THE MAN WHO ORGANIZED NATURE

The Man Who Organized Nature

THE LIFE OF LINNAEUS

Gunnar Broberg

Translated by Anna Paterson

PRINCETON UNIVERSITY PRESS
PRINCETON & OXFORD

English language copyright © 2023 by Princeton University Press
Originally published as *Mannen som ordnade naturen : en biografi över Carl von Linné*
by Stiftelsen Natur & Kultur, Stockholm, Sweden © 2019

Princeton University Press is committed to the protection of copyright and the intellectual property our authors entrust to us. Copyright promotes the progress and integrity of knowledge. Thank you for supporting free speech and the global exchange of ideas by purchasing an authorized edition of this book. If you wish to reproduce or distribute any part of it in any form, please obtain permission.

Requests for permission to reproduce material from this work
should be sent to permissions@press.princeton.edu

Published by Princeton University Press
41 William Street, Princeton, New Jersey 08540
99 Banbury Road, Oxford OX2 6JX

press.princeton.edu

All Rights Reserved

Library of Congress Cataloging-in-Publication Data

Names: Broberg, Gunnar, 1942- author. | Paterson, Anna (Anna Tora), 1942-translator.
Title: The man who organized nature : the life of Linnaeus / Gunnar Broberg; translated by Anna Paterson.
Other titles: Mannen som ordnade naturen. English
Description: Princeton : Princeton University Press, [2023] | "English language copyright © 2023 by Princeton University Press. Originally published as Mannen som ordnade naturen : en biografi över Carl von Linné by Stiftelsen Natur & Kultur, Stockholm, Sweden © 2019"—title page verso. | Includes bibliographical references and index.
Identifiers: LCCN 2022040781 (print) | LCCN 2022040782 (ebook) | ISBN 9780691213422 (hardback) | ISBN 9780691248196 (ebook)
Subjects: LCSH: Linné, Carl von, 1707-1778. | Naturalists—Sweden—Biography. | BISAC: BIOGRAPHY & AUTOBIOGRAPHY / Science & Technology | HISTORY / Modern / 18th Century
Classification: LCC QH44 .B87513 2023 (print) | LCC QH44 (ebook) | DDC 508.092 [B]—dc23/eng/20230109
LC record available at https://lccn.loc.gov/2022040781
LC ebook record available at https://lccn.loc.gov/2022040782

British Library Cataloging-in-Publication Data is available

Editorial: Eric Crahan, Whitney Rauenhorst, and Barbara Shi
Production Editorial: Kathleen Cioffi
Jacket Design: Katie Osborne
Production: Danielle Amatucci
Publicity: Alyssa Sanford and Charlotte Coyne
Copyeditor: Karen Carroll

Jacket illustrations: The sexual system (of plants) after G. D. Ehret, 1736. Photographed by M. Hjalmarsson, Uppsala University Library.

The cost of this translation was defrayed by a subsidy from the Swedish Arts Council, gratefully acknowledged.

This book has been composed in Miller

Printed on acid-free paper. ∞

Printed in the United States of America

10 9 8 7 6 5 4 3 2 1

CONTENTS

Preface · ix

Who Was He? 1

PART I   A GREAT MAN CAN COME FROM A
         SMALL HOUSE, 1707–1741   13

    1   The Guardian Tree   17
    2   Studies in Växjö and Lund   30
    3   The Academy in Uppsala   45
    4   In a Mythical Landscape: Lapland   61
    5   *Diaeta Naturalis*   82
    6   In the Mountains and under the Ground:
         County Dalarna   91
    7   In the Land of Tulips   103
    8   Nature's Order 1   119
    9   A Stockholm Interlude   130

PART II AT THE HEIGHT OF THE AGES OF MAN,
         1741–1758   143

   10   Uppsala and Enlightenment   147
   11   Three Programmatic Speeches   153
   12   Provincial Travels on Behalf of Parliament   158

13	A Language in Which Everything Matters	176
14	Flora et Fauna Svecica	181
15	Family Life 1: Scenes from a Marriage	186
16	In the Garden, at Herbations, among the Collections	194
17	Ex Cathedra	207
18	What Is More Precious than Life, More Pleasing than Health?	211
19	Academic Amusements	223
20	Appetite for Work, Weariness, Communication	228
21	When Linnaeus Wrote, Salvius Printed, and Tessin Bought the Books	238
22	Linnaeus, "the Sexualist"	247
23	Curiosity-Driven Research	253
24	Nature and Culture	257
25	Entrepreneur and Economist	264
26	To Describe the World	271
27	Nature's Order 2	279
28	*Homo sapiens*	294
PART III	THE OLD LINNAEUS, 1758–1778	301
29	Honors	305
30	Among Students and among Senior Academics	309
31	Family Life 2: Hammarby	317

32	Friends and Enemies	327
33	Problems	337
34	A New Synthesis?	346
35	A Philosopher of Science or a Scientist?	359
36	The Back of God and God's Footsteps	365
37	*Nemesis Divina*	373
38	Solomon on Growing Old	380
	Epilogue I. Family Life 3: Mother and Child	394
	Epilogue II. Linneanism	402

Abbreviations · 411
Notes · 413
Sources and Literature · 437
Index · 459

PREFACE

FOR SWEDISH SCIENCE, Carl Linnaeus is a flower in the buttonhole that has stayed fresh despite the passage of time. By now, over a century has passed since the publication of Thore M. Fries's massive Linnaeus biography, two volumes adding up to a work unbeatable for detailed, factual knowledge and sheer love of its subject. Fries however didn't try to answer post-Linnean questions, not even about evolution, a topic for debate at the time he was immersed in writing. He also shied away from the darker aspects of his biographical narrative.

This book, too, is a biography, but discussions of the history of science will recur throughout. One unusual feature is my emphasis on Linnaeus as an old man. I have been trying to work out how he thought and from which sources he drew his ideas; catch him in mid-step and hear him speak; visit him at home. Yet another aspiration of mine is to inject new life into old texts, so that his story is to a large extent told in his own words and those used by his contemporaries. This should give us opportunities to get closer to the man and his times, although Linnaeus's determination to control his posthumous reputation means that his words must not be accepted without question. It is worth noting that in his autobiographical writing, he refers to himself in the third person. One needs a keen ear when listening to the voices from various rooms of the past. The way in which spelling and meaning have changed are matters I have not been able to do anything about: for instance, that to describe a dish as "rather good" meant it was "excellent" or that "worm" once was spelled "wyrm."

Like most Swedes, including Fries, I use the ennobled version of my subject's name: Carl von Linné. It was his proper name for barely twenty of his seventy years of life and in anglophone countries, he is known as Linnaeus.

Linnaeus is a man for our time. As I write this in May 2019, an intergovernmental panel of experts on biological diversity[1] has declared that, out of a global count of about seven million species of plants and animals, one million are at risk of extinction within the next few decades.

The list of references, pruned but still long, has been placed at the end of the book, as have the endnotes. Page references to Linnaeus's travels have not been included.

Acknowledgments

I am grateful to Anders Bergman, Agneta Engström, Annika Lyth, the seminar series in the history of ideas at the university in Lund, Eva Nyström and the Linnean Correspondence project, and Lynda Brooks and Isabelle Charmantier at the Linnean Society (London), Gunnar Eriksson (for inspiring coffee breaks), Ulf Marken, Staffan Müller-Wille, Nils Uddenberg, David Dunér, Jakob Christensson, Carl Gustaf Spangenberg, Richard Wahlgren, Annika Windahl Pontén, Anton Härder in Stenbrohult, Erik Hamberg, Åsa Henningsson, Karen Beil Magnusson, Agneta Helmius, and, for grants, I thank Kungliga Patriotiska Sällskapet and Längmanska kulturfonden.

Gunnar Broberg

THE MAN WHO ORGANIZED NATURE

Who Was He?

A PROFESSOR OF LITERATURE wrote this appreciation: "Carl von Linné was a great man, and a remarkably happy man. For his happiness he could thank his harmonious nature, and an optimism drawing on his Lutheran piety, learned in the rectory that was his paternal home. His mind was not prey to intellectual anxieties; he did his work, and gathered in its fruits, with humble, grateful joy. . . . Von Linné strongly held the opinion that the grand style was not for him. Instead, only half aware of doing this, he created his own style, straightforward but occasionally with poetic overtones, sometimes striking notes of biblical conviction and, at other times, of Rococo idyll."[1]

It is worth pointing out that practically every element in this oration could be turned into its opposite. Who, then, was the man whose earthly remains lie under the tombstone just inside the entrance to the cathedral in Uppsala?

We know more about him than almost anyone else alive in Sweden at the time. A quick overview serves as an introduction to his appearance and personality, as for instance in this verbal self-portrait: "Linnaeus was not tall, nor was he small. He was thin, brown-eyed. Light of foot, walked quickly, did everything promptly. Had no patience with tardy folk; he was easily moved, sensitive, working continually as he was incapable of husbanding his strength. Ate good things with pleasure, drank good drinks; but was never excessive in such things. He cared but little for the *exterieure*, believed that the man should make his clothes suit him, not vice versa."[2] To make his point come across, he described himself once more. This passage has been translated from Latin: "short of stature, rather tall than low; neither thin nor fat, with some *musculeuse artus* and large veins, ever since childhood." Furthermore: "His head strongly curved inward at the back of his neck. Hair a snowy white in childhood, later dark, then graying in old age. His eyes are brown, lively, very sharp, excellent power of sight. Old age has left his forehead deeply furrowed. On his right cheek, a feebly growing wart, another on the right side of his nose. Teeth poor, caries having developed from inherited toothaches in his youth"—another study of himself, covering just over a page and obviously composed with the previous self-portrait in mind.[3] This intrusively physical description might surprise some readers, but similar passages are found elsewhere in Linnaeus's writings. Here, for instance, he describes his father-in-law: "Of middling height, he had grown tall, evenly set and straight. His bones and joints were strong, surrounded by firm, solid flesh, tendons, and blood vessels. His face was manly, his eyes quick, with a frank expression, his beard grew black and his skin tended to a dark shade; in a few words, he looked a fine figure of a man." Naturalism makes him mention his own slight warts, one on his right cheek and another on the right side of his nose—features that can be seen in the painting of Linnaeus used as the cover image of this book.

Johann Beckmann, a German historian sometimes spoken of as "the father of technical history," has described Linnaeus several times. Beckmann wrote: "The nobleman von Linné was of short stature and, as he walked with a slight stoop, appeared smaller than he actually was. His was a liberal heart, and his passions were lively and strong. . . . He loved joking, happiness, and, in every way, good living. His urge for glory was boundless. He cared nothing for his neighbors' opinions, only for the judgments of learned men."[4]

The editors of *Vita*, his autobiography, observe that "Linnaeus is one of those writers who never discovered the danger of superlatives."[5] At times, he seems unaware that "self-praise is no recommendation." Writing to his old friend Carl Fredrik Mennander in 1762, he mentions that he has been writing his own eulogy: "finding that *propria laus sordet*, yes, indeed, I would say it stinks. I would never show this to anyone in this world unless were it to one man only, my benefactor ever since my years of struggle. If, my Dear Friend, you would consider taking words out, they would be those which attracted attention as they came from such a source."[6] We know nothing of Bishop Mennander's reaction, nor what Linnaeus actually sent him—yet another autobiographical outline? He would probably have had the relevant *curriculum vitae* notes at hand when the matter of a national reward was decided. As late as in 1770, it was still in the cards that Mennander would write, or at least edit, Linnaeus's autobiography. A farmer had been dispatched to hand the text to Abraham Bäck, an old friend, who was in turn to pass it on. It was the second version of Linnaeus's Life—*Vita* number two—and "penned at diverse times, it will also be diverse in thoughtfulness." Mennander was urged to freely change the order of things. (An earlier, overruled note however instructs that, after the author's death, the *Vita* should be made available to professor Magnus Beronius to use for the encomium and afterward to Linnaeus's widow, "insofar as she will be well," for her to have it printed.) Mennander was to submit the manuscript to the Academy of Sciences, of whom "my wife will demand" that the institution fulfill its promise to have it printed. Does all this mean that *Vita II* is the most authoritative? Was it really sent on to Mennander and then back to Bäck? While *Vita II* stops at 1750, the following autobiographical text, *Vita III*, goes as far as 1776. Bäck's tribute to his friend seems based on *Vita III*:

> His stature somewhat below middling height, neither fat nor thin, of a solid and full build with, since childhood, prominent veins, a large head protruding at the back with a deeper furrow separating the frontal and posterior parts of the head; brown, fiery eyes, sharp vision, good hearing but not for music, quick and easy on his feet, an excellent memory well into his sixth decade by which time first names began to escape him; ability to learn languages was however not one of his gifts, so that he was less than content to be with foreigners who did not know Latin. In Latin, he expressed himself swiftly, easily, precisely, and more briefly than anyone else, when it concerned descriptions of natural phenomena, but for other matters he did not trouble himself much as long as what was said fitted with what was observed. When writing to his patrons, he instilled in his language uncommon and captivating turns that cannot be mimicked. His few speeches displayed the author's wit, daring, and great learning, and could not, one would fancy, have been delivered by any man other than Linné.

With little more than a year of his life left, but still keen to control his posthumous accolades, the old Linnaeus labored over his autobiography. Yet another one; depending on how one cares to count them, it is the fifth or the fourth or possibly the sixth. True, these texts are probably better seen as lists of his qualifications, set out to suit this or that academy preparing to salute its great former member. He was making bids for himself. Attempts to outrank all other claimants make some of his self-assessments read like job applications with the applicant's perfect suitability described in such terms as to make rejection impossible. Below, he outlines his personal qualities for a presentation by the Patriotic Society, which had not approached Linné until as late as 1775—a noteworthy fact. The final words are familiar: "He [Linné] was of somewhat less than ordinary height, his build solid and full, and his eyes brown and fiery."

Of these fiery eyes, the notorious gossip J. G. Rothman Jr. observed: "When Linnaeus intends to utter a malice, he narrows one eye markedly more than other." He added: "His stature is rather less than ordinary, he was neither fat nor thin, had a large head expanded backward, brown eyes full of fire, sharp vision, and good hearing although not for music.... Always wore boots and, when at home, dressed mostly in a short nightshirt and a velvet cap. Used to excess both Coffee and Tobacco. With foreigners he spoke only Latin, in which he was not well versed. Cared only a little for medicine."

Berge Frondin, who at the time was an admirer of Linnaeus's wife, characterized her husband thus: "He was easy and light on his feet. His temper could flare—he was jovial, witty, and spoke well—always wore boots and walked about in his own house in a nightshirt and a velvet cap— with tobacco and coffee abundant. He was much amused by card games." Staying with his physical status and its problems, he was plagued by toothaches "ever since in my mother's belly" and smoked tobacco, probably to excess, to deal with them. In 1772 he wrote: "I hold back on him [tobacco] as far as I dare, not to break utterly with *antiquam consuetudinem, quae in naturam transit*"—an old habit that has become part of one's nature.[7]

More fiery brown eyes: J. G. Acrel recalls the man, as seen in 1796: "In his not unpleasant face, one noticed his quick and fiery brown eyes, a little short of sight and narrowed, not so much by nature as by habit and work on the investigation of matters in hand, practices that also caused him over the years strong wrinkles around his eyes due to the pulling of the muscles." He continues:

> Walked somewhat stooped but had otherwise in his earlier years a light step which more and more changed so that in his fifth decade he had begun to shuffle his feet forward instead of lifting them.... His disposition was quick and easily moved to grief, joy, and wrath but also quick to regain calm. In his youth, he was full of joy, in his middle age always cheerful, witty, and easy with words, and when in cheerful company liked to share laughter with others, an inclination that followed him into his last year. He did not speak much but liked to listen to others and interjected at times his own brief but always interesting anecdotes, with especial preference for events that belonged to his own lifetime, or to his disciples or friends. When in his professorial chair, he showed a

singular and distinguished eloquence, which, although not supported by a notably strong and pleasing voice nor by an elevated way of speaking (as he still had a county dialect), never failed to beguile his audience to the greatest degree.

There are many who have attempted to diagnose his condition, or pronounce him a genius—or both. However, the diagnoses vary. Nils von Hofsten has spoken about Linnaeus's mind and how it was possible to trace signs of incipient physiological decay. Depressed moods and happier states of mind came and went at lightning speed, sometimes seeming to coexist: "He was definitely showing signs of cyclothymia and, at all times, his emotions were labile"—in other words, he had a bipolar disorder. Von Hofsten wasn't a nobody, but the long-standing chairman of the Swedish Linnaeus Society and, furthermore, the expert advisor to the National Board of Health in cases of sterilization.[8] No one among Linnaeus's contemporaries seems to have made the usually close-at-hand connection between genius and melancholia—for instance, as written up by Samuel-Auguste Tissot in *De la santé des gens de lettres* (1768). Linnaeus never refers to his own state of health other than in the context of physical illnesses.

He could fit into a modern diagnostic category such as ADHD (attention deficit hyperactivity disorder), which includes people who can also be characterized as quick to take initiatives, energetic, creative, curious, stubborn, impatient, and ceaselessly active. Toward the end of the eighteenth century, the Scottish doctor Alexander Crichton grouped such symptoms into the syndrome "mental restlessness," which could well be compared to Linnaeus's description of himself in *Vita*, quoted above. A list of other creative geniuses with this diagnosis is supposed, rightly or wrongly, to include Wolfgang Amadeus Mozart, George Bernard Shaw, Edgar Allan Poe, Salvador Dalí, and Thomas Alva Edison. If medicated in some way—would their special brilliance have disappeared? There are other sources for the biographer to try: in 2013 the *Diagnostic and Statistical Manual of Mental Disorders* (DSM-5) included "compulsive hoarding." Linnaeus was in his early twenties when he mentioned in a letter (dated 1730) that his insect collection had reached four thousand specimens. His collections would continue to grow steadily, and he later claimed that his herbarium was the world's largest. His library was also very extensive. His contemporaries were all much given to manic collecting, and there might have been an epidemic of compulsive hoarding. It wouldn't be totally unlikely since it is a not uncommon pathology which allegedly afflicts about 2–6 percent of the present-day population.[9]

Today's visitors to the Linné Museum in Uppsala often stop to contemplate the bed where Linnaeus died. "Was he really that small?" the child asks the adult, who, of course, knows the answer: "No, you see, they slept sitting upright in the eighteenth century. Besides, look, you can pull the bed out." But Linnaeus was short, so one might surmise a certain sense of inferiority—a "Napoleon complex"? He and H. C. Andersen were alike: both men had lowly origins, became internationally famous early in life, had huge imaginations, and were famously prickly as well as showing evidence of paranoid traits and fears of death.

Linnaeus quoted God's promise to David: "I was with thee whithersoever thou wentest, and have cut off all thine enemies out of thy sight, and have made thee a great name, like unto the name of the great men that are in the earth." He continued in his own words: "*No one* at our academy has with greater diligence practiced his profession and had more *auditores*. *No one* versed in the ways of nature has made more observations about natural matters. *No one* has had more robust insights into the three realms of nature. *No one* has been a greater *Botanicus* or *Zoologus*. *No one* has with greater skill worked on the understanding of the natural history of our Great Country, its Flora, Fauna or done as many travels in it. *No one* has written more numerous works from his own experiences, nor as neat or orderly. *No one* has reformed an entire science and created a new epoque." And so on, through a further ten "*No one*" assertions.[10] Linnaeus was *somebody*—not least in his own view. His constant flow of self-praise sounds almost incantatory. He doesn't trust Fortuna and her unpredictable handouts, from the Creator's happy rewards to Nemesis's dark retributions.

Using the words of William Blake, Linnaeus could "see a World in a Grain of Sand / and a Heaven in a Wild Flower." As he writes himself: "My greatest labor has consisted in being an attentive *observator*." By now, another question should be considered: Was he a scientist—and, if so, what kind of scientist? A bright spirit, of course, but not to be compared with the brightest lights of the Enlightenment. At times, and notably in the work of his old age *Nemesis divina*, his mind moved in the deep shadows where ghosts are lurking. He disparaged folk beliefs but was never quite free of them. His was a "genius of the eye," but it is perfectly reasonable to portray him as a scholastic list maker and a traditionalist who never got around to using the microscope.

Often, he acted like a pragmatic utilitarian, but at other times he seemed to see himself as one of God's elect. His intellectual awareness begins in the late Baroque period and ends in the early Romantic period—but was he ever a man of the Enlightenment? It is, alas, only too easy to equate the eighteenth century and the Enlightenment, but Linnaeus seems not to have read any of the contemporary French philosophers. To him, "Science" was the prime source of Enlightenment; it was a favorite theme of his. In 1759 he said, in a lecture held in front of the royal couple: "Without Science, we would place our trust in priests from Rome and medical men from the French town of Monspelier. Without Science, *huldras* could still be concealed behind every bush; ghosts, and apparitions emerge from every dark corner; gnomes, trolls, river spirits, and all others in Lucifer's battalions share our lives, like gray cats."

In Linnaeus's New Year's greeting for 1749 to the architect Carl Hårleman, he refers to two kinds of admirable men—those who, like Hårleman, "have made great discoveries and carry tall, heavy torches" and, on the other hand, "the Imitators who have small candles, quite *ordinaire*"— like, perhaps, the writer of the letter? Next, a surprise: "In that very moment, I caught sight of myself in the Mirror, walking with a quite small torch in my hand but, looming over me, stood a large and strong hero"; the protective figure was Hårleman.[11] It could be that Linnaeus was not all that surprised because he normally didn't conceal that he saw himself as an "inventor" and "auctor"—an originator—rather than a "compilator." He wrote frankly about flattery to the journalist Carl Christoffer Gjörwell: "For your generous praise I thank you most humbly; ... should I pretend to detest praise, I would lie as would all who said likewise, for whoever hates their own flesh? Love of yourself is the foundation for all that is good. My Dear Sir, you inundate me with eulogies such that, did I not know myself too well, I might have become proud."[12]

He had a stock expression when disapproving of a state of affairs, which was "dense and barbarous." "More than 70 years have passed since Swammerdam, Lister, Blankard et al. opened the eyes of the peasants, hitherto blinded by the monks." To him, monks, the Middle Ages, and "darkness" added up to a set of linked concepts: "Indulgences could be bought if they frequented church diligently to hear some absurd Mass being said [or] if they visited a miracle-making cross, or images to do with Mary so that their world filled with poems, dreams, and monkish fairy tales." No one had any learning except for the priest, who was actually uneducated but for clichéd book learning. His outburst probably didn't exclude the Lutheran Church of his own country.

One might compare his stance to that of the poet and wit Johan Henric Kjellgren in *The Enemies of Light*, writing in defense of the Enlightenment. Unlike Kjellgren, Linnaeus didn't attack esoteric teachings, and sects such as Hermeticism and Freemasonry, but directed his ire at the erroneous or plain superstitious beliefs of simple people he had encountered and noted down in the travelogues. As he speculated in *Miracula insectorum* (1752), lacking knowledge about nature is "the most prominent reason why many imbecile superstitions are held to be true and become the cause of such vain terrors."[13] Still, Linnaeus didn't always believe himself to be "superior to the common man." Although his life and his time on the international academic stage coincided with the period called the Enlightenment, "offstage" he personally followed different lines of thought. Was he a Renaissance magus? Such labeling attempts always bring complications in their wake. It is true that, for all his belief in the scientific method, he was intrigued by the occult. Marie-Christine Skuncke, author of a Carl Peter Thunberg biography, provides wall-to-wall coverage by one of Linnaeus's most adventurous pupils—but also sighs that her hero "still remains to me an enigmatic figure."

Linnaeus's scientific work can be summarized under several headings:

1. Creation of the first complete, systematic schema for classifying, in principle, all living organisms;
2. Standardization of the descriptive methodology and terminology for living organisms;
3. Classification of thousands of animals and plants according to his system;
4. Establishment of the principle of binominal nomenclature by naming an organism by just two attributes: its genus and species;
5. Demonstration that human beings should be classified as an animal species;
6. Creating, through the travels of his pupils, a basis for a global natural history;
7. Teaching his own, as well as later generations, the value of knowing how nature works.

Linnaeus was a physical, sensuous man, who responded with all his senses to the signals from his surroundings: shapes and colors, sounds, tastes. He saw everything. His acute sensory responses were essential to his approach to science and seem to have formed his experiences of man and nature. Instead of being discrete about sexuality, he recognized it in himself and lectured about it to others. Censorious voices have loudly claimed that Linnaeus never made a single discovery! Of course, the man himself saw it differently. In his curriculum vitae, the twenty-seven listed entries under the heading *merita et inventa* included his classification system based on sex organs, the binominal nomenclature, and his invention of the "flower clock."[14] He might well have added a few more: the 100-degree temperature scale, dendrochronology, and perhaps his version of evolution. His insistence on standardization as an important feature in many different contexts is rightly influential. How "discovery" is defined matters: surely methodology counts, as well as perspectives, ways of thinking, and recognition of contexts. One might well add his influence on his contemporaries—his "impact"—though some could view this as a demerit.

Another line of criticism takes its cue from the darker aspects of his persona. Not just the shadowy night of *Nemesis divina*, but what was said about him, for instance that "he was one of the least generous of men; what he could reach, he wanted for himself and, in his naive vanity, he hoarded his worldwide reputation like a dragon its gold treasure."[15] It is easy to portray Linnaeus as a careerist, a man with an outsize ego who enjoyed power and liked being the head boy in the class, but easier still to describe his charm, capacity for hard work, inspirational teaching, and lasting influence. The charismatic young scientist, the aging authoritarian, the writer, and the patriot—the man with "a genius of the eye" and an acute ability to perceive larger meanings in small observations, who had a global outlook and a drive to make natural history available to everyone—women, too. All these characteristics and traits must, of course, also be seen in the context of his life, which was from time to time difficult, plagued by poverty, poor health, and a heavy workload. Linnaeus was naive, yes, and had a monumental sense of his own worth, but he was also an outstanding observer and a vivid writer.

To get a grip on who Linnaeus was, it is not enough to contemplate *one* image—neither of the old man with his gentle gaze, nor the young one with his alert eyes. That simple truth should have become clear even in this lightning-quick introduction, and will be a theme throughout the biography. For instance, we must not allow ourselves to be taken in by the bent old man in the portraits by Per Krafft or Alexander Roslin. Linnaeus had a merry, charming side to his personality but could descend into brooding melancholia. He worked hard but sometimes lost touch with his surroundings. He had a remarkable capacity for grand overviews of systems but would spend days and nights pondering details. His curiosity hardly ever faltered even though the tables he compiled to show the diversity of Creation can be wearisome reading.

As he paced around in the natural world, Linnaeus inspected it with a field marshal's eagle eye; one might argue that he superimposed a soldierly hierarchy on natural history. The kingdom of plants became structured in an orderly manner from the top down to the single soldier—or plant. This

FIGURE 1. Bust of Linnaeus, hailed by Greco-Roman gods and goddesses. On the left: Asclepius with his serpent-entwined staff. Then, the winged Cupid, Flora, goddess of flowers, and Ceres, goddess of harvest. From Robert John Thornton, *The Temple of Flora*, 1806. Uppsala University Library.

was one of Linnaeus's pedagogical principles and, speaking abroad, he even used analogies such as infantry men, centurions, and decurions.[16] In a famous review of contemporary botanists, he classified them by imaginary military ranks. In some respects, he might easily have joined Sweden's legions of conscientious civil servants; he actually went on to become one of their models.

His perception of the role of numerical order in nature, as Bach's in music, made him believe that God used mathematics to construct his palaces. Linnaeus had probably never heard of Bach and wouldn't have cared for his compositions (he always said "I have no ear for music") but, like Bach, he was attracted to the mystery of numbers and completions of series. *The Well-Tempered Clavier* and *Systema Naturae* both contain twenty-four variations on a theme: preludes and fugues in all 24 major and minor keys, and plants sorted into "classis 1 to 24." Keyboard instruments as well as the medical practices needed tuning to be fit for the great harmony of the world. Bach's *Art of Fugue* and Linnaeus's corresponding synthesis *Clavis Medicinae Duplex* were separated by little more than a decade.

Both men directed their polyphonic creations to their Lord in ecstatic gratitude. Linnaeus loved sequencing—trying out runs over the keys of nature—and excelled in variations based on themes and schemata discovered in nature. Both were systematizers, but they also shared a taste for the esoteric. They wove their signatures into their compositions: one used the notes B-A-C-H, the other his personal logo, the twinflower or *Linnaea borealis*, to show off to the watching audience—and ultimately, to God.

PART I

A Great Man Can Come from a Small House

1707–1741

FIGURE 2. *Linnaea borealis,* introductory illustration in Johan Palmstruch's (1770–1811) *Svensk botanik* (A Swedish Botany), 1810. Uppsala University Library.

The Modest Twinflower (Linnaea borealis)

The wild flower the Swedes call "linnea"—the twinflower—was known before Carl Linnaeus adopted it. Before his time, Swedish natural historians had recorded its existence, as seen for instance in an illustration in *Acta literaria upsaliensis* (1720) by Olof Rudbeck the Younger, where the linnea is named *Campanula serpyllifolia* or "small bell(flower) with leaves like wild thyme." Linnaeus took note of it, wanted it formally to bear his name, and wrote: "This Lapland plant . . . is of low stature, despised, flowers only for a short time and so is like Linnaeus."[1] His Dutch friend Jan Frederik Gronovius acted on this wish in his work *Genera Plantarum* (1737).

Linnea borealis became emblematic of the man Linné, of Swedish science, and, indeed, of Swedish culture. In Linnaeus's *Flora Svecica* (1745), the twinflower is the only plant to be portrayed and, among his thirteen known personal seals, ten are engraved with an image of "his" flower. His eminent contemporary A. F. Skjöldebrandt commented: "The very sight of this plant makes you remember the man after whom she was named."

The identification of the man with the flower is almost complete, and explains why an image of *Linnea borealis* introduces the first part of this biography. It stands for Linnaeus's modest origins and his expedition to Lapland, which was to become the takeoff point for his career.

In 1908 the twinflower was adopted by County Småland as its signature flower.

The story of Linnaeus begins in Småland.

CHAPTER ONE

The Guardian Tree

A BIOGRAPHY BEGINS long before the subject's birth and does not end with his or her death: the lifespan it charts is inextricably linked to time and natural surroundings. A human life depends on weather and wind as well as guardianship. The personal narrative is given shape by parents and by those others who remember or care enough to investigate, to follow the tracks in the grass and in the archives.

We live in the shade of trees. In ancient times, Nordic people believed that the World Tree was an ash and the protective guardian tree a linden—a *Tilia*. The biography of Linnaeus should surely begin with a linden. In the late sixteenth century, a huge, split specimen with three trunks grew in Jonsboda in the Småland parish Vittaryd. Three local families were said to stem from the tree and to have taken their names from it: the Lindelius, Linnaeus, and Tiliander families. The great linden had once been declared dead but is still alive. This is Linnaeus identifying himself: "God let him spring forth from a stubborn root, replanted him into a distant place and, praise be, allowed him to soar and grow into a worthy tree."[1]

On the 23rd day of the lovely month of May, the newborn boy, who would be called Carl, opened his eyes to see the world around him: "The most beauteous spring, when the cuccu hailed the summer between *frondescentiae* [coming into leaf] and *florescentiae* [flowering]." His place of birth was the rectory in Råshult, County Småland and near the border with County Skåne, which within living memory had belonged to Denmark. Carl was born in 1707, during the night: "Between the 12th and 13th day by the Gregorian style of reckoning, at one in the morning, in the realm of Sweden, the parish of Stenbrohult, the village of Råshult."[2] As this quote shows, the "old-style" calendar was still in use. Sweden had not yet adopted the new order, so was "behind the times" and would, from then on, try in all things to catch up with the international lead.

The year of his birth fell during the somber final stages of the Great Northern War. Just two years later, the Swedish army would lose the battle against the Russians at Poltava and, by then, the country was exhausted from never-ending warfare. In 1709 the Danish government set out to reoccupy the Swedish territory lost after the peace of Roskilde. The Danish army landed near the coastal city of Helsingborg in November that year. The governor general of Skåne, Magnus Stenbock, retreated to Växjö in central Småland but ultimately led the defensive forces to victory.

The witch-hunting hysteria, particularly feverish in the north, had reached Småland by then. A few generations earlier, Johanne Pedersdatter, a distant Norwegian relation on Linnaeus's mother's side, had been convicted of witchcraft and burned at the stake in Stavanger.[3] Linnaeus seems never to have referred to this relative and may not have known about her. He also makes no reference to the plague that followed in the wake of the war and caused at least a hundred thousand deaths, nor to other forms of contagion made worse by poverty and starvation, nor to the King's taste for war games. A freer, more sensible world was on its way, though. Little Carl's arrival might have been seen to counterbalance the unrest of the time, or as heralding better days to come.

Carl was the firstborn child of his parents, Nils Linnaeus and Christina Brodersonia, but several siblings were to follow. Nils was a minister in the Lutheran Church and, as the son of a clergyman, Carl was to follow in his footsteps and, ideally, succeed him. It was not to be; the baton—or, rather, the hymnbook—went to his brother Samuel.

Both Linnaeus's early homes, first in Råshult and later in Stenbrohult, burned down, but a look-alike house and garden have been created in Råshult, lovingly cared for and popular with tourists.

In his several autobiographical works, Linnaeus writes of his parents with warm affection. His father was a farmer's son, born Nils Ingemarsson and a man who "walked slowly throughout his world, finding his pleasure in the ordering and care of his garden with its several and sundry plants as, in such matters, he found all his peace." A few years after Carl's birth, Nils was promoted from curate to rector in Stenbrohult parish. He now had a home near the parish church as well as his house and grounds in Råshult. The document confirming his position as rector was properly signed by the king on 12 August 1708, though Charles XII was at the time somewhere near Mogilev in White Russia (Belarus). His wife Christina, Linnaeus's mother, was the daughter of the former rector of Stenbrohult.

She was "heedful and indeed so industrious as to never give herself time to rest. She feared God greatly and was the mother of 5 children. . . . She was a beautiful young girl. . . . The boy was nursed, suckling his own mother's breasts." This last remark reflects a significant element in Linnaeus's later instructions about natural nutrition. Christina's stepmother was a harsh, difficult woman, which might be why the younger woman accepted Nils's proposal—"although otherwise, she had not been thus inclined."

Carl's parents are both described as "of middling height"—that is, "short" by present-day standards—but the differences between them are more striking. Nils was heavily built and she was slender; furthermore, "his spirit slow to anger, even-tempered, and good, hers sharp-tongued, quick, and workaday."[4] Linnaeus is mostly rather silent about his mother, which might suggest secret reservations.

"For her, the day of 13 May 1707 was a day of mourning, as she gave birth, with the greatest difficulty and danger to her life, to a well-formed son: this despite her wish that the child would have been of the gentler sex. . . . The man was however made happy indeed and his gladness atoned for her grief. Thus, they joyfully christened this child, their firstborn, on the 19th day of that same month." Here, the writer is Samuel, Carl's younger brother by eleven years; he is addressing the Småland "nation" at Uppsala University after his eminent brother's death in the New Year 1778.[5] How is it that Samuel knows about his mother's regrets? Why mention it in this context? We don't know. One explanation might be found in the "tradition of conservation"—a form of social support based on the rule that the young clergyman should marry the dead pastor's widow and support any children. The pastor's grown daughter would be free to marry "out."

Linnaeus speaks of his mother only once more: "At 6 o'clock [6 June 1733] after midday, my most dear and pious mother departed, causing me in my absence an ineffable anxiety, grief, and harm." There is nothing more. Nils wrote about his wife in her book of remembrance: "She always feared God and ordered her home well, always diligent and cautious, generous and heedful, and gifted with fine understanding."

"A great man can come from a small house," Linnaeus remarked, referring to himself. Physically, he was a small man, even for his time (estimated height about 153 cm), and seems fascinated by the tall and strongly built: "Those living here [in Stenbrohult], as well as in most Småland parishes, incline to be larger than elsewhere, as is true for both sexes, for the probable reason of belonging to the old tribe of Göthaland, as strangers are seldom seen here and a farmer rarely has his daughter marry anyone not born in the parish."[6] Linnaeus's notes contain occasional references to

FIGURE 3. "A great man can come from a small house." Råshult, Linnaeus's place of birth. The original house has burned down. Uppland Museum.

a tall Finn called Daniel Cajanus, and one note mentions that the Sami are shorter than himself. Johan Lång (Tall) was a member of Linnaeus's student "nation"—the Småland nation at Uppsala University—and his height caught the attention of its recordkeeper: "Master Tall from Tall Lycke village is tall—a tall man's tall son."

In 1703 Stenbrohult parish had 206 inhabitants but, by 1729, the local population had increased to 578. Local authority was embodied in the rector, who kept the parish records of births and deaths, went to people's homes to make sure that they knew their catechism, and would advise in day-to-day matters. People understood their country in the terms of the Protestant exposition of the relationship between the state, the working people, and the church; this text was printed and distributed with hymnbooks and catechisms. Its fundamental thesis was Luther's teaching about the three hierarchies or estates: church, political establishment, and household. These three entities were also defined respectively as the learned, exploitative, and nourishing estates, and their roles illustrated by a dozen short passages from the Bible.

Nils was a practical man, well able to restore the decaying parish church and keep the rectory in good repair. He was a man of learning but also a farmer who knew how to speak to other farmers. The parish supported their rector generously and helped him to construct a morgue and an ossuary near the church, and also to restore the rectory after a fire.

The rector paid for guest rooms, farm laborers' quarters, and a bathhouse. Everyone was forthright and trusting, the old soldiers as well as other villagers. What was said of the farmer Åke Kvick in Råshult was more or less generally true: "Well versed in the Bible, a patient man who never let the world weigh on his spirit."

Linnaeus writes in *Spolia Botanica*: "Stenbrohult is a parish found some 30 miles from Wexiö, toward the border with Skåne in the municipality of Allbo, and which, compared with all other places, in appearance is like a queen among sisters; preferred to others even in the location of rare and wondrous herbs not often to be seen elsewhere in the country. Indeed, the very rectory here is as if adorned by Flora herself; I would doubt if any space in the whole world could present itself more pleasantly. Surely it is not strange that I have reason to lament along with the poet: 'Nescio qua natale solum dulcedine cunctos / ducit et immemores non sinit esse sui'" (Our native land charms us with inexpressible sweetness and never allows us to forget that we belong to it).[7]

Much later, when Linnaeus reluctantly agrees to describe landscapes, he declares that he is like a lynx abroad but a mole at home, and knows more about "what is produced by Virginia in America, Cap de Bonne Sperence in Africa, and Zeylon in the East Indies, than in my own native land which I left before I had properly woken and rubbed the sleep from my eyes."

He had hardly seen more than Stenbrohult, the village where he was born, and Växjö, the city where he went to school. He had left these places before reaching adulthood and had, ever since, hardly seen more of any one place than a wandering goose in its flight—a migratory bird. "What I saw in my youth, or through the eyes of old Cubae or of Arfvid Månsson Rydaholm, I still recall as if it were a dream and, now, it is all I can speak about."[8] He goes on, however, to describe uplands and high hills, forests, meadows, fields, marshes, and lakeshores before getting around to the plants, the animals, and the minerals—in fact, he turns out to be a man with an excellent memory. Far from being a blind mole, he was a lynx also in the forests of his childhood.

Göran Wahlenberg, a colleague of a later generation, provides an explanation of Linnaeus's scientific approach in the 1822 issue of the yearbook *Svea*, based on what he believes was a natural kind of cultural simplicity. "A beautiful and solitary situation, in a setting of low hills pierced by many streams and cloaked in vigorous birch woods. The scarcity that affected Linné in his first youth here appears in its way to have made him stay close to his natural surroundings, indeed to unite his lively spirit with them and thus develop in him that natural sensibility, which to such a great extent nourishes the imagination and so becomes a source of the finest qualities."

In other words, Råshult and Stenbrohult were the right places to foster qualities such as characterized Linnaeus the man. Wahlenberg develops his theme and argues that the long, low ridge that runs through Uppsala has helped to change the university town into a source of creativity—of the "imaginative power" so cherished by the Romantics. Linnaeus would later admit that his home "appeared as if adorned by Flora herself. Here I have with my mother's milk been infused by the shapes of a multitude of plants."[9] When he grew up, the enclosures were still in the future, and the village was surrounded by woodland meadows, a form of cultivation typical of southern Sweden. Species diversity is an outstanding characteristic of these hay meadows, and in Stenbrohult they were "richer in tree canopies and fuller of flowers than in any other province. . . . When you sit there in summertime, hearing the song of the cuccu and also songs of many other birds as well as the piping and humming of insects, and you see the bright and splendidly colored flowers, you cannot but be astounded at so excellent a creation."[10] Where the Taxås ridge enters Lake Möckeln, the views from the top of the steep rock face leave you with a sense of a great and spacious world.

Impressions of the landscape's striking natural beauty were instilled in Linnaeus from childhood. He had learned to recognize all the flowering plants around Stenbrohult: viper's grass, common milkwort, quaking grass, yellow rattle, heath spotted orchid, wolf's bane—the poetry of the names alone was wonderful. The church and the rectory stood in the middle of the village. The rectory had a garden and an avenue ending at a circular drive in front of the main house. The guardian tree grew next to the house, and behind it stretched a kitchen garden and a somewhat wild "pleasure park." These were quite large cultivated areas looked after by hired workers rather than the pastor himself. The practices and layout followed ancient patterns, perhaps going back to monastery gardens. Most of the inspection records reflect such traditions, including cabbage patches, hops cages, and ponds stocked with Crucian carp. At Råshult, they had an unusually well laid-out cabbage plot as well as an herb garden and several hop cages—hops for brewing the dark local beer. In his *Travels on Öland*, Linnaeus describes the garden of his childhood home: "It had many more species of plants that any other garden in Småland had and did with my mother's milk inflame my spirit with an unquenchable love of plants."

The rectories around the lake formed part of a distinctive culture. The inhabitants were related, and shared interests and tasks. The care taken of gardens and plants is documented; one of the clerics actually owned *De plantis* (1583) by Andrea Cesalpino, the great Italian botanist.[11] Families, neighbors, and at least some members of the local gentry bonded with each other by being godparents when children were christened—as was carefully noted down.

Rectory kitchen gardens were "sermons made real." Clergy who traveled abroad brought back ideas and plants—lilacs, for example, and herb gardens and arbors. We know about the gardens at Råshult from Samuel's letters and later from Linnaeus's *Adonis stenbrohultensis* (1732), in which the plants were already ordered by their sex.

His father, Nils, was impressively knowledgeable about plants. Carl was barely four years old when, one afternoon when house guests were resting on the grass, Nils held forth about Latin plant names: "The little one found listening to this a heart's delight. Ever since, the lad would give the Father no peace." Once, he had forgotten a name and was scolded by his father, but "since that moment, the boy's entire will and thought was to remember the names and never cause his father displeasure."[12]

Nils had an interest in botany that went well beyond even what might be expected. Linnaeus writes: "How come the boy fell in love with Flora I do not know. What I do know is that his Father had always loved the company of plants." As a student in Lund, Nils had learned the Latin nomenclature of plants, and collected and pressed some fifty specimens. Linnaeus continues: "It was known that when the boy was troubled and could be soothed in no other way, he would soon fall quiet the moment he was given a flower to hold. This, what I believe to be his innate delight, was heightened when the boy listened to his Father speaking of some characteristic of a plant that seemed noteworthy." Young Carl was given "a garden *en migneateur*, where in a small space he grew samples of all that was found in the garden."[13]

Samuel Linnaeus comments in his letters on his father's keenness for gardening and how it captivated his older brother. He also describes how Nils went about planting: "In this garden, my dear departed father had with his own hands created a round, elevated area like a *table* around which plots with herbs and shrubs represented the guests, and groups of flowers, the dishes served on the table. Our mother often went to see it: this was at the time when my brother was conceived." The layout was recreated in 1982, the round table "set" with wild thyme, sweet William, lavender, feverfew, and musk mallow

Samuel tells another story about Carl at play with his siblings during school holidays: "At the slightest suggestion of someone's ailment he would

FIGURE 4. The young Linnaeus? Unknown artist, oil painting in the art collection of the Småland Museum, Växjö.

palpate the sufferer's pulse, make as if to use a thumb lancet of wood (for bloodletting), and search for herbs with which to cure his sisters."[14] As a boy, Linnaeus played at medicine and, by his own system, he was then at the third age of life "when, by running hither and thither in constant preoccupation, the child practices his body incessantly, day after day"—as he would later describe it.[15]

How does an interest start? How does it become an obsession? Linnaeus speaks of a vision: "I was ill in 1718 from winter until Whitsun, then came out into the greenery which appeared to be not of this world but of Paradise."[16] Then:

> I believe there are persons who, when stepping outside, see the ground in front of them as green with some other colors, the cloud he sees like shadows and the sun like a bright disc, so enthralled is he by economic, political, fanatical, arrogant, lustful, mercenary, vengeful, etc. concerns and impulses that he cannot see further into what comforts our Creator has provided and placed us in the midst of. . . . I admit in my own case, one summer in my youth when ill with a strong fever, I did not look at nature from *medium Martii* until *Juli* and then, when I was allowed outside, I saw the world in a very much changed manner, different from before and all spread out in front of me, being so high, so beautiful. . . . Then consider Adam and Eve: perfectly made, in their finest, most healthful flourishing, free of prejudices, and so shown hills and green valleys with rivers running wherever the most temperate weather might be (in *Mesopotamiae terras* or *alibi*), everywhere clad in grass, plants, trees, all green and also with every kind of flower of great loveliness, divers form and *coleur*, animals who run about, birds on the wing flit through the air, singing and calling, fishes silently wander in the pellucid river gently flowing forth, insects seated on flowers and trees like small jewels, wings shimmering like a *pocader*, everything together in *migniatyr*, then would it not be the proper moment to admire the Creator as if only then had they been given eyes to see with and in their joy could not decide where to turn to see more, to observe sun and moon, stars, sky, and wander in the night. As their ears hear the murmuring of the weather, the sounds of animals and songs of the birds, would they not easily mingle with many.[17]

The Fall from Grace awaits . . .

Linnaeus came home to Stenbrohult to visit in the summer of 1728 and at Christmas in 1731 and 1732. In spring 1735, he stops by on his way to Holland and stays from 19 March to 15 April. All his siblings are there, and his old father: "Mother missing as she had died since one was last at home. House in confusion."[18] On 15 April: "Finally, after a month of staying at home, one must *valedict* one's sweet *natale* Stenbrohult with one's *Patre* in his sixties and 4 siblings. . . . My elderly father commended his *Biblioteque* and also my youngest sister to my protection should a fatality come to afflict him."[19] Linnaeus returned for a few weeks in the summer of 1738, again in 1741 on his journey to Öland, and in May 1749 on his way to Skåne. It adds up to about eight visits, surely evidence of what he calls his *nostalgia*.

When he visited in 1749, his father had died and the rectory had burned down. He wrote: "Here, I found the birds dead, the nest burned, and the young ones dispersed. I could hardly recognize the room where I had myself been hatched and felt as if present at *campum, ubi olim Troja* [the field where once stood Troy], the place where my dear departed father the rector Nils Linnaeus planted his garden that formerly glowed with the finest plants in Sweden but was utterly destroyed by fast flames before time took him away on 12 May of the past year. The pleasures of my youth, the rarest plants once growing wild in this location had not yet emerged. I, who 20 years ago knew every person in the parish, could hardly find 20 of them now; those who were stout lads in my youth had gray hair and white beards, their lives were done and a new world had taken the place of the old."

This passage bears witness to his sense of alienation. Linnaeus had been born in an impoverished country where little was thrown away. Hordes of starving, homeless people drifted along the roads in the years when the harvest failed. Nowadays, visualizing such scenes makes us uneasy; they are very different from the glamour we believe characterized the last half of the eighteenth century.

Linnaeus tells us of his own poverty: "He could lie awake all night as he lacked the money for evening meals when he was a student." He had to "incur debt to afford food, had no coins to have mended the soles on his shoes but must walk with his bare foot on some paper that he put inside the shoe."[20] Later, he would try the natural foods eaten by the Sami as well as share sumptuous meals in the homes of the wealthy but still find his years of hardship difficult to forget. In *Nemesis divina*, his book of exemplars, Linnaeus shows compassion for the poor, as in this example: "The poor farmer labors all year, has barely the straw to rest on, and is paid but little; *sic vos, non vobis* [thus you work but not for yourselves]. Consider the poor slave, at work while you sleep. You would say of him, he ploughs my field, it is my farm and I decide. I tell you: nothing is yours. God has lent you all."

In such times, a healthy child might well be the best insurance and old age pension. The injunction to "be fruitful and multiply" was taken seriously not least by the reverend clergy. It went without saying that Linnaeus had siblings. In 1730 his eldest sister, Anna Maria, married *magister* Gabriel Höök, appointed rector in Virestad in 1742. Her son Sven Niklas painted a group portrait of Linnaeus's only son Carl and his sisters. In 1749 the next eldest sister, Sophia Juliana, married *magister* Johannes Collin, pastor in Ryssby. Both women gave birth to a dozen children. Once widowed, Sophia was destined to live in poverty, and Linnaeus pleaded with the diocese on her behalf. Her daughter Anna Sophia married *magister* David Widegren, also rector in Ryssby. Carl's younger brother Samuel (born 1718) studied in Lund, where his doctoral thesis was supervised by Sven Lagerbring, professor of history. He fathered twelve children, of whom only three reached adulthood: two daughters who married into the clergy and a son, Carl Samuel, born 1778. The boy had made Stenbrohult the subject of his doctoral thesis in Lund but died at the age of twenty-two. Carl's youngest sister, Emerentia (born 1723), married in 1749 to the clerk to the local authority, a Mr. Branting from Virestad. This overview indicates the social circles of country clergy.

In deepest Småland, spirits roam and can be glimpsed in the moonlight: "In Stenbrohult, when the moon shone at night, everyone could see three dancing white-clad ghosts, around three musket shots away. I denied their existence and laughed at those who believed therein but they promised to show me. One evening, when Dr. Rothman, the local judge, was present, a farmhand came with the message: 'Now they are dancing.' We went outside and saw them with our very own eyes. Then Rothman took me and my father to the place where a boulder striped with white moss seemed to shift in the moonlight." Later, he wrote self-confidently that "Spectra are seen in Smoland every night.... I believe there are as many in the world as there are in just Smoland." Rothman and Nils and Carl Linnaeus undeniably come across as a trio of enlightened men in a dark place.[21] Even so, Linnaeus was always uncertain about the validity of many popular beliefs.

Linnaeus insists on the importance of birthplace. Rather than an astrologist, he is a "topologist." He would instruct his students under the subject line *Solum natale* that your health is always at its best in the place of your birth and where you were brought up because you are used to the air. Migratory birds, returning in spring to their native homes, are examples of this principle. "In this context, an illness called *nostalgia* [homesickness] is found, which principally causes the sufferer to be weak, anxious, *cacheticus* [cachexia, a wasting syndrome] and look as though having contracted pneumonia." Swedes have found that traveling to Holland can bring on this condition: "The symptoms afflicted also *regius medicus* [royal physician] Linnaeus when he was in Holland and contracted the ague before he left the place where he stayed and arrived at the sandy heathlands of Brabant where the air was clean and, on that day, he was well again." Linnaeus has another example at hand: an Inuit girl from Greenland who lived briefly with the late queen of Sweden [Ulrika Eleonora the Younger]. He observes that no *Lapp* living high on a hill is a *melancholicus*, but the opposite is the case with those who live in forests. The place where you live and the air you breathe are critically important, and it follows that homesickness is very strongly felt by people born at high altitudes who have ended up living in lowland areas, for instance Swiss people in Holland.

Linnaeus also stressed that inheritance matters, for instance to be conceived by healthy parents, and points to himself as living evidence. True, his temperament at least was very different from that of his placid father, who was content to live quietly in his own small world. Linnaeus was perhaps more like his mother, with her "quick" mind.

When he died at the age of seventy-four, Nils Linnaeus was celebrated in learned journals for his beautiful garden but above all for instilling a ceaselessly active mind in his eldest son. Fredrik Hasselquist, a naturalist who had traveled widely, writes in praise of the son as much as of the father: "His son sprung from his root / and brought his country honor; as Nature's spokesman he stood out / peerless in the Nordic lands."[22]

What were Linnaeus's thoughts about childhood, his own and more generally? He insisted that babies should continue to suckle mother's milk for a few years because he saw it as important on moral and practical as well as medical grounds. Worm infestations can follow drinking cow's milk but not mother's milk. The baby consumes milk at will, without forcing, and the mother need not chew the food for her little one. A child must never be deprived of rest and sufficient sleep; for boys, he recommended sleeping twelve hours out of every twenty-four. Parents often urge their children on: "[They] think it will give their children quicker minds, but instead of catching fish they get frogs. . . . Is it not so that every day you can observe the children of the wealthy families who, from the age of 4 years until they are 24, have been made to study and tutored daily whereas the poor farmer's son who begins with books 10 years later, has come much further in his studies and also in his *ingenio* although having taken the same length of time while the other's youth had passed in silence."[23] The physical effects are dire, too: "To force the children to sit all day over the books in *scholis* distresses them and makes them smaller and thinner."[24]

He summarizes: "All that is done to bring up a child is directed toward this: to teach them good habits." On the other hand: "Habits are of the Devil. God help whoever has been given a bad habit. . . . As when children are told to fear the villain who lurks in the darkness beyond the windowpane so that they must not step outside at night and must be quiet. . . . As for me, I did not dare go outside alone until my twentieth year and, in some places, I shudder even though I know better. . . . *Caveat hinc* [beware here]: If you instill a belief in a child, it will stay in his mind for far too long a time."[25]

What were the factors that contributed to shaping the young Linnaeus? He grew up in a rural-agrarian Sweden, in a Lutheran pastor's home, and experienced its natural surroundings, but also in wartime, with all its consequences of poverty, pestilence, and a high mortality rate. Oscar Levertin, who has written a classical account of Linnaeus's childhood, ends with the following passage: "On one of the innumerable pieces of paper on which, in his high old age, he noted down his solitary thoughts, is written in the shaking, perhaps stroke-troubled hand of an old man, a single word under the heading *Nostalgia*: Stenbrohult."[26] True, Linnaeus discussed *nostalgia* in his large manuscript on dietetics and also in his taxonomy of illnesses. We know that, for him, childhood is bathed in a blessed light. As for that "piece of paper," no one except Levertin has ever seen it.

The image of Linnaeus as a child is an integral part of the cult of the man. As the "child of nature," he gained by instruction in natural learning of a kind approved by Rousseau. He was, by definition, a child of nature, and his authenticity implies that he shared ancient wisdom and natural religiosity. Even as an adult, he was thought naïve, innocent, and, hence, charming. There are risks, though, when you gild a childhood like his—especially as he himself was lending a helping hand.

CHAPTER TWO

Studies in Växjö and Lund

STENBROHULT AND THE DIOCESAN CITY of Växjö are about fifty kilometers apart. The slender, elegant cathedral spires towered over the small city with its population of little more than a thousand souls. On 10 May 1716, Carl was taken to be initiated into the first year at Wexiö School, "where coarse masters, using equally coarse methods, instilled a thirst for knowledge into children."[1] There was a war on, and Linnaeus was almost exactly nine years old. Later he would advance to the Grammar School, still in the same building: the young boys were taught on the ground floor and the scholars on the first floor. The building, barely a year old, had been located next to the cathedral and given a handsome doorway under an ornamental pediment where heraldic lions flanked the monogram of King Charles XII and the date 1715; it became known as the "Karoliner House." The building still stands, despite the fires that have ravaged the city. Linnaeus was taught in the ground floor rooms from 1716 to 1723, and upstairs from 1723 to 1727—ten years, important for his development, which by now has been keenly and usefully examined.

These years were also important for the whole nation. The hour of midnight had struck for Sweden's era as a Great Power, and the hand was moving "from 12 to 1" when Queen Ulrika Eleonora the Elder, after one year of ruling from the throne, handed over the power vested in her by her dead brother Charles XII to her husband Fredrik I. A new age—the Age of Liberty—had arrived and brought with it the abolition of absolute monarchy. Arvid Horn, the cabinet secretary, was the strong man leading the country. The opposition against Horn led to the formation of a politicized group, the Hats—the name referred to the tricorn hat worn by officers and gentlemen—which later became an opposition party with foreign wars on its agenda. Horn's peacekeeping policies were supported by another party, the Caps. When the 1738–39 parliament finally got rid of Horn, power shifted into the hands of the parliamentarians.

For Linnaeus, going to school meant being absent from his family home for long periods and boarding in an unknown place with new daily rhythms and demands. Sixteen terms in Växjö would have made the boy travel 52 times along the 50 kilometers between home and school: 2,600 kilometers in carriages or sledges, or on foot. He eventually learned to recognize the territories of every plant along the roads linking Stenbrohult and Växjö.

Like all the little ones, Carl was placed under the tutelage of an older boy, or praepostor, one Johan Telander, who was a harsh taskmaster: "Instead of cajoling, thrashing was used to teach him to read."[2] Telander ended up as the rector in Korsberga parish north of Växjö. In Linnaeus's school, the daily timetable was quite closely based on the 1724 statutory guidance on education: the day began at 6 a.m. and ended at 5 p.m. including times dedicated to prayers and hymns. The week's forty-seven study hours were mainly devoted to Latin (17 hours), theology (14 hours), and Greek (4 hours), a scheme that in the main stayed fixed throughout his school years. Still, learning how to dance was also taught, and he enjoyed that.

The "coarse masters"—Linnaeus's tutors, one of whom was also a relative—were Johan Telander and Gabriel Höök. There were sixty-three pupils, some of them of relatively advanced age, enrolled in his year. The collegium of teachers had members of international standing, such as Johan Rothman, one of the senior masters and a doctor of medicine. Some colleagues were immigrants. This small, remote city with its oddly un-Swedish name was the home of men with even more un-Swedish names—for example, the music teacher Ernst Zeidenzopff and the cathedral organist J. C. Zschotzscher.

Johan Rothman taught logic and physic—"natural sciences"—and was also the province's chief medical man. He gave Linnaeus private lectures and lent him books to study. It has been argued that Rothman contributed more than anyone else to the development of the boy's mind. It was he who introduced his pupil to an essay on the sexual life of plants, *Sermo de structura florum* by Sébastien Vaillant (1717). On Rothman's death in 1763, Linnaeus wrote to the widow, describing her husband as "my only teacher in the elements which have become my profession and all my happiness."[3]

Linnaeus's memories of his school years are illuminated by Samuel Ödmann's book *Om disciplinen och lefnadssättet wid Wexiö skole och Gymnasium in til 1780-talet* (On the discipline and living conditions at Wexiö junior and grammar schools until the 1780s). Ödmann left Växjö in

FIGURE 5. The young Linnaeus. Romantic, mid-nineteenth-century image. Engraving based on a painting by Frederic François Joseph Roux (1805–1870). Uppsala University Library.

1768, forty years after Linnaeus had left, but probably little had changed and life was still tough for the juniors. Ödmann wrote: "You would see honest men's offspring, often no taller than their broom, occupied with sweeping all Saturday afternoon." The school had its own penal code with established types of offenses having punishments specified in terms of the number of strokes with a cane. Pupils were given special tasks, like waking the teacher in the morning and lighting his candle. A solemn ceremony marked their transition from junior to senior school. Ödmann also comments on the special use of language in the school: you "journey" from the school to the academy but at all other times, you "travel." Pupils about to set out on their "journey" would gather to sing and play music below their teachers' windows. The teachers addressed the younger boys as "you" but the seniors as "Monsieur."

Their living conditions were, according to Ödmann, to be seen as "remarkable links in the history of Swedish Scarcity." He tells us of how boys of all ages were eating from the same dish "with utter equality," that all card games were forbidden but play and music making encouraged—unlike *amourettes*, which were not permitted. Smoking tobacco was not tolerated in the school, but you should nonetheless have mastered it in time for arriving in the academy.

Pupils who couldn't raise the school fees went begging in a local parish, an accepted practice known as "walking the parish." Linnaeus "walked" at least once, in 1717, when his name is found on the list of participants. The walking, accompanied by singing and playing instruments, was done twice every term and most enthusiastically at Christmas. Walkers were allotted to a parish, some of which were better value than others; Linnaeus apparently did the rounds of Stenbrohult.[4]

The 1724 statutes prescribed that every grammar school must have seven senior teachers, at least one of whom must teach mathematics and another one logic and "physic." The math teacher had to cover arithmetic, understanding of the globe, historical time as taught by the church, geometry, and geography; "however, he is also to undertake the duties of demonstrating in the summer to his Disciples while in the open the *praxin geodeticam*, and in the winter *Constellationes* and more such as pertains to Astronomy." The other science teacher lectured in logic, philosophy, and allied disciplines and "ought also to propose to his Disciples the *Physicam* of the *Suicero*, in particular of such as pertains to Anthropology, as he is most earnestly obliged to offer up the light which this science offers, due to the dedicated work and investigations carried out by newer adherents, so as to prevent Youths from addressing to *Philosophia Scholastica* various obscure and unfitting questions."

It is, of course, dubious whether the new edicts were turned into actual teaching and whether Växjö was among the grammar schools that could muster the right teachers. Nonetheless, the new statutes that redefined the education of Linnaeus's generation contained some remarkably modern ideas. He was forbidden to study scholastic logic, which was thought to be the basis for his ideas for a taxonomic system. On the other hand, his studies gave him useful knowledge about geodesy and anthropology. The anthropology textbook had been elaborated by Andreas Rhyzelius, bishop in Linköping, and printed there in 1725. It was entitled *Suicerus Erotematicum physicae Aristotelico-Cartesianae compendium* and was deemed "dry and lacking in interest as it intends to explain all in a philosophical manner and knows nothing whatsoever of studies in nature itself."[5] Latin was practiced with Cornelius Nepos's *Lives of Illustrious Commanders*, and theology with Hafenreffer's *Compendium*, but the textbooks for other subjects are not known. They have not been found in Linnaeus's library and possibly depended on what the teachers happened to have available.

Nils Linnaeus had given his son five literary works, all matching his own particular interests: above all, a folio edition of *Historia animalium* by Aristotle. On the endpaper, young Linnaeus wrote in his childish hand: "That Alexander is said to have donated to his preceptor Aristoteli 80 Talenta for the books he composed on divers parts of nature and properties, and that makes 60,000 silver thalers." It is a note that tells us something about a poor boy's wish for glory. He also owned much more modest texts, for instance *A most usefull book of herbes* by Arvid Månsson, a farmer and mill owner from Rydaholm in Småland, and other similar works by Johannes Palmberg, Olaus Bromelius, and Elias Tillandz. His copy of Aalborg's *Medicine or a book of healing* is full of notes, often in Latin—Linnaeus was apparently determined to become a doctor. A note on the inside cover reads "Who on his health doth set a highly price / will for his carelessness pay dear / which is what he should always fear." The dates found in this and other books help establish the development of Linnaeus's education from about 1725 onward.

While at school between 1725 and 1727, Linnaeus entered his botanical finds in his own Book of Herbs, still kept in the Växjö library, and only recently published in English translation as the *Carl Linnaeus Notebook* (2009). A book of herbs was a botanical miscellany on the roles of plants in food, medicine and childbirth, beauty treatments and poisoning (the frontispiece shows an alchemical tree complete with mineralogical symbols; a tree, shorn of any symbols, is the logo of Kalmar's Linnaeus University in Småland). The entries in the book include older botanical knowledge and works of Swedish and international literature as well as notes on a very wide range of subjects—tulip mania, the potency of plants, aphrodisiacs, the legacy from Galen, lines from the Bible urging us to "consider the lilies," quite a lot of information about brewing beer but also about the use of cumin for deworming and rhubarb as a laxative. He had added extracts in the style of the early medieval encyclopedia, *Etymologiae* by Isidore, but also told curious anecdotes drawn from the local lore—for instance, this story about the effects of black henbane. Pastor Flintsten, his wife, and his organist were sitting down for a meal in the Virestad rectory. They fancied some horseradish and ordered the maid to get some from the garden. By mistake, she brought them henbane roots and, after a while, the pastor "began to fix his gaze on a pane in the window for well nigh an hour and then he turned wicked and hit out on all he saw. The pastor's wife began to sing at the table, then to dance, holding on to whoever she saw. The organist wanted to jump out through the window."[6] Someone rode off to Växjö to fetch the doctor and the matter was sorted out.

One note concerns a rare mallow, the "Chinese mallow" or *Malva verticillata v. crispa*: "About this plant I have heard it said that if a maid who is no longer a virgin sniffs its powder it will cause her waters to break at once and against her will." Some of the longer entries appear to be based on his own experiences: "Cheeseweed, whose flower if much smelled by a maid (who is not a virgin) makes her come down in a faint. Probatum est 22 Julii 1724. Maid, Stina Trolle from Skatelöf" (the details about the girl are added in upside-down mirror writing). Sex features in many of the different pieces of advice about plants—this is about an orchid: "Old Age Succor. Bed Comfort. This herb invigorates male nature, brings on manliness. When a young desirable and keen woman has to suffer an old laggardly and slow man, she must prepare its root, make him take it (since it is true that the old do not have such urgency in bed as do the young); truly, by doing so she turns the half-dead man vigorous anew and he will be enjoyed as well as would a young urgent and jolly man, and so give succor in his old age and also help him make himself a young son. Bed Comfort is for the old, to have them comfort each other as the herbe makes the man as well as the woman desirable." All of which goes to show that the world of plants had been sexualized in various ways long before the introduction of the sexual system of classification.

Animals and nature clearly played major roles in the life of Linnaeus. "I had a dog that was very dear to me when I was only a lad. He never left the farm with me away at school for ¼ year but the day I was to come home he went out to wait, yelped all day until I arrived when he ran to meet me with 1,000 joyful affections."[7]

Some older accounts contain the claim that, since no one had any better ideas, one plan was that young Carl might well become a shoemaker. Without taking the idea literally, it should be seen as an option for someone different, possibly defective in some way or otherwise peculiar. For instance, the mystic and Lutheran theologian Jakob Böhme was a shoemaker. Young Linnaeus didn't care to join the clergy and instead wanted to become a "medicus," which also entailed being active as a "botanicus." At first, his father was taken aback, but he later came around to it. Regretfully, with his eyes full of tears, he said: "May God give happiness, I shall not force you to do that for which you have no inclination."[8] His mother was a more difficult proposition, and he delayed telling her for a whole year. When, in the end, he finally broke the news, she almost had a stroke but took comfort in her husband's promise that their younger son Samuel would be kept out of that dangerous garden.

In the summers between 1724 and 1727, Carl stayed in Stenbrohult, occupying himself with reading, and collecting plants and insects. He put the plants between the leaves of his books, and the insects, mounted on pins, inside a small cupboard. When all was said and done, his mother was pleased that Carl didn't go in for any lewd activities, and "the father supported her in these thoughts." She also enjoyed "Carl's quick wit."[9] What this means isn't clear, but it presumably refers to talent or intelligence.

There was a moment when everything could have come to an end. Much later, Linnaeus told his students a story about himself: "Once, when the archiater was at the Vexiö gymnasium and had gone to the lake, where to cleanse himself, he stood on a stone, but some of other fellows knocked him over. In the water, he became giddy, his eyes spinning, and *anxietas* gripped his heart. Then his mind went black and he sensed nothing. However, his preceptor, a final-year boy who had missed him, pulled him out and blew life-giving breaths back into him."[10] This event might well have taken place on the shores of Lake Löje, where people went in the spring to wash the winter off their bodies. The lake was shallow near its banks, Samuel Ödmann has explained, but was widely believed to be under the rule of Näcken, the capricious water spirit. This myth had a built-in warning of the abrupt depth and the rock that marked the danger point, the Djekne stone.

Oddly enough, Linnaeus doesn't mention any near-drowning incidents in his autobiographies, which means that his biographers don't mention it either. A very brief version of the story can be read in his *Lachesis Naturalis*, here in translation from the Latin: "When I nearly drowned in 1722, I experienced a spinning sensation, anguish, giddiness, blackness, swooning."

Linnaeus was fifteen years old on the day he risked not getting any older. He never learned to swim and seems in later years still to be frightened of water and journeys by boat. His inseparable friend, Nils Rosén, had an episode of apparent death in his childhood, and his future fate was in the balance. A miraculous event was an expected feature in the lives of great men; in a lecture, Linnaeus mentioned three further examples apart from his own. Besides, "apparent death" was an issue in mid-eighteenth-century medicine: how many had been unnecessarily lost to society?

Later, Linnaeus himself successfully used the mouth-to-mouth method on his daughter Sophia, who had appeared lifeless at birth. His son, the naturalist Carl Linnaeus the Younger, would refer to this method in his lectures on dietetics and believed it to be an ancient skill developed by the Egyptians: "My sister Sophia was born dead. She showed no signs of vital consciousness and lay there quite dead for a while. Father called for an old woman who would blow into the child by laying her mouth over the little one's," he continued. "It matters that the blowing is not too strong, because the air that enters the other has lost its *electricum*."[11] The early biographer

D. H. Steever has his own version of the narrative, in which Linnaeus, faced with the unconscious Sophia, cries out: "She must not, shall not die! Pulled her up into his arms and gave her of his breath—and she came alive!" This is only one of many contemporary scenes linked to the terror of being buried while only apparently dead. Linnaeus Jr. also tells of how medical students in Uppsala dissected dogs, but their owners tried to blow life back in their animals, and also of how he had himself, as a youth, a much-loved dog that fell off a table and died but was successfully revived by the mouth-to-mouth method.

In their Sweden, so rich in open waters, death by drowning was relatively common because so few ever learned to swim. Vital signs were hard to establish, and being buried alive was a realistic risk. When, at Christmastime in 1731, Linnaeus visited his parents, his sister Emerentia was ill in bed with smallpox. She was cured by being bedded inside the still-warm carcass of a recently slaughtered sheep. Her brother insisted, "This pulled my little sister out of death's grip," and added: "If someone is weak by blood loss, such as with almost all the blood having flowed from a cut limb, then nothing will restore and revive other than speedily dissect a living creature and place the sick inside it so that the spirit of life is attracted. Archiater Linnaeus assayed this with his own sister when she was but little and so weakened by variolation that you hardly saw any life in her; he laid her in her entirety inside a sheep and after a while she was helped and came to once more."[12]

Linnaeus finished at the gymnasium in 1727, where he was ranked as eleventh out of the seventeen school-leavers. A party—"a pleasing diversion"—was held in the home of the merchant Isac Munthe. Next, the twenty-year-old Linnaeus, clutching the headmaster's testimonial, set out on the 150-kilometer journey to his new station in life: Lund University. It had been founded only sixty years earlier and then held back by the war. As far as learning went, his luggage consisted of Latin, some Greek, biblical stories, and classical history. His rank in the final-year Växjö cohort did not point to a very promising future, but Rothman's private tutoring had impressed him so much it determined his academic choices.

In Lund, as in Växjö and later in Uppsala, he was in the presence of a towering cathedral. The great church cast a massive shadow over the ancient city of Lund, by then badly damaged by fire and warfare. A few hundred students came and went, a large minority in a local population of about one thousand. The university was surrounded by a wall that set it apart, a state within the state. The library with several thousand volumes was kept in the Lundagård House. It was a good place for medical students to work, thanks to a donation by the polymath Christopher Rostius, the first holder of a chair in practical medicine. Lund's status symbols included remnants of the city wall, a pharmacy, a printing press, a botanical garden, and, of course, the cathedral. It was then, as it still is, a city of gardens. The cultivated land was on either side of a central street, joined by so-called cow wynds—grassy, narrow alleyways, deep in manure at all times.

Linnaeus brought his headmaster's testimonial to Lund—the young man is said to be "God-fearing and studious"—as his entry permit to higher education. On 19 August, he matriculated at the university but did not join a students' "nation." He might have intended to end up in Uppsala all along but, if so, why not go there at once?

Now he had become a *civis academicus*, a citizen of academe, with all that entailed, including limits set by time. Much later, he recollected: "Now, studies became for Linnaeus as joyous as until then they had been vexatious. Truly, the aversion for *humaniora* instilled at school contributed greatly to Linnaeus's failings in *exercitium styli* at the academy and that he did not join *collegium humanioribus* or any of the sciences but solely medicine, which he lately has endlessly regretted."[13] Is it his oft-mentioned lack of languages that Linnaeus refers to in this remark?

It didn't take them long to find each other: one, a brown-eyed, quick-witted student, the other a sickly, one-eyed man. On 21 August 1727, only four days after his arrival, Linnaeus had moved in with Stobæus. "Made it so that I had a chamber in the house of *Doct, et Practicus nobilium Scaniae* Kilian Stobæus, whose *privatissimus* I was for as long as I remained in Lund. Here was a fine opportunity to acquaint oneself with minerals as the subject he prized most highly was indeed Mineralogy, of which he had collected many specimens together with many other Curieusities. Within his privat. colleg. he studied with his own *collegium physiologicum* and also *sciagraphia* but *privatissime* in the Valentinii Musaeorum. For the love he had for me I will be greatly obliged to this Gentleman for as long as I live, as he did not love me as a Disciple but more as his own son."[14]

Who was this man? Linnaeus's intimate characterization of "Tobus" puts great stress on his otherness: his family was common, he was self-taught and had many physical defects, walked with a limp, spoke with a stutter, was one-eyed and fat, with a birthmark on his lip. Linnaeus has seen "a sadness" in him, which could suggest a manic-depressive tendency, or bipolar disorder in present-day terminology—a strange man and a genius. He was an adherent of Stahlian medicine, that is, he believed in waiting for nature to do its best before attempting a cure—a better idea than most practiced at the time, though different from Linnaeus's own views. Stobæus's practice was large; as is quoted elsewhere, he was said to be poor in looks but held in high regard.

Stobæus was an armchair traveler. The world had to come to him, not the other way around. He needed his specimens for teaching, but his collections also led to contacts with other learned men. Cabinets of curiosities supplied topics for research, and were displays that attracted visitors as well as being useful capital investments. There are many examples of how similar collections became the bases of great museums. In Stobæus's case, his "specimens" mattered because they formed part of the core collections of the Historical Museum in Lund, and its stores of geological, botanical, and zoological (especially entomological) materials. It is hard to stress enough how important they are for scientific investigations. If assembled on reasonable principles, inventories and their sorting and classification were the foundations of science.

The pupils taught by Stobæus were a remarkable year: Carl Linnaeus, Nils Rosén, Sven Lagerbring, Johan Leche, Nils Retzius. Linnaeus wasn't the only favorite. During the summer holiday in 1732, Stobæus had taken the waters at a spa and wrote: "[I was] in the company of my most preferred and curious *physicos studiosis*, whom I, and throughout the entire year at home as well as away, serve as a humble tutor in the Knowledge of Nature and *historia naturalis studium*, about which I can truly declare that I was the first at this academy to address and inspire." A German medical student, Samuel Koulas, lived in Stobæus's house. Linnaeus, who had studied physiology under Rothman, tutored Koulas in the subject and was in return lent books "out of Stobæi library" overnight. This is a popular anecdote: Stobæus's elderly, sleepless mother came across Linnaeus reading at two in the morning. She warns her son about the country lad, but Stobæus is impressed by his young student's industry and hands Linnaeus the key to the library, telling him to read as he pleases. This story has been told many times but without anyone mentioning the likely reason for the old lady's reaction: a common, justifiable fear of fire. Linnaeus is offered a bed and a seat at the table at meals; the childless host wanted him to inherit.

To be seen means recognition that matters for everyone, but some people seem incapable, even with two functional eyes. Stobæus saw "the singular man" in the youth in poor clothes who had no money for books but a deep knowledge of botany, and allowed him to borrow his books and study in his library—a story that Linnaeus often told, and it has even been turned into a poem by Herman Sätherberg. Like many anecdotes, it can be taken as an argument in a discussion about virtue. The relationship between patron and patronized is not unlike that between father and son. If we chose to stick to Linnaeus's own accounts, Stobæus took him on as his "beloved Linnaeus," "a son" as well as "a disciple, taught *privatissime*." The same could be said about other students, for instance Koulas, who also lived with Stobæus "as a son." In this context, "son" has nothing to do with genetic links, just as the present-day German expression "Doktorvater" (lit. "doctor[ate] father," that is, thesis supervisor) signifies a relationship that once also existed in Sweden, and was much closer than "supervisor." Linnaeus has also claimed that he acted as Stobæus's secretary: "Stobæus calls out to have Linnaeus come down to help him pen a letter in reply to a sickness; but Linnaei writing was uncivilly treated and rejected."

By then, Linnaeus had moved into "the uppermost floor, uppermost in the gable" and had access to Stobæus's library of some seventeen hundred volumes. Lundagård House, home of the university and its library, natural history collections, and dissection rooms, was a long stone's throw away. The cathedral loomed just next door, but if Linnaeus ever visited it is not known. Generally, his years in Lund are documented only in occasional notes. For instance, in May 1728, he was somewhere between Lund and the neighboring city of Malmö: "As the sun rose, I observed 2 earthworms which had their holes close by and lay with one end in each other's hole. . . . One had a small penis that lay on the other and its origin was at the ring and the penis was inserted into the other where its ring was. But I was not able further to observe them accurately as they were quick and leaped out of sight, each into its hole."[15] An intriguing note that gives us a vision of the young student as he crouched to watch one of nature's many wondrous scenes—this time, how hermaphrodites go about vital matters. His interest in reproductive functions became a recurring trait. One is also reminded of Darwin and his fascination with earthworms.

One memory that would always stay with him was a persistent bite inflicted by the terrifying "*Furia infernalis*." He was on a botanical excursion in the spring of 1728, and the day was warm enough for him to take off his coat and waistcoat. Next, the *Furia infernalis* stung his right arm, which swelled into a rigid trunk. He had to take to his bed, but the inflammation grew worse. Stobæus, by then fearing for his dear Linnaeus's life, had to leave the young man in the hands of the barber-surgeon. The surgeon, a man called Snell, opened up the arm from elbow to shoulder and bandaged it. The "Furia" is mentioned in *Miracula insectorum* and included again in *Systema Naturae*.[16] Later, the apparition spooks several entomological works. No one is sure about the identity of this monster insect, which by now has a small writerly genre devoted to it. Linnaeus had, one way or another, escaped the jaws of death on three occasions: when he was ill in bed in 1718, almost drowned in 1722—and, in 1728, this attack!

In *Nemesis divina*, Linnaeus recalls that his brother Samuel, then at school in Växjö, was thought to be very bright. "I was held to be dumb, had just arrived in Lund. Everyone had spoken of my brother as if he were a professor already and said that he would become a professor." A fortune-teller had called at their home and heard of the brothers: "She who had seen nothing of us asked to be shown clothes worn by us and said of Samuel's clothes he will be a clergyman and of Carl's he will be a professor. Our mother changed the clothes over but the answers remained the same."

Then, a ghost story: "I was quartered with arch. Stobæus 1728 in the topmost floor at the gable high up to where no Lund night watchman's staff could reach me. Blows 2 times 3 fall so hard that I wake, with fear in my heart, full of dread that it is for me. After 2 days the archiater receives a melancholy notification of the death of one he was to cure and had been a most notable personage. The state of the mother alters, she contracts cholera and dies the next day. This is all I have heard and know for myself."

Linnaeus explored the riches kept in Stobæus's library and carried out field studies in the countryside and on the city walls. Stobæus was on hand to instruct his pupil in how to make an herbarium. The student's early notes record an older series of lectures by Stobæus on Lange's *Collegium* (1723), and private tutorials with Stobæus on Valentini's mighty catalog of collections, the *Museum museorum*. Stobæus owned the 1704 edition of Valentini's work, generously illustrated with copperplate engravings, which must have made a strong impression. On 25 September, Linnaeus buys in Lund Johrenius's *Vademecum botanicum* and inscribes it with his motto *Ens Entium miserere mei*—"Being of Beings, have mercy on me." The volume is full of annotations, often about plant locations in Skåne and Småland, and much about it indicates that it had been used as a textbook. Linnaeus's educational progress can be followed by studying his notes.

Visiting libraries was part of his studies and he borrowed books; as on 27 February 1728: "from the Doct. Stobæi Library, the *Historia indiae orientalis* by Pisoni, also *Aldrovandi ornithologiae et lib de quadrupedibus*," both large, important books. On 8 March, his university library loans included Rembert Dodoens's medical herbals in three volumes and Rudbeck's *Campus Elysii*. He also borrowed Caspar Bauhin's famous *Pinax theatri botanici* (1623), at the time still the standard work of botany. The stocky Linnaeus must have been quite a sight, leaving the Lundagård House weighed down by Aldrovandi's enormous folio. A year or so later, Nils

Retzius dragged this awkward burden back to the library, plus Aldrovandi's just as weighty *Ornithologia* and a few others of Linnaeus's loans.

Who was the professor of medicine in Lund at the time? "The archiater and knight when he studied in Lund listened as a Professor lectured an entire year *de spiritibus animalibus*. The student took it all down with care but now finds it not well put together." The professor was the much-disliked Johan Jacob Döbelius, to whom Linnaeus also referred in another note: "Doebeln lectured on Miscellanea and whatever else but to no profit for us." The young Carl Christoffer Gjörwell says about Döbelius: "Döbeln, German, good *practicus*, voluminous talker, looked just like a German. He once lectured to us in the physiology and spoke about how those who had high foreheads were possessed of great wit. As he spoke, he now and then pushed his wig back so that by the end of the lesson he had acquired twice as high a forehead as he had had before." Nowadays, a portrait of "Döbeln" has become the most frequently seen of all at Lund University because it graces the label of the bottle of Ramlösa mineral water, which has refreshed innumerable committee tables.

Next, a sudden change: "Stobæus is very vexed that Carl has left him." Stobæus had only learned of Linnaeus's departure for Uppsala when Linnaeus needed credentials and also through a letter from Koulas, who was in Uppsala supported by a grant from the Piper trust at the time. In October 1728, Linnaeus received what must have been a disappointed and very displeased letter (not kept) from his former teacher. He replied on 8 November 1728, admitted to having "repaid good with evil which is a devilish act" and begs forgiveness for his crime. In a letter dated 19 November, he is still distraught: "Therefore, my Master and Doctor, I pray you find it in your always so noble mind to forgive me who doggedly begs you thereof and wishes to be absolved from my crime"—to the best of my knowledge, this is the only occasion that Linnaeus asks forgiveness in writing. As he recalls on 12 October 1731, Stobæus never treated his student as a preceptor would a disciple but as a gentle father would his only son.

Linnaeus's letters included bits of information about his new university. He described its gardens as having been allowed "to run to waste"—probably hoping it would please Stobæus—but also expresses delight at Rudbeck's famous illustrations of birds. "How I would wish that you, my dear Sir, could see these images as they would most certainly be to your taste." He often included field reports and questions concerning mineralogy and petrified material. And, as if Stobæus were still responsible for his welfare, he added items about his finances.

Early in 1732, he visited Stobæus "to see his earlier so dear Stobæus and inform himself in *mineralogicis*, though it may be in vain as the Stobæi method may well not agree with Linnaei head." This visit, ostensibly about mineralogy, was likely an attempt to explore the situation. He would have found that others had moved into his place or were about to; in any case the appointee was a member of the Stobæus clan—the great man's nephew Kilian Stobæus the Younger. Linnaeus might have hoped to be offered a lecturing post in medicine, but Johan Gottschalk Wallerius was finally successful. Still, in a later, rather rough encounter in Uppsala, Linnaeus would win over Wallerius.

What drove Linnaeus to leave Lund? Was it Rothman's persuasion? Rothman had set about convincing him and his parents that the young man should be allowed to go to Uppsala—a large library and many sources of grants. "He listened." Why did he? Because scientifically Uppsala was preferable, or because the opportunities of making a career in science were better? Was he guided by botany? Did it bother him that his friend and rival Nils Rosén had already made the move? Among many likely reasons, another could have been that Kilian and Florentina Stobæus had no daughter for him to marry and so advance as a member of the family. Would the outcome have been different if he had started out in Uppsala and then moved to Lund?

Yet another possible explanation: a year or so later, Linnaeus wrote in an autobiographical note: "Committed a *faut* that was understood uncharitably by ill-disposed folk and caused great annoyance." There are no corresponding entries in the "official" autobiographies so we are left no wiser about the "error" Linnaeus referred to. It surely cannot be the reading at night in Stobæus's library—the outcome had only been benign. But in 1730, he wrote to Stobæus: "Alas, how numerous are now the occasions when I rued my follie in Lund, where otherwise there was such good prospect to learn of stones that I will never find the likes of." It might have been a case of rowdiness. It is strange that Linnaeus made no mention at all about the undisciplined student activities in Lund: they drank, fought, and generally misbehaved; "quite an abnormal state of affairs" as the historian Martin Weibull commented after going through court protocols from the time. The city burghers and the university got together to write a joint report, castigating the disorder. The fact that Lund's most profitable industry was beer brewing didn't help, nor that the consumption of aquavit was out of all control. Perhaps the dutiful student Linnaeus didn't participate in the boozing, or got into woman trouble although such things happened. We can only conclude that the young man's "faut," whatever it was, may have caused his parents and Rothman to force a change of environment.

Interchange between the Swedish universities was limited and dominated by a few families. In Lund, the leading names were Stobæus and Bring, while in Uppsala, three members of the Celsius family played key roles. Within the systems of patronage, important trade-offs were made between celebration of eminence and learned publications on the one hand, and various forms of support on the other. This was how ministers of the church were appointed, for example. Curiously enough, there is no study of the relationships that Linnaeus may have cultivated with mentors, patrons, and sponsors, nor any systematic analyses of how he chose his career. At that time, once he had excluded the church as a future employer, his choices were between academic and other, less settled forms of work: medicine or mineralogy—or, alternatively, combining the two by taking on a post as a medic in the copper-mining town of Falun, which thrived on its mineralogical wealth.

CHAPTER THREE

The Academy in Uppsala

LIKE LUND AND VÄXJÖ, Uppsala had grown from being an ancient site of a heathen cult to a city with a cathedral at its center. Uppsala University had been founded some 250 years earlier but stayed closed for a large part of the sixteenth century. The university still bore the scars of the fire that ravaged the city in 1702. The finest buildings clustered around the cathedral and the Gustavianum, the main university building with its anatomical dissection theater under the roof, library, and lecture halls. Previously, the university had been housed in the medieval building Academia Carolina (completed in the 1430s), which occupied a site between the cathedral and the Holy Trinity Church. The old academy was demolished near the end of the eighteenth century.

Traveling northward and eastward, Linnaeus probably arrived in Uppsala via Stockholm and the small town of Enköping. Like Lund, Uppsala consisted mostly of single-story unpainted timber houses. In the autumn term of 1728, Linnaeus matriculated together with some two hundred other new students. Most of them brought spare clothing, bedding, and food packed into wooden chests, all of which was meant to last them as long as possible. Linnaeus had taken his herbarium and insect collection with him. The protocols of the university court record the rowdy, disorderly student life, how they misbehaved in the streets and were "running *grassatim*," their rowdiness disturbing the sleep of the good burghers. Only a few years earlier, when the Russian navy had landed to burn and loot, the university's precious book collections had been hidden on barges moored in the bays of Lake Mälaren.

The city was divided by the Fyris River into western and eastern halves, with the academy dominating in the west and commerce in the east. The castle on its hill was in poor shape and the cathedral in need of repair, but both were soon to be restored. A miasma of smoke often filled the air, and Linnaeus would later observe that "in Uppsala the air is noxious due to the marshy ground." Vegetable patches on the edges of the city grew mostly cabbages, the streets lacked pavements and were used not only by people but also by household animals such as horses, pigs, dogs, and cats. The human population was a few thousand strong, not including the students, and supported a post office, a bookstore, and an academic printing workshop.

The youth of the students was a striking feature of university life, and many of them had to be supported when away from their parents by being placed in good lodgings and joining their fellow students in the nations, where the inspector served as a paternal figure—but this was not for Linnaeus. Each student was supposed to bring a supply of butter and a side of pork, food that would go rancid with time. The students came from varying backgrounds and included aristocratic youths but also farmers' sons and a handful of truly poverty-stricken outsiders. Of this last group, two have featured in Linnaeus's notes: Backman from the local county, Uppland, and the perennial undergraduate student Mats Westerberg. Backman, who lived mostly on milk (an early lacto-vegetarian, it seems), was described as "cendré—with sandy hair. He was of ordinary build, Linnaeus writes, but had in infancy been brought up on milk and, from the age of four, on milk and bread. He also acquired a liking for fish (twenty items listed). Westerberg presented himself in his "Description of a Life" (*Levernesbeskrivning*), an eccentric work in every way.[1] Westerberg got bored with the *gymnasium* in Västerås and traveled to the fairly nearby Uppsala "with the intent to take up my so far fallow studies," but he wasn't allowed to matriculate on account of his record of previous misbehaviors. Finally, after "much begging, bowing, and running around," he was accepted as a student. He was the kind of person Linnaeus couldn't help noticing: "Then Westerberg during ¼ year consumed meat, salmon, herring, butter; periit," in other words, he ate one foodstuff only for a quarter year, with a fatal outcome. If Linnaeus had lived under a less benign king "then I would have become Westerberg and Westerberg, I."[2]

Linnaeus had a hard time in Uppsala. For a poor student, every day brought new worries. "The parents had little wherewithal and the monies Carl received were soon gone. *Professores medicinae* lectured little or not . . . so in Upsala he frequented a barbarian medicine. He began to suffer privations, could not pay to have his shoes soled. . . . He had wished to go back to his good Stobæus, but for such a long journey the money in his purse did not suffice."[3]

Then, his existence lightens thanks to a friendship with the botanizing cathedral dean, Olof Celsius, Anders Celsius's uncle. Their relationship led to a meeting with the professor of medicine, Olof Rudbeck the Younger. Linnaeus's life story is full of such singular encounters. He met Celsius outside the botanical garden, and the older man became his "paternal friend." Celsius introduced him to the Rudbeck flora in and around Uppsala, which he knew better than anyone else, though his interest never turned into a completed piece of work. The good God had delivered another Stobæus to Linnaeus—and also money to buy a new pair of shoes. Employed as the tutor to three young Rudbeck offspring, he also had access to a splendid library and could hope to join in learned conversations. From the spring of 1730 and throughout 1731, everything looked promising. He clearly knew how to make himself noticed: first in Växjö by Rothman, next, in Lund by Stobæus, and in Uppsala by Olof Celsius, and, later, in the Netherlands by Herman Boerhaave.

Now, Linnaeus is on the move. In May 1729, he and Johannes Humble travel north of Uppsala to Dannemora, long famous for its production of iron ore. The students wanted to examine the mines and their flora but could also inspect "what was the rarest sight, the fire machine [a steam-powered pump] as drawn by Triewald." Linnaeus is observant: he spots big things but small ones, too; the shrews in the mine "are as tame as dogs so that they ran toward people to eat from their hands. They are thought to be sacred and no one cares to hurt them." On several occasions, he travels to the capital: "In Stockholm 1729, lodged near Norrbro in an upstairs room; by 7 o'clock of a summer's night many ran quickly home like [word indecipherable] and by 9 o'clock some would come home on weak knees, then at 12 o'clock others would sneak this way and that. The first being collegians, next drunkards, 3ième, whores." He attends six dissections of the body of a woman in January–February 1729, his only hands-on experience of anatomy. He thought it essential: "A blacksmith, carpenter, etc. know their *subjectum*, so how could we opine on our *subjectum* without having made ourselves familiar with its nature." It was perhaps with these dissections in mind that Linnaeus told his students much later: "Your Archiater and Knighted Gentleman while on a fasting stomach once went to the *nosocomium* in Stockholm where he fell down and had to be carried outside." Linnaeus never took to anatomy. On 23 June 1729, he applied for the post as *hortulanus* or academic gardener at the botanical garden in Uppsala. It had been held by Erik Winge, who had recently died. Linnaeus lists his qualifications: "In childhood spent with my dear parents I had every opportunity of practice in the care of a garden." At Lund University, medicine had been his chosen subject, and he had notably devoted himself to botany. He owned an *herbarium vivum* and observes that it would be easy enough to find "a German garden underling or rustic cabbage planter" but that a man who had studied would be better able to carry out a species inventory, like that done by Olof Rudbeck the Elder. Olof Rudbeck the Younger was to be in charge of the appointment, but he was away for a spa cure and the post went to someone else.[4]

A break for country travel: "In the year 1731, and in the company of some twenty students, I took a journey by water aboard Rudbeck's packet boat, and after much hard labor and, as the air was quiet, with much pulling at the oars, we finally arrived at the small island of Kofsö in a bay of Lake Mälaren." The crew and the students soon went off to rest, "but I walked on the island along and crosswise, back and forth, one of my paths ran hardly further from the earlier one than by two feet almost as if ploughing... and I declare that hardly a herb or plant escaped my hand, or *muscos*." How many have vanished and how many have been added? There are 88 species on Linnaeus's list and, in 1953, the corresponding number is 105.[5]

Two professors, Olof Rudbeck the Younger and Lars Roberg, were responsible for the medical faculty at Uppsala University. The two men, both getting on in years but exceptionally able, were otherwise different in most ways. Rudbeck was of noble birth, the father of twenty-four children and, in the portraits, always wears impressive wigs. Roberg, a bohemian bachelor, is shown with his own thinning hair and his shirt unbuttoned. Rudbeck held the chair in theoretical medicine, specializing in flora and fauna, and directing the compilation of encyclopedic works: *Fogelboken* (The Book of Birds); *Campus Elysii*, drawings to their natural size of more or less all the plants in the world; and a huge project, aimed at deriving all the world's different languages from that spoken by Adam and Eve—the *Lexicon Harmonizans*. Roberg, professor of practical medicine, excelled in publishing small works, among which the linguistically eccentric *Likrefningstaflor* (Corpse Dissections) from 1717 deserves a special mention. Roberg worked in one of the city's hospitals and cared for the sick during the first plague year (1710). Rudbeck took pride in the title "archiater" or chief royal physician. An outstanding man in many ways—but Roberg was a true nonconformist. It escaped no one that the two men didn't get along. In principle, Roberg might be expected to have taken Linnaeus on for mentoring, and he did but at a later stage. At first, Roberg favored the student Nils Rosén and Linnaeus became Rudbeck's protégé.

Linnaeus paints a grim picture of medical training in Uppsala: "None of us saw any Anatomie nor Chemi, and Linnaeus had never heard any lectures in botanical matters. Neither *publice* or *privatim.*" Such statements by Linnaeus should be taken with a pinch of salt as he was very prone to talk up his judgments. In this case, he probably intended it to be understood as a contrast to what Uppsala came to stand for after his return: his light shone more brightly against such a dark background. He had, however, nothing but praise for Rudbeck's lectures on Swedish birds during the academic year 1728–29. The drawings of birds attracted his attention: "All very well and nicely portrayed with their *coleurs* precise as if the art of man."[6] Oddly enough, Linnaeus seemed to have missed the opportunity—was he not allowed?—to study the great work of botanical illustration, *Campus Elysii.*

Anders Celsius testified that medicine was in a bad way, indeed had hit the floor, because neither Rudbeck nor Roberg was lecturing. "Botany alone is flourishing somewhat thanks to a student called Linnaeus, who is from Lund and receives my Uncle's support as he is poor; he has an incomparable inclination for this *studium.*" He added that this young man knew the authorities—for example, the works by Tournefort and Ray. Celsius had learned that Rudbeck was teaching his favorite student in the garden, "but it is a pity that Prof. Roberg, in the way he always mistreats decent folk, pursues him quite frequently."

Linnaeus wrote to Stobæus on 5 April 1730: "Prof. Roberg is badly stricken with disease as he has a prodigiously swollen *bubo* between the shoulder blades; that the Good Lord release him from this world, to enlighten us and to praise Him. But should he recover this time I fear he will never die."[7] It was only much later, after Linnaeus's journey in Lapland, that he began to appreciate Roberg. So far, Rudbeck was Linnaeus's protector, the man who arranged the Lapland expedition and wrote the first entry in his book of patronage—a kind of memory book with contributions from friends and supporters. Roberg wasn't mentioned and didn't make an entry in the book of patronage. Linnaeus and Rosén were competing for Rudbeck's approval.

In Sweden's history of science, the rivalry between Nils Rosén and Carl Linnaeus has become a classical conflict that began early in their careers. They were almost the same age and both were the sons of the manse. At this point, Rosén was a few paces ahead in the race. He had made his journey abroad, all the way to Montpellier, and completed his doctorate in Harderwijk—as would Linnaeus a few years later. Back home, Rosén was offered a medical teaching post as the "medicine adjunct" after the early death of the previous incumbent, Rudbeck's son-in-law. His position was rather more secure than Linnaeus's but, with time, they would both be appointed to professorial chairs and their double-act at the top of the medical faculty would raise it out of its mediocrity.

Linnaeus wrote: "Once I had learned enough as to the extent which R[osén] pursued me, lied about me, suggesting that I would never reach the top of the pile, and so caused me much pain, it sometimes came to mind that I would ruin him in the manner of Samson [presumably, by violence?], sometimes calling to GOD for revenge, and sometimes seek to pay with an eye for an eye, but nothing offered comfort as I perceived God let him act so as to oppress me. Then I saw it all as a game for children for as long as I earned enough to feed myself; knew it all to be in vain as we in the end rot and be thrown together."[8] Later, their relationship would become less tense and normalize. At Rosén's funeral, Linnaeus spoke well of his colleague, but in his autobiographies he still lambasted his old enemy. Linnaeus's great biographer Thore M. Fries does his best to write a calm account of their rivalry and counter the rumor of hostility and an eventual duel, but it was a serious conflict.

In the summer of 1730, Linnaeus earned his keep by tutoring private pupils, perhaps as many as a dozen. They paid for the teaching partly in goods. A youth from a noble family gave him a bag wig, another four feet of black cloth, shoes, and gloves. Carl Gustaf Warmholtz, son of a pharmacist and a bibliophile in the making, presented his tutor with perhaps the most valuable gift in the long term: the botanist Caspar Bauhin's book *Pinax*. A letter to Stobæus, dated 12 September 1730, speaks of the many members of the nobility in the audience when Linnaeus lectured. It was rare for the professors to attract an audience of more than 80, but Linnaeus "most of the time had 200 à 400 *auditores*," a good proportion of the 1,000 or so students at the university. He reported that his collection of insects by then had reached more than 4,000 specimens, "all collected by my own hands." On 5 April 1730, he had twelve stuffed birds and "fishes I have also begun to preserve."

Ingratiating flattery was part of good manners at the time and, in 1730, Linnaeus presented his own celebratory poem for Olof Rudbeck's name day (29 July). The verse is full of compliments: "So at Your feet I cast / this small note of adoration," and, furthermore: "May Your sun long shine / May Your light not be doused / so the brilliance that is Thine / will light me from my house." As a thank-you for having been given the job of tutoring Rudbeck's children, Linnaeus one year later congratulates his employer on his name day with a "Humble Offering": "Could I bring You the fruits of *ananas* [pineapple] and *musa* [banana], my gifts would be but flavors that amuse for minutes and then flee like dreams. Therefore, I will find a more agreeable gift. . . . And so, I offer on the altar raised in your honor, a tender plant, a transient herb but one that will bear Your great name to its invaluable glory, the RUDBECKIA"—on and on, line after line. The use of military metaphors is remarkable: "Thou hast shown Thy bravery by raising recruits in distant Lapland, a place to which no Botanicus has earlier dared to journey. Like a gallant Commander in Thy youth, you lined up and led an Armée with sagacity. Thus, it is the strongest reason that Thou here finally triumph and that this soldier stands by Thy door garnished with Thy victorious livery. This Soldier, I here mean the herb, stands tall."[9] For all the flattery, he was soon dismissed from his post.[10]

By 1731 the Rudbeck home had become the stage on which was played the tragedy of Greta Benzelia, a hopeless tangle that Henrik Schück managed partly to unravel in his essay, appropriately entitled "An Uppsala Scandal." The fruit of previous sins, a daughter was born to Bishop Erik Benzelius and his wife née Swedberg/Swedenborg. The girl, Greta, turned out to be "rather frivolous" and was married off to a librarian, Mr. Norrelius, who was "a learned ass" according to later notes by Linnaeus. Young Mrs. Norrelius "walks out at night with students, cannot stand her husband, teases the students to beat up her husband, ruins the poor man's finances." One act in this real-life play took place in Olof Rudbeck's home where Greta "a *hysterica*, swoons," and Nils Rosén was called only to forget himself: "In the morning he is found in her bed." Greta became pregnant, but her condition was blamed on a man called Gerdessköld. Asked about a witness, she replied that witnesses were not usually part of such encounters. "A convicted whore, Norrelius is now well rid of her"—a divorce, in other words. We learn that, in 1743, Greta, the bishop's daughter, made a living as a common whore with the Russian troops, then ran away to Norway. Her life ended there in miserable poverty. Her child, a son, was described as "fast": he became a forger and was extradited from Denmark to Sweden, where he was sentenced to death by hanging.

Linnaeus sounded strangely well-informed about this story but still did not write a complete account. Seemingly, he was also caught up in this mess. In an obscure note, he wrote, referring to Greta: "In discomposure I do that which I otherwise could not: raise high, strike. . . . Rashly, as I would not, I hit. Rashly, catch hold of the girl, such as I would not in sound mind. *Emitto semen* rashly, unnoticing as I would not do unless agitated."[11] This was presumably why Mrs. Rudbeckia dismissed him from his post as the children's tutor. Linnaeus had nothing good to say about the mistress of the house: "In his third marriage [Rudbeck] found for himself a housemaid, a virago, and a rancorous human being. She is so strict with

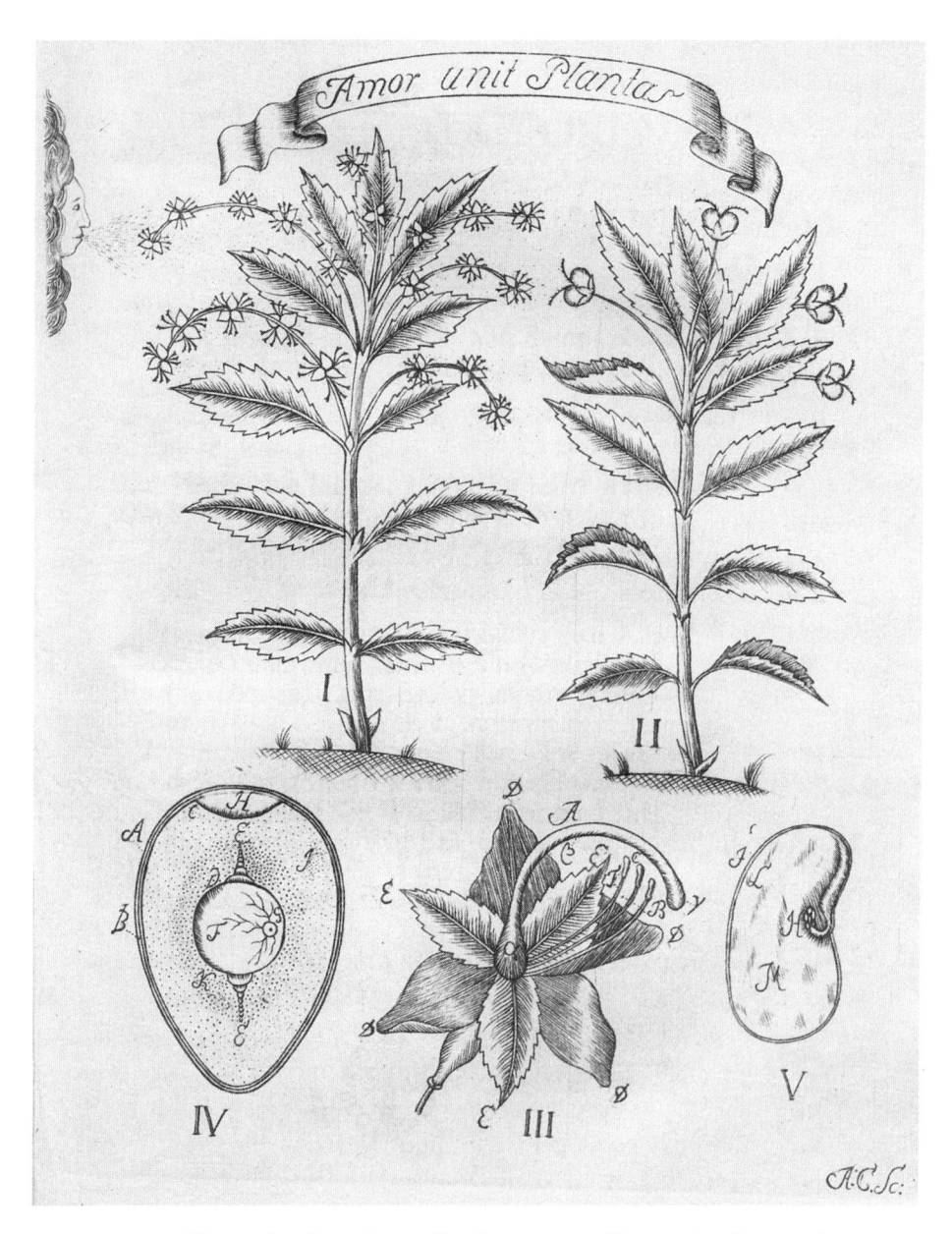

FIGURE 6. Illustration from *Sponsalia plantarum*, a dissertation from 1746 on reproduction in plants. We are shown the male and female forms of dog's mercury, *Mercurialis perennis*. Part III shows the parts of the flower, and IV and V compare eggs and seed, the animal and vegetable equivalents, all under the title "Love unites the plants."

him he tires of quarreling with her. She is a Devil in her own house."[12] She will receive a well-deserved punishment.

In a much lighter mood, he praises the power of love in *Praeludia*, dedicated to Olof Celsius for New Year 1730. It begins: "In the Spring, as the bright sun reaches our Zenith, she stirs the liveliness in all bodies which throughout the cold winter have been still and smothered.... Look, as all the herbs leap up and all trees burst into greenery although bedraggled in winter, yes, as mankind itself springs back into a new life.... The joy that the sun instills into all lives is beyond expression; the grouse and capercaillie can be seen at their dances, the fish at play, yes, all animals are in heat. So, love conquers the very herbs and among them both *mares* and *feminine*, yes, even the hermaphrodites, hold their weddings. Such is the scene that I have now set myself the task of describing."[13] This is vivid, sensual, and fresh, so just right for reading to the sound of Johan Helmich Roman's celebratory music for the royal wedding and bedding-down (1744). In 1731 Linnaeus introduced *Praelectiones Botanicae Publicae* in the same spirit: "Now the sun with its warm radiance begins to warm our land. Now the sun begins to bestow on all creatures a livelier spirit, now small critters and insects wake up from the chilly beds where they have lain under cold winter's cover. Now, trees adorn themselves with beauty and herbs creep out of the soil so that all in concert can praise their Creator, displaying such artfulness that no artist could imitate nor the power of Salomon accomplish."[14] Linnaeus wrote like a devotee of pagan sun worship, praising Venus and the Creator separately, or together in the shape of the Creative Venus—*Venus creatrix*. In the early 1730s, he practiced using older botanical systems and tried out his own variants—for example, in *Spolia Botanica* (1729), *Hortus Uplandicus I-IV*, *Praeludia sponsaliorum*, and *Adonis Uplandicus*, and in lectures commissioned by Archbishop Carl Fredrik Mennander (1733); in all, he spent almost four years testing ideas. Later, he would locate his system for the sexual reproduction of plants to the summer of 1730. The *Hortus Uplandicus* was rewritten three times in one year—four times, if you count *Adonis*. He wrote *Praeludia* around New Year's 1729 as a gift for Olof Celsius. In *Hortus Uplandicus I* and *II*, which date from 1730, he refers to the approach of Joseph Pitton de Tourneforts in parallel with his own. *Hortus II* was dedicated to Olof Rudbeck and *Spolia Botanica* to Lars Roberg. In 1730 Linnaeus also wrote *Fundamenta Botanica*, a collection of aphorisms and commentaries. This is a quote from his introduction: "Hence, my little Book, step forth and submit yourself to the judgment of the entire world," and this, a quote from the postscript: "And I can assure you of this, namely that with time you will come to inhabit the palace of *Botanices principum*, yet often act as the secretive guardians' advisor." Meanwhile, Linnaeus found the time to collect insects and teach students in his post as temporary *botanices* demonstrator. In *Animalia per Sveciam observata* (1736), he announces that "*Insecta vere fuere summae meae deliciae*; this was *arsit mea Juventus*"—it translates as "Insects have truly been my source of happiness. In my youth I burned with enthusiasm for them."[15]

Linnaeus did not discover the reproductive mechanisms of plants but built on the work of two earlier authorities: the German professor of medicine Rudolf Jakob Camerarius and his French contemporary, the field surgeon and botanist Sébastien Vaillant, who studied under Pitton de Tourneforts. Actually, there is also a Swedish amateur botanist with a claim to have discovered the sex life of plants. Andreas Hesselius, at the time the minister for a Swedish congregation in America, wrote this in a letter (1712): "The mulberry tree has a rare property, namely it can be distinguished by its gender. The tree that I believe to be the Male bears no fruit but only a flower similar to a berry but which consists of nothing but dust so that if it is shaken, the dust scatters and vanishes into the air. The tree which is She has her place not far from her Spouse and should I be allowed guesswork it appears as if Nature has so ordered it that the dusty flower impregnates precursors to fruits that the other tree possesses."[16]

At this time, the elements of a system describing the sexual relationships of plants were suspended in the air like pollen. Such ideas, together with the development of microscopy, had led to the sex life of plants joining a more contemporary and wide-ranging discussion concerning animal reproduction. Olof Celsius, who might have picked up the idea from Linnaeus, wrote in a letter—annoyingly, it's undated—to Georg Wallin, commenting on Wallin's *Nuptia arborum* or The Wedding of the Trees (1729): "In this very moment, Sir, I received your fine Doctoral thesis on *nuptis arborum*. Vaillant has posited that he can demonstrate this for all *stirpibus*, and had he not died so promptly, it could have become a new *methodus cognoscendum plantarum*. I have received from him recently [his work?], and communicated with him at that time, *ob paritatum argumenti*. . . . In my *Flora uplandica* I have in the passing observed which trees and herbs are *mas* and *foeminae*. I pen this in greatest haste."[17] Had Celsius borrowed this from Linnaeus or made the observations himself? In the biography by Thore M. Fries, the essential passages in this letter are cited but without further comment on what "the new method" entails.

The following comment by Andreas Browallius must, however, have been based on a misunderstanding. He had made notes about Linnaeus's botany, among other subjects: "As the notary Johan Elvius much favored this [division of plants by gender] and had thus understood it must be possible to classify herbs by their sex, then their very aspects would surely signify their strengths and actions. As the sexes are unlike each other in appearance and also in constitution, it will soon be recognized what they are good for [!]."[18]

Linnaeus's "wedding of the herbs" had parallels in the extensive verse writing that was part of wedding celebrations and has been much studied by literary scholars. The time for weddings was not typically the spring; most of them seem to have been arranged for the autumn. There was, of course, nothing to prevent poets from linking love to spring and claiming that it must be the right time for one and all to wed. So, for instance, Johan Tobias Geisler, at a wedding in September 1721: "The best time of the year to marry all considered." As public interest in natural history grew, it was reflected in wedding poems such as this: "When the Ruffe shows off most quickly; When the Salmon fearlessly goes upstream / When the Perch begins to play; When the Bream caresses his Bridal Bream."[19]

Concerning the sex life of plants, there were also plenty of predictions and antecedents. Throughout these conversations, leading arguments were often based on analogies: God had had an overall plan for what He created "in Heaven so also on Earth"—the same plan could be discerned in the macrocosm as well as the microcosm, above and below ground, on land, and in water. However, it apparently didn't occur to Linnaeus to draw up a sexual system for animals. He distinguished between *vivipara* and *ovipara*—that is, between animals born alive and from an egg.[20] What about the "stones"? They "grow" and are entire objects, Linnaeus pronounced, not pure elements or mixed chemical substances.

Reproduction: "We have noted it as proven that even plants are *viva*. Thus, it follows from the general proposition that they by necessity *per ova propagari*. Observe that the seed is the egg of the plant. Then where the seed is must be the location of its *organa genitalia* as there is no *ovum* that will bring forth a *vivum* with less that it has been made *foecundus*. Neither *mas* nor *foemina* has in itself the entire *rudimentum foetus*, as is clearly demonstrated by such animals as are generated of diverse *specierum* blended together and display one part from the father and the other from the mother, as when to a horse and a donkey is born a mule and so forth. In Lund only a few years ago, piglets were born to a sow and the newborn were in much similar to their mother but had heads and feet like those of a hare as they were offspring of *patre lepore*."[21] In other words, Linnaeus endorses neither ovism nor animalculism—that the egg or the sperm contains the whole embryo—but believes that both egg and sperm must be combined. We are going to consider the hybrids again.

Around New Year's 1731, Linnaeus wrote to Stobæus: "*Nuptiae plantarum*, a part of what I refer to as my *nova methodus plantarum* but not my foolish reasoning concerning the same matter of a year ago afore I conferred with nature by autopsies.—[The method] which I shall never present in a common discourse but would demonstrate what it can entail should I ever be granted the joy." Linnaeus had discussed his ideas in front of the scientific elite who, to a man, thought him mad at first but, once he had explained a little more, "restrained their mirth" and promised to help.[22] That Linnaeus visited Stockholm for some time just after midsummer, and afterward tried and failed to find a publisher for some of his findings, is by now thoroughly investigated. To persuade possible sponsors, he had to be on the alert to advertise his "method" while at the same time prevent it from becoming so widely known that a rival might steal his thunder. He outlined publication plans to Stobæus (23 October 1731): "I made final changes to my *Hortus Uplandicus* at the end of July in Stockholm, and should from a Book Press that undertook to frame it up receive only 100 copies but he has not yet begun the printing. I planned to interest the Germans in my remaining *Lucubrationes* [nighttime products]."[23]

His growing library of medical texts reflected how his interests had evolved. It contained some two hundred works by the end of 1729. Many of the books dealt with subjects that Linnaeus thought of as relevant to medicine but by now no longer are, such as mineralogy, botany, and the art of gardening. The "materia medica" included the classical works by Hippocrates and Aristotle, but much of the rest was by then too arcane. Linnaeus used the word *hyperphysiologia* for such works—for example, Cornelius Agrippa's *Opera magica* (ca. 1550) and Martin Mylius's *Hortus philosophicus* (1597). Both belong to the world of Paracelsian thinking. A religious work such as Johann Arndt's Paracelsian *True Christianity* was very widely read, almost certainly also in Linnaeus's childhood home. It seems to have left a profound impression on him and his way of regarding nature as a book that can be read and its "signatures and powers" interpreted. This has given rise to speculations about Linnaeus's "hidden philosophy" and suggestions of possible links to esoteric brotherhoods such as the Rosicrucian Order. None of this proves that Linnaeus was an anti-empiricist, and instead only that he shared with Francis Bacon, a man he admired, a fundamental Hermetic belief that everything emanates from a creative God.

To test his "natural philosophy," one might try to connect Linnaeus's great, systematizing ideas to the leaps of thought that took place during the Baroque period. There is a possible link between the direction his thoughts took and the German form of Enlightenment rationality called "Wolffianism," which spread triumphantly in the late 1720s. In Uppsala, Wolffianism on the one hand turned into "sensible" notions about how to achieve heavenly rapture, and on the other, a mania for classification. Apart from this, no notable links to philosophy have been identified. There are echoes of Descartes in his Foreword to an early, unprinted version of *Fundamenta Botanica* (ca. 1730): "Here I offer you, benevolent Reader, my first *meditationes* concerning the manner of rescue for our vitiated botanique and on what pedestal it ought to be placed.... As I at first ransacked nature herself and observed her striving against the opinions of *autores*, so I let go all *praejuducier*, became a *scepticus* and doubted everything, and first then did my eyes open, first then did I perceive the truth."[24] And, again, elsewhere: "To enter the mighty forest of the world of plants without an Ariadne's thread to follow would be as if to travel on immense oceans without a rudder. Thus, all botanists worthy of that great title have attempted to twist together a thread, calling their efforts a method or a system.... But each time the inexperienced man sets out to follow his methodological thread, he finds a tangle of knots, all more difficult to unravel than the Gordic."[25] Methods and systems are essential. Once more, impressions from the Hermetic tradition seem evident. We will have reasons to return to this discussion.

Linnaeus tended to note the provenance and date of purchase of his books and also systematized his books according to subject matter. Here, the subjects and the bracketed numbers in the list are Linnaeus's own: "Anatomici & Physiologici (26), Pathologi (8), Practici (25), Particularium Scriptores (5), Botanici (25), Mineralogi (5), Zoologi (6), Diaetici (6), Pharmacopaedi (11), Chymici (13)"—and so on. It is clear that his choices have been weighted toward medical studies. He had acquired at least a dozen books in Växjö, among them gifts from Rothman. Some are heavily annotated, as for instance the interfoliated *Vademecum botanicum* by Johrenius. Several books were of interest for various reasons, including Niels Aalborg's *Laegebog* (Medicine or the art of healing) of 1633, Cornelius Agrippa's *Opera*, Herman Boerhaave's *Institutiones medica*—a copy that belonged to Linnaeus's father-in-law Moraeus—and Johan von Hoorn's *Anatomes publicae*. He owned classic works such as Santorio's *De Medicina Statica*, and Ramazzini's *De mortis artificium*, and also Cardanus's *Contradictionium medicorum liber Venetiis* (1545)—there are no annotations in these. On 18 February 1726, Linnaeus bought Thomas Sydenham's *Opuscula* and Verheyen's *Corporis Humani Anatomia*, and annotated the latter on a total of five and half pages. The dietetics texts included works by Santorio and Keill, and the medical textbooks Friedrich Hoffmann's *Fundamenta Medicinae*, with Linnaeus's *Vita Caroli Linnaei* up and until January 1729 added at the end.[26]

For an impoverished student, his library seems surprisingly large. It might also be thought remarkably out of date, given the number of works on the occult and also what is missing by way of authors and centers of study. It is unmistakably the collection of a medical rather than a natural history student.

As has been pointed out quite often, Linnaeus was to a large extent self-taught, but good mentors supported him and his teaching tasks had been useful. Another boon was meeting like-minded fellow students. At some point just before 1730, he came across the Petrus Artedi or Arctedius, who came from Nordmaling on the northern Baltic coast. Apart from memories noted down in *Vita Caroli Linnae*, there is little on record about their friendship except for the presumably polished formulations in Linnaeus's introduction to Artedi's *Ichthyologia* (1738). Generally, Artedi's profile is low, especially as there is no known portrait of him. However, no one can doubt his ability even though he oddly enough is absent from the documentation of the professional quarrels between Linnaeus and Rosén. He might have been a more or less explicit adherent of radical pietism: his father was a clergyman who had been admonished by the diocese. In 1724 young Artedi enrolled as a student of theology but changed to medicine after a couple of years. He seems to have been something of a loner; Carl Tersmeden, the impressively built admiral-to-be who met him in the Netherlands, called him "the invisible man." Linnaeus compares his friend to the English pastor-naturalist John Ray: both were tall, thin, and dark. He also saw Artedi and himself as a pair, like Ray and his college friend, the mathematician and ichthyologist Francis Willughby. They had been taken on by different professorial mentors, Linnaeus by Rudbeck and Artedi by Roberg, and agreed to divide their work on describing nature between themselves: the botanist Linnaeus took on the birds while the fish expert Artedi carried out systematic studies of umbelliferous plants and of furry animals, about which he wrote *Idea Institutionem Trichozoologiae*. Artedi's interest in alchemy was another characteristic that set him apart, although one he seems to have shared with Roberg. Much later, Linnaeus's son Carl writes of Artedi that "he was a *solitair*, who went out drinking from 3 until 9, worked the night through from 9 until 3, and then slept until 12."[27] In a somewhat obscure comment about the two students, Linnaeus's trusted friend Abraham Bäck writes that "a common inclination for natural history instituted their friendship although there was modest affection."[28]

The lack of portraits, the paucity of manuscripts, and limited biographical information combine to make Artedi one of the most enigmatic figures in the history of Swedish science. What would have become of him, had he lived longer, is a tantalizing question. It has been suggested that Artedi inspired the sexual system of classification since he had developed a similar methodology in his study of umbellifera "whereupon it came also to Linnaeus's mind to draw up a new method for all plants"—as Linnaeus himself has written.[29]

Where in Uppsala did Linnaeus live? We know that Olof Celsius helped him find lodgings at Östra Ågatan 11, at the corner of Järnbrogatan. In June 1730, he left to live in Rudbeck's estate to tutor "the little masters Rudbeck." He moved again by Christmas 1731 but spent the holiday and the first month of 1732 in Stenbrohult, before setting out for his travels in Lapland. After his return to Uppsala, his digs were at the corner of Nedre Slottsgatan and Drottninggatan. Johan Browallius visits him and describes the room:

> The ceiling he had embellished with trophies taken from birds. One wall likewise with Lap clothing and curiosities. The other wall with larger plants and a collection of mussels; the two remaining ones were handsomely equipped, partly with books generally on subjects in medicine and natural history, and partly with scientific instruments and stones. In a corner of the room stood tall tree branches in which he had taught tame birds of almost thirty species to live. Finally, the inner window nooks were set aside for larger clay pots containing soil as nourishment for the rarest plants. It is my hope that we will soon see the three kingdoms of nature illustrated on maps or drawings, printed under the title *Geographia naturae*.

Why all these birds? The interior fits with Linnaeus being at work on an ornithological system in 1731. His idea was to classify birds according to their beaks but, as we have noted, he also tested a possible variant on a sexual system by sorting them according to numbers of eggs laid: Dispermia, Tetraspermia, Pentaspermia, and so on.[30] It seems likely that he also hoped that gendered classification would apply to all the animal kingdom—a shared basis of reproductive functions in nature.

How far had Linnaeus come at this stage of his life? He lacked professional training and was still poor but far from shy—on the contrary, he was up for competitive challenges and had impressed important mentors and patrons. We get glimpses of his many uncertainties concerning the specialty he should devote his life to: medicine, mineralogy, natural history,

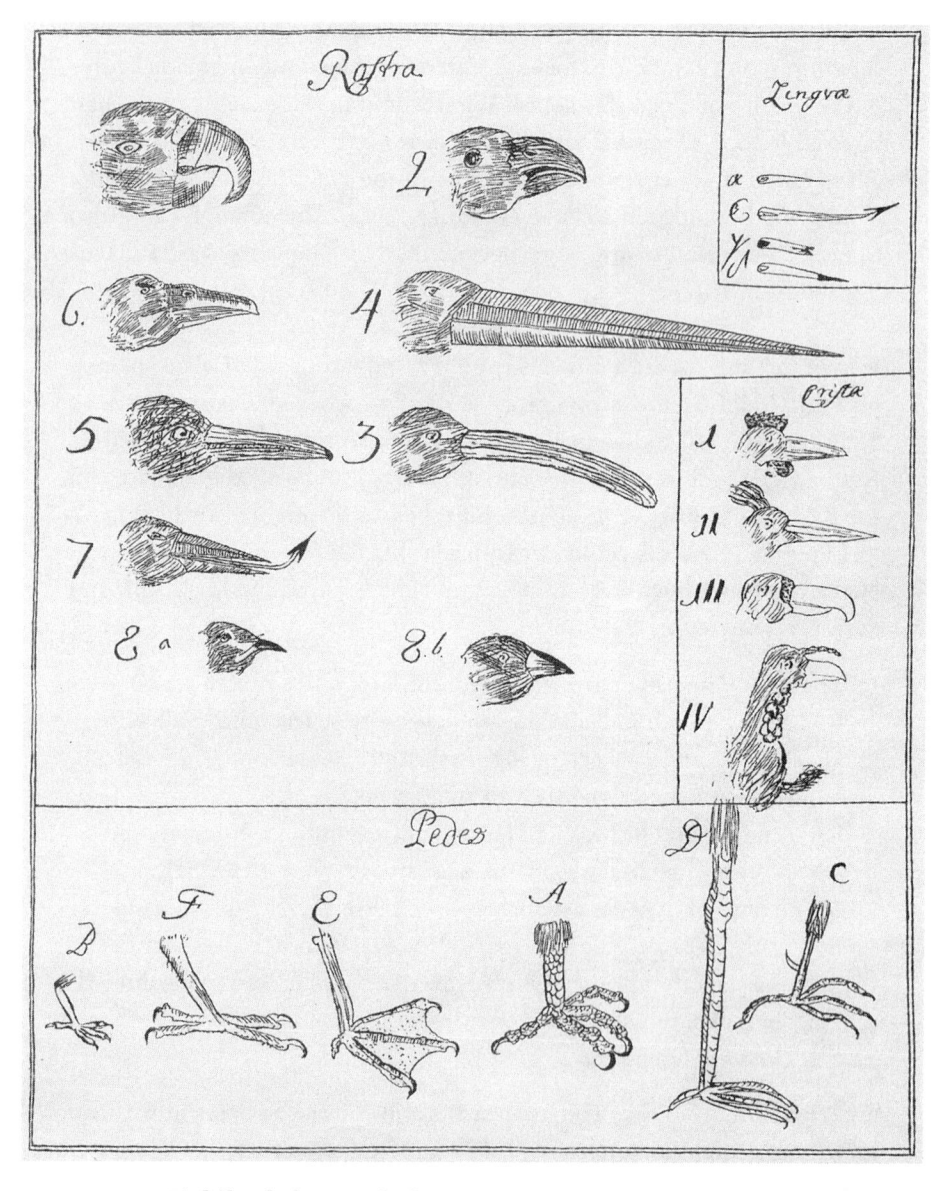

FIGURE 7. Birds' heads from *Methodus avium sveticarum*, ca. 1730. The illustration shows the division of the bird into *rostra*, *lingua*, *crista*, and *pedes*—beak, tongue, head crest, and claws.

or, especially, botany or all of these—which is how it would be. As it would be with his philosophical search for ideas in Aristotelian and Hermetic texts, and in the works of Bacon.

He needed a change of air after a few years, which had served both to overheat and to cool his mind.

CHAPTER FOUR

In a Mythical Landscape

LAPLAND

THE GREAT ADVENTURE in Linnaeus's life was his journey to Lapland in 1732. He was commissioned by the Royal Society of Sciences in Uppsala as part of a drive to extend the country—to investigate the hidden secrets of the high hills and glens and try to grasp what the benefits of missionary activity and colonialization might be. Of course, sources of knowledge about northern Sweden already existed, notably the still influential pages about the North in the great ethnographic work "A History of the Northern Peoples" (*Historia de Gentibus Septentrionalibus*, 1555) by Olaus Magnus. It was further explored in Johannes Schefferus's *Lapponia* (1673) and had been inspected by King Karl XI, who in 1695 saw the glow of the midnight sun in Torneå. Rudbeck the Younger had accompanied the king and planned a sequence of accounts of his "gothic" travels, starting with *Nora-Samolad* (1701)—however, he didn't get any farther north than the Dal (Dalälven) River. A few years after Linnaeus's journey north, the French Academy of Sciences was sending an expedition led by Pierre de Maupertuis, and with Anders Celsius representing the Swedes, to measure the degree of flattening of the North Pole.

The motives, as well as the routes followed and landscapes examined, can all be understood in different ways. It is reasonable to think of it as an "imaginary world"; that Linnaeus moved then, as well as later, through the geography of his imagination, a perspective that overwhelmed detailed observation of the natural features and day-by-day writing up of the journey.

Linnaeus's application to the society contains a passage dealing with the qualifications they would expect from the prospective traveler: "1. To be

VIRO NOBILISSIMO ET CONSULTISSIMO
D: GEORGIO CLIFFORTIO J. V. D.

FIGURE 8. Frontispiece of *Flora lapponica*, 1737: altogether too high mountains, midnight sun, a reindeer, Linnaeus outside of a Lapp tent. The stems of *linnea* (twinflower) are winding in front of him, on one side. Royal Library, Stockholm.

a native-born Swedish man; 2. That he be young and light so that he with all the more agility could run up the steep mountainsides and again down into the deep valleys; 3. Healthy so that he . . . with enjoyment could carry out his daily tasks; 4. That he be resolute and not after soft living become comfortable; 5. Unmarried so that he dares meet dangers without fears for his impoverished children; 6. That he was *Historicus naturalis* and also *medicus*." This is how the applicant, who met all these requirements, formulated his task in *Diaeta Naturalis*: "The Royal Acad. of Sciences has designated, in order to elucidate *historiam naturalem Laponiae*, that C. Linnaeus, for some years in attendance at Professor Rudbeck's public lectures in *Horto Academico*, should undertake that journey to the country of the Lapps recording not only stones and soils, etc. but also trees, herbs, grasses, mosses and insects, birds, and animals, in addition to the diets of the natives and their lives."[1]

The instructions issued by the Society of Sciences have gone missing, but judging by the list the academicians had compiled, they were especially keen on what we would nowadays call curiosities. The Sami people aren't mentioned until item 4 on the list (it concerns their physiognomy, build and height, skin color, and ages) and when they turn up again in item 16, the focus is the same: "[If they are] short, well built, thin, light, have a broad head, etc." although the last of these entries has a line through it, possibly because it was repetitive. Pretty meager, given that the list contains some eighty questions, often referring to oddities: was it true, as stated in an Italian book, that the "water snakes" in Lake Inari are "four feet long"? (item 51); could "a maiden turned into stone" be found? (item 54); how about "the spots on the reindeer"? (item 59). Nils Rosén, Linnaeus's rival and at the time working on a textbook of anatomy, had entered this question: "How is it that the Sami can be so quick on their feet?"

The route, with a few excursions, followed the Baltic coastline as far as Luleå, then swung inland northwestward to Jokkmokk and Kvikkjokk, and crossed the mountains into Norway. After reaching the Norwegian Sea, Linnaeus turned back, and instead of completing a suggested trip even farther north, he traveled back to Luleå, followed the coast around the Gulf of Bothnia and in Finland to Åbo, before crossing the Baltic via Åland to return to Uppsala. The fate of his Lapland travelogue is worthy of a separate essay: in any case, it was not out in print until 1811, in an edition translated into English and published under the title *Lachesis Lapponica*.

"Never would I have undertaken such an unpredictable journey and one along a road with so unaccountably many death-dealing abysses, had I only been able to recognize the truth. Emphatically, no!" Linnaeus wrote to Lars Roberg.[2] Linnaeus had indeed begun the journey quite unprepared for what was waiting for him. Nothing suggests that he had read up in the fundamental accounts by Olaus Magnus and Johannes Schefferus, or in the older Olof Rudbeck's inexhaustible *Atlantica*. Olof Rudbeck Jr. was his mentor, and Linnaeus will have known his writings on the subject and presumably spoken with him about it. The *Nora-Samolad* (now in the library of the Linnaean Society in London) might well have been in Linnaeus's luggage; insofar as Linnaeus owed his journey to someone, it had surely been Rudbeck, a man listened to in the Society of Sciences. The society's secretary Anders Celsius, Linnaeus's professorial colleague-to-be, was also a strong supporter; as we know, Celsius joined the Maupertuis expedition to the Polar Circle.

Linnaeus was just twenty-five years old when he rode through Uppsala's northern customs gate on a lovely day in spring 1732. His luggage contained:

> A light coat of unlined Västgöta cloth with narrow cuffs and collar. Breeches well made all of leather, tailed wig. Wrist supports in green hemp, sturdy high boots for the feet. A small bag about one foot long, somewhat less across, the toughened leather with loops on one side to tie her firmly and hang her on my body; inside her was placed 1 shirt, 2 pairs half-sleeves, 2 nightcaps, ink well, pen box, microscope, field glasses, a cap with a net veil to exclude the mosquitoes, this protocol book. A bundle of papers stuck together between which to place herbs. Both these in folios. A comb. My *Ornithologie*, *Flora Uplandica*, and *Characteres generici*. A *hirschfängar* hunting dagger at my side and a small carbine between my thigh and the saddle, an 8-sided stick with *mensurae* inscribed along it. A wallet in my pocket with my pass issued by the chancellery in Upsala and the Society's recommendation.[3]

Petrus Artedi, the northerner, had advised Linnaeus about the best places to stop for the night along the coastal road.

As he sets out, his notes are rich in the ornamentations of his time: "Now Nature itself began to rejoice and smile, now beautiful Flora comes to sleep with Phoebus."[4] Linnaeus brought the mythological goods of the Baroque period and Sweden's Great Power period along on his travels. Here is a passage from another sample of his writing, planned to be an account of *Oeconomia Lapponica*: "Indeed I could not at any time have dreamed of making a journey to Lapland until some fourth part of a year before I set out. Dreams I have always held in contempt as the greatest phantasies but I can still state that I, just one year before I traveled far away on *ipsa natalitia die* 13 March, dreamed how I was in a landscape or forest where some ancient women lived inside a quite low house and the door was so low that I . . .".—the writing ends abruptly at this point.[5] We can safely assume that the women Linnaeus is about to meet are not any old biddies but are called Clotho, she who spins the thread of life, Lachesis, she who measures the thread to the allotted length, and Atropos, she who cuts the thread. The Fates were guarding Lapland, that quite low house that he was about to visit. He didn't allow the dream to end, but the landscape he traveled through is unmistakably mythical.

Iter Lapponicum, the Latin title of Linnaeus's Lapland narrative, is a complex text. Not only did its author fail to complete it, but it describes a world that has changed between then and now and in language that is difficult to interpret, a stylistic blend of Latin and Swedish, where the Latin is used for exact scientific observation and the Swedish for general and speculative writing. There are times when his own text compares badly with the ideal manner of speech he ascribes to the people of the North: "Oratory, flattery, exaggerated, and distracted talk is not heard here." What Linnaeus implies is that, in these distant parts of the realm, everyone speaks his or her mind just as he does himself. Actually, the language in *Iter Lapponicum* is typically a macaronic jumble, with the additional complication that Linnaeus addresses different groups of readers so that passages can seem at cross-purposes. One element in his mental geography is the influence of the classics and, above all, Ovid, so well loved by him and his contemporaries. That his mind was much preoccupied with economic issues was characteristic of his time, even though he was personally fascinated by natural history, by the life cycles of animals and plants.

Why did his travelogue remain unfinished and unpublished? Many of the best descriptions from the journey were incorporated into *Flora Lapponica* as notes on ethnobiology, such as the native footwear and the types of grass used to line them, or the role of the great hummocks of haircap moss as beds in the land of the Sami, or "the obnoxious practice of stuffing the nose full of snuff."[6] *Flora Lapponica* with its exotic subject matter contributed more to raise Linnaeus's profile in the world of science than any of his other publications at the time. On 13 January, Olof Celsius wrote to Linnaeus: "Herr Doctor, *Flora Lapponica* strikes me as perhaps unparalleled among your works. For myself, I will subscribe to a copy and it may cost what it must. I had fancied ridding myself of my botanical books but now I will reconsider a little."[7] Another reason why the full account was never published might have been that the route taken did not agree well with the report Linnaeus submitted to the society. What was the whole story of his stated northwestward digression into the high mountain ranges beyond Kvikkjokk? What was he up to during that long stay in Kalix—forty-four days documented only in a few scattered notes? He surely couldn't have made it to the high glacial lake of Torneträsk, even though the dateline of a letter to Roberg is "14 August, Svappavaara," which suggests he was on his way. He reported to the society that he had traveled along the Torne River in the company of the region's governor, a man called Höijer, and he told Roberg that he was taken by *ackja*—reindeer sledge—some 70 kilometers into the mountains, both items that are not found in his dairy.

He claimed to have made another digression, covering 800 kilometers in four days, on the way back from Norway, which implied that he traveled about 1,350 kilometers in a fortnight (a feat noted by his biographer Mrs. Gourlie, who exclaims: "Telling fibs is naughty!"). Starting from Torneå, the outward journey went along the Torne River up to Vittangi, and the homeward one followed the same route. He was penniless when he returned to Uppsala and wanted his sponsors to pay for all his expenses, but because he couldn't produce receipts, no money was forthcoming. Linnaeus's inclination to "tell fibs" is perfectly understandable, as is his need to impress the members of the Society of Sciences. The present-day Skytte Society, based in the Norrland city of Umeå, has questioned his own estimate of having traveled 6,890 kilometers and suggests it would have been in the order of 4,400—quite a difference, to which one explanation, with reservations, could be Linnaeus's "relatively common carelessness with figures."[8] Understandable it may be, but also regrettable.

There is a gap in his diary for the period between 20 and 30 August, but it is at least to some extent accounted for in his report to the society.[9] He needed a break: "I could go no further and must retreat some 1,000 furlongs. Then I journeyed hither and thither among the isles in the archipelago. Once in Calix, I could accompany (27 aug.) the governor of Öfwer Torneå, and from there I went by boat to Kengis (4 sept.)." This period can be understood in various ways: that he needed rest "after a strong and lasting fatigue" or that he hoped, as already mentioned, that more time spent on the road would earn him more cash from the society; other ideas are that he stayed because of a love affair, or to learn how to analyze metal-containing ores, or was working on the manuscript of his travel diary. If we stop to consider the last option, our attitude to the final text will change, as it can no longer be thought of as a record of immediate inspiration, not even as a diary. Many, Knut Hagberg among them, have

FIGURE 9. Linnaeus's drawing of a short-eared owl, *Strix ulula*; he shot it late one evening near Skellefteå. *Iter Lapponicum*. Uppland Museum.

seen it as a moment-to-moment record: "It reflects what Linnaeus thought of the Lapps, and the swamps and mountains of Lapland, what the weather was like on the day, if the air was mild or if icily cold wind cut through his clothes."[10] In the expense account that Linnaeus submitted to the society, he includes "3 b[undles] Paper for Journal; 3 books," which gives the impression that he did much more writing than what has been kept. But his entry does not match what he included in his list as he set out, where his luggage is said to hold "this protocol." Furthermore, it can't be excluded that he reported the time spent in Kalix in another folder of paper and that this was lent to someone. The editors of the book entitled *So Why Does Linnaeus Travel?* (*Varför reser Linné?*) conclude that there is no trace of an earlier record later turned into a clean copy.[11] The errors in his text indicate that he was in a hurry. So their arguments can be summed up as "the diary was written by his hand'—agreed, but when, and who were the intended readers? Did he not admit that, once back in Uppsala, he spent "not only many days but also waking nights of labor to conscientiously compile the observations made during my journey."

Just because the manuscript pages remained unpublished it would be a mistake to regard them as a collection of private notes; for one thing, they were to be presented to the society. The text had some features of a subjective, spontaneous notebook, but this aspect has been exaggerated. His own views have actually been given more space in his medical notebook, the *Diaeta Naturalis*. It is reasonable to think that Linnaeus had sponsors other than the academicians, but it should be stressed that the diary only rarely—hardly ever—suggests that the Swedish Church might be among them. It is worth establishing at this point that religion has no more than a minor role in Linnaeus's writings, here or elsewhere. True, he sometimes speaks about church matters. On his travels, he would stay overnight with the local clergymen, be treated to food and drink, enter into discussions, and once, in Jokkmokk, quarrel irritably with schoolmaster Malming and curate Högling, neither among Norrland's brightest—but he does not explicitly refer to religion, nor does the paganism of the Lapps trouble him. He attended one church service (in Lycksele) but wrote little about it. He can exclaim with awe at the designs by the almighty Creator but doesn't develop the theme any further. He wrote in a passage about the religion of the Sami that it respects three forces—thunderstorms, the devil, and the dignity of the dead—but without lecturing on the true faith.[12] Linnaeus experienced a pre-Christian society, still ruled by Pan but becoming subject to the claims of the Swedish king.

The same context appears in a couple of famous passages. The first one describes an encounter with a scary woman who made him recall Styx—the River of Death.

> 3 June. We waited long into the day toward 2 o'clock after midday for the Lapp sent for us and he came with time but very tired as he had searched so many *stabula* in vain. With him came a personage, I could not say whether man or woman; to my mind the Poët had never so exactly pictured a furia, such as this being represented. One would not without reason have to assume that she was come from Styx. Her aspect was so very small, the face utterly brown, blackened by smoke, its eyes brown, shining, black eyebrows, hair black as tar wrapped around the head and on top sat a red low cap, the skirt was gray, and the chest was covered with skin like that of a frog and with long, lanky, brown teats but hung with brass chains all around. Around sinus she had a belt and on her feet boots. I was afraid at my first sight of her.[13]

A "furia"—a Fury—came to meet him in an earthly hellhole. He explained that "to my mind the Poët had never so exactly pictured a furia" and, although for him Ovid is usually "the poet," this might have been a nod to Virgil's account of the Cumaean Sibyl; if so, Linnaeus is another Aeneas about to found Rome. The Fury spoke to him with solemnity, as if from another world. He was attempting to cross the Swamp of Misfortune, an extremely arduous route. He was on his way to Sorsele, the planned starting point for his first attempt to enter the country of the Lapps. In this manner, Linnaeus emphasized that he had been crossing the boundaries of another world, just as in *Nora Samolad*, his mentor Olof Rudbeck had turned his journey through Uppland's Älvkarleby and his crossing the Dalälven river mouth into a splendid mythical scene, complete with a ferryman—the Älv-Charon. Rudbeck devoted twenty-odd pages, heavy with learned references, to this flight of fancy; in contrast, Linnaeus's narrative stands out for its brevity and concrete detail. Besides, the Fury provided cheese made from reindeer milk, and the creatures he met on the other side were friendly, if pitiful. It stands to reason that Linnaeus took pains to write to suit his chief sponsor.[14] He knew, of course, of *Nora Samolad*, recalled Älv-Charon already when he himself crossed the Dalälven River, and later links his narrative to the frontispiece of Rudbeck's work. However, though the Fury may well be an homage paid to Rudbeck, she is also a conventional emblem; the scene can be understood as a meeting of old and new, a boundary marker on mythical ground.

A few pages further along in the diary, there is another significant episode: a description of bog rosemary (*Andromeda polifolia*), which is a classic in Swedish literature—not only emblematic but also a striking example of how the Baroque ideas influenced Linnaeus's thinking. He was unmistakably proud of having composed these passages on the Fury and Andromeda, because he copied both in *Flora Lapponica*. His account begins with a quotation from Virgil (*Aeneid, Book VI*): "Divine Law prevents it, and the sad marsh and its hateful waters bind them, and ninefold Styx confines them." The plant is compared with the sun at midsummer, with the myth about Perseus, who defeated the monster and rescued the princess Andromeda. Other passages that come to mind include Rudbeck's exposition about the red neck of the swallow as a reflection of Minerva's golden apples or as a letter in a chain of images. The subject of the following analogy is the tall plant, King Charles's scepter, "whose flower likens to a helmet / with the yellow sheen of gold / its pale mouth bloodied and blood-speckled leaves." Further, concerning Andromeda:

> Chamaedaphne Buxb. or Erica palustris pendula, fl. Petiliolio purp. Now stood in her finest flowering and conferred to the marsh an excellent ornament. I saw how she, before she opens her petals, is as red as blood. But when she floweres the leaves become all over a shade of violet. I would doubt if any painter exists who can show such pleasing appearance on a maiden's image and such beauty to clothe her cheeks. No cosmetic has yet had that effect. When I saw her, Andromeda as

portrayed by the Poets came to my mind. The more I thought of her the more the princess seemed to agree with this herb. Had the Poet made to describe her in a mystic sense, no one from the Poems would have suited her better. Andromeda was described as an extra ordinaire maiden and woman on whose cheeks beauty still lingers. This loveliness remains hers only for as long as she is a virgin (such as is the way with women), id est until she has conceived and that will not be long now; for she stands a bride. This herb is confined in the middle of water, always on a tussock in the wet marsh as if bound on a rock in the sea, in the water up to her knees above the root. Unceasingly she is surrounded by poisonous dragons and creatures, that is, evil toads and frogs who in the spring as they mate blow water onto her. So she stands, her head bent in sorrow.[15]

The figure of Andromeda represents a positive contrast to the Fury and also a biographical dimension. Why did he recall the story about "an extra ordinaire maiden and woman," chained to a rock in the sea and surrounded by poisonous dragons and evil toads, the part of his drawing that is *ficta* but is compared to *vera*—that is, the plant he observed? It seems relevant to recall that Linnaeus, as the tutor to Rudbeck's son Johan Olof, wrote a boyhood thesis about King Charles's scepter. Now, the bog rosemary became Linnaeus's very own contribution to political interpretations of natural history. Extend the line of thought further, and the bog rosemary could change from being just a beautiful maiden into— perhaps—Sweden itself, now threatened by commercial developers rather than by the tsar and the Russian army. The knight riding to the rescue of Andromeda stands before her, the young student Carl Linnaeus.[16] Sir Carl conquers Lapland, a narrative that would surely please the learned gentlemen in the Uppsala academy. But one might also consider his possible identification with Ovid in exile.

A passing comment: Was Linnaeus color-blind? The question points to the Latin name of the plant *Andromeda caerulea* (current: *Phyllodoce caerulea* or blue mountain heath). The species name "caerulea" means blue but a lighter, less vibrant shade than indigo. What might the reason be? Was Linnaeus writing up a pressed specimen? No, on this journey, he described live plants. Anyway, the question has been asked. "When I by midnight—if I am to call it night as the never setting sun illuminated the Earth—searched for a Lapp's tent . . . I do not know what it might be that after nightfall in our mountains so perplexes our sight that we even though the sun is still shining can no longer discern objects as clearly as in the

IN A MYTHICAL LANDSCAPE: LAPLAND [71]

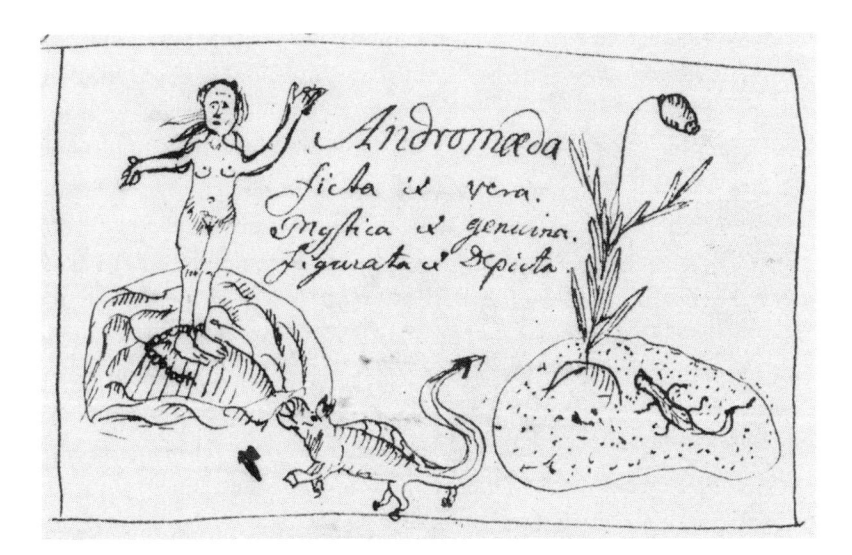

FIGURE 10. Andromeda threatened by the dragon but about to be rescued by Perseus. Nationalistic botany? Bog rosemary, from *Iter Lapponicum* (facsimile ed.).

middle of the day."[17] Linnaeus had by then managed to pass the Swamp of Misfortune, crossed a boundary, and rescued Andromeda from hostile territory. He climbed farther up the Vallivaare mountain near Kvikkjokk on his own or, rather, guided by a Sami who also carried his bags. Once he crossed the tree line, he was "in a new realm," unsure whether he was in "Asia or Africa" as everything seemed alien to him. "I was now in the mountains. All around me stood snowy peaks and I walked on snow as if in the strongest winter"; he felt "as if guided into a new world."[18] The way he expressed himself must have been meaningful because almost exactly the same words recur in his presentation to the Society of Sciences.

What was Linnaeus referring to with the phrase "into a new world"? A new kind of natural environment is the obvious interpretation: he describes what he saw as changeable between white and green, winter and summer, sterile and fertile, and close to the sky. When, in July 1746, he had traveled westward in central Sweden and reached the coast at Marstrand, he wrote about botanizing on the bottom of the sea "as if in a new Sweden." The "new world" could also have been his way of referring to "the New World," as had the statesman Axel Oxenstierna, when he spoke of the North of the country—Norrland—as Sweden's Caribbean islands. However, Linnaeus isn't thinking about colonies at this point. He wrote to Lars Roberg that "when I had arrived into the mountains I knew not where I was, whether in the east or west indies, so unlike anything I knew was the world around me, so many curious objects occur there." But experiencing the unexpected can also mean something almost alien, visionary. Linnaeus felt as if "surrounded by the very clouds and not ever alone," according to one autobiographical note. Walking in the high hills, he felt "inordinately well" and recalled Ovid's vision of the Golden Age—"if someone would only take the snowdrifts away." And again: "As soon as I went up into the mountains, somehow I was given new life and it was as if a heavy burden had been taken off me." All in all, this new world could offer him a heightened sense of being alive, of happiness. In another passage about another place, he remembered: "It appeared to Linnaeus that he in Fahlun had come to a new world where everyone loved and supported him."[19] This brief overview shows many different ways for understanding him, and that his outbursts of emotion have literary and intertextual content rooted in his wide reading.

The journey to Lapland displays some of the ideas that Linnaeus had made his own. Observations during his travels provide new impetus for his special area of medical interest, dietetics and, generally, in how to live well. Again and again, he expresses himself in a "pre-Rousseauian" mode, following classic primitivistic tradition and mixing it with homespun "Gothicism," which perhaps isn't very original but rarely done as vividly as here. The idea of a golden age also emerges, linked to the Christian myth of Paradise. The Gothic idea may seem rather backward and belligerent, but it can also be used, as it is here, in an almost forward-looking sense, mixed with utopian hopefulness about peace to come: "Ah, thou secure Lapp, thou fearest nothing, be it war or hunger or priest though they can in a short time ruin other lands."[20]

Primitivism, chronological as well as geographical, was unhesitatingly combined with Gothicism, religion, folk wisdom, Hesiod, ethnography, ideas about the "noble savage," and words such as "natural," and pedagogics, often influenced by Rousseau. In the eighteenth century, primitivism is alive, encapsulated in notions such as the "great chain of being" and democracy, and also as an ingredient in travelogues, cultural analysis, climatology, and health advice. This idealized version of the primitive is contradicted by the theories of progress and economic growth, luxury consumption, and "the unclean native." For example, in *Diaeta Naturalis* Linnaeus wrote about how giving birth comes easily to the American Indians, that they tie the infant to a board, so it can be carried on the mother's back, and that the child is breastfed until the age of four. "They have the acute sense of smell of animals such as dogs, and often live for more than one hundred years." They also believe in a life after death in a world "where prey for hunting and fishing is plentiful."[21] "As for the spouse of the Lapp, once with child she is given no peace by her husband or husband's mother until her time has come so that she will find it easier to give birth. The opposite is true of the highborn as they are enjoined after conceiving to stay still and so they suffer great pains in their bed."[22] The dream of bright, innocent eyes and free, uncontaminated minds was related to another dream of human beings in a state of innocence, and of the wild man, noble and untainted.

These ideas emerge in the introduction to Linnaeus's unfinished treatise "On Lapland and the Lapps" (*Om Lappland och lapparne*). His travels among the Lapps had allowed him to "delve into their well-secured *republique*" and appreciate the peaceful, contented rule, and the good life it offered the people, free of illness, warfare, and spying. The Lapps were happy and healthy and not threatened by crop failures and famine. Science, indeed medicine itself, were "alien birds," manual work was shared alike between man and wife, and the blacksmith's craft was the only missing one. "I would be content to exchange all that now pleases me for a mountain Lapp's refuge, his *bene latuit*, had it not been that *Frigus e teneris non adsvetus*"—if only he had been accustomed to the cold from childhood. This little quotation serves to illustrate how Linnaeus added gothic mythmaking to his traveler's tale. The narrative of a land of peace without fortresses and war, and far from the world's noisy traffic, continued. Comparing it with Rudbeck's vision is unavoidable; as he had discovered the lost Atlantis at the end of the world, Linnaeus had found a happy population of shepherds like the stories told by Ovid and others. However, the similarity between mentor and mentee only goes so far. The Lapps he met were no gothic grandees, nor the northern "wild men" of the usual allegorical portrayals. His northerners were described in levelheaded language, placing them within a toned-down gothic context, tired of war and embracing peace. They were healthy, lived well, and were therefore models of moral beings.

It has been said about Linnaeus that he felt a "deep, genuine admiration for the Lapps" and that he was someone who "loved the Lapps."[23] These quotes refer to the Sami people as an idea and an ideal reality, but the expedition also had glaringly obvious pragmatic goals. At this time, Lapland was the new Peru, its resources praised by Rudbeck and its role seen as part of a nationalistic ideology. The driving force behind Linnaeus's journey was a royal command that made the Society of Sciences responsible for sending someone to explore Lapland. Christopher Polhem, a leading scientist and member of the Society, compiled a list of twenty experiments that were to be carried out in the hills and valleys of Lapland.[24] Utilitarian notes were struck already in Linnaeus's application to the Society in December 1731, when he wrote that Lapland is still in the grip of "what seems the cruelest barbary"—"barbary," here, is used to mean "ignorance" and refers to the need for schooling but also to lack of cultivation—matters that should be taken in hand.

Linnaeus cared for the mountain Sami: "I believe that, given the natural conditions of their land, the Lapp has adapted in the best way the manner of his household's livelihood. Thus, their animal herds tolerate the climate well, find enough to eat, and supply their herders with both food and clothing. Where in these parts would sufficient fodder be found for horses and cows during such long and stubbornly constant winters? Where would be a suitable place for sowing in soil always subject to frost? New settlers could surely survive, but unless, by trading goods with the Lapps and by fishing in the many waters and hunting in the forests, they could acquire a more assured profit than by tilling the fields, I dare pronounce that their lives would be most wretched."[25] He is blowing hot, then cold.

Categorizing Linnaeus's different descriptive passages requires a scale between the extremes of accurate record and ideological statement. One apparently strictly neutral account dealt with how a Lapp urinates, worth noting because he can't get his penis out through the fly on his trousers: "instead he pulls his breeches down and over his thighs as their shirt does not prevent it. He pisses along the line ahead and *et sic mingit* [thus he urinates]." It does not seem ideological at all but, at the same time, clothes and the body are meaningful to Linnaeus for economic and political reasons. The man's clothing is made from skins and fits easily around the waist, he has no use for underwear, and his shoes don't have heels. All of this adds up to a theme that is given plenty of room in Lapp narrative. At times, seeming to argue in favor of colonialization, Linnaeus stressed the kindness shown by the Lapps toward the settlers from the South, and also how rough the response often was:[26] "The poor Lapp, now much reduced himself to live off fish and with it to feed his household, has now hardly more than one or 2 fishes left. I asked him why he would not complain but he answered that he had once raised such matters with the authorities (the Governor) who said it was *lapperie*, with which he meant 'of no consequence.' . . . New settlers who come to sit down with the Lapps are well loved, and showed without holding back where they may gather hay not needed by themselves, as their reindeer would rather feed in dry places; and they can buy flour and milk from the Lapps." Linnaeus knew and seemed ready to accept that the composition of Lapland's population

FIGURE 11. A Sami woman on the mountainside. Lapland, seen as fertile and suitable for cultivation. In the lower edge of the image, a man plants a tree (a palm?), with the caption *posteritate*, which would become the motto of the Academy of Sciences. Drawing from the journey through Lapland. After Thore M. Fries's 1913 edition of *Iter Lapponicum*. Uppsala University Library.

would soon change: "In Lapp territory, they [the new settlers] are permitted to live where it takes their fancy if the soil can be worked. There can be no question: where Lapps once were will soon be farming land."

There must have been other, still unanswered questions in Linnaeus's brief. He wrote in Latin: "Problem—the question of how the mountains can be cultivated. The whole region is very cold." He listed several, mainly climate-related issues and went on to a conclusion: "Persons who have occasion to travel here will find these parts exceedingly healthy, and also to drink snow water rather than lie in the sultry air of spas by boggy wells."[27] Rudbeck was interested in this aspect, too, and Linnaeus reported at about this time on the Swedish spas in a pamphlet called *Najades svecicae*. It reads like a fusion of advertisements for mountain tourism and recreational visits to sanatoriums, showing a role for the author as the guide to a new era.

Once back home in Uppsala, Linnaeus turned his journey into politics. He planned to write *Oeconomia Lapponica* as a report on the utilitarian aspects of the nature in the region and the Sami land management; in Linnaeus's view, there was nothing to stop him from arguing about the "economy" of a native population. His promise to offer "propositions to how Lapland through all three realms of nature could be improved" was presumably intended to support the settlers. His submission to Gabriel Gyllengrip, lord lieutenant in County Västerbotten, was an extreme case of the project enthusiasms of the time. In 1733 Norrland had been hit by a famine, and Linnaeus insists that something must be done. His letter can be regarded as an expression of patriotism but also acceptance of central government intervention. Linnaeus proposed cultivation of winter cereals (or marram grass), hardy varieties with little or no need for proper soil but added that "to in this instance attempt to persuade the Lapps, the Settlers, or Farmers, would all be in vain as they never give up the practices of their ancestors," so this reform must be imposed by the governor. Once completed, a good outcome was granted as over time it "would earn our great country some barrels full of gold." These ideas suggested that fast, almost alchemical transformations were possible. Gyllengrip sought funding for Linnaeus from the Collegium of Commerce, but without success.[28] However, these ideas stayed with Linnaeus, who a few years later developed them further in a proposal made to Nils Reuterholm, lord lieutenant in County Kopparberg: "I have received projects as to the mountains and how they could best be cultivated with root vegetables of which Nature has offered me several exemplars through a new *autopsie*." What he means by "new" seems clear from how he continues: "Alas, could one only travel throughout all the provinces of Sweden, best each summer, how many things could one not invent for the good of the realm! How much would not a province learn from what another had facilitated. I profoundly hold the thought that such a post in Sweden would profit it more than Poetic, Greek, and Metaphysical professions at some *academia*, as *publicum* could establish such a post with much less expenditure than any of the latter."

It would become a reality. Also, the small drawing of a figure planting a palm on the slope of a mountain, sketched by Linnaeus during his journey in Lapland, would be developed into the motto of the Royal Academy of Sciences: "An old man digging" and the caption *posteritate* translated into "For those who come after us."[29]

The anticipated changes were sometimes to be fast—"metamorphoses"— and sometimes slow transitions to benefit "those who come after us." The traveler prepares the way for the modern, which in Linnaeus's case included the powers of naming and identification, and a capacity for abstraction that was a basis for his technocratic appreciations of nature and people. He described a small event that seems symbolic: "I showed a Lapp some drawings in my book. He at once was fearful, took off his cap, curtsied and shivering with his hands against his chest mumbled a little with his mouth as if with the greatest veneration, and would perhaps swoon. Many have swooned on seeing something extraordinary such as . . ."" and here, Linnaeus had drawn a toy—a snake leaping out of a container like a jack-in-the-box. This is probably the phenomenon he referred to in this note in *Diaeta Naturalis*: "The Lapp if he has never seen a snake will swoon at the mere *pictus*." He might, of course, have misunderstood the reaction—perhaps it was the creature in the drawing that scared people. He also noted that when walking in the hills, his Sami companion became irritated whenever he wanted to examine a plant: "The Lapp was galled when, here and there, I stopped for the herbs I saw."

Linnaeus was young and free but still inescapably a creature beholden to the central power, as a medical student and a grant holder. One of his roles was to act as investigator and prospector on behalf of the expanding nation-state. Classical, medical, and political attitudes mingle, myth and modernity meet in his mind; the ancient myths are mobilized in support of present-day motives. One might argue that the old concept of euhemerism applied to him and to his time in general (cf. Euhemerus, ca. 300 BC, argued that the gods of popular religion could be identified with dynamic personages who were worshipped after their death). In other words, euhemerism equated gods with real heroic figures and was, in its way, uniting scientific and humanist traditions. Rudbeck worked alongside etymologists while Linnaeus used a taxonomic net to capture forms of life recorded in classical texts. To Rudbeck, the pygmies of antiquity corresponded to the "gothic boys," that is, the Sami. This kind of discourse places classical texts as concrete models and thus confers charisma on reality while, at the same time, it makes the past seem more real. Linnaeus discovers or reveals the hidden identifications offered within history as well as in nature.

He traveled as a natural philosopher and also as an economist, with no distinct boundary between the two disciplines. The task was framed as an investigation of nature. His application to the academy doesn't emphasize economics, but after his return the premises seemed to have changed. He was annoyed at the society's lack of interest in his three submitted practical overviews: "So this is how *oeconomica* is cultivated in Sweden!" he wrote—including the exclamation mark. When the poorly funded Society became unable to meet his wishes, he was told to petition its honorary president, who in his outside role was the most powerful man in the land: Arvid Horn. Linnaeus had been evicted from academe and sent out into a more worldly society. Tensions inside the older institution had almost certainly driven the founding of the Academy of Sciences, originally proposed as the "Oeconomic Society." By the 1730s, new ideas were bubbling inside such groups. From Linnaeus's point of view, during his travels it was self-evident to turn to the new officeholders, as for instance the governors Gyllengrip and Reuterholm. The latter in particular would play an active role in founding the new Academy of Sciences. Carl Gustaf Tessin once described Linnaeus as having the mind of "a brave and vigorous county governor who rather resolves 2 times than waits for one resolution," an ideal that often went down well with the reformers of the time.[30]

The memories of the journey faded with time, and the bursts of patriotic emotions were replaced by utilitarian projects—insofar as there was a clear distinction between them. Linnaeus will go on to pronounce again and again on when, where, and how the Swedes should set about planting crops, trees, and shrubs; the Lapland project was finally shelved twenty-two years later, in an essay published as part of the 1754 Proceedings of the Academy of Sciences. Thankfully, the marram grass was forgotten by then, but the bare mountains were still to be clothed with trees and other plants. The climate would eventually instruct the cultivators about which species were the most appropriate, but some were thought of as safe bets; Linnaeus hopes, for instance, that the Swedish dinner table will one day enjoy the tang of Norrland-grown saffron. Irony comes easy when facing an overindulgence in captivating projects.[31] Did Linnaeus ever realize that he had brought back a veritable Pandora's box, full of destructive powers? As much as he emphasized the lifestyle of the Sami as ideally adapted to their land, at the same time he outlined ways to go about ruining their culture.

In 1733, Gabriel Gyllengrip was appointed to the post of governor in County Västerbotten. He had one year to go before the introduction of the rule that governors had to submit an annual report, and traveled widely in his huge province, of particular interest to the state for its possible mining potential. Gyllengrip corresponded with Linnaeus about options for cultivation and often had two friends as traveling companions, Seger Svanberg and Jonas Meldercreutz, a young mathematician and owner-to-be of a mining company in Tornedalen. The group's efforts were successful: the rich ore findings in Kirunavaara and Luossavaara were identified.

Gyllengrip wanted to persuade Linnaeus to return to the North, but he didn't. Instead, Meldercreutz, who had heard of "the excellently skilled Herbalist Herr Lenaeus," came across a cereal variety that could be cultivated locally. He also described what they saw after climbing a mountain: "It appeared to us as if by ascending to the highest mountain ridge we were in another world from where we observed, as if on a stage, a large part of our land" and also found signs of "all 4 seasons of the year at the same time."

This passage seems utterly Linnaean, and it could well be that Meldercreutz had read at least some of *Iter Lapponicum*. Could it be that Gyllengrip had been given a copy during the long stay in Kalix? A little later, he remarked that "Hr. Lenaeus has elucidated that the same kinds of herbs are found in the northern mountains as in the Alpe ones." Such comments are reminders of our fragmented knowledge of Linnaeus's activities and the historical development of *Iter Lapponicum*, while they also demonstrate that, as a visionary explorer, he was not alone.[32]

Evaluations of Linnaeus's travels in Lapland have been changing and, indeed, swung right around. Recent hypotheses include that the text was penned by a poorly informed enemy of free trade, which fails to take Sweden's postwar finances into account and instead turns history into a lesson for the present. The notion of cultivating the northern mountains has led to smiles all around, but the Swedish ambitions look more normal in the context of the many other attempts to acclimatize plants and animals from elsewhere on the basis of geographical similarities. Some students of his work have become so engaged in discoveries of Linnaeus exaggerating his achievements that they have forgotten or undervalued both his physical application and the many valuable outcomes of the journey. Criticism has been based in part on the philosopher Michel Foucault's analysis of power relationships, and insists on the arguably trivial fact that Linnaeus represented the authorities. Another issue seen as practically compulsory is the ideological content of the Linnaean discourse on "the Sami Other" and whether it is to be rated "modern." It is a large, possibly inevitable question but one which must not be allowed to overshadow all others. It seems clear that, as a traveler, he represents a mobile changeable—modern—people, in contrast to the Sami with their nomadic but nonetheless stationary lives.[33]

The art historian Wilfrid Blunt wonders how daring journeys of this kind actually were, and feels that Linnaeus's risk-taking has surely been talked up too much. Really? Given the circumstances, would Mr. Blunt have set out? As for unmasking a presumed authority, this can easily lead to generalizations. In her 2015 book on the problematic attempts to categorize nature (*Animal, vegetable, mineral?*), Susannah Gibson comments on Linnaeus's travels in Lapland: "He later produced journals, fantastic charts and maps detailing his adventures. Sadly, most of the documents were fakes. . . . But this fraud wasn't discovered until many years later, and Linnaeus' successful career continued."[34]

There are good reasons to emphasize the wealth of factual content of *Iter Lapponicum* but also to open up the debate about its more far-reaching interpretations. The book is not "a source of authentic information," true—but are there any of its kind that can count as wholly authentic, even today? Practically half of Olaus Magnus's "History of the Northern Peoples" (1555) consists of quotations from classical texts, but it is nonetheless the main source of our knowledge about the Renaissance in the Nordic countries. That Linnaeus could be less than truthful and occasionally tell lies or express himself obscurely, but was still taken on his word, arguably says more about the historians, for so long uncritical—deliberately or carelessly—than about the writer. His flexible attitude toward facts was not remarked on by people such as Thore M. Fries, Knut Hagberg, Carl Otto von Sydow, or even the critical Sten Lindroth: to all of them, *Iter Lapponicum* is a source as pure and sound as a mountain stream.

Karl Bernhard Wiklund, an expert on Lapp culture, noted in 1925 that Linnaeus had failed to give a proper description of the physical characteristics of the Sami and so, obviously, was a rather mediocre anthropologist. Wiklund's conclusion: "His imagination runs out of control and allows him to write what is nothing but fantasy, especially so in his account of his travels for the Society." Wiklund had been in touch with Herman Lundborg, the head of the State Institute for Racial Biology, who at the time was busying himself with measurements in Lapland.[35] Actually, Linnaeus was attempting another and more inclusive narrative. His scheme for classifying what he saw among the Sami demonstrates again that his journey came to breed new forms of texts and that he reached out to different audiences. He addressed the narrower scientific community, of course, but also the collectors and the taxonomists.

Johan Browallius assures us in his long essay of the journey in Lapland:

> Linnaeus however joyfully took on the task, although arduous and entailing great dangers, without hope of monetary gain or glory and instead enthused only by his lively desire to shed light over the entire span of the three realms in natural history.... The expense of the journey was met by the Society of Sciences. This payment was the only modest reward sought by and ever given to Linnaeus, so great was the value he placed on learning to know nature.... He came to spend all of six months among the Lapps, a wild race. He was often in danger of losing his life: once, climbing Skjul Mountain, the snow betrayed him, [once] a bandit fired his gun at him.

The image of the dashing explorer-scientist recurs in the "Journey in Dalarna" (*Dalaresan*), which has Linnaeus clambering about in mine workings. As for taking notes, Browallius assures us that upon Linnaeus's return to Uppsala "he worked conscientiously to finalize his account of the observations made on the journey, not only for many days but also during many wakeful nights."

There is still much to be said about this journey, the most classical traveler's tale in Swedish history. Take the mosquito plague for example. This quotation, signed "Linnaeus," is now on display in the high hills of Norway: "The Lord made the mosquitoes mainly for the North so that its mountains would not compete with Paradise."

His journey is comparable with Darwin's on *The Beagle* a hundred years later, but this is not the place for us to speculate about how the comparison would work out.

CHAPTER FIVE

Diaeta Naturalis

INSPIRED BY HIS TRAVELS in Northern Arcadia, Linnaeus began to work on *Diaeta Naturalis* not too long after his return. It was to be a collection of medical commonplaces and aphorisms, and in the main amounted to one long celebration of the Sami and their happy lives. Its stated date is 1733, but it was in preparation for at least five years—printing took longer still: 1957. It is a medical thesis in the Hippocratic tradition, with emphasis placed on the importance of the peaceful state of mind that was under threat from a new social order. The criticism was mostly directed toward urban life, but the court and "Frenchification" were also vigorously targeted.

He wrote with a personal touch at times: "In my early childish years I admired *primi parentes* Adam and Eve that they could have grown so old, and then it came to me the saying by Svedenborg that the year was shorter then and so [they] had not been living much longer than now." He was probably referring to [Emanuel Swedenborg's 1714 treatise?] "About the Motion of the Earth and the Planetary Movements and Relationships" (*Om jordenes och planeternas gång och stånd*, 1714). Linnaeus also pondered about the limitations of human life: "When I considered the healthy Lapp in Lapland, some principles came to me that, through whose useful application, man could increase his age twofold without illness, *ex principibus naturalibus*."[1] He explicitly rejected his own trade's professional interests: "Do not wonder, my Reader, how come I, a *medicus pro primo*, took it upon myself to write about how the common people should escape illness although such advice is against the profits of myself and all *medicorums*. . . . Young *medici* forbid all things, but what I want is otherwise."[2] Linnaeus is uncompromising in his insistence on reinstating a natural form of medicine, based on the physical characteristics of the body and how surroundings and culture might affect it for the best. Everyone should go back to how he or she began—was created to be.

His arresting declaration was intended for Gustaf Cronhielm, the chancellor of Uppsala University and the dedicatee of Linnaeus's work. He hoped that these thoughts would open the doors for a stipend, perhaps a post, and even a restructuring of medical studies. It seems that his treatise never reached Cronhielm, who was preoccupied with his epoch-making legal reform that in 1734 resulted in a new Swedish law book, which still provides the basis for current legislation. Linnaeus addressed a readership unversed in medicine directly in *Diaeta Naturalis*. Its long title was "70 rules that with explications, and personal observations that have led to proof, taken from the principles of zoology, in that the human being is an animal and should therefore live like an animal." Or, as he put it in one of the lectures to Mennander: "A human should not be considered as a machine but as an animal and none is more a relative to the human than the wild *apian*."[3] Then again, he says in his fast-moving, rather incomplete introduction: "This book I would not recommend for reading by the *praejudicieuxe* as to those it will seem like nothing but a daft caper. It is for unlearned folk that I have written."[4] Despite his populist intentions, the text is full of Latin; an arrogant youth held the pen, ambitiously sketching out a possible utopia by creating conditions for a human rebirth that could save the nation—but only by sticking to traditions and simple habits, and unprejudiced sympathy for the ways animals feed themselves. This sounds like Bacon; a book not suited to the nonempirical mind. At first, Linnaeus drew up fifty rules, but the number later increased to seventy; later still, additional rules were added and taken away, but his six main elements of dietetics remained recognizable: fresh air, rest and sleep, activity and stillness, ingestion of food and drink and excretion of waste, external sensory stimulation and peace for the soul. Linnaeus didn't trust medics and their cures and instead urged his readers to go in for prophylaxis.

The Lapp should be our teacher, Linnaeus explains. His views might be categorized as "medical primitivism," given the sheer quantity of his advice on hygiene and diet linked to Lapp practices.[5] "Oh Contented Lapp, concealed on the outermost edge of the world where you live well, cheerful, and innocent. . . . You know not of any blows other than the strikes of lightning bolts from thunderous Jupiter. . . . You live in your forests as the birds do, who sow not nor do they reap but are nonetheless bountifully fed by the almighty Lord."[6] He alludes in several contexts to the Greek philosopher Diogenes and the large pot he apparently sheltered in; Linnaeus refers to it in *Diaeta Naturalis*: "The soulless animals live without possessions. The Lapp is in his small *goahti*; in a moment, he can change and pull it up. Diogenes was content with his pot and his crock but then he saw a youth drink from his hand, and threw the crock away. And yet the great Alexander said, 'Were I not Alexander I would wish to be Diogenes.' But we do not content ourselves unless our houses are filled with affectations and wish for still more castles of foolishness."[7] Two factors primarily shape the lives of men: the climate, which no one can escape, and habit, made up of cultivation and culture; the distinction between them is far from clear. To prove his point, Linnaeus compared a series of statements about the variations between people from different parts of the world, and also about how children and animals are formed by the upbringing, exercise, and food they are given, and the clothes children learn to wear.

Until the present day, history tells us of everyone's endless walking. It was a necessity but also recognized as a good thing to do. Linnaeus agreed: "When I had walked in Lapland some 1,000 furlongs and came home, my legs had acquired twice as much meat as before."[8] Walking was also a fruitful exercise for the mind, as thinking is activated by the body's movements. Among the twenty-odd questions handed to Linnaeus by the Society of Sciences, one concerned the Lapp's running speed. Linnaeus had the answer: "We build up our shoes with large heels, females even more so, with the aim of walking more lightly; it is a wonder that the Creator lacked the wit to add high heels to us as He was so merciful and did everything in its entirety such that no more could be desired. It is known that to run barefoot is more easily done than *ocreatus*, indeed with Lapp shoes or boots it is as easy as barefoot because without a large heel, the foot has its whole length for its use, but with a stuck-on heel no more than half."[9] Linnaeus is advocating a future of shoes without heels; fashion was at all times a perfect target for his irony.

We must also learn from the animals. Linnaeus argues against current conventional wisdom: "If we inquire into what separates man from animals, it is not sensible reasoning."[10] True, what he quoted as proof is not all that convincing: "As Councilor Lund sat asleep in Skänninge courtroom, when aroused by the Mayor he reached for the inkpot, saying 'cheers, bottoms up' while it dribbled down him." Linnaeus goes on: "I tremble when I observe the advantage God gave the human over the animals and how she has abused it and [lost a God-given place] on the scale of perfection . . . as the insect is less wise than the bird, the bird less so than the animals; the ox less than the wolf, the wolf less than the ape, and the ape than man."[11] He speculates further: "*In bestiis non seviendum* [do not take out your anger on the wild animals]. *Theologi* state that man has a soul and that the animals are no more than *automata mechanica*, but I believe they were better counseled to concede that animals have a soul with distinguished nobility. Were they nothing but *pura* machines, why should we feel compassion for them?"[12] Of course, were we essentially machines, how could we feel compassion for anything?

Habits make up our second nature—this is one of the main themes in *Diaeta Naturalis*—so it follows that our habits must become our friends.[13] Again and again, this principle is made the subject of a variety of expositions, single-mindedly but often amusingly. The topics can be clothing, exercise, or food and drink; the medical moralist Linnaeus is always in full flow. One problematic question concerns how dominant one's habits should be. Flexibility is necessary since how else would progress be possible, how else would moral improvement be an option? However, in principle, habits run counter to nature, and habit alone can defeat nature. Culture is also capable of changing us in endlessly complex ways. If and how an infant is swaddled determines its place in a taxonomic system. We may feel none of this makes sense, and may well be right. "Should a stranger arrive in Leyden, he will hear a set of tinkling bells like our *spiel* here in Upsala. He who is unused thereto must clap his hands over his ears, but once used to the sound he makes nothing of it." Another example, the final and rather odd one: "A Stockholm nobleman could not abide hearing of *peapods* so his people began to say *teapots* until he became so habituated that he himself spoke of *peapods* without being moved thereby." In *Morbi mentales*, he used the same example to illustrate how habitual resentment can become attached to minor irritations: "A baron would never expose himself to the word *peapods* as he vomited at the sound of it. Why, the archiater himself could not for a long time tolerate the sound of certain bells in Leyden."[14]

He paid keen attention to food.

> The first human beings fed on vegetables and fruits as is clearly said in the Book of Genesis. Until the time of Noah, such was the diet of all mankind. Here we should stop to consider that such were not the fruits as nowadays grow in our Northern lands.... We are convinced of this as are our hands, teeth, and stomach, etc.... When much more time had passed and mankind multiplied even more, men began to plant grain; who the first man was to invent the planting of grains is not known. However, we have read in the Book of Genesis that Cain was the first man to till fields.... Gymnosophistae, Brachmanni, or Sapientes Indiae eat only *vegetabilii*, a practice which is said to increase their genie.[15]

Linnaeus valued the Sami highly, but they were of a different kind. Who were they, then? He didn't write of anyone by name and only discussed "the Lapp" as a type or member of a group. Linnaeus traveled not only as an ethnologist and a cultural anthropologist but also as a taxonomist, a role he indicated by using Latin for precision and to reach a wide audience. He raised questions of how nature made us and to what extent culture has changed our natural state, for better and for worse. Considering animals, wild men or "natural people," and children, he observed: "The small newly born child, recently coiled inside its mother like a squirrel inside its nest, once born is swaddled so that the bone might break or become all straight; the American binds its offspring to a board, the Lapp is laid in the almost loosest cradle and thus the American walks as straight as a post and the Lapp leans a little, walks with his feet this way and that as God decrees, not by the instructions of the dancing master."[16] He would soon present his proposition for the classification of mankind. We are made by nature but *formed* and *grow* as shaped by culture and climate.

He praised the lifestyle of the Sami, as is seen in this mosaic of quotations: "Oh thou holy minister in Särna, said Governor Renstierna of Falun, as the war ravaged. Oh, thou secure Lapp; thou fearest naught, nor war, hunger, pestilence, and such afflictions that in a short time might destroy a people elsewhere. The Lapp knows no more worry than what is caused by God or the weather." "The Lapps live together alike, without *luxuria; hinc* are not slaves of others." "Someone who becomes *maniacus* is like an animal with but little soul, his being so like one as were they the same. She freezes not and acts as she fancies, untroubled by desires, can be vicious and rude, glum, etc. but is healthy, strong, *æstimates* no *medica*." "All *theologi* want that nature should be destroyed as we always accustom ourselves to what is *malum*. This is so, because our parents and upbringing and then all conversation instruct us to indulge in things that are unnatural, to gluttonize, dress colorfully, be mean with money, etc. Is this seen in the poor Lapps, Eastasians, Americans?"[17]

Consequently, his next step is contempt for the aristocracy. "The farmer and the Lapp do not know what it is to purge. . . . The court gallant even better at it than the doctor." "Thus all wealth is drawn into the grasp of certain families who eventually are given the name of nobility; are however mostly like bee drones in a honeycomb as they suck out of the realm that which the farmer and the merchant like honeybees have been assembling; would that the realm was without the nobility, and ennoblement not a birthright but the virtuous reward for hard work, if so, the country would be a better place."[18]

The climate, part of nature and yet a singular phenomenon, was useful as an explanation of what seemed still inexplicable, but as a part of an argument, climate was difficult to handle. Linnaeus would eventually come to see the world as subdivided into concentric climate zones, its population distributed with the Lapp at the northern extreme and the Hottentot at the southern. In this scheme of his, northern peoples benefited from positive effects of the climate, which became increasingly negated the farther south they lived. Cold doesn't soften "the fibers," he thinks, but rather toughens and firms them up; cold keeps us healthy and energetic. Discussions of this kind are, of course, much older than his near-contemporary Montesquieu's, and Linnaeus remains close to Gothic-Rudbeckian ideas. Many stages in the argument are bewildering: for example, contrary to all expectation, southern peoples don't grow larger than northern—instead, a hot climate makes you lax, and your body sponge-like. Every aspect of the Nordic climate directs the individual's development in a good direction and, by that logic, the most northern Swedes, the Lapps, are the best—first among near-equals. Real patriots should reject any negative responses to his reasoning.

The dietetic debate of the day provided advice in matters such as rhythms of work and leisure, and food, drink, and dress. When Linnaeus wrote about the beard as "natural," he had been beaten to it by combative men of the church, notably Jesper Swedberg and Jacob Boëthius, who disapproved of the new clerical habit of shaving. In portraits of Linnaeus's father, he has an untidy but suitably devout beard—as his son wrote: "The beard distinguishes men from women, a mature *judicium* from one who is immature and unjust, and a fine, upright man from a lad, not yet fledged."[19] During his travels in Lapland, he must have sprouted a fine beard—soon enough, though, he would be clean-shaven and bewigged. As for clothing, supply issues amounted to a nationwide financial problem. The desire for change, or fashionable style, was also problematic. In this, Linnaeus seems to lean toward the Baroque attitude of deploring change, which only leads to dissatisfaction—as opposed to the older ideal of steadfastness, bringer of security and happiness of the right kind. He was a sworn enemy of fashionable manners and all things French, including food, which also had political overtones. Linnaeus was above all a patriot at this stage; later, he became a utilitarian.

The elevation of the Sami was to some extent a literary device, made to express the fundamental notion of the superiority of animals and the natural world in general. It also expresses a preference for positing oppositions, another feature of Baroque thinking: nature is positive, culture negative; animals are wiser than men. Among "natural" peoples, the Sami

FIGURE 12. Linnaeus in Sami costume—the cap is not genuine. Engraving after Martin Hoffman. The art collection of Uppsala University.

are good, in contrast to the Hottentots—but studying both is instructive. It is, however, hard to make out where his real experiences, such as they were, give way to intellectual construction.

It is intriguing to think that Linnaeus might have identified himself with the Sami—with the Other. He wore a Sami outfit in Holland—could it be that he also developed a Sami persona? Does passing the Styx change you? Writing about his travels in Lapland, his tone shifts between levels. There were moments when he felt in touch, as suggested for instance by his fleeting interest in clothes. He observed that among the Sami "their caps are like those worn by women in Stenbrohult"—comparisons between harsh Småland and Lapland came easily. He didn't hesitate to dress up like a Sami, as in the portrait by Hoffman. Is it true that "the clothes maketh the man"? He described his physical appearance in one of his autobiographical sketches as "brown-eyed. Light of foot, walked quickly, did everything promptly." Under the title *Crede mihi, qvi bene latuit, bene vixit* [Believe me, whoever lived in concealment, lived well], Linnaeus wrote: "I would have wished to be a wealthy farmer if habituated thereto from childhood but better still, had it not been so cold, would be the life of a wealthy mountain Lapp."[20] Linnaeus allows himself the option of a classical retreat into the countryside—an *otium* in Sami land—but backs away from having to put up with the cold (although known to be very good for you).

No noticeable distancing occurs between the writer of *Diaeta Naturalis* and Rudbeckianism but, generally, the opposite: "Regarding Rudbeck *pater*. Is there any young scholar who, having attended the academy for a couple of years, would not in his Thesis ascribe to him all he hoped to achieve in honor of his country, but also *in exteri* admitted that he cannot achieve it? How else could he not represent the academy"[21] (a note of identification here). But he is by now abandoning the Gothic themes in his journey in Lapland in favor of the economic ones to be pursued in the pending journey to County Dalarna in central Sweden. He is also about to shift his dependence on the younger Rudbeck to Nils Reuterholm. Still, the Gothic thought patterns remain: Linnaeus is known for his celebrations of the sun and of springtime, but he could also grow lyrical about winter: "The hare, the ermine and the squirrel clothed in white coats ... at night the Northern Lights played and shone across the heavens."

Some have commented on the introduction to *Diaeta Naturalis* to the effect that "it has a certain quality of melancholy and bitterness." The backstory is that he wrote it during "the most dejected period of his student years, so full of both worry and joy." He had studied for seven years and still had no secure post. In addition, he would soon leave for foreign parts. Instead of "dejection," however, one might focus on the reforming spirit and Baroque ideals that characterize his book, and its generous exposition of how to live healthily, with its many riveting anecdotes and examples. The tough style, its recurring Gothic motifs and references to classical ideals of good living, was present in his lectures on dietetics even though his own practices suggest adaptation. The same question could be addressed to Linnaeus as to most others: Did he betray the ideals of his youth?

In 1731 he described the usefulness as well as the happiness he had found in the career choices he had made: "Who could find greater joy than he who stays in the countryside and also has a good understanding of the *botanique*. Should he become a pastor, the farming folk will hold him to be Aesculapius, becomes he *politicus* he will be thought *curieux*, and one who understands more than others do, which even so gives him less pleasure than the joys he himself finds every day."[22] Is this Linnaeus the medical moralist or the ambitious climber in his chosen profession? He had at least three career options: one, stay in Falun as a doctor, possibly also a mineralogist; two, somehow get himself a professorial chair in Uppsala; three, become a new kind of mobile expert in regional economics.

An unexpected journey would make him postpone a definitive decision.

CHAPTER SIX

In the Mountains and under the Ground

COUNTY DALARNA

IN 1733 LINNAEUS went home to Stenbrohult after his mother's death but then returned to Uppsala. It was the year he began work on *Diaeta Naturalis*, but overall little else is known about it. His early biographer Sven A. Hedin saw it as a time of desperation and makes a great deal of this conclusion, as did Thore M. Fries; the cause was thought to be that Linnaeus still had no proper post, despite seven years of university studies. Hedin declared that, after Linnaeus's journey to Lapland, "Upsala expelled him from her womb."[1] Still, on 16 November 1732, Linnaeus applied for and was granted a stipend from the Wrede foundation. In 1733 he enjoyed teaching Mennander, the archbishop-to-be, though the joy was tempered as he wrote to his pupil in 1734: "This year, thou, my dear Brother, is the only one to have accompanied me through all the extensive realms of Nature. And also the first one to have done so. Thou are likely also to be the last. This, as not all heads are cut out to practice *methode* and admire Nature's mastery, nor are there many of those who are ready to soak their brains in observations and experimentations on small piteous insects when clothes as well as food can be yours in exchange for pietous sermons."[2] He went on to make more fun of the clergy and stressed how poorly you are paid for studying the natural sciences. In 1733 (no date), he began a letter to the chancellor of Uppsala University, Gustaf Cronhielm, on a similar note: "My brief *tragoedia*, I would insinuate with only few words as I would not care to expand on this subject,"[3]—only to carry on with a list of his qualifications.

What of his plans for the future? He had set aside April 1734 for sorting and categorizing his collection of minerals, and May for beginning to run a private botanical collegium or seminar. This plan was blocked by Nils Rosén, who by then had married into Mennander's family (his sister's daughter) and "so joined one of the most esteemed families in academe." Linnaeus comments: "It was woeful to see him at this time somehow abandon his self, he who had wished to be *Academicus*." Rosén, who realized very well that Linnaeus could be an influential rival, visited him in his lodgings and asked to borrow his manuscript. He used threats to force Linnaeus to give him the first part in the end. Once Rosén had copied the text, he demanded access to the second part, or he would not return part one. "To Linnaeus, it was as if lightning had struck because his entire system and collection would be ruined." He decided not to bend to Rosén's demands and set his heart on revenge—like Samson—only for "a professor Ölreich from Lund, at that time a Linnaei disciple in botanicis, who counseled against, as revenge was for the Lord." Then Governor Nils Reuterholm's commission arrived, and Linnaeus "changed his Mind, praised his persecutors and left his fate to the Creator."[4]

The year 1734 is the year of Falun, the place where his future would be forged. The year before, Linnaeus's friend Claes Sohlberg had gotten him invited to spend Christmas with Nils Reuterholm. He told the party about his journey to Lapland; it is possible that the connection to the governor in Falun went via his colleague Gyllengrip in Västerbotten. Reuterholm's son Axel Gottlieb kept a diary and recorded for 24 January 1734: "A student, Linnaeus called, a small, invisible but *speculatif* man is said to possess great knowledge in the *botanique*, also in Medicine and has been in Lapland (re his *lectus* on haircap moss)."[5] The winter months were taken up with visiting mines in the district of Bergslagen. In the early days of the summer, when Linnaeus was back in Uppsala, Reuterholm sent him a commission to journey in County Dalarna to chart its resources and also a bond to pay for his expenses.

This task was quite different from the Lapland adventure: the brief didn't include describing a mythical-poetic landscape with academic turns of phrase. The Reuterholm circle joined the expedition, each man funding his own costs: geographer Reinhold Näsman; doctor and also secretary Carl Clewberg; mineralogist Ingel Fahlstedt; botanist Claes Sohlberg; zoologist Erik Emporelius; ethnologist Petrus Hedenblad; and economist Benjamin Sandel. On their return, the journal was produced—or perhaps a copy? Fourteen-year-old Gustaf Gottlieb Reuterholm writes in his diary on 29 March 1735: "In the afternoon, I began Linnaei Dahl journey to copy, is 8 sexterns long." By then, Linnaeus was teaching the boys in mineralogy and ore-testing methods, and included their tutor, Johan Browallius, in the lessons. Browallius, who had "never studied *historia naturalis* made the same journey which showed his fire."[6]

This time, Linnaeus didn't travel on his own, and the journal was correspondingly less personal than his "Journey in Lapland." It gained by the reporting being better structured and replete with facts. The local flora, mostly consisting of typically middle-Swedish varieties, disappointed; Linnaeus had been spoiled by the novelties found in the Lapland mountains. When *Flora Dalecarlia* was finally published almost fifty years later, it described around three hundred spermatophytes (seed plants), in many cases also including their uses. He sees his own plant, the *Linnaea borealis*, several times but is less excited by it.

All the same, it is Linnaeus who holds the pen, and we recognize him already in the introduction to his travel diary. It rained a little, and "between the eye and the farthest mountains rose an exhalation, as were it smoke. The sun shone warmly between the clouds: finally, toward the evening, we were awarded a clear sky." They traveled on horseback, following first the East Dal River northward to the Norwegian mining town Röros, and then the West Dal River along its course. On 1 August, the night was spent "in the wild forest. The night was very hard as first we were given no peace from the mosquitoes and then, toward the dawn, rain began to fall strongly. Our quickly erected hut soon was less bearable than the rain itself as raindrops were smaller outside than inside. Wet through we went on our way and ever more wet we became."[7] This is just one of several entries showing how changeable the Swedish summer can be. The ways and means of men are often as disagreeable: "I would rather not dwell on the long distances when the road twists and turns for the sake of the fencing.... *Contraire*, I once observed in Småland that to raffine the road it was drawn straight through the house and across part of the hop-growing plot where a large fruit-bearing pear tree was cut down so that the road would be without bends."[8]

This is the description of their arrival at Mount Swucku on 27 July:

> When at 12 o'clock we commenced scaling from the root of Svucku, the thunder was soon upon us. The higher we climbed the clearer it became and the harder the weather blew. When we arrived at the highest part of the mountain, everywhere between the peaks stood in thick smoke. At a few places, rays came from the heavens as if it had rained and within them glowed manifold colors as in a rainbow, from which scene our guide foresaw rain toward the night. In the east, a black cloud

of darkness such that we could not well discern the nearest mountain, even more of this here and there in other directions. That dark sky, as black as death, was threatening us who stood upon the bare, high rock, warning of lightning, clap and peal heard from around us on all sides.[9]

As expressive as an impressionist watercolor and, this time, shorn of classical mythology.

What follows, page after page, is a catalog of facts: natural specimens, mining techniques, manners of dress, history, dance, churches, markers of time and season: "At midday, the mountain addresses the field-man that spring is in the offing, so that like cow and calf, from the same mountain as from a wise *prognostico*, all understand that the fields should be made in readiness; as, in spring-time, the snow lies on this mountain for longer than anywhere else."[10] He drew a choreographic scheme of a dance, possibly a quadrille, and comments on local ceremonies for greeting: "Largely seen over the entire county of Dalarna, and particularly among those working the mountain-sides, a somewhat excessive welcome was offered when a visitor arrived for the first time." A large vessel—buffalo horn, coconut shell, flagon, or tankard would be filled and "must be drained, as otherwise is inexcusable, although one's stomach explodes, head bursts, and health, all life's pleasures, valedict."[11]

They used the local clergy as chief sources of information; one of them was the poor and knowledgeable curate in Transtrand, Lars Dahle. He told them about what was eaten in hard times: "Mosses boiled with water and milk, be it peat moss or reindeer lichen." A comment noted that the barley sown on mountain slopes by an eccentric vicar did not turn out well: "It is passing strange that Sweden owns land with mountains of perhaps a hundred thousand miles and as yet no one has made it his business to cultivate the same. In the valleys there is plentiful rich dark soil that well supports the *vegetabile*." They join a lively wedding party, and everybody drinks to the couple's future happiness "on site and in house." Linnaeus is shocked that people in the Mora and Insjön communities don't hunt down the hundreds of overwintering whooper swans: "Swans float all winter long on Lake Orsa in the river currents where the waters do not freeze so they are difficult to shoot or catch, particularly by the farmers who believe God and nature gave them privilege."

A comparison between his two travelogues in Lapland and in Dalarna would be interesting, especially if it were possible to use the copies of the originals that were apparently made. In the account of Dalarna, there is little or no trace of the Baroque style with its embroidery of mythological motifs. There is one well-known exception, a few lines about the great mines in Falun: "Never could any Poet describe Styx, *Regnum Subterraneum* and *Plutonis*, nor any *theologus* Hell so abominable as what is seen here. Thus, outside rises a poisonous, stinging sulphureous smoke that for a long distance around adds poison to the air and makes it harder to reach the place. It so corrodes the soil that no herbs can grow around it."[12] He examined the ground around the mine keenly and observed that the vegetation had been stripped away by the mine workings.

In the Rättvik community, Linnaeus notes the prevalence of a form of chronic breathlessness then known as "stone dust lung," now as silicosis. "Notably, it should be remembered that these people are in a miserable condition because they thought how to best conserve their lives rather than destroy it. All who attempted to attain their finest station in life by handling stone grinding rarely reach more than 20, perhaps 30 to 40 years of age.... In church, one saw a couple of men gone old and gray, and was told that their professions were tailor and shoemaker, and not to go into the mine." A matter-of-fact comment pointed out that grinding was anyway unprofitable as "the workers eat up to 3 à 4 times their income."[13] There were no local doctors; he noted the lack of medical help also in the Idre area in northern Dalarna: "*Medici* up here are there none, not even an *empiricus* or *anus medicatrix* [he means "healers or wise-men"] but only the clergy who are consulted only in the most difficult of passions as they are clever studious men even though they have never read, understood or heard any *medicum* as he carried out his profession." Linnaeus gives some examples of the cures and bloodletting practices that were tried; "As soon as someone falls ill, he must endure having the chisel applied to his artery.... The people are plagued by bleeding and the flux [dysentery] all die with inflamed lung sacks."[14]

The county governor Nils Reuterholm was content, seated at his dinner table in Falun, and the conversation was taken down by a young student, Johan Browallius's younger brother Andreas:

> S [the governor] praised tremendously Linnaei account of his travels all over Dahlarna and said he knows of no one who could have done something like, as there are observations from all the realms of nature although it could be remarkably expanded and so end with being a peerless work, if one with accuracy added still more observations then it would be known and valued in all languages for being peerless and so it would grow, as in such a brief time as this journey not everything that occurred could have been so well noted.... The man who is to make

such a journey and observe all around, he must not travel in comfort but be unknown, ride or walk on foot through the forest.[15]

At Easter 1735, the governor was once more praising Linnaeus as he had

> through his own attempts come quite so far and in practically all, followed such a road as no one before has known or thought of. He makes sure as he walks that he can fit all he sees into a system; has also made some. In the botanique there is no one in Europe like him and none either in orderliness of mind, nor in much else. The wonder is how quickly he has come so far. He is notably attentive and in his final reflections takes all into account, so that he cannot possibly fail. He is not accomplished in languages; holds out against all that is not of reality as nothing as being a waste of time and empty science. His poverty has driven him to become advanced in his experiments and work exceedingly hard. In Medicine he is also said to be without peers.[16]

Recurring praise for having an orderly mind and for being "peerless." Nils Reuterholm was ready to understand Linnaeus because he, too, came from a simple background. His grandfather had been a tenant farmworker on the Baltic island of Åland, and his father had married into the clergy—both good instances of functioning social mobility.

Not everyone was positive about the upstart Linnaeus: "One who has risen fast in society so that his life has quickly altered, cannot become other than too proud, and such a speedy change-about could easily break a man's neck, as one might say of Linnaeus who, when he first arrived in Fahlun had neither clothes, food, nor friends to trust and then in a flash got all this in excess. Hence, he became conceited, quite *capricieux* and minded his habits terribly. He was as if estimating every word he uttered, should each one too many mean a loss of 10 ducats."[17]

In Linnaeus's absence, Browallius stood in for him—although the texts were probably written by Linnaeus. In 1735 Browallius wrote in his magazine the *Swedish Patriot* (*Den swenske patrioten*) that the author had long sought in vain for a practicable way of classifying minerals:

> I was already tired from my searches when I came across considerable help and enjoyment in a humble dwelling. Here were specimens from the realm of stones collected and set out in natural order. The cabinet was paid for not in wages but in painstaking. It lacks any ornamentation on the outside but the order and its sound basis pleased so much more. Here I was told about stones that served as markers of its province, parish, village, and house in the whole of this realm of nature. The

owner allowed me to attempt on my own how one should distinguish between the stones in the fire by their kind, and the metal ore by the kind of stones, and correct their smelting accordingly. We journeyed through all kinds of salt—Sulphur—metal; all species of soils—sands, gravels, and so forth of the interior secrets of the realm, and so I concluded in the end that, instead of some 1,000 the varieties counted to little more than 60 or 70.[18]

Straightforward classification rules, unpretentious display, the scientist as an investigator! A later issue launches an introduction to botany, represented by Flora, queen of the kingdom of plants. Her enemies worry her, but next she was told of one of her courtiers who is a friend of Botany (we can guess who). She mustered her supporters. "'Just folly,' said an onlooker, Mr. Fop—'Leave grass to the cattle.'" A pharmacist's lad pointed out herbs for remedies can be brought in. Flora told the mobilized crowd to be silent and gave a fiery speech about the multiple uses of plants, and how clergy and bureaucrats should know about such things. Botany should be taught to top students in senior schools and at universities, ahead of philosophy and other subjects.[19] Throughout, military terms are used—muster, a solider of botany, and so forth. The pointed phrases about fops have a Gothic tone, and the bid for botanical supremacy in the end sounds like a plant breeding program.

The visit to Dalarna in January 1734 also offered welcome entertainment breaks. As Linnaeus revealed: "The sirens of Dalarna have, contrary to expectation, tempted me to forget all—friends, cares, sober thoughts, annoyances, even home, study, and time. I must not let my ears stay open to their song though they sing that worry comes with time no matter how I hurry away. They have taught me this and that, to take one drink and then another, yes indeed more than that, to drink myself sober. This is no joke! I have felt extraordinarily well here." And this was just for starters. Next, a well-known romance began when Carl met Sara Lisa Moraea, the daughter of the mine company doctor Johannes Moraeus. Linnaeus's courtship moves were recorded in his almanac pages for January 1735, day by numbered day: "2. Visited Sara Lisa wearing Lapp outfit, 3. Similar, parents absent, 10. Visited Sara Lisa, brought a little game, 20. Wrote to her father, done, 22. Visited, offered the ring, 23. Correspondingly [yes] from her mother." On 20 February, he left together with Claes Sohlberg. A period apart to allow for Linnaeus's going abroad to study was the next step, a condition that had been set by his putative father-in-law. Linnaeus had composed a poem to say farewell, dated 20 February 1735: "Morea My

FIGURE 13. Linnaeus. Portrait, probably painted by Lars Roberg some time prior to 1735. From *Linnéporträtt* (1907) by Tycho Tullberg. Royal Library, Stockholm.

friend / Linnaeus Your lad / Hopes Your Days be glad." Quite a long time would pass before they met again, but they agreed that Browallius would act as the go-between.

In a letter to the Swiss botanist Albrecht von Haller (a critic of Linnaeus), dated 12 September 1739, Linnaeus summed up the events of early 1735: "He [her father, Johan Moraeus] had a beautiful daughter—also another, but younger. A certain baron was after her, but in vain. I saw her, amazed I was aware of my heart filling with strange sensations. I was in love, she finally vanquished by my courtship and pleading, loved me in return and told me: let it take place!"[20] About the leave-taking in Falun, on 18–19 January: "magister Browallius assured me of his assistance, as and when needed.... Dr. Moraeus wrote in my book of patronage. S.L.M. gave me, in oath and in writing, assurances of her faithfulness. She was born in 1716 on 26 April."[21]

In Falun, the partying continued, recorded in the diary kept by the governor's second youngest son, Gustaf Gottfried Reuterholm. He included the key to the names of the thirty-odd members of the circle: Sabina (Sara Lisa), Amaryllis (her sister Anna Stina), Hermidorus (Lisa Ehrenholm), and Adamar (Johan Browallius). Gustaf Gottfried took notes of conversations in the governor's home: "My brother said that Linnaeus was engaged to Sara Lisa Moraea and would come to live in Falun as the doctor after Moraeus, of which my father approbated." His approval probably mattered a great deal. For another ten months, there were no new messages from Linnaeus. On 1 March 1736, he wrote that he was installed in Holland, liked staying with Clifford, and had been offered a post as professor in Utrecht but that the pay was poor. In the margin of his diary, he added: "200 thalers in Swedish copper coins in Linnaeus travel purse, nothing given to him by Doct. [Moraeus?] and so he was well broke before he became more in fashion."

Sara Lisa must have received letters now and then. She seemed to take the separation from her Carl in stride, and flirted with Carl's rivals, among whom the diarist mentioned Daniel Tilas and himself. The group of young people went to dances, played cards, fell a little in love with each other and whatever else. The diary entry for 7 December 1736 read: "Hermidorus is not at all bad-looking, has polite enough manners but is malicieux, ambitieux, has not got a good reputation, is apt at passing on gossip, in which arts Sabina is also tutored as she is Hermidori complete ape and seems a clever coquette. Amaryllis is said to be designated the spouse of Adamar but he is flattering Sabina." In clear text: Browallius was flirting with Sara Lisa though linked with her younger sister and, meanwhile, fathered Lisa Ehrenholm's child, a scandal in the company town.[22]

Johan Browallius and Carl Linnaeus skirmished about Sara Lisa. Linnaeus believed that "during my absence, my best friend B regularly dispatched the letters from my beloved by post. It was a task he carried out conscientiously." However, by the time the third year of his stay abroad had passed, "it seemed to B that he was next in turn. He had been made professor, and that on my recommendation. He soon opined that I would not ever return home and tried to win my fiancée. He might have succeeded, had not another come between them by revealing B's falsehood. He was eventually punished by a thousand setbacks."[23] The helpful another might have been Mennander, but it is not known for certain, nor is it clear what he meant by "my recommendation."

They later settled back into a normal relationship, but something had changed. Johan Browallius fought with Linnaeus against the botanist Johann Georg Siegesbeck, a German based in Petersburg.[24] Once the German Russian became known to have exclaimed: "Who could believe that Gud created such wanton behavior to allow plants to multiply? Who could countenance teaching the young such a lecherous system without causing scandal?" Linnaeus went into counterattack by publishing a vitriolic response under Browallius's name, gave a sticky plant the name Siegesbeckia, and—later—ranked him as a "noncommissioned officer" in his order of botanists.

Thank-yous were due to some of his friends, and especially to Carl Fredrik Andreas Browallius: "I thank you for your frankness, dear Brother. You are the one who by being honest has proved himself a true friend.... Thank you for the news of Browallius, but for God's sake do not show yourself as my friend but rather as if you dislike me so that B. can be told everything and vexations are avoided. I am writing an *aequivoce* letter to him now. Can the system [*Systema Naturae*?] gain, a copy will be offered." This might be the copy with the owner's signature *Andreas Browallius*, now in the diocesan library in Linköping. Linnaeus continues: "Fear not Siegesbeck, I shall respond to all myself if only you, dear Brother, would lend me your name, as did once Chomel to Tournefort, [but] only if it pleases my Brother to so transfer."[25]

He planned his writing for 1733: *Bibliotheca Botanica, Systema botanica, Philosophia Botanica, Nomina botanica, Characteres botanica, Species*; as well as *Flora Lapponica, Lachesis Lapponica, Historia insectorum, Aves sueticae*, and "also *Diaeta Naturalis* according to the principles of zoology." On 12 July 1734, *Hamburgische Berichte* announced via a letter from the librarian in Greifswald, a Herr Nettelbladt, that a print run was planned for *Fundamenta Botanica* including two tables to demonstrate both *Systema vegetabilium sexuale* and *Systema vegetabilium calycinum*. However, the two latter systems did not materialize. A similar overview, probably compiled on his way to the Dutch university in 1735, is also available; it lists some thirty works, including the classical botanical texts, *Flora Lapponica* and *Diaeta Naturalis, Sponsalia plantarum, Nuptiae plantarum, Spolia Botanica, Geographia Regnorum botanica, Aves sueciae, Insectae S,*

Ceres noverca, Flora Dalekarlicum, Flora Uplandica Celsiana, Adonis Uplandicus, Najades Sv, Aesculapius Extemporaneus, Vulcanus docimasticus, Pluto Svecanus, Pan Europaeus, Iter ad Superos & Inferos, Iter Dalecarlicum, Oeconomia Lapponica, Diss de Febribus intermittentibus, Diss de Sceptro Carolino.[26] More lists have been found in his undated 1733 letter to Gustaf Cronhielm: *Adonis, Ornithologia, Fundamenta Botanica* and its eight parts *Insecta, Flora Lapponica, Ceres Lapponica, Lachesis,* and *Diaeta Naturalis.* His journey in Lapland is not listed, unless it is hidden behind the title *Lachesis. Systema Naturae* appeared first in 1735. He clearly counted titles in different ways at different times, and later conflated several of them. People often register surprise at the number of texts that Linnaeus published during his years in Holland, but it seems to have been only a small part of what he had planned.

Even if problematic, it is worth examining Linnaeus's publication plans. The texts he mentioned were pulled together during a fairly short time—from 1732 to 1735—but differ in format and, to some extent, in message, indicating the wide range of his interests and the possibilities latent in his themes. The self-assured tone of his publication lists probably didn't so much reflect his state of mind as what was expected from one of the men in the new sciences and the social reformers of the time. Society would be improved through the application of medicine, natural history, and economics. For example, he said in 1736 that the "Journey in Dalarna" (*Dalaresan*) was about to be published by Johan Browallius, and such plans were mentioned again later. Apart from *Flora Lapponica* he hoped to see in print corresponding works on mineralogy, zoology, economics, and "physics"—that is, probably medicine.[27] Why this didn't happen could be explained as due to a lack of time but probably also to an inability to sort out his many themes. Hence, what was made public at the time was actually just *Flora Lapponica*. That the printing plans did not become reality is, however, not a sign of failure but rather that he tested the wind and his own ability and examined what he wanted to say.

On the list is an entry of what Linnaeus called *Iter ad Superos & Inferos*—"Travels to the high and the low" or, more freely translated, "Travels in heaven and hell," written in "the taste of the old Poets and also their manner and choice of words" and composed "more as repose and for my own gratification than aimed at especial usefulness." In it, he describes and compares "the incredible delight and happy lives owned by mountain Lapps while mine workers labor like slaves, their lives daily endangered thereby." A comment states: "The Lapps' great happiness and the hellish conditions of the mine workers are described within and in the manner of a Novelist and in a style all his own."²⁸ Scientific rivals trying their hand at descriptive writing come to mind, notably Albrecht von Haller's famous *Die Alpen*, a pioneering work on nature in the Alps. It seems likely that there are traces of that text in Linnaeus's letter to Roberg, written in Svappavaara on 14 August 1732: "I have seen *solem inoccidum* [the never sinking sun] in the deepest winter. I have frozen in the strongest winter so that my face and fingers felt as if falling off although in summer it could seem that the heat would fry turnips on the slopes of mountains." And, again: "Thou great Creator and protector of all things Who has allowed me to come so high up in the mountains of the Lapps. In the Falu mine You let me descend deep down, While in the Lapp mountains I saw *diem sine nocte* [day without night], in Falu mine I saw *noctem sine die* [night without day]—a promise of all things that You have created, from the beginning to the end."²⁹ It seems obvious that Linnaeus had ambitions to produce fine writing.

He was riding several horses and might have wanted to mount his own Pegasus. The role of the state as taskmaster was a given for his travels through Swedish landscapes in the 1740s, but here similar financial lines of thought were applicable. Linnaeus was being trapped ever more firmly in the economic stall. Despite his attachment to tradition, Linnaeus was a man of the new era in Sweden, a reformer and aware of a new scientific and moral context. His mind was on the move away from often lyrical patriotism, toward often prosaic utilitarianism. Even though his proposals for an agricultural reorganization of Lapland should not be regarded as innocent impulses, they are representative of a much gentler approach than official attitudes later that century (1789), documented in the responses to a competition launched by the Patriotic Society concerning the best way to deal with roaming Lapps.³⁰

"Before my 23rd year [1730] I had conceived of it all."³¹ This was before he had traveled abroad or, rather, before he could have been influenced by foreign colleagues. He was the sole "conceiver," which suggests creativity. It is worth adding that before his 28th year (1735), he had conceived still more.

CHAPTER SEVEN

In the Land of Tulips

LINNAEUS WAS PLANNING to travel abroad as early as 1733. His official reason was to acquire a doctorate in medicine, a qualification only available in some non-Swedish universities. Another overt ambition appears to have been to get the maximum of text in print. During the seventeenth century, Leyden could easily be taken to be Sweden's second university with about a thousand matriculated Swedish students. There was, in those days, only a thin line separating spiritual freedom and brassy commercialism. "Tulip mania" had been driving the country's economy for too long and ended in a market crash. On the other hand, in the late seventeenth and early eighteenth centuries, Dutch science and medicine had included many learned celebrities: Jan Swammerdam and Antonie van Leeuwenhoek, Bernhard Albinus, Frederik Ruysch, Albertus Seba, and the remarkable Herman Boerhaave. Pieter van Musschenbroek, a mathematician and physicist based in Utrecht, had also written a 600-page textbook in natural history. In both Utrecht and Leyden, Cartesianism was widely supported.

On 24 November 1734, Linnaeus took an obligatory examination in theology set for those who wanted to study abroad. The examiners were Olof Celsius and Georg Wallin—two theologians who knew him and granted him *attestatum*. It was the only Swedish academic examination he ever attempted. The result had to be presented to the chancellor's office before it issued a student travel pass. His friend and traveling companion Claes Sohlberg must also have passed the examination, which cost the candidates 18 thalers in Swedish copper coins. Between 1732 and 1735, ten students took the exam, and Linnaeus was among the very last to have their knowledge of theology tested before being allowed to travel to complete their studies. What did the examiners ask about? What were Linnaeus's answers? Was the idea to protect the young against impiety?[1]

A 1735 almanac was annotated day by day during the journey: On 12 March, Linnaeus had a meal with bank assessor Rothman; on the 14th with General Koskull; on the 15th with his brother-in-law Höök; on the 16th with accountant Bergman—then, again, Rothman on the 17th and Bergman on the 18th. On 19 March he traveled home to Stenbrohult, and on 26 March they visited Möckelsnäs. Rothman visited him on 31 March in Stenbrohult; they left for Diö on 3 April; then, on 15 April, Linnaeus said farewell to his family and Stenbrohult and on 22 April boarded the ship in Helsingborg harbor. On 25 April, "the Germanic land was within sight," and on 29 April he toured Hamburg. It is noteworthy how much time Linnaeus spent with Rothman and, according to the almanac, his first letters were for his father and Rothman. He noted no letters to or from Sara Lisa—until, yes! on 28 July *"literae ad uxorem"* (letter to the wife!). On 19 August, more letters to inspector Sohlberg, Browallius, and to Sara Lisa. The records of the conversations in the governor's drawing room give us glimpses of what he wrote about.

This, from *Iter ad exteros* (The Journey Abroad): "Setting out from Stenbrohult in very fine weather, the rye was growing, birch leaves were bursting from their buds, and birds made the forest ring with song like Paradise."[2] The journey aboard the ship began with crossing the Öresund, and the experience made Linnaeus remember the Sami: "O contented Lapp who is not afraid of the waves when the ship leans sideways and every gust of wind is fearsome. The sea is merciless, never dilatory, without art or reason, not subject to human arguments, *sed purus casus fortuitus, vel sola gratia Dei* [only chance or the mercy of God]."[3] As the ship passed the Danish island Møn, a sense of insecurity was creeping into Linnaeus's mind—the sea scared him. He was full of vitality when climbing mountains, but a victim of anxiety aboard a ship.

As they approached Hamburg "it seemed the frogs barked 3 to 4 times more loudly than in Sweden, perhaps had their own language. Some sang so that life might have been born again and I know not of the wishes desired, but some so grimly that one might die of *melancholie*." In "the marvelous city of Hamburg," he observed that "all Jews are bearded but cut like the Lapps do." He visited the synagogue in Altona: "The priest sang rather well, thrilled, held his fingers in his ears stamped first one then the other foot as if needing a piss. So did all the other Jews who read."[4] Neither at that time nor later does Linnaeus seem to have picked up any sense of Jews being singled out for notice, approving or hostile. They were alien to him and hence compared several times with the Lapps in *Iter ad exteros*.

FIGURE 14. The frontispiece of *Hortus Cliffortianus*, 1737, drawn by Jan Wandelaar. The intricate symbolism includes the revelation of Flora by Apollo/Linnaeus, standing on the dragon/hydra of idolatry in Hamburg—a forgery that Linnaeus had exposed. The central panel is flanked by the garden's fruit-bearing banana tree, and, to the left, by representatives of different parts of the world, all of which had supplied the garden with exotic specimens. In the lower right-hand corner, little putti point to the 100-degree thermometer—indicating that Linnaeus invented it. Uppsala University Library.

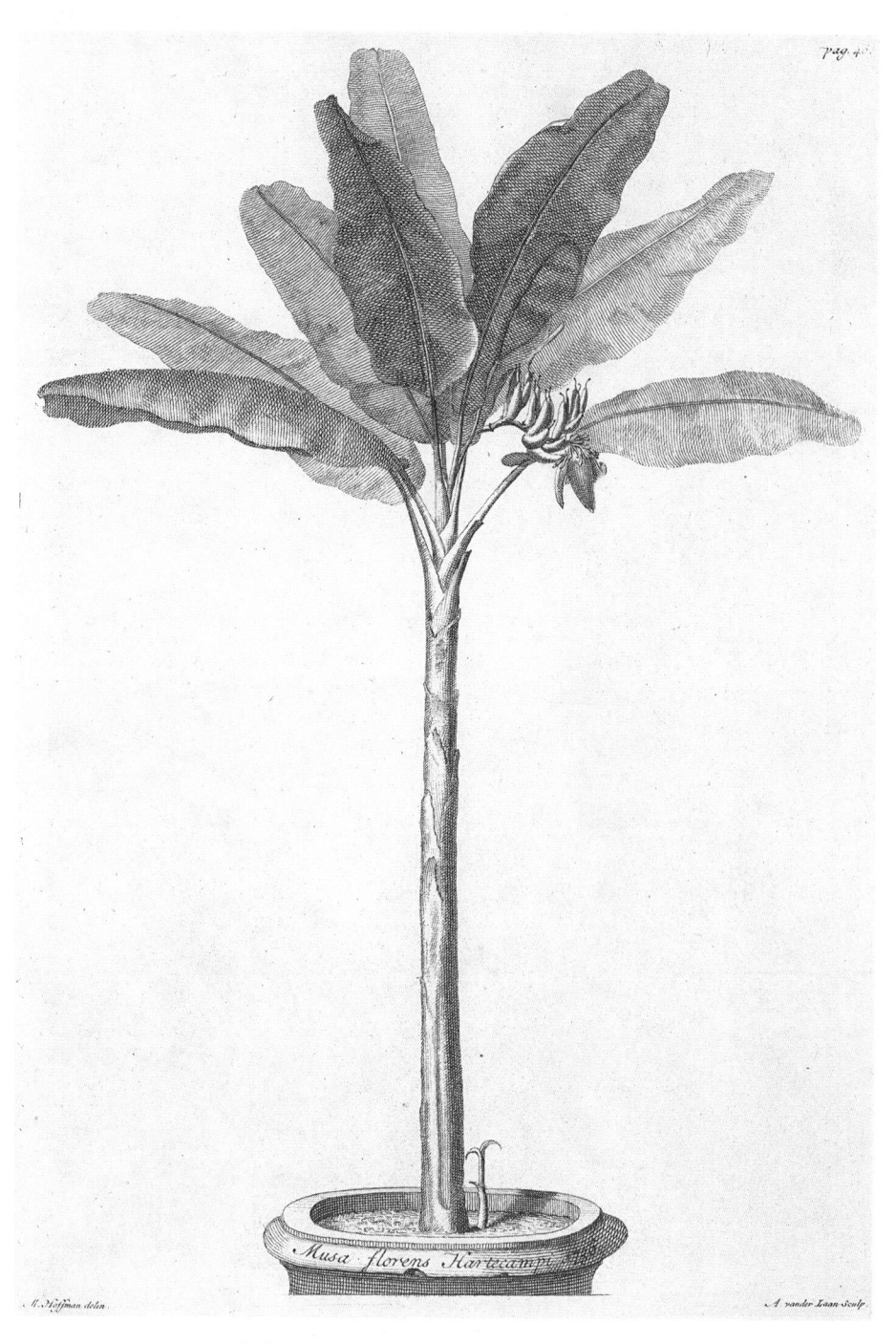

FIGURE 15. *Musa Cliffortiana* by Martin Hoffman, 1736. Uppsala University Library.

The exposure of "the hydra in Hamburg" has often been written about as an outstanding feat.[5] In one of his autobiographies, he insists that "Linnaeus was the first who could reveal this *falsarium*."[6] The story has been told on the basis of Linnaeus's version as printed in *Hamburgische Berichte*, including how the widespread agitation had turned him into *persona non grata*. He had noted that a "hydra" was being exhibited, allegedly with seven heads all like a dragon's, and a body built from parts of weasels. This sensational showpiece was owned by a well-connected merchant called Spreckelsen—he was related to none other than the mayor. The place was beginning to feel too hot for Linnaeus. In Holland, he could also expect a hard line being taken, notably by the influential pharmacist Albertus Seba, who had drawn the hydra to include it in his huge work *Thesaurus*. Linnaeus had been warned by everyone to stay in Hamburg. The hydra affair is worth studying as one skirmish in the battle against the old natural history, for instance in *Paradoxa* and on the frontispiece of *Hortus Cliffortianus*. The hydra in Hamburg can also be connected to the Andromeda episode described by Linnaeus in *Iter Lapponicum* and, more generally, a political form of natural history with roots in the Baroque emblem books; a Swede might be reminded of the political relevance of Bernt Notke's sculpture of St. George and the Dragon in Stockholm's Storkyrka. Linnaeus probably retold his adventure many times; Olof von Dahlin referred to the hydra in his "A Song of the Maiden Natura and Herr Lodbrok."[7] In the telling of the saga of Knight Carl of Linnaeus, this event is a necessary element.

Stories of the Sami went down extremely well abroad. As early as in 1733, Linnaeus tells of his journey in Lapland in *Hamburgische Berichte*. He is presented as "Carolus Linnaeus, a learned gentleman in the subjects of medicine and natural history." He must have been helped by a translator for pieces like his account in *Hamburgische Berichte* on 10 June 1735 of the Lapp shaman or "nåjd" playing his drums and falling to the ground in ecstasy. "Herr Linnaeus told us that no Sami could sing, instead they give off sounds that as they came out are like the howling of dogs."[8]

During his journey, Linnaeus is said to have dressed in Lapp outfits and even yoiked, (traditional Sami singing style); he wore Sami clothes on the well-known full-figure portrait by Martin Hoffman. It is hard to avoid feeling that he was "dramatizing the exotic" and even comparing him with the fraudster Nils Örn, who had traveled around Europe a couple of decades earlier, claiming to be "prince of Lapland." Linnaeus, of course, was soon to become known as the prince of botany; it is not known how many times he tried the effects of dressing up, or when he stopped trying. As a matter of fact, he owned Örn's brief autobiography.[9]

There were, of course, already plenty of Swedish travelogues from the land of tulips. Arrivals from the desolate North were impressed by the open land and the many cities. "Leyden is costly, a very large and fine city," the student Sven Bredberg noted. Another student, Samuel Pontins, allowed a critical note in his comment: "Seemingly buttermilk runs in their veins rather than blood." Some were put off by the country's commercial drive. Emanuel Swedenborg, on his way to Paris in 1736, briefly stayed in Amsterdam and observed grimly that the city breathes nothing but the profiteering spirit. The historian of ideas Johan Hinric Lidén in 1769 listed three elements of Dutch life that he disapproved of: contempt for religion, the arrogant ways of wives, and the careless upbringing of children. As for learning, he thought it "practically extinct," at least in Amsterdam: "The academies lie in deep sleep, despised and without encouragement." Contrary to all this gloom, Carl Peter Thunberg a year later found the room for private enterprise invigorating. As for Linnaeus, he doesn't mince words about the way women dress: "In Sweden the females are deranged. Here they are raving maniacs. Among them there was still nothing that pleased less than *papilionacei* [like butterflies], very wide sideways, all black. These were the *domine* [wives]. So, the botanist leaps to study the flora with all his energy."[10] But he also wrote: "In Holland, the aim of *conversie* is not drinking but speaking together."[11] So, voices for and against, the judgments probably saying as much about the visitor as the actual circumstances.

Linnaeus found the freedom to worship a strength: "When you travel and come to such places where the people are free to practice their religion, what thoughts go through your mind? In Amsterdam, I went to the churches on Sundays, one after the other, observed how people worshipped each in their manner so that the Jews shook and howled; the Papists crossed themselves, sprinkled, and fell on their knees; Quakers, Anabaptists, Reformed folk, and Armenians all engaged in diverse habits, all with great reverence for their own God, all certain theirs was the right one."[12] "Theology. In all places where your thought is free and you can write of what is on your mind, studies will flourish. Where theology rules, where there is none of such freedoms, those are bad places."[13] The natural sciences needed freedom to worship because "no one makes it harder than theology as it will propound the absurd theory of the old concerning *generationem aeqvivocam insectorum minimorum* [the dubious generation of the smallest insects]."[14]

In the merchant cities, everything was of interest; the vicar Jacob Wallenberg observed that "in Holland, the cities are like dolls' houses, all are neat, clean, and properly made. At home, Gothenburg gives us the nearest city of that kind. The people there know a little more than just digging a ditch and fishing for herring." Linnaeus's very close friend Abraham Bäck kept an extensive diary starting in the early 1740s, and his entries draw attention to other aspects of Holland: one is about Clara, the rhinoceros shown at Het Loo in Amsterdam, another about when he attended a lecture by the anatomist Bernard Albinus. He also inspected the Dutch facilities for sinful goings-on. On a visit to the brothel in Haag, he paid an entrance charge, climbed two flights of stairs, and entered a large room where he paid special attention to four of the whores:

> dressed all in white and well-polished, reaching out with their hands between the grid bars and caressed the male visitors lustily. A long hole was there where cunning bachelors could better handle them. I looked on as she noticed one who was eager to grasp that which had earned her a place in this room; soon she ran for a small stool to stand by the previously mentioned hole, stepped up on it, and lifted *geschwindt* her petticoat and shift, then as he fondled, she touched and moved about her behind. He said no liquor had come nor did his hand smell any different from before. . . . In a house along from this one some were said to be common *canailles*. One who looked quite good but was lusty, demanded a kiss from me. I answered her that I had promised keeping my kisses for another. They called themselves *filles de plaisir* [and] were not ashamed to test your faults and display their breasts. Some sat there for 2 years, some just 1.[15]

The Swedish traveler was attracted by the sinful life in the Batavian region and paid as willingly as those who turned up to watch the mad men and women in London's Bedlam. The Bäck account was written just a couple of years after Linnaeus had left Holland and at the same time as when Swedenborg, visiting the same country, records his erotic fantasies in his *Journal of Dreams and Spiritual Experiences (Drömboken)*.

What were Linnaeus's impressions of this foreign country? Some memories were auditory: "In Amsterdam from 7 o'clock until 8 in the evening the sounds are nothing but shouts and screams as one hawker is heard above the next, one worse than the other, one more crooked than the next, all calling out 'buy mine,' then you hear horns sounding, from time to time the horse-drawn cart taking the uncleanliness out of town."[16] On music: "Who is so earnest as not to be moved when hearing beautiful music. Vocal music however rises even higher. Music is the wisest cure for a melancholic, music can gladden one entire crew and sadden another one." He writes about music and dance in *Lachesis Naturalis*: "Minuete, Gigue, keep a straight body, polite *gestus*. Walk like a magpie."[17] More sounds from the street of Amsterdam: "clop-clops of steps, clinks of cups, hubbub and rumpus &c. As the evening draws in, their shouting is the worst."[18] And "a bell in Leiden, it tolled morning and night, the noise was so bad Arch. Linnaeus at first had to cover his ears with coverlets and pillowcases."[19]

Next, olfactory memories: "At the seaside, there is the stink of Fucus [algae, possibly seaweed]. County Skåne plains smell badly of using peat and cow shite for wood. County Österbotten of soured whitefish and smoke."[20] "I saw an *apinia* [?] in Amsterdam. All that she found she smelled on and rejected unless it sufficed to eat when she put it in her mouth but never before she had smelled it."[21] More on the sense of smell: "It is held by most people as the simplest *sensus organum*, as it seemingly matters little; but did she only know how taste and smell correspond with each other and that when the smell vanishes so does the taste and also appetite, their unborn offspring, then she would surely wish that she had kept him even as she grew older."[22] Furthermore: "The longer *nasus* is, the stronger smell sense, *hinc canes*, as the air runs through for longer and also has harder affinity for the *membranam*, as that gives a finer smell and less affected by snuff."[23]

"All heavy snuff-takers will in the end lose their sense of smell."[24] "The greatest wonder I have encountered anywhere is that tobacco, a stinking, nauseous, and toxic thing, is used by all peoples in the world; no nation is plain living enough not to have failed to do as nature bids." Indeed, Linnaeus did himself, and not in half-measures. More on how the upper class is the source of bad habits: "Tobacco first smoked in Stockholm became ill seen because country folk learned to smoke. Instead, snuff was taken, then country folk learned it also. In 1734 smoking was back in fashion as the plebs took snuff."[25]

Linnaeus didn't take long to prepare his doctoral thesis on the causes and prevalence of ague (malaria) in Sweden, and defended it in the small fishing town of Harderwijk on 23 June 1735. An ironic old Dutch rhyme defines Harderwijk: "Harderwijk is een stadje van negotie / Men koopt er booking en bullen van promotie" (Harderwijk is a trading town / It sells smoked herrings and doctorates). Under the title *Dissertatio medica inauguralis in qua exhibitur Hypothesis nova de Febrium intermittentium causa*, the slim thesis was presented by Carolus Linnaeus Smol-Svecus; stipendiary *Wredian ad diem 23 junii*. Johannes de Gorter was *preses*; that is, he presided over the examination. The examinee's argument was that people living on wetlands were inhaling clay and this in turn caused the febrile fits. The brief text reflects at every turn a mechanistic approach; the examiners also gave themselves time to interrogate the candidate on a few Hippocratic aphorisms concerning "dropsy" or edema. The ceremony was concluded swiftly and the entire process took a week. Linnaeus could have gone back to Sweden, but that is not what happened.

Linnaeus was the last of the thirty-odd Swedes, including Rosén, who had passed their doctoral examination in Harderwijk. The practice ended because, from 1738, medical students could acquire their doctorates in Uppsala. In Leyden, Linnaeus followed Herman Boerhaave's lectures in botany and chemistry. As a medical man, Linnaeus is usually described as a "mechanist" and an adherent of the leading iatrophysical or iatromechanical (also called iatromathematical) school: bodily functions should be interpreted according to the laws of mechanics; illness was a sign of dysfunctions in the machinery. Linnaeus often argued in rigidly iatromechanical terms, as, for instance, in his thesis on the mechanism of ague. Others, including Boerhaave, were more restrained in their defense of one or another medical system. For a long time, questions had also been asked about the mediating role of the vital spirits, possibly in a refined, almost soulful sense.

A fact worth emphasizing: Linnaeus was not alone in his claim to greatness in the natural sciences. For one thing, he had mentors, notably Rothman, Stobæus, Rudbeck, and Boerhaave and, as we will see, eventually people in positions of power who acted as patrons and sponsors, like Carl Gustaf Tessin, Carl Hårleman, and (Queen) Lovisa Ulrika. He also relied on colleagues and assistants such as Celsius, Gronovius, and Bäck, and his own pupils and disciples. The big idea was that Herman Boerhaave, the most prestigious figure, would accept being his *patronus*, but the roles changed so that Linnaeus himself became the mentor and patron of Johann Bartsch. Bartsch was in many ways a disciple and also a martyr: he went to Surinam on his patron's errand and died there in 1738. Linnaeus honored him by naming a flower genus after him; the Bartsia is in *Hortus Cliffortianus*. Other helpful contacts included a couple of country governors; for instance, Linnaeus dedicated *Classes Plantarum* (1738) to Nils Reuterholm and Gabriel Gyllengrip, "*maecenatibus maximis*." His relationships with his mentors—as seen in his recurring praise for the "peerless" royal couple—his disquiet whenever Höpken made a face, and his courtier's role in the company of Tessin—were tempting Linnaeus toward grand plans and the politics of the royalist party, "the Hats."

After leaving Harderwijk, Linnaeus went immediately to Leyden. He failed in his first attempt to meet Boerhaave because he didn't grasp that the maidservant who was to usher him expected a guilder for her trouble. However, the professor of medicine in Leyden, Johan Frederik Gronovius, helped with an introduction. It was difficult to get an appointment with Boerhaave, who was usually either busy or unwell. They then agreed on a meeting at 2 o'clock, but Linnaeus had understood it to be 3 o'clock. He was late, of course, and the old man had left. The next time it worked out, only for Boerhaave to begin with interrogating his visitor about the plants in his garden. Linnaeus passed the test with flying colors, so Boerhaave recommended that he should see the botanist and medical man Johannes Burman. Once more, Linnaeus impressed with his superb knowledge of botany. When Burman later realized that the young man was staying in a very basic hostelry, he offered him free board and lodging. The consistent testing element is strange—almost like in a fairytale.

Boerhaave seemed to remain irritated, possibly because he had read in *Hamburgische Berichte* that Linnaeus had been corresponding with him—which was untrue. Whatever the reason, his entry into Linnaeus's book of patronage was a brief, impersonal greeting, and the great man also refused the offer of writing an introduction to *Flora Lapponica*. Earlier, Claes Sohlberg had remarked to Tersmeden: "Curiously, Boerhaave can't stand Linnaeus." It rather contradicts what we know in general about how people tended to trust Linnaeus. However, Lindeboom, the Dutch historian of medicine, has stated: "Something about Linnaeus's personality made Boerhaave dislike him. I find it relatively easy to understand as the two men were each other's complete opposites. I may be forgiven for arguing that Boerhaave was the nobler."[26]

Despite his aversion, Boerhaave realized that the Swede was exceptionally able and supported his application for a post with Clifford. He also penned a letter of recommendation to the collector Sir Hans Sloane. In 1737–38 Linnaeus stayed with the medic Adriaan van Royen, who tried to tempt him into taking on the task of managing the botanical garden and rearranging it according to the sexual system. Linnaeus backed off, aware that Boerhaave would disapprove. As a reasonable compromise, he outlined a system, which van Royen later used in his *Florae Leidensis Prodromus* (1740) without mentioning that it was Linnaeus's idea.

When Boerhaave suggested that Linnaeus should travel to South Africa and also take up a well-paid job as a doctor in Surinam, he wisely turned the offers down. The reasons may well have been mixed: fear of traveling by sea, his promises to Sara Lisa, his unease about foreign languages, different plans—whatever it was, he didn't want to go. Johann Bartsch, his replacement, only lived a few months before dying of a fever.

Linnaeus describes the sad farewell in 1738, when Boerhaave was ill: "When Linneai took leave of Leyden, Boerhaave was already in the grip of his *Hydrops Thoracis* [pleural effusion], which caused a strong breathing difficulty so that he had long before forbidden any visitors.... So Linnaei was the only one allowed to enter, kiss the hand of his great tutor with a grief-stricken *vale*, and then the ill old man had still enough strength in his hand that he could hold Linnaei hand and bring it to his mouth and kiss the back of her, saying: I have lived my time and my years, done all I had been able to, may God save you, as it all remains for you, what I have loved in the world I have given back but she is loved still more than that by you. Farewell my dear Linnaeus. The flow of tears allowed no more. Once back in my quarters I found he had sent me a fine copy of his Chemistry."[27]

Linnaeus held one permanent post in Holland as the head manager of George Clifford's garden at Hartecamp, a position for which Boerhaave had recommended him. After once more undergoing a test, he began work in mid-September. He reported back to Falun and then wrote again on 1 March 1736: "This man Clifford is counted as the third in rank among the greatest capitalists in Holland."[28] The introduction to *Hortus Cliffortianus* reflects his awed impressions of the wonders he had seen: "My eyes were at once enraptured by so many masterpieces, artfully improved yet created by nature, the tree-lined roads, border plantings, statuary, ponds, and artistically constructed mountains and labyrinths. My heart delighted at your menageries, so full of every animal, such as tigers, apes, wild dogs, Indian deer and goats, South American and African boars; with their sounds mingled those from hordes of birds.... I was overwhelmed when I entered the glasshouses, full as they were with such numerous herbs that a son of the North must perforce feel enchanted and bewildered as to which continent he has been transported."

Linnaeus's dedication to Clifford begins with an account of Paradise and Genesis, and continues: "You who are other animals, you move about, taste, smell, hear, and see without knowing what it is that reaches your senses, as animals see the green carpet but investigate no one single element in it." Then, human beings were created within this most wonderful paradise; ought they not wonder at the Master's Creation and worship him? "Such is the first and natural law.... But harder times followed the Golden Age and mankind moved from nature to art, from innocence to tyranny. Luxury takes over from human health as is still true today when people are slaves to divers pleasures. But none could be more innocent than that embraced by the first human beings. Hence, let me find my pleasures among the plants. Health is the most precious gift of all. Medicine demands two kinds of knowledge, of the body and of what cures ailments. The virtue of plants can be founded and judged on botanical learning. No field of study is larger than botany, none more difficult. Therefore, sponsorship is necessary." After listing a line-up of sponsors, this overview ended with Clifford and his garden, which had stunned Linnaeus by its number and variety of animals and plants. "Captivated as I was by these splendors, bound to the rock by these sirens, you asked me to reef my sails and come ashore. I have now spent two years in utter innocence, forgetful of my native land, friends and family, forgetful also of my future and unhappy past."[29]

From September 1735 to October 1737, with a few additional months for convalescence, Linnaeus lived in Clifford's Hartecamp, situated between Haarlem and Leyden. He had fallen ill with a fever in the spring of 1738, and it was followed by what he called cholera. He rested in Clifford's place and set out for Paris in May. As soon as he crossed the border into Belgium, he felt well again—or, rather, this is the story as it is told in Linnaeus's *Vita Caroli Linnaei*. No other material backs it up, and he naturally recounted it differently from time to time in his lectures.

These years were full of intense work. Visits by Swedish guests are on record, as for instance when Petrus Artedi met up with Linnaeus in Amsterdam. Artedi had been in London, and Linnaeus came from his desk where he was working on *Genera* and *Bibliotheca Botanica*. A few weeks later, around one o'clock in the morning between 27 and 28 September 1735, Artedi drowned in a city canal or *gracht*. He had been dining with the wealthy pharmacist Seba and was on his way back to his plain lodging house. No one knew exactly what had happened as the only records are the account by Linnaeus, who wasn't there, and a couple of more or less reliable notes elsewhere.

Artedi's landlord apparently took his dead lodger's manuscript as security against a bill for his board. "When Artedi had passed on to a better world in Holland, as fatally as Hasselquist with his great collections, his manuscripts were sequestered against three hundred and twenty guilders." Linnaeus paid the landlord in cash, which reduced the debt to one hundred guilders, "monies I loaned and thus rescued his writings which are now the choicest work on Ichthyology that the world has ever seen." The county governor in Falun was quoted as saying in October 1736: "When from Browallius much was spoken of Linnaeus . . . the *ichthyologie* that he has communicated from the work of the great *ichthyologus* Artedius, who died not long ago without any light falling on his observations. However, Linnaeus has pledged for all his manuscripts as much as 99 guilders that makes 228 thalers in silver coins."[30]

We cannot know whether Artedi committed suicide, but his situation was desperate, and going by what was said about his mindset makes it seem likely enough. At the time, people speculated about his being the victim of a crime. The state of Petrus Artedi's *Ichthyologia*, the fundamentals of knowledge about fish, is also unclear. Linnaeus writes about it in his autobiography: "This work would never have seen the light of day had not Linnaeus vindicated it, bought and written a clean copy of it, put it together, and seen to its publication so that we do not know if we should more regard as its author our Artedi or our Linnaeus."[31] In fact, the person who prepared the clean copy was a young—barely twenty years old—medical student, Tiburtius Kiellman, who would soon afterward go to Surinam and meet his death there. It is as impossible to work out how much Linnaeus had paid in the end, as to ascertain what Artedi's original script looked like—it has disappeared.

Admiral Carl Tersmeden recalled late in his life how he had once met his old companion from Uppsala, Claes Sohlberg, at an eatery somewhere—Tersmeden seemed most concerned about choosing the food. Sohlberg introduced their fellow Swede Linnaeus, and someone else turned up "just a little later, also from our country, a small thin and invisible man called Arctelius. The pleasure at meeting 3 men from Sweden was not negligible. . . . My handsome livery stimulated in the other guests . . . a curiosity to find out who I was and much whispering on all sides. . . . As I saw they did not order any wine I asked to be allowed to serve them from my *bouteille* of Rhine wine and for the servant man Jean to give them glasses."[32]

One thing led to another. "On 17 September [1735] we were wandering around 10 to the *schuit* [canal barge], when Sohlberg alerted us, saying 'You have forgotten, we are to take supper with Linnaeus.' . . . There, we moored and strolled forward to the gate. The gardener's lad opened up and told us that Herr Linnaeus was with the head gardener in the orangery so he led the way and once there we were not so terribly well received by a severe Linnaeus. . . . At 1 o'clock we went to the house where we were served with 3 dishes without any meat but otherwise good food: water sauce, a dish of globe artichokes and cauliflower, good butter, cheese, and bread and with it red wine but still both my brother and I, as we are not *botanici*, found little gave us pleasure." A vegetarian meal—why not! In *Diaeta Naturalis*, Linnaeus discussed eating vegetables and referred to ascetic Hindu groups, such as gymnosophists and Brahmins, but also to what he calls "a beautiful book," the vegetarian *De esu cranium* by the Neoplatonist Porphyry. Besides, in Paradise, human beings did not eat meat. He never arrived at any definite conclusion but clearly cared a great deal.

The time spent with Clifford, however, left Linnaeus with a sour taste in his mouth. He spoke about being served "a pâté made of cocks' combs and how every comb cost as much as the entire cockerel."[33] At Hartecamp, servants brought him tea, coffee, and regular meals; he lacked for nothing and spent his days in the wonderful garden but, left on his own, he felt melancholic and wanted to exchange his life in Holland for the hard life of the Sami.

Johan Frederik Gronovius, a medical man and a senator, was an unselfish supporter with deep pockets. He was very patient with Linnaeus, a moody protégé on whose behalf he had to intervene with Clifford more than once. Gronovius's help deserved every recognition, but Linnaeus doesn't mention his debt of gratitude in his autobiographies except for one reference to his mentor publishing *Systema Naturae* at his own expense. A mention in a 1736 letter to Olof Celsius is his only known reference to that particular act of generosity: "D. Gronovius works night and day, year out and year in, with the proofs of my works, which he believes does him credit." They consulted each other frequently, going up and down between Hartecamp and Leyden in canal boats, "4 hours there and 4 hours back." We can now follow their work. Gronovius warned him against Seba and his friends Hans Sloane and Philip Miller, but Linnaeus took no notice. The printing took almost a year; Gronovius also helped with the printing of the important *Genera Plantarum*.[34]

Linnaeus joined a club for scientists in Leyden, where his fellow members included Johan Frederik Gronovius, Gerard van Swieten, Isaac Lawson, Johann Nathanaël Lieberkühn, Johann Kramer ("a lively and disorderly German" with an outstanding memory), and Johann Bartsch. A quotation from Gronovius suggests that Linnaeus was the luminary: "An excellent society with Linnaeus as president. . . . We have made such great advances that, consulting his tables, we can refer each fish, plant, or mineral to its genus and hence to its species even though no one has seen its like before. In my opinion these tables are of remarkable usefulness and everyone should have them mounted on the wall in his study like maps."

A few years later, when Abraham Bäck was staying in Berlin, he learned from Lieberkühn of the hostility that had flared up between two members of the society, Linnaeus and Kramer, on who was likely to be Clifford's favorite:

> That is to say, L feared that Clifford would feel he had put too much faith in L so when Clifford had once come to watch one of Cramer's experiments and how he went about first making the Capell, then

L had thrown salt into the *materia* and still assured the silver would not blacken. But Cramer had noticed the chicanery and so utilized another Capell. Then the Experiment went as he had said and Clifford railed at Linnaeus for his prediction. To avenge himself, Cramer set himself to be very friendly toward Linnaeus and, for as long as the wine lasted [?], which he had received from Clifford, L held back and pretended that he would write against Siegesbeck on Linnaeus's behalf in the first place.[35]

All this is very confusing, but what Lieberkühn meant was that Linnaeus was not "in every way admirable. How he constantly whispered in Dr. Lieberkühn's ear that he should not continue with the club begun by van Swieten, Gronovius, Lawson, Kramer, and professor Linnaeus."

In the summer of 1736, Linnaeus traveled to London, and the following autumn he set out for Paris, but we know relatively little about these journeys. Clifford paid for the London trip because Linnaeus was to buy seed. He visited Hans Sloane and Philip Miller at the Chelsea Physics Garden and also examined William Sherard's collections. He had little time for either: Sherard's were moldy and Sloane's very disorderly. While in London, Linnaeus stayed with the Swedish pastor Tobias Björk but spent most of his time in Oxford with Johann Jacob Dillenius. As usual, he felt himself to have been unkindly—"disdainfully"—received at first, but was then made to stay on for the whole month until his hosts "at last let him go after much tears and kissing."[36]

He is more circumstantial in a letter to Olof Celsius: "My *genera plantarum* came at last presentable from the printers. D. Dillenius was given half part of it, at his request. When he saw how I had gone about mastering his *genera*, this gentleman turned on me as angrily as he had previously been cordial. When I came to him in Oxford, he would scarce bring himself to ask me to step inside and then burst out, carping and putting on contemptuous faces. I was with him in the city for 3 days but hardly got to see more than a few plants. I paid for the carriage in his presence in which I would leave the following day and then he looked at me in a kindlier manner." Eventually, Linnaeus gained the trust of his host, and "when I finally made to leave, he was in tears as he let me go."[37]

We have already learned that he fell ill with a severe fever as he was about to set out for Paris. He also got an unpleasant message from home at that time—presumably the one about Sara Lisa and Browallius. Clifford warned him that he was too exhausted to travel, and made him stay on at Hartecamp to enjoy all its comforts. However, he never saw "a day of regained health ere he had said farewell to Holland." He felt better first when he had reached Brabant; there, he was "refreshed by the air and freed of a heavy burden."[38] Continuing on his way, he broke the journey in Leyden to see his fellow scientists Adrian van Royen and Gronovius. For safety, the case containing his own books was sealed in Valenciennes. Antoine and Bernard de Jussieu received him in Paris, and he met Mademoiselle Bassaport, the lady appointed as court artist by the king, an encounter he liked to recall. He had contacts now in the Académie des Sciences.

A few brief notes from his time in Paris: "When I left from Fontenblau to Paris, I rode without gloves. My hands became swollen and burned so that I had to stop at a farm and ask for some vinegar. Had I not smeared this on I might well have contracted a fever."[39] His son, Carl von Linné, has noted an otherwise unknown visit: "My father went to see a madhouse in France and a man had come with him who when he entered laughed at them all with wide-open mouth, then one came up to him and asked why his Companion carried a fencing sword as if he were a hero. The madman laughed and so did all the others, then asked what he wanted with that saber and asked my Father why he wore a Peruque. All were laughing at us by then and thought us the mad ones."[40]

In August, he traveled to Rouen and, from there, homeward to Helsingborg. He went to stay with his father and rested for a few weeks before going directly to Falun, where his beloved had been waiting four years for her dear Ulysses.

CHAPTER EIGHT

Nature's Order 1

ORDER WAS A QUALITY everyone needed after the mess created over many troubled years—years of war and hunger, and radical doubts about everything, intellectualized by Descartes. At the governor's table in Falun, order and systems were discussed. Johan Browallius: "Wolff, together with Leibnitz, are the biggest names the Germans can boast, but to set against them we have Linnaeus and Klingenstierna." The mathematician Samuel Klingenstierna was thought to be Sweden's leading scientist, a match for Leibnitz. Wolff's range of knowledge was more universal than Linnaeus's, but "it would be wished that each and every one took on no more than one task and worked on it with dedication as Linnaeus now has sacrificed himself for the people, so that, in every science, a system was worked out, and then in 3 years as much would be done as in an entire lifetime."[1] Klingenstierna was actually a Wolffian, and perhaps Reuterholm had also been impressed by him, but there is no evidence for any influence on Linnaeus.

When Linnaeus went to Holland, his case was packed full of manuscripts; some of them can be seen piled up next to him in the full-figure portrait by Hoffman. The most significant was *Systema Naturae*, the first major overview, ordered according to scientific principles. It was to serve as the backbone for future inventories of the kingdoms of biology. An early, undated version was the so-called *Manuscripta Mennandria* with twenty-three classes—polygamy is missing and a variant, also undated, manuscript exists in a preprint state, kept in Stockholm's Hagströmer library.[2] Instead of *Systema Naturae*, the title at this stage was wonderfully grandiose: *Geographia naturae*. The printing was a complex project in its own right and turned out to require help.

An account of the structure and content of *Systema Naturae* could easily become excessively technical. Here, a few translated passages will follow to stabilize the grand structural design. From *Observationes in Regna III. Naturae 1*: "As we observe God's Creation, it is more than sufficiently obvious that every separate being has its origin in an egg and that each egg generates a progeny who will be in every way like its parents. Therefore, in present times, no new species will emerge; 2. Through reproduction, the number of individuals will increase. Hence, the number of individuals within each species is greater in the present day; . . . *15. The Stones* grow, *The Plants* grow and are alive, *The Animals* grow, are alive and sensate— hereby are the boundaries staked out between these realms; . . . *17.* I have here presented a general overview of the system for natural bodies so that the curious reader can, with the help of this table used as if it were a map, learn what course to follow through these huge kingdoms."

From *Observations in Regnum lapidum*: "*5.* That rocks, common stones, fundaments of rock and mountains, were not created in the beginning, is proved by their components, and that not all were generated in the Flood is confirmed by many findings of such stones as are incessantly formed; . . . *11. Petrified objects* are as sirens leading on present-day authors, who take great delight in classifying them into as many genera as there are species, in the same manner as gardeners classify their plants. They imagine there are as many species of tulips, hyacinths, anemones, etc. as there are variants of the plants. Truly, all petrified objects can be sorted into seven genera."

From *Observationes in Regnum vegetabile*: *17.* "I assume that botanists will by now argue that my method entails too great difficulties as it requires examination of the smallest parts of the flower which are barely visible to the naked eye. *I respond*: if the truly curious person brings a magnifying glass, a very essential instrument, what more is required? Even so, I have examined all these flowers with my unaided eyes so without any use of a magnifying glass. The last class would however appear to have been excluded by the Creator from the theory of stamens; . . . *19. The effects of plants* are investigated by a botanist, where one is at hand, according to scientific theory or with the support of experience. The person who can interpret the information that comes from either direction will truly understand the powers of plants. All plants within a natural class agree with each other and are similar to each other in the nature of their effects."

From *Observationes in Regnum animale*, much abbreviated: "Zoology, being the noblest component in natural history, is much less cultivated than its other two components; . . . *2.* If we scrutinize the zoological texts by many authors, we find mostly little else but fabulous tales; *3.* Therefore, I have made an orderly system, supported by such observations of my own

/of the four-legged animals, teratology, based foremost on the teeth, in ornithology the shape of the beak, in entomology the antennae and wings, etc.; *4*. In ichthyology, I have not worked out a method of my own, but the greatest ichthyologist of our time, the Famous Swede Herr Petrus Artedi . . . has communicated his to me."

Short extracts such as these allow selective stress on certain features, but it is generally true that Linnaeus used the first person to state the correct take on a subject, and that his self-awareness is very prominent. He connected his arguments to principles he would always adhere to—for instance, his denial of any form of generation of life other than from an egg, or his adoption of a *Scala Naturae* in the style of Aristotle—"stones grow, plants grow and are alive, and animals grow, are alive and sensate." He still used a geographic model and referred to maps, though his grand work was by then entitled *Systema Naturae*.

What did Linnaeus include in the concept of species in *Systema Naturae*? "We count as many species as the Eternal God has created at the beginning of all things." For him, a genetic concept linked to the denial of all spontaneous generation added more stability. Linnaeus emphasized immutability, strikingly so in his *Philosophia Botanica*. However, the diversity he observed challenged a stable origin of all species. As a working hypothesis, he would in his old age still insist that only one species had been created within one genus, and that variants had occurred by interbreeding; any questions were left for future natural scientists. In his view, species and their genus are the outcomes of natural processes, while culture creates the variants, and orders and classes are formed through combinations of natural and cultural processes. As a scientist pointed out two hundred years later: "While the practical applications of the Linnaean concept of species were continuing unchallenged, its theoretical basis was given scant attention. Their common origin was quietly accepted as a given, although pious Christian scientists would emphasize it more."[3]

The constancy of species and the once-and-for-all act of creation were both protected by the sentence *omne vivum ex ovo* (all that is alive comes from an egg), a rephrasing of William Harvey's *ex ovo omnia* (all out of an egg), a motto on the frontispiece of his *De generatione animalium* (1652). A drawing depicts Jupiter opening Pandora's box with a difference: all kinds of critters leap out. That the formula spoke of "the egg" rather than "the seed" is an indication that the animal kingdom is the model. In 1725 Lars Roberg had actually been *preses* at the examination of a thesis discussing the similarities between the egg and the seed. As the seed must originate somehow, the conclusion that all living things have their origin

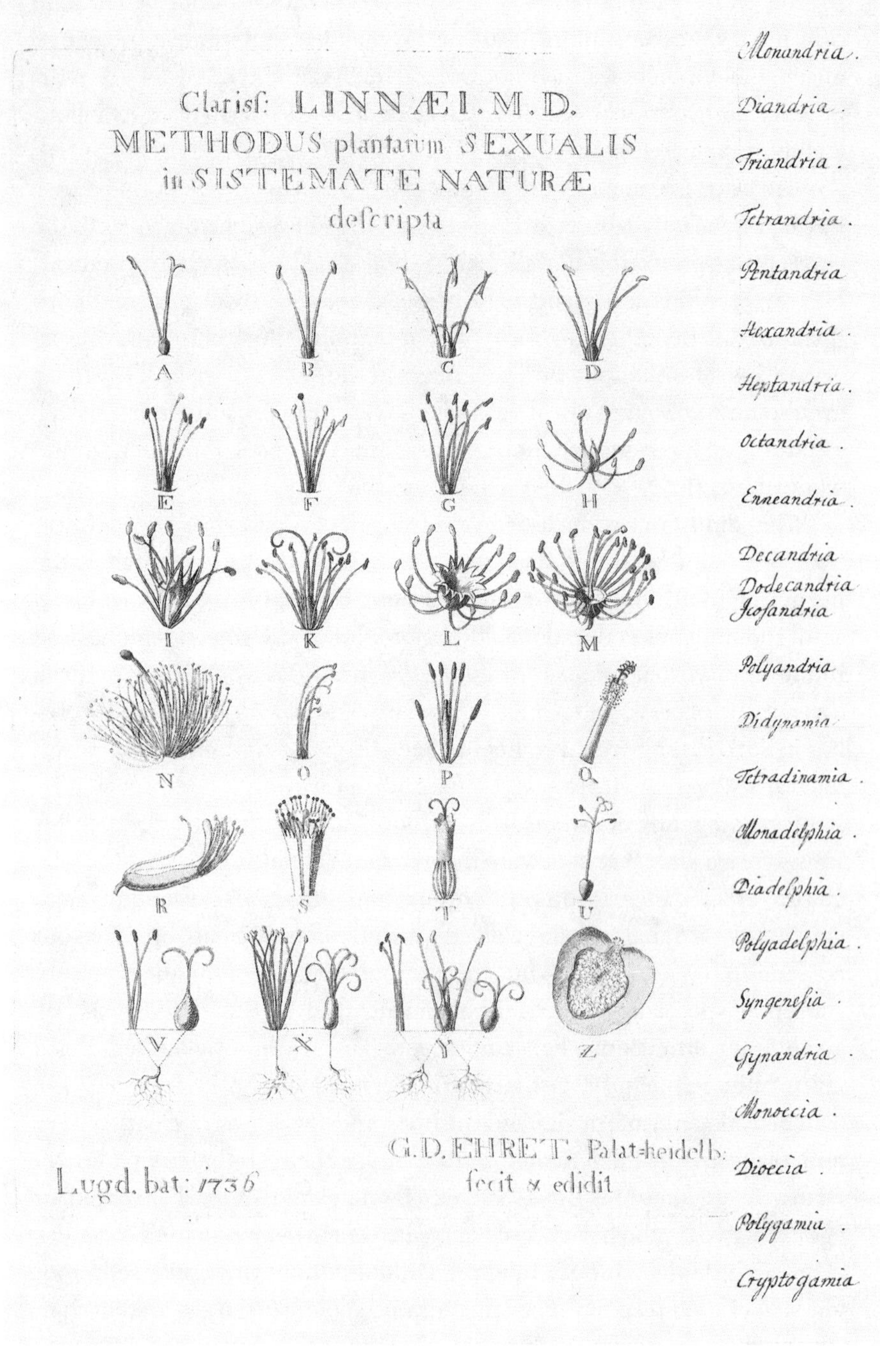

FIGURE 16. The sexual system (of plants) after G. D. Ehret. Uppsala University Library.

in an egg guaranteed the idea of plants having a sexual life. This was critically important for Linnaeus because his task would become more or less meaningless if there was no firm limit to nature's inventiveness—if it could produce new species obeying new laws. Of course, he had to readjust soon enough because new species were indeed found to emerge. Then, there was the tricky behavior of the cryptogams—who wed in secret—to whose "seeds" and their discovery he devoted a thesis (*Semina muscorum detecta*, 1750). Just how much eggs mattered to him became clear when the ennobled von Linné had a row with the head of the College of Arms about the design of his coat of arms. The egg turned up also in his seal and was carved into the chairs of his son's room in the family's Hammarby summer house. The egg was the symbol of nature's continuity, and used by mystics, including naturalists such as Swedenborg, as a representation of the world.

In Linnaeus's system, there were five or six taxonomic ranks: (kingdom), class, (order, family), genus, species, and variety. His focus was on the genus, which was defined in plants as their reproductive structures; certain species were placed in a genus because of similarities in "habitus." In nature, the species was the fundamental rank, and distinctions within a species were made on the basis of numbers, position, and overall shapes and proportions but not including scent, life span, or other features. Naming a plant correctly was the essential start of an examination. It was an open question from the very beginning whether the sexual system of classification was "natural," but it was left for future biologists to answer. Throughout, there is an unmistakable ambition to reduce the diversity of nature to a simple schema that uses the similarities in the reproductive activities of human beings, animals, and plants—everyone weds a mate. Nature is pervaded by lust: its simplicity expresses the Creator's rationality and leaves it open to interpretation by all investigators of nature.

In its first edition, printed in 1735 by Haak in Leyden, *Systema Naturae* consisted of seven sheets of which eleven pages were printed with Latin text; the format is large folio (53 × 42 cm). Only 29 copies of this precious rare book are known to be in existence worldwide. However, the Swedish Medical Association (Svenska Läkaresällskapet) owns a volume with the note *exemplar auctoris*—the author's copy—a unique find, printed on only one side of each page. Every sheet is marked by holes made by nails probably used to hang it up to dry, perhaps after a test printing. The printing was generally a complicated process—for instance, the right-hand side of the kingdom of plants had to be redone three times. The large format suggests inspiration drawn from the work of the older Olof Rudbeck, as seen in other Linnaean contexts. He had surely seen the illustrated volume of *Atlantica* (1679), with its tables on the two final full-page spreads showing the chronology of the Creation of the world, starting with Japheth some 1,670 years after Genesis and ending with the Swedish king Charles XI, 1,670 years after the birth of Jesus. Not that Linnaeus necessarily believed in it all, but it did demonstrate what was typographically possible—as well as suiting his Rudbeckian ambitions.

In 1736 a separate printing on one sheet was published. It was entitled *Methodus* and is often linked to *Systema Naturae*. It is divided into seven subheadings aimed at directing the botanist as he describes a species from its name to what class, order, and genus it belongs to, where it grows, and what its structure and general characteristics are. That last item aims at superstitions and use in poetry, which signals a slimming down of descriptive criteria compared with earlier natural history, and has led to speculation about the influence of Artedi on Linnaeus. He in fact has not followed his own scheme, as set out in *Musa Cliffortiana*, but it was often used, for instance in several theses. In summary: "The *methodus*, which was to apply to descriptions of anything in natural history, was a remarkably concise table. It proves, contrary to doubts expressed elsewhere, that Linnaeus did indeed live up to his own classification rules."[4] *Methodus* was reprinted in *Philosophia Botanica* and set the standard for scientific investigations.

Linnaeus was untroubled by having to shed the classicist inheritance as well as old folk beliefs. It is enlightening to examine his stance in relation to older learning in natural history and what is known as "crypto-zoology": in the first three editions of *Systema Naturae* (1735, 1740, 1744), he takes up some zoological curiosities under the heading *Paradoxa*. In the middle of the two-page spread showing the animal kingdom in the 1735 edition of *Systema Naturae*, ten paradoxes, with brief attempts to explain them, were inserted into the brief column for *Amphibia*, where there was space to spare. The seven-headed hydra is first on the list, and followed by *rana-piscis*, an act of transformation described by Maria Sibylla Merian. Next, number 3—the unicorn which he dismisses as another name for the narwhal—that is, Artedi's *Monodon (monoceros)*. The old tale of the pelican's offspring suckling their mother's blood is explained as a misunderstanding of the bird's oddly shaped beak, and the satyr as a kind of ape. *Barometz*, "the vegetable lamb of Tartary" was said to grow on trees but is just an artifact, as is the image of a sheep's fetus growing stuck to its umbilical cord. The magic Phoenix is a linguistic mix-up with the date palm and *Bernicla*, the Scottish geese, are not born from barnacle shells growing on tree trunks that have fallen into the water.[5] Like the hydra, the dragon was a fake, composed of bits and pieces from, perhaps, lizards or rays and left to dry. The ominous ticking of the death watch in your home is made by the larvae of a small woodboring beetle chewing the woodwork (Linnaeus called it *Pediculus pulsatorius*). During his travels in Dalarna, Linnaeus had noted its presence all over the rectory near Lake Hormund: "It ticked as if 50 trouser pocket watches hung on the wall." In the 1740 edition of *Systema Naturae*, the list of paradoxes, now under the title *Animalia paradoxa*, is placed after the animal kingdom and lengthened by four more items: the manticora with its multiple rows of teeth; the creature called an "antelope," which corresponds to an animal observed but oddly described; the lamia, which is either a dragon or an animal/woman composite; and the siren, another composite creature straight out of the fables. In total, fourteen of nature's "paradoxes" are discussed in *Systema Naturae*.

Linnaeus was staking out the boundary between science and fantasy; in a solemn mood, one might argue that his critical analysis of the "animal paradoxes" marks the beginning of modern zoology. In earlier works, from medieval herbals to the encyclopedias of the Renaissance by Conrad Gesner, Ulisse Aldrovandi, and John Jonstone, references to these and other fantastic creatures are common. They were not understood as nature's exceptions but, in Swedish literature, too, "obligatory" as some of the most well-loved tropes of Baroque thought. Linnaeus's list can arguably be seen as a rejoinder to Haquin Spegel's widely read *Guds werk och hwila* (God's Work and Rest; 1685 and 1705), an extensive, versified, and very enjoyable retelling of the story in Genesis. If you accept its premises, it is also the first Swedish book of natural history, but it still included them all: basilisk, boramet (*sic*), dragon, hyena, phoenix, pelican, unicorn, and many other wonderous animals, described for the reader's instruction and enlightenment.[6] Two distinctive minds confront each other. To Spegel, such creatures were not "weird" but were ready to inform us lucidly—if only we would listen—of Divine intentions. To Linnaeus, God used a better defined and regulated grammar. In 1748, when the sixth edition of *Systema Naturae* came out, the *Paradoxa* were not included, which has been understood as signifying a battle won. Still, the borderline between science and fantasy could still be crossed, as demonstrated when, a few years later, Linnaeus introduced *Homo troglodytes* and other freakish notions that he had decided to trust.

The *Paradoxa* category may have been inspired by John Ray, author of a systematic study of quadruped animals—*Synopsis quadrupedum* (1693)—who wrote in his foreword that he would include only descriptions given by people who have personally seen the animals. Hence, Ray rejects "[the upalim], the catoblepas, a gorgon-like buffalo, and the Indian satyr or the famous Orang Hutan" described by the Dutch doctor Jacobus Bontius. The gulo or quickhatch, now known as wolverine (Gulo gulo), was also suspect, following the pictures and lurid accounts by Olaus Magnus in his *Carta marina* and also in his great *Historia* (1539, 1555): to get more room inside to feed, it empties its gut by squeezing between two tree trunks, which makes it an easy target for the hunter. The wolverine is in fact missing from several systematic accounts of animals until Linnaeus's *Fauna Svecica* (1746). The wolverine and hyena (Linnaeus's "filfrass") were mixed up and thought to "dig up the corpses in cemeteries." In his lectures, he remarks how shaming it is "that we have in this country such a rare animal and yet no Swedish person has described it." His claim to have once dissected a wolverine during his time in Lapland is exaggerated or worse (yet one more fib about Lapland; it can be found in a letter to Stobæus, 12 June 1732). He waits all his life in vain to see the rare animal, which is finally given a descriptive overview in the 1773 Acta of the Academy of Sciences. So, the wolverine in Olaus Magnus's version endured for more than two hundred years, an example of a "persistent beast" and proof that some tales are long-lived because they are such instructive and excellent stories.[7]

He studied widely to prepare for the *Paradoxa* list, and had questioned several other composite creatures before. *Pan Europaeus*, Linnaeus's zoological notebook from his early student years, had a corresponding category that begins with the lamia—a quadruped with a human face and a maiden's breasts—and continues with the manticora, the antelope, and the rhinoceros, which looks more like a unicorn with the body of a horse, the feet of a predator, and a spiral horn. Another precursor to the *Paradoxa* list was included in *Diaeta Naturalis* (begun in 1733) and linked to a Latin motto to the effect that the study of nature is an antidote to prejudice. This time, the entries are the date palm, the phoenix, the heavenly deluge of algae, the flying dragon, ghostly fire, the cobra, the basilisk (the danger that kills by midday), the pelican, northern lights, zodiac signs, and the hydra. Linnaeus remarks: "As the systems of the world are best ruled by a God, why should the bodies not be thoughtful?"[8] Indeed, why do people fail to grasp that this is a rational world organized according to uniform laws? His preparatory inventories show that the *Paradoxa* category was no sudden inspiration but an element in an intended reform of natural history.

It is tempting to express the same idea in another way by adding a quotation from Linnaeus's Lapland journey: his conversation with the curate and the schoolteacher in Jokkmokk. "I was taken aback in my admiration of such great haughtiness and ambition in such a simple soul, that such lack of understanding paired with such solid stubbornness and simplistic reasoning should be found in a clergyman, thought by simple people to be a man of some learning"—using variants of "simple" three times suggests his utter contempt. "Herr Curate began to explain the skies over the Lapp wilderness, how the skies sweep across the mountains and thus remove stones, trees, and various creatures to carry them away; I, as seemed plausible, attributed this to strong weather and stated that at other times the sky cannot lift; he smiled at me, saying I had never seen the skies; I have so, I replied, when the fog is down, then I walk inside the sky, when the fog is low soon enough it rains underneath it. But he had to smile a *risu sardonico* at such precious stitching. . . . The other (the *pedagogus*) admonished me as so much time and effort is wasted on this worldly and farcical display while neglecting spiritual matters and thus many are ruined by their rummaging in various learned studies."[9] Both were wondering why the society had decided to rely on just a student. A few years later, the two daft Jokkmokkians could have read—though one feels pretty certain they didn't—that the "mere student" had published in the Proceedings of the Royal Academy of Sciences an explanation of the notion that certain animals fall from the sky, such as lemmings: "[They] are conceived and born as are other mountain animals, leave their hillside in certain years."[10] Nothing mysterious and no punishment for sins.

Linnaeus's examples of "paradoxical animals" might seem randomly chosen, but his choices were partly dependent on his own experiences, as we have seen in some cases. He had himself investigated the seven-headed hydra in Hamburg and had been able to show that it was a fake. The wealthy apothecary Albert Seba had staked some of his prestige on this creature by including a handsome engraving of it in his recently published *Thesaurus*. Gronovius, Linnaeus's friend, had warned him against challenging Seba, and there had been time to consider the matter as the printing of *Systema Naturae* took longer than planned. He insisted, despite the risk of causing offense—and just the following year, Seba passed away.[11]

These are the initial stages of Linnaeus's work in zoology; he was sure of himself and insecure at the same time. Nature is endlessly varied but certain universal laws must not be broken. The animal kingdom was harder to define precisely than that of the plants because he had a firm grasp of the sex life of plants. The matter of boundaries was important; for example, *Barometz* could be understood as an organism straddling the boundary between plants and animals—literally a "zoophyte." *Systema Naturae* contains several aphorisms intended to exclude the so-called theory of spontaneous generation and insist that nature relies on sexual reproduction. His argument was based on the laws of reproduction that have been in operation ever since the genesis of all things, and also that nothing can come into being spontaneously—*generatio aequivoca* is rejected in favor of *univoca* as eventual cross-breeding would fail to produce vigorous offspring. All creatures are like their parents so that, for example, geese can't arise from wood or barnacle shells. His entire system depends on sexually mediated reproduction functioning continuously from the beginning and nothing extraneous creeping in at any time. Linnaeus's task was to cleanse Augeas's stables, get rid of the accumulated muck, and be seen as the individual who represented the new, scientific zoology that stood against the historical version, crowded with more or less ancient fables, philological oddities, and fantasies. Hence, his task included enlightenment of the people and a disenchantment of nature. He is also promoting a new view of science itself and of the scientist as someone who opens up to truth and enlightenment. Linnaeus, the dragon slayer, steps forward.[12]

He was, however, not immune to all reports of strange natural events. As the very last item in *Systema Naturae*, he added a note about "Microcosmus"; if one follows his references, it turns out to be the sea monster Kraken of Scandinavian folklore. The one thing people tend to know about Linnaeus is that he believed swallows spend the winter underwater—a widespread misconception, probably related to swallows flying close to the water's surface in the autumn. "They fly not away from us during the winter but lie on the bottom of lakes, still it is not known whether all species do this or which ones it might be,"[13] he explained to his students. The watery realm remained a mystery to him—most things can be found in or around it, like the sirens.

The siren is the last entry in Artedi's *Ichthyologia* (unlike Linnaeus, Artedi moves from the simplest to the most complex animals). The manatee was placed among the whales, and the fish species *Plagiuri* remained in the mammalian class. The publisher, Linnaeus, might have added the manatee. One clue could be that the same printer's error is found in both *Ichthyologia* and in Linnaeus's *Manuscripta medica*: *Massiliam* is used instead of *Brasiliam*, suggesting that these freakish creatures are found off the coast at Marseille rather than Brazil—where it could well have been seen as the dugong who had lost its way out into the sea.

We can already attempt a summary of Linnaeus's internationally important contribution: (1) He had created a system that included plants and animals; (2) he had constructed a new terminology; (3) he had placed sexual reproduction as the basis for classification of plants; and (4) he had made an inventory of the plants in a previously unknown part of the world (*Flora Lapponica*). As he summarizes his work: "Linnaeus has changed everything and thus it must be admitted that he is the greatest reformer in this branch of science (*botanique*). All this, Linnaeus had conceived of himself and before he was twenty-three years old, all had been worked on abroad before he returned home."[14] More was in the offing.

In fourteen published works, Linnaeus discarded earlier ways of naming plants and established his own rules. These works include, for instance, *Critica Botanica* and *Classes Plantarum*, with their orderly tabulation of all systems known to date, and then a draft version of a "natural" system. He explained the botanical terminology, but "victory was not won at once. Many took exception to his rather dictatorial manner and were especially offended by his assumption of a sovereign right to restructure plant nomenclature," as his biographer Arvid Uggla put it.[15] A distinct irritation can be detected in people such as Gronovius, Clifford, Lieberkühn, Ehret, and Haller. Nonetheless, he was mostly successful in avoiding polemics. In the case of Haller, the objections were probably due to jealousy because his own pioneering ideas had not received the same recognition.

By now, knowledge of plants had been given a solid ground from which to take off. "Now must I, too, begin to speak botanical," his friend Olof Celsius wrote in 1738.

CHAPTER NINE

A Stockholm Interlude

AFTER THREE HARDWORKING but truly successful years abroad, Linnaeus was on his way home. Nothing much is known about that journey other than that he arrived in Helsingborg and traveled on to Falun, after a rest in Stenbrohult.

Just as he had done before leaving for Holland, after coming home Linnaeus made a list of his planned publications: *Species Plantarum* in two volumes, *Sponsalia plantarum*, *Philosophia Botanica*, *Flora Svecica*, *Ornithologia Svecana*, *Bibliotheca Botanica Aucta*, *Insecta Sveciae*, *Diaeta Naturalis*, *Ceres noverca*, *Iter Lapponicum*, and *Iter Dalecarlicum*.[1] By now, the list had been slimmed down, and the titles had become well known as *Fauna Svecica* and so forth. The last four works were published in relatively recent times. *Hortus Agerumensis* by Carl Linnaeus and Johan Rothman (1739) was the first Swedish printed text to set out the sexual system of classification.

In traditional accounts of Swedish history, the country's time as a great power ended with the loss of territory by treaties in the early 1720s, and was followed by the so-called Age of Liberty, which lasted until absolute monarchy was reinstated by King Gustav III in 1792. The gloss of the "Age of Liberty" was marred by trade in civil service posts, bribery, and party squabbles, but its reputation improved once it became seen as a first, fundamental step toward democracy. This was after all the period when public discussion was served up in coffeehouses, but above all by an ever more varied press. By midcentury, keeping up with population statistics had become a function of a department of the state, and health care reforms had been initiated. Laws ensuring freedom of the press and abolition of censorship were voted through in 1766. The political debate prioritized utility and industrial enterprise, and consequently also economics and science.

Swedish life was influenced to a degree by France but increasingly more so by developments in Germany. Depending on where the emphasis is placed, the country either was becoming enlightened or sinking into obscurity. The Age of Liberty—paradoxically crowded with rules and regulations, and to that extent "modern"—might also be seen as an age of peace, though there were belligerent episodes such as the war against Russia and the Pomeranian War. It was after all a period between two variants of royal absolutism, as exerted by Charles XII and, by the end of the century, by Gustav III. But it could also be called the age of Linnaeus due to the wave of inspiration that swept through the nation and lasted a long time.

A new generation had set about changing the understanding of science in Sweden: Linnaeus and his students Carl Fredrik Mennander and Johan Browallius, and furthermore Anders Celsius, Samuel Klingenstierna, Nils Rosén, Johan Ihre, Pehr Wilhelm Wargentin, Abraham Bäck, and Johan Gottschalk Wallerius—a group of clean-shaven young men who daily took control of nature's rebellious urge to grow. A fashion note: there are no portraits of Linnaeus or the other young scientists sporting a beard—but how did they look on expeditions such as the Lapland journey? There is a later description of Linnaeus by Johann Beckmann, who met him in an Uppsala bookstore and saw a bearded man: "Getting on in years, not large, with dusty shoes and stockings, a long beard, and an old green coat on which his medal hung."[2] Generally though, these men shaved off the beard of orthodoxy.

A new society was being built, supported by new ideologies. Philosophy changed from the old academic topic to a physical theology, incorporating the principle that nature's order proved that a god must have created it. In addition, the main ideas of the Enlightenment had crept in, and a sense of national pride was still extant. True, Linnaeus was overstating things when he wrote in an application to the Academy of Sciences: "Lapland is a possession of Sweden, which has become *literatissima* [very highly educated]." He was himself an upstart who now belonged to the new elite, felt that his career was taking off, and knew his mission was important. The emerging generation of the 1730s was the driving force behind what can be thought of as a Swedish miracle—a remarkable transformation. They were, of course, very different individuals. Among them Linnaeus seemed the closest adherent of the Baroque ideals, and hence the man with the richest range of references to both old and new learning. There were other, older scientists and thinkers who should be counted as members of the group, like Emanuel Swedenborg and Christopher Polhem. A know-it-all, like Linnaeus, Polhem's wide-ranging mind produced prescriptions for sorting out every kind of national problem and was always ready with quick solutions. The welfare of the people depended on the circumstances, chiefly the climate, and the availability of food and clothing. The Swede was to be shaped the right way, his wit and wisdom encouraged, and so the nation would flourish.

Arguably, 1732 is the year when the transformation began: it is when Linnaeus set out on his journey to Lapland and Dalin published the first issues of *Then Swänska Argus* (The Swedish Argos). The *Argus*, together with the laws set out in the Civic Code of 1734, laid the foundations for the modern Swedish language.

The reputations of some of the young reformers were tarnished by political rivalries; Johan Browallius is a case in point. He was an uncompromising champion of limiting royal influence by a transfer of power to the four "ständer"—a representational system based on the four estates of farmers, merchants, clergy, and nobility. He ended up being known as "the worst brat in the country," mainly because he favored his friends with the best clerical posts. In this and his many other faults, he was a man of the times. For a while, he was actually thought of as a candidate for the archbishopric, but he died when he was just forty-eight years old. As we have noted, Linnaeus picked up a rumor about his flirtation with Sara Lisa and was jealous of him. Browallius was a difficult character and his posthumous reputation has not been the best, but at least his name is not found in Linnaeus's roll call of the righteously punished sinners in *Nemesis divina*.

In September 1738, Linnaeus arrived in Stockholm. The Swedish capital was home to about sixty thousand souls, distributed between the old town or Gamla staden, an area to the east, and Söder, an island to the south of it. The parliament of 1738–39 would be decisive for the politics of the following two decades. Arvid Horn was fired and two parties formed: the Caps were the party of choice for Horn's supporters, but the parliamentary majority were Hats. That majority was largely made up of young noblemen, almost all officers whose goal was a Swedish revenge on Russia. They soon got to try their luck in a war (1741–42) and lost it comprehensively. Both Caps and Hats wanted a mercantile future for Sweden, but the Hats were more extreme. Linnaeus, known as "the Caps' own archiater," made remarkably few comments on the foreign policy fiasco—including a passing note in *Nemesis divina*—but would become all the more engaged in the practical implementation of industrial and trading policies. Stockholm had some fifteen coffeehouses that offered opportunities for conversation and the exchange of ideas.

The young Axel Reuterholm recorded moments in his diary: he went visiting, took cups of coffee in cafes, met the librarian Dalin (still "Dahlin" at the time), and a few times Doctor Linnaeus, whom he, of course, knew from Falun. He also took an interest in the House of Nobility debates and listened to Count Horn, who gave a fine speech. Axel was a Cap and didn't miss out on hearing God's words in church, but then we meet him again just wandering about "without utility or pleasure." Next, he went to a party and tried to charm a girl called Doris. He noted down his meals and the menus. In 1738, on 29 September—the feast day of St. Michael—Doctor Linnaeus, who had arrived a month earlier, "passed the peace of the meal quite enjoyably with tales from his journeys in foreign places."[3]

At first, Linnaeus was treated as a stranger in his homeland. As he remembered it: "Linnaeus aspired to earn his keep as a doctor, but as he was unknown to all, no one dared to place his dear life in the hands of an untried doctor, indeed, not even that of his dog, and he came to doubt his advancement in this country. He who had been honored everywhere abroad as a *Princeps Botanicorum* was at home regarded as a 'Klimius' [Niels Klim is the accident-prone traveler to alien worlds in Ludvig Holberg's 1741 novel], recently emerged from the subterranean realm, so had it not been true that Linnaeus was in love, he would infallibly have traveled outward again, leaving Sweden."[4]

His life improved soon enough. On 1 December, he writes to Mennander in Åbo: "I have unaccountably found myself in such demand as a practitioner that from the morning at 7 o'clock until the evening at 8 o'clock I can hardly spare the time to take a brief midday meal. All of which will help with putting pennies in my purse but also to such an extent consumes my time that I can hardly find an hour for rest or for my dearest friends." Carl Gustaf Tessin, the nobility's exceedingly well-connected parliamentary spokesman, stepped in to support Linnaeus with contacts and a pension. Linnaeus remarks that he had been "taken in as a member of Tessin's household." It meant that he had access to fine collections of art, books, and natural objects and was suddenly connected to the Swedish cultural elite, but as one among many in the Tessin palace, from the house steward to footmen, maids, and scullions. He slept in Österlånggatan 33, in the civil service building known as Räntmästarhuset. Just down the street from there, the bachelor could choose between several places serving food with names like The Dragon, The Dove, The Riga, The Star, Three Kings, and The Dutch Thun.[5]

In a letter of 21 January 1740, to the botanist François Boissier de Sauvages in Montpellier, Linnaeus spoke of having been granted permission to examine the corpses of men who had died in the navy hospital: "Before now, I could not do this. If I live for another year, I will find the answer to what is the most immediate cause of the fevers but no sooner; I shall not posit any hypotheses but only eternal verities."[6] He was prepared to begin to dissect bodies and study medicine seriously; Swedish history could have turned out differently. He worked on another group of patients, too: people with syphilis, whom he tried to cure with Rhine wines or with mercury. It brought him new and valuable contacts; *Nutrix noverca* (1752), his treatise on breastfeeding, includes an anonymized case history about a wet nurse who had in her care the children of a captain (Augustin Ehrensvärd) and had given them not only milk but also syphilis. Their mother sought help from a local healer, but her cure did not work, so a doctor—Linnaeus—was called. He immediately suspected venereal disease, prescribed mercury, and fired the wet nurse. The story doesn't quite end there: the father was probably the artist who created the frequently reproduced etching of Linnaeus (1740) and signed off with *amica manu*—"by the hand of a friend."

Linnaeus next took over from the engineer Mårten Triewald, who had been giving lectures at the House of Nobility—Riddarhuset. The new series of lectures in botany was held in the Auditorium Illustre and began in May 1739. Linnaeus's lectures became very popular and were advertised in *Stockholms Post-Tidningar* in 1739, tempting the public to attend on Wednesdays and Saturdays between half past six to seven o'clock. Free tokens were available for those who wanted to come along to the next presentation—"Besides, once a week, the same Doctor will accompany attendees on the grounds and on flowering meadows," the object being to make everyone "familiar with such herbs that grow in this country."[7] As Linnaeus put it: "We, more observant than the lynx, calculate time and years, know what has been and plan what shall become, are in our own nests often as blind as moles. Thus, although we reckon to have procured the world's sciences, we must surely admit to knowing little more of the origin of our food than do the wild animals."[8] Linnaeus expounded on the pleasure as well as the utility of learning about "herbs." The excursions began on the grounds just beyond the customs post Hornstull, then on to the large island Kungsholmen and other areas around Stockholm.

The new generation made the Royal Academy of Sciences their chief forum. The academy was founded in 1739 by a group that included Linnaeus. The institution was connected to the king and the capital city, and its role was defined by utility, science and economics, nature and the new society. Linnaeus made few references to the old heroic history and kings like Gustav Vasa or Karl XII, but referred all the more frequently to the queen, Lovisa Ulrika, and, later, to her son King Gustav III. The academy provided new income streams but also increased his workload—Linnaeus often took on the sensitive role as scrutinizer and censor of new findings in natural history. The number and range of his contacts also grew and came to include people from all walks of life—senior civil servants from various departments, mining experts, clergy, publicists, counts, pharmacists, post office functionaries. Crucially, the new institution cut across the old class structures: academicians were knowledgeable and the nobility interested. Internationally, London's Royal Society (1660) was the first model, followed by Académie des sciences (1666) in Paris and Preussische Akademie (1700) in Berlin—all enjoyed royal patronage. Linnaeus must have remembered the small academy he had joined in Holland and might also have been inspired by his visit to Hans Sloane but didn't mention this in his account of how the Swedish foundation came about.[9]

On 2 June 1739, the academy decision was made in the House of Nobility at a meeting of Linnaeus, the physicist and engineer Mårten Triewald, Count Anders Johan von Höpken, Earl Sten Carl Bielke, and the merchant and agriculturalist Jonas Alströmer. They cast lots for who should be the first *preses*, and chance chose Linnaeus, the youngest but an instant central figure. They elected others—politicians, entrepreneurs, and scientists and, by 1750, the academy had around one hundred members.

Initially, the members met frequently—42 times in the first year, with Linnaeus presiding at 37, and in 1740, the following year, the frequency was 48 and Linnaeus's attendance 36. Linnaeus toured the provinces and took up a professorial post in 1741, and he attended barely half of the 41 meetings. The minutes of the meeting on 30 September 1741 states: "Herr Linnaeus came and explained that he now intended to travel to Uppsala in order to remain in that place but would appeal to the academy that he would be informed of all botanical observations that may be presented."[10] These are some extracts from the academy's minutes in 1739–41: Herr Ahlström (later ennobled to Alströmer), von Höpken, S. C. Bielke, C. Linnaeus, and M. Triewald met on 2 June 1739 in the House of Nobility in Stockholm and debated the initiation of a planned society "dedicated to 'the upholding of education and the spreading of knowledge in Sweden of mathematics, natural sciences, economics, the useful arts, and manufacturing.'" They talked about what to call their creation: "Society of Oeconomic Sciences" or "Guild of Scientists"; both testify to how much the practical aspects mattered to them. By the end of the summer, Linnaeus emphasized on 5 August: "The academy should be encouraged, with every tender care, to express itself in pure and neat Swedish," and texts should be copied out in either Germanic factura or in classical Roman letters. At a meeting on 19 September, the founders discussed the academy's motto, and Linnaeus suggested a Swedish phrase—not a Latin one—meaning "For those who come after us."

In the autumn, their meetings dealt with a variety of technical matters; for instance, on 6 October 1739 they discussed a certain type of leather and whether it could be manufactured exclusively in Sweden. The surveyor Jacob Faggot thought not, but Linnaeus disagreed and claimed that the plant source of the essential tanning agent, marsh rosemary (then *Ledum palustre*, now *Rhododendron tomentosum*), was widely distributed and "found only in the Nordic countries." On 10 November 1739, they argued about the possibility of building an "eternity machine." Linnaeus believed it had to be possible since nature itself was in perpetual motion, and the heart inside the human body likewise. Triewald and Faggot objected that life—the vital principle—drove the circulation of the blood. On 28 November, everyone approved of Linnaeus's paper on horseflies and other clegs, and "it was also decided that Herr Salvius should read through the text before the treatise is sent to the printer."[11]

Linnaeus pronounced on economics: "No science in the world ranks more highly, is more necessary, or of more utility than oeconomics, as the welfare of all mankind rests upon it. The conclusion must be that this science is practiced and applied with utmost diligence, also the learning of physics and natural history as no oeconomics can last without these subjects."[12] However, their shared hopes for Swedish silk and tea production were disappointed.

It is interesting to note which candidates Linnaeus backed for academy membership. Browallius was not approved the first time around but is on record expressing his thanks for being elected in 1740. "Herr Professor Svedenborg was by Herr Linnaeus put forward as a candidate for membership, as was Herr Pastor Westbeck by Herr Faggot."[13] Linnaeus didn't like Westbeck, who was a rival for carrying out landscape inventories. The minutes of the meeting on 9 June 1739: "*Praeses* proposed further the academy should consent that the travels from Stockholm on certain commissions, which he of necessity must soon undertake, would preclude acceptance of further members."[14] He repeated this on 20 June—but what were the "commissions"? His wedding? A trip to Uppsala, or perhaps a job application? Usually, it is believed that his promise to Sara Lisa was the cause of his travels, but the impression remains that Linnaeus was in no hurry and perhaps hesitated between untroubled work time and the baby-minding duties of a married man.

The marriage had been postponed for one more year to give Linnaeus time to consolidate his position. The wedding took place at Sveden, the mining director's manor where the Linnaean romance had first come into flower. According to the Linnaeus biographer Thore M. Fries, the celebrations were all "traditional joy and togetherness," but we have no proof one way or the other; Linnaeus himself is silent on the subject of his wedding. We know that it took place on 26 June and that the groom stayed for a month before leaving for Stockholm in order to contribute to a good start for the Academy of Sciences. A celebratory wedding poem has been kept. It referred to the sexual system of classification by calling Sara Lisa "monandric lily." The church records were lost in a fire, but the guess is that Eric Björk, a friend to Johannes Moraeus, officiated. These days, Sweden boasts the finest of outdoor privies. It bears the date 1692, and it seems reasonable to assume that Linnaeus, perhaps also Swedenborg, had at some point been seated in it, deep in thought.

On 26 January 1741, Linnaeus was at work in Stockholm and Sara Lisa at home in Falun. That day, Linnaeus wrote her a letter about a most important event: "My very dearest mother and wife. How impatiently I awaited the post, and how moved I was by the announcement it brought.... However deeply it pains my heart that your work, my dearest mother, had been so very hard, I kiss the merciful hand of the Almighty that it was not harder as He blessed us with a well-shaped son born without any debility." He added, "I hope also you my friend will forgive him who confesses himself guilty of your agony though without evil intent." He entered on the date of 20 January in his 1741 almanac: "Born, my first son *Deo Laus* at 7 o'clock in the morning." Another entry on 23 January: "Christened, my first son Carl."[15]

A piece of biographical entomology from his paper "Notes on the Practices of Ants": "When I, in the year 1740 in the month of August, was at Sveden near Falun, for some days kept there by heavy rains, I had nothing with which to pass my time.... Then came to my mind the words of Salomon, who commanded the idle to study the ants ... § 13. As, apart from the common unwinged working ants, I saw in each ant heap two kinds of flying ants, one larger and one smaller, I took to postulating that these flying ants belonged to a quite different type. Thus, I became convinced that the flying ants were without doubt one and the same species as the worker ants ... so with every certainty I could now deduce as follows, (a) that the larger flying ants are all females. As is the practice with the bees, (b) all the smaller flying ants are males, as are the drones in the beehive ... (g) that unwinged ants are geldings (*neutra*), as are the

FIGURE 17. Linnaeus's wedding portrait. Painting by Johan Henrik Scheffel, 1789. Photo by Michael Wallerstedt. The art collection of Uppsala University.

FIGURE 18. Sara Lisa's wedding portrait. Painting by Johan Henrik Scheffel, 1789. Photo by Michael Wallerstedt. The art collection of Uppsala University.

worker bees in the beehive. Once these begin to emerge, all the ants in the heap build with much zeal such high and vaulted passages that males and females who will soon celebrate their wedding feast may within these, sheltered from harsh weathers and burning sun, have freedom to play and amuse themselves in leisure and lust. In this way it happens that males become fathers, and females, mothers. Lastly, I should ask what is the cause that ant heaps reproduce and become many more? Is it that the ants swarm? Or by which means does it take place?"[16] This almost autobiographical piece of writing ended abruptly at this point. Did the final questions concern demographic policies?

An even more direct hint is found in his lecture on the curiosities among insects: "Observe the unceasing work of the ant but also how the males carelessly enjoy their brides inside for them built arched passages while the worker ants must continue to slave in the home all day, bring in winter feed and building wood, and carry swaddled babes into the sun; but the honeymoon does not last long as, once the wedding is all done, father and mother must leave the home where the gelded slaves have made their place until their offspring are fully grown and begin to think about their engagement."[17]

The time had come for a dramatic competition for two professorial chairs in Uppsala as two elderly occupants were being pensioned off. Rosén wrote to Linnaeus on 28 September 1739: "At last, it looks like the venerable archiater Olof Rudbeck and ditto professor Roberg are to demand leave from *consistorio et al.* . . . they however protest *hautement* and refused in particular the first-named. . . . I was also called up but R would from time to time have it noted in the Protocol that I was guilty of letting the Garden go to ruin for which task I had never been hired by the venerable Archiater? . . . Rudbeck claims that he in 1730 was one among the most distinguished *hort[ulani] Botanici* in *europa*, and it was through the hard work of Martini he said and the ven. Doctor can bear witness thereof. Ergo keep an eye out should he write to the ven. Doctors. . . . It would be good if the ven. Doctor gave him true principles and ideas in this matter." The interpretation of all this is uncertain; Rosén was clearly agitated, and the overall message is foul play! Linnaeus was to stand by him.

The three main candidates for the two posts were already prominent and have been called "the father of" something: Linnaeus of botany, Rosén of pediatrics, and the third man, Johan Wallerius, of agricultural chemistry. All three took part in the creation of a golden age in Swedish life sciences. The voting by the *consistorium* (university court) for Rudbeck's chair is on record: the other retiring professor, Lars Roberg, placed Rosén first, Wallerius second, and Linnaeus third; Johan Ihre's ranking was Rosén first, then Linnaeus, and last, Wallerius; Petrus Ekerman's rank order was Rosén, Wallerius, Linnaeus; Anders Celsius made Linnaeus his first choice (the qualities he stressed were experience of garden management and publications of several books in Holland), then Rosén and Wallerius; Olof Celsius also chose Linnaeus for first place (same arguments), then a professor Spöring from Åbo, and finally Rosén. It was a close race, but the final outcome was clear: by sum of ranks, Rosén had won while Linnaeus and Wallerius were neck and neck.

The appointment to the Lars Roberg chair was nothing short of a scandal. The qualifications required by the candidates made Linnaeus's merits look less appropriate than before. As he wrote himself: "Every possible circumvention by Uppsala to stop Linnaeus from getting the second professorship."[18] The Caps on the appointment committee demanded proof of Linnaeus's ability to write in Latin, but this was turned down in view of his published works. He was supported by Anders Celsius, who reserved his opinion and persuaded the chancellor, Carl Gyllenborg, to refrain from asking for a Latin test. However, the rival candidate Wallerius had to do his test. He had already, possibly assisted by Rosén, penned a fiery dissertation directed against Linnaeus's thesis on the causes of fevers. Among the twenty statements in Wallerius's polemical publication, seven were critical of *Systema Naturae*; one, the first, denied the classification of the human being as a quadruped animal. Several objections followed, mostly to the role of clay in the causation of fevers. Wallerius had "mustered all his forces" to refute Linnaeus but failed nonetheless.

The students, most of whom were rooting for Linnaeus, stamped their feet, laughed, stood on the seats, and tore Wallerius's pamphlet to pieces. But Linnaeus was also strongly backed by people in power. There are also the recommendations in *Orbis eruditi judicium de Car. Linnaei scriptis*, probably printed in Stockholm in 1741—it was produced anonymously and is very rare. It records twenty-odd statements testifying to Linnaeus's qualities, taken from letters or printed forewords. Boerhaave endorses *Genera Plantarum* as it "demonstrates to the ever more astonished reader a labor of boundless exactitude, uniquely consequential reasoning, and incomparable learning.... Centuries will continue to praise it." From Montpellier, Sauvages wrote: "If I tasked myself to express in words how great the pleasure has been as I devoured these works, a mere letter would be insufficient." From the report by the secretary to the university, Carl Klingenberg, it is clear that Nils Rosén supported Wallerius.[19]

It took thirteen months to complete the appointments to the two chairs in Uppsala: the process lasted from April 1740 to May 1741, when the conclusion was to Linnaeus's advantage. The competitive atmosphere was an indication of how narrow were the opportunities for the new generation. The source material from this period is scarce: a couple of almanacs, mainly listing places where he ate and so forth while pregnant Sara Lisa stayed in Falun until the birth. Linnaeus moved to Uppsala on 6 October; he had swapped the specified professorial posts with Rosén so that, in practice, Linnaeus became Rudbeck's successor. However, this meant that he took on more subject areas than prescribed: botany, chemistry, semiotics, dietetics, and medicine while zoology, for example, which he also taught, is not mentioned in the chancellor's letter of 21 January 1742.

As has been shown, Linnaeus's formal higher education was well-nigh nonexistent but his capacity for independent studies all the more impressive. Now he had arrived at last. The road to the professorial chair had been long, and he had already reached the age of thirty-four.

PART II

At the Height of the Ages of Man

1741–1758

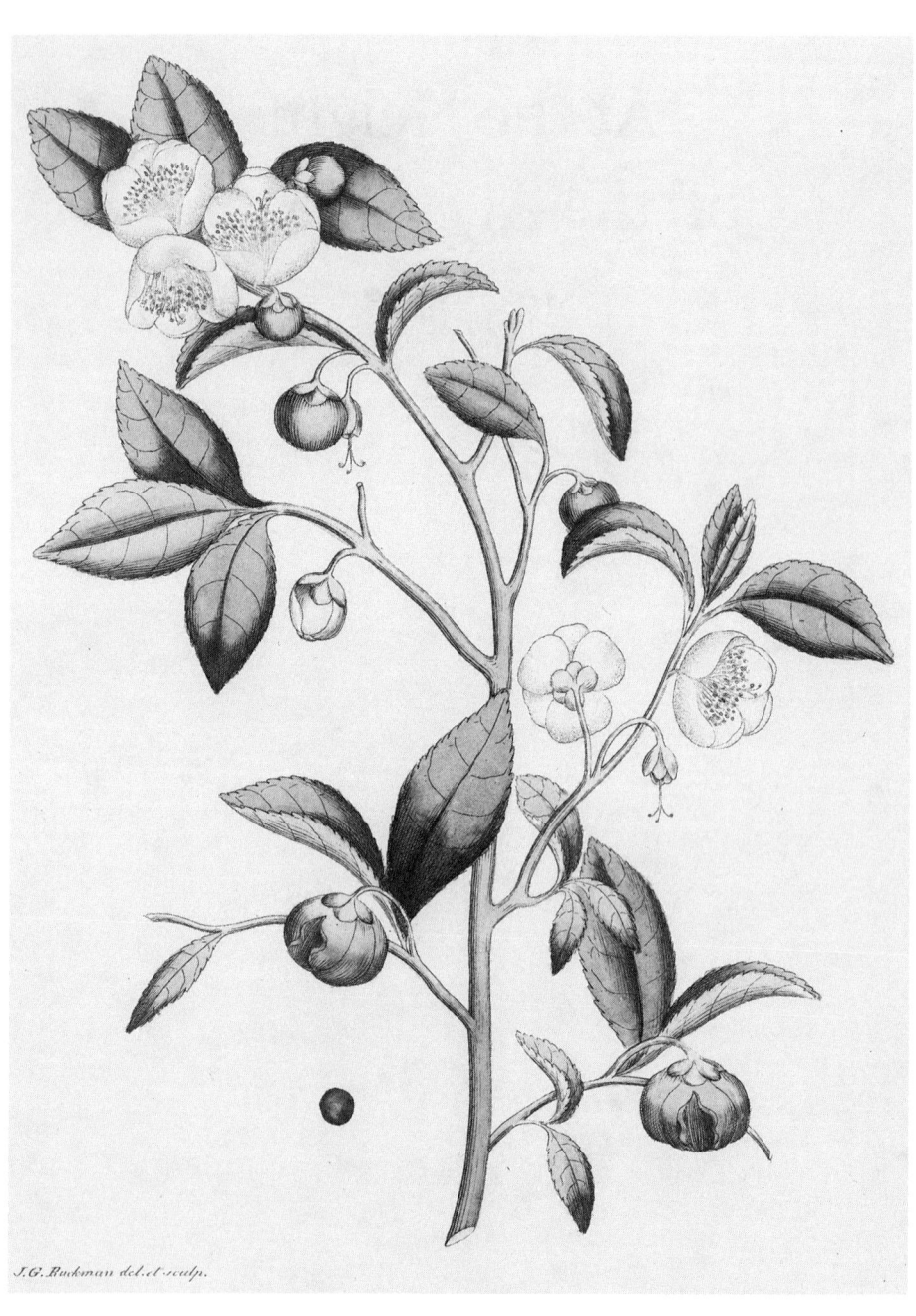

FIGURE 19. Tea bush by Johan August Pfeiffer (1777–1842). Published in *Magasin för blomsterälskare och idkare av trädgårdsskötsel* (Magazine for lovers of flowers and practitioners of gardening). University Library, Lund.

The Desirable Tea Bush

Linnaeus wrote in *Diaeta Naturalis*: "Thée and coffe are novelties, always to be drunk hot so that by drinking we wash out our stomachs as they are truly rendered lax thereby; these concoctions bring nothing good."[1] Later on, and with the authority of a professor, he would change his mind: for one thing, he was becoming something of an addict himself, and his new message was that, as Sweden could be self-sufficient in most things, producing Swedish tea should at least be attempted. He notes in his almanac of 1746: "Let us then bring the Thée-tree from China to our country! Here is the same soil, same Sun, same water, same air. The Winters are truly sharp here in Sweden, the single circumstance that could be a cause of damage to the Thée-bush." The tea bush was a top priority on the list Linnaeus handed to Christopher Tärnström, the first Linnaean acolyte sent off to explore China. It would be easy to find many more expressions of hope centered around tea cultivation, which makes the desirable bush a good emblem for this period in Linnaeus's biography. During the 1740s and '50s, many similar projects were pursued, with Linnaeus as the expert advisor.

When, in 1763, Captain Carl Gustaf Ekeberg finally managed to transport a handful of tea bushes to Uppsala, Linnaeus was over the moon: "Thus God had blessed him even in this matter and allowed him the honor of observing and also introducing thée. . . . He could conceive of no higher goal than to close the gate through which leaves all silver in Europe."[2] It did not work out all that well, though; two years later, only two specimens were left, and the last survivor expired in 1781.

CHAPTER TEN

Uppsala and Enlightenment

LINNAEUS HAD BEEN ABROAD and learned about the big world outside of Sweden; he had spent several years in Holland and visited London and Paris. For what it was worth, he had also been living for a couple of years in Stockholm. What did he bring to Uppsala? His reputation as the new star among naturalists, his knowledge of how to run an advanced garden project, and many foreign contacts; above all, perhaps, his successful role in the creation of the Academy of Sciences and evangelical convictions on precisely how the natural sciences could drive the growth of the nation's wealth.

The city, still small and easy to cross on foot, was in a state of vigorous restoration after the fire in 1702. The man in charge of the rebuilding, Carl Hårleman, was the leading architect of his time. By 1723 a new city courthouse had already been completed at what was then known as "New Square"—Nya torg—on the eastern, mercantile bank of the river. Hårleman cleared the old castle precinct and, in 1745, had a pair of low domes built to replace the fire-damaged castle towers, topped with spires, dating from the era of the Vasa kings. He gave a new shape to the archbishop's walled compound and to the building where young lads were taking their school-leaving exams, a place ominously known as "Duds." And, naturally, Hårleman and Linnaeus both had a hand in the design of the elegant orangery. We continue the walk, past Anders Celsius's observatory and Anders Berch's Theatrum oecononicum at Gamla torget (Old Square), down the hill to the Theatrum chymicum on the western bank of the Fyris River. The central network of streets was being transformed, straightened, and ordered.

All this was documented by Johan Benedict Busser in his proud history of the city (1769), and much of what was built by then still stands. The city was not a healthy place and the Fyris River was an open sewer. In 1754, the year that the "Uppsala fever" reached its devastating peak, the city's thick air stank. People who had climbed Slottsbacken (Castle Hill) saw the infected fog hanging like a pall over the Uppsala valley. Also, the danger of fires was lurking, as Linnaeus revealed: "Here is terror *panicus* for fire and firestorms, people frighten each other with lies and analogies; however, each and every one is so fearful for their own that they will hardly dare leave to enjoy their holidays."[1]

The Age of Liberty has been considered the city university's peak period, an era of truly engaged authorities, perhaps even too much so. The university chancellery was very effective; the Hats were keen to determine policy and for royal personages to carry out ceremonial visits. The outcomes, however, included interference with professorial appointments, directives about the range of study choices, and the introduction of "useful" subjects. The brief of the new Education Commission was to regulate the academic community in a bureaucratically satisfactory manner. Its proposals have been compared to the reforms of Sweden's tertiary education in the 1960s and '70s; both attempts have been found just as regrettable. In 1751 the university court responded to the commission in factual, and hence all the more devastating, terms. "No one could have been more insistent than Linnaeus. Normally loyal towards authority, he shuddered at the thought of what awaited the university," as Sten Lindroth has written. Still, it was good to see chairs funded in new subjects: economics, physics, and chemistry. "For all the restless activity, despite party-political intrigues and tampering, the 1740s signified the initiation of the university's high summer season."[2] The lineup included Rosén, Linnaeus, Klingenstierna, Johan Wallerius, Torbern Bergman, and later, Carl Wilhelm Scheele, but also a controversial theologian in the prominent person of Nils Wallerius, an ardent follower of Wolffian philosophy. Readers and lecturers were added to the academic staff but with poor salaries. Among the professors, Johan Ihre was wealthy, and Linnaeus by then was well off. In the 1730s, the number of students exceeded one thousand, but it later fell again to about four hundred in the 1780s. It took time for the Enlightenment, in the Continental sense, to catch on. Students were instructed to read foreign newspapers in the 1720s and '30s, but despite this modern idea, neither Locke nor Montesquieu had been added to the teaching schedules. By the 1770s and '80s, the changeover to a new generation of academics meant that the golden age had come to an end.

Of course, enlightened thought had taken root to some extent. When Anders Celsius was on his deathbed, the pastor spoke to him about the eternal life to come, and he is said to have replied: "Is that so, Herr

Magister? Well now, this is the time when I'll soon be in a position to find out whether it's true or not." This apparently shocked his contemporaries.[3] Another sign of new ideas gaining ground was a promptly forbidden pamphlet by a disciple of Linnaeus, Peter Forsskål: "Thoughts on Civic Liberty" (*Tanckar om den borgerliga friheten*, 1759), which Linnaeus, ordered by the rector of the university, had to gather up to prevent its dissemination (how effectively Linnaeus tracked the copies down is not known).

There was a symbolic break with the past in 1753, when the Swedish state decided to go along with Europe and introduce the Gregorian calendar. As with many reforms, the Academy of Sciences drove the change. Other events took place in 1755: the ruling Hat party struck a new medal for silkworm cultivation, a redoubtable home cook and housewife Cajsa Warg had her book on "how to run a household" published, and care of the sick in the hospital became better organized.

Measured out on the clockfaces on the church tower, on the clocks displayed outside watchmakers' shops, or the watches in men's waistcoat pockets, time itself was a reminder of modernity. Young voices spoke of it: "When mankind invented how to measure time, they invented a notion of prodigious utility for the commons; although time in itself is no matter, it is a fact that all living beings are nonetheless under its rule; we hold for simpleminded, even barbaric, such people as do not know how old they are; when we are in the country, without a watch, under a dull sky and unaware of the time of day so that the evening ensnares us unexpectedly, we find it insufferable."[4] One could live with nature but must avoid being a stupid barbarian. The style might seem familiar, but this was not Linnaeus's writing—it was his fifteen-year-old son's, writing in the introduction to his dissertation "The Flower Clock" (*Blomsteruret*, 1756).

The outline of Uppsala was a rectangle surrounded by greenery, with some 4,000 inhabitants living within its boundaries. Customs posts controlled access by outsiders. In the year 1756, the city also housed 253 horses, 412 cows, 487 pigs, and 64 sheep; in other words, manure was produced in bulk. On one of the city streets, you would find Lokk's pharmacy, where the young Scheele was once employed. Russworm's, later Kiesewetter's, bookstore was even more remarkable and modern; it sold, among much other printed matter, "periodicals published in Upsala." The city had several inns, a regional market held in February, Lunden's coffeehouse on Drottninggatan, and assorted tradesmen including the globe-maker Anders Åkerman.

The Fyris River divided the city into academic quarters on the west bank and mercantile quarters on the east, a division that still holds true.

FIGURE 20. Uppsala city: Appearance from the south side. From *Draft of a description of Uppsala* (1773) by Johan Benedict Busser. Uppsala University Library.

FIGURE 21. *Arithmetica, or the Art of Counting, thoroughly demonstrated by Andreas Celsius, Printed in the year 1727.* The drawing depicts Gustavianum, the central university building at the time of Linnaeus. The onion dome is placed above the anatomical dissection theater, built by Olof Rudbeck the Elder. Uppsala University Library.

The university was a citadel, but the students were also part of the crowds in the streets and lanes of Uppsala. They could be youths of noble birth, shepherded by their tutors (children of wealthier commoners often had similar minders), or the sons, usually in their twenties, of clergy or farmers. Among them were the odd lost souls who had found in the university a long-term haven, the so-called old boys; a few underage offspring of noble birth; and a scattering of foreigners.

The street scene also included elderly academics like Lars Roberg, the old professor of medicine. The well-informed journalist Carl Christopher Gjörwell, still a relatively young man, was often seen at the journeyman's inn in Läby, wearing his traveling clothes, which "consisted of a coat of green wool so worn that it was halfway yellowed." Another notable was "professor Meldercreutz, who wore on his mathematical head the remains of a hat, its brim not pinned up as it was torn to shreds. Should it be that he has chosen for his party the nightcaps, as have most of his gentlemen colleagues in Upsala, he still seems to take his hatred of hats too far." Other tales were told—for instance, about the immensely fat jurist Christer Berch, and about the notoriously unwashed polymath and naturalist Daniel Solander: "In his house, dogs and squirrels were his company." He was, however, a very learned man, as was Lars Hydrén—learned in the old scholarly ways—but he "lacked taste." The brilliant linguist Johan Ihre and Anders Celsius—who died far too young at forty-three—were outstanding; Celsius was the university's most brightly shining light of learning.

And then, of course, there was Linnaeus.[5]

CHAPTER ELEVEN

Three Programmatic Speeches

IN THE THREE YEARS that followed his return to Sweden, Linnaeus indicated how he saw the direction of his future work. As the first and, by then, retiring president of the Academy of Sciences, he addressed an audience in the House of the Nobility's Auditorio Illustri on the subject of "Curious Features of Insects" on 3 October 1739. His theme was set in the quotation from Job 12:7: "But ask now the beasts, and they shall teach thee; and the fowls of the air, and they shall tell thee." The speech was, of course, about entomology and the usefulness of insects as well as the damage they can cause—which was *not* what Job was after. Linnaeus began by showing the connectedness of all things, as "the soil becomes the foodstuff for the plants, the plants feed the wyrms, the wyrms feed the birds," and so on. His point was that nothing is created to be sufficient unto itself: tigers, lynxes, and bears have to be in their skins, as the hen has to feed us with her eggs and the cockerel to wake us in the morning. Human society works on the same principle: "And shall some prodigious and great task be accomplished, it must be with the help of many," which is why every being has been given different talents. Mankind cannot compete with animals in many things but has the power of reasoning, which we should apply and also practice. Hence, expect even the smallest critter to have notable talents. "Observe the minute aphids as they haul themselves along in the shade under the leaf. Could you have truly believed that when the mother has bred with her male, then the daughter, daughter's daughter, yes, daughters until the fifth generation, can give birth derived from its fourth grandmother's intercourse needing no other male assistance?" Forty-odd "observes" and "considers" later, all new insights into the lives of insects: "Here you can contemplate a field of study for all those who are *curieuse*, those who long to see wonders and new things such as no one has seen before." Linnaeus's science, more than most, is *seen*—a visual art.[1]

He had been appointed to professor on 5 May 1741, the same day on which he set out to travel to Öland and Gotland—journeys that will be discussed shortly. Linnaeus's induction didn't take place until 17 October 1741: his inaugural lecture was given in the larger of the Carolingian lecture halls, and its title was "About the Necessity of Investigative Journeys in Our Native Country." He promoted journeys in Sweden as useful exercises before venturing on peregrinations around Europe, with the temptations this would entail. The extensive topographical literature already in existence demonstrates the widespread optimism about the nation's resources. By then, Linnaeus had new experiences to speak about and, as a medical man, he felt that investigative travel could be particularly useful. True, medicine also required many other things, notably a botanical garden and good hospitals. Visits to London and Paris were very important for updating hospital practices, but he emphasized the native country at every other turn. This is what he had to say about his travels in the mountains of Lapland, the forests of Dalarna, and the woods of Gotland, all full of natural wonders: "I would venture to insist that there is in this world no country richer in birds and insects than our Sweden."[2] He went on to point out that by traveling, doctors can learn much about diseases and their cures, and there is also money to be gained. Attentive travelers will learn much about agriculture—and have much to teach.[3]

Throughout, he juxtaposed the national and the international, only to find his own country with the advantage. He mixed Gothicism with utilitarianism: "We demand that our pharmacists shall procure from abroad and it costs them." Our well-endowed nation, he argued, would be able to provide all we need. It was only a matter of discovering the resources, which was Linnaeus's task on his tour of the Swedish provinces. The common man's treatments of illnesses should not be disregarded, nor should his understanding of farming. Everything that grew would do well in Sweden, especially plants with origins on similar latitudes; all this reads like something more than a hint of the tasks that he would give his pupils—his disciples to be. Any such journeys should be done while the traveler is young, so "before family, home, a beloved spouse and household concerns have taken too large a place in his heart." Linnaeus ended his lecture by greeting and honoring the memory of Lars Roberg, "my teacher and predecessor."

Linnaeus's opinion of the Bible as a source of wisdom tended to be very liberal but, at times, could be peculiarly literal. He often allowed himself to speculate along experimental lines. One case in point is the speech he gave in 1743 at the ceremony when he was granted the title "doctor of medicine": "On the Growth of Habitable Parts of the Earth." His talk was an eccentric spin on the idea that all land had been arranged in concentric circles around Paradise and that "the withdrawal of the waters" caused the available landmass to increase, an idea drawn from a then-current debate. His theme can be understood as a development of a core concept in *Fundamenta Botanica* (1737), set out in paragraph 132: "We count today as many species as were created in the Beginning." There was more to it, though: the impact of the Genesis narrative is clear, but so too is the discussion of land elevation and the geographical spread of plants, followed by animals and plants at different latitudes spreading eastward or westward, and acclimatizing. Consider, he argued, that from the beginning, the world was an island with high mountains in the center. It could mean that Adam had flora and fauna within reach and could therefore name all things alive without becoming overwhelmed, especially as all the climate zones would have been present nearby, from low-lying ground to mountaintops. Once Genesis was complete, the world was a tightly populated area, analogous to Noah's Ark. What happened next, probably after the Flood, was a successive decrease in the amount of water in the oceans, laying bare more and more land. At the same time as the water level was falling, various mechanisms favoring the spread of land-based life took over and allowed flora and fauna to take hold everywhere on Earth. Linnaeus discussed examples of such mechanisms with true insight, making most of his talk a presentation of plant geography. Linnaeus didn't seem to believe in the Flood and never referred to Noah's Ark, but he conceived of nature as developing from Genesis onward through continuous change. Based on what he already knew, he extrapolated about the state at the beginning by moving backward through time.[4]

The speech was not taken entirely seriously. The botanist Pehr Kalm, who had been in Uppsala cathedral for the ceremonial address, commented: "The Lord alone knows what truly happened. I find it hard to believe; maybe it was just a small *lusus ingenii*"—a play with ideas. However, Immanuel Kant became interested and linked Linnaeus's theory about the steady expansion of land with the variations in the song of nightingales and finches among different regions. A century later, the botanist Elias Fries wrote: "Linnaeus who, more ultraconsistent than any theologian, interpreted the first chapter of Genesis and, in addition, spoke as a Neptunist, as he sought in a truly inventive manner to prove that all that is generated by nature, animals as well as plants, were present in Paradise in their original parental forms.... This theory has been contradicted far more frequently than it deserves."[5] Thore Fries, Elias's son and an authority on Linnaeus, considered the theory "bizarre, almost puerile." But the speech, given in Latin, was translated into Swedish, so it must have been thought to be of general interest.[6] The concept had indisputably been preoccupying Linnaeus for a long time. In *Observationes in Regna Naturae*, he argued that the landmass had been increasing and, as time passed, so had the numbers of plants and animals. Looking back, each species would be represented with ever fewer individuals, until reduced to a single pair living in a relatively small area of land. This original area must by definition offer a wide temperature range to let every kind of species thrive. Linnaeus based his thesis to a large extent on "reduced water levels," later understood to be the land rising.[7]

The first of the exactly one hundred paragraphs in his inaugural speech declared: "Not only Holy Scripture but also common sense assures us that all the world has received its origin by the hand of the eternal Master Builder." Linnaeus claimed that he had foreseen the objection that the plants would not have had enough time to carpet the world in greenery given the short period available (according to the old, deep-rooted belief in a time scale of six thousand years) but attempted to disprove this by analyzing the different ways of reproduction and geographical spread: roots, shoots, wind, water. He reminisced about childhood play in his father's garden and the small bed he had dug, where among other plants he grew a thistle variety. Every year, his father hoed it up as a weed, but new plants emerged all the same; clearly, seeds can survive for long periods in the ground.

The Flood had been replaced by the reduction in water levels, as had been observed by Linnaeus's friend Anders Celsius, who might well have inspired the idea of growth in available land starting with Paradise Island at the equator. Could it not have been Cyprus, where Kiöping had described the peak Pico Adam—Adam's burial place? Then again, one might imagine an Atlantis Island, emerging Rudbeck-style. Linnaeus's thesis stated that species were constant but their numbers increased in parallel with the available land surface.[8] The solid knowledge presented in the speech consisted of examples of plant proliferation and not, as Elias Fries said, presumptions about the original animals—for instance, that "tigers and lions originally lacked the appetite they demonstrate in our times."

In terms of geological history, Linnaeus was apparently a Neptunist—someone who believes that the solid land has been shaped by the power of the oceans rather than, according to the Volcanists, by the forces operating from inside the Earth. On 24 November 1749, he wrote to Bäck:

> My Dear Brother, do believe in God and not in Moro. God created nature in His likeness, the same today as it was yesterday; as Gotland grew yesterday so it grows today, and recall that in the old days there were but 2 islands, Torsburgen and Hoburgen. My dear Brother, explain how an underground fire could carry seashells from America to our beaches. Underground fires are surely rarer than is thought; as she can never hide herself but must smoke through the tops of fire-erupting mountains. My dear Br., I wish we could talk on such matters together for a whole night and the next day."[9]

CHAPTER TWELVE

Provincial Travels on Behalf of Parliament

SWEDEN WAS SMELLING OF MANURE, smoke from wood fires, and sweat, sometimes also of perfume and powder. The rustling of winds in forest trees filled the air, interrupted by the peals of church bells; people speaking in foreign languages were heard in many towns and cities. Your affiliations were shown by what you wore: not just hats or caps, but everything about dress signaled the wearer's place in society. Small groups of travelers followed the potholed roads. In terms of time, distances were longer then. You had to find your own way, and to arrive before darkness fell. The cry of the night watchman was heard in the towns: "Nine o'clock has struck, may God keep us all / from fire and thievery as the night will fall."

Linnaeus's writing makes landscapes come alive, rich in interest, but he made many journeys that we know nothing about: for instance, his trip from Lund to Uppsala in 1728, and later, from Holland to Paris, from Holland by ship to Sweden; also, travels to and from his mother-in-law's house in Falun. His provincial journeys in the 1740s are, however, very well documented. The state had recruited widely to complete an extensive inventory of Sweden's resources. Academics had produced several papers on topography, and the Academy of Sciences published in its Proceedings of 1741 the land survey specialist Jacob Faggot's lengthy memorandum with 145 questions about the country. Projects had been initiated in certain provinces—for example, by the bishop and politician Jacob Serenius in Södermanland. Charting the home nation was a priority, but the framework for economic advancement widened when disciples went out to explore.

Fact-finding "in the field" had precedents. Kilian Stobæus was a direct source of inspiration, and not just as a deskbound scholar. In 1729 Stobæus drafted a set of points for his pupils Mathias Benzelius, Nils Retzius, and Johan Fjellström to investigate during their excursion in Skåne. Linnaeus made a note of these instructions in his *Manuscripta medica* and singled out some of them: irregularly shaped plants likely to be charms to ward off nightmares, parasitical plants, the circumscribed areas where certain plants grew, and horticultural plants; also, common names for local flora and fauna, for waterways, fish, and wells. They should record the tales told in different localities, stories about ghosts and little people in the mountains as well as about caves and events linked to caves, mines, and minerals. Notes must be kept of *lapides figurati* (fossils), precious and semiprecious stones, and also characteristic kinds of local stones; furthermore, of different types of useful soils and sands, of meteorological conditions, and notions about the weather. Ancient monuments should be charted and drawn in detail. The same went for communal workplaces such as mills, blacksmith's forges, and other workshops.[1] Overall, the emphasis was on locally held beliefs and old sites of worship—such as sacred wells—as well as natural features.

In Linnaeus's personal history, the 1740s were his years of traveling, compared with the previous decade, and in marked contrast to the next three, when he didn't travel at all unless we count the to-ing and fro-ing between Stockholm and Falun. His explorations were based on the economical ideology of his time, as Eli Heckscher was the first to elucidate. His analysis served as the springboard for Lisbet Koerner's influential thesis *Nature and Nation* (1999).[2] Heckscher sees Linnaeus as "a Yankee," a tough entrepreneur in tune with his enterprising times, while Koerner stresses his naïve and unrealistic side. The journeys around his homeland were a search for balance between occupations, between the territories of the church, the state, and the private lives of the people.

The parliamentarians were full of utilitarian ideas. Probably, it was one of the country governors—either Reuterholm or Gyllengrip—who came up with the provincial travel project. In the spring of 1741, the Trade and Manufacturing committee looked to establish sources of plants and raw materials, and had picked locations of interest: the large Swedish islands in the Baltic Sea, Öland and Gotland, and areas in County Västergötland— Kinnekulle, Halleberg, Hunneberg, Mösseberg, Ålleberg, and Billingen. Linnaeus was thought to be the right man to lead the investigation. The parliamentary manufacturing fund was to pay out three thalers per day in remuneration and also pay for three horses per day. Three parliamentary estates—the farmers were excluded—voted in favor straight away.[3]

Later studies of Linnaeus's reports have been focused on specifics: particular plants, or parish customs, or local expressions and dialects. Still, a seemingly inexhaustible list of practical questions remains: Did he move freely or follow a pre-prepared route, perhaps at least booking ahead in certain inns? How did he find the horses? How far could he travel in a day? Did he ever backtrack because he had forgotten his toothbrush? How did any traveler find his way? Did he depend on maps, milestones, and signs? How was the daily journey decided, who made the decisions, and who undertook the planning ahead for the night? Anyone traveling was required to carry an "inland passport" (obligatory until 1860). People generally had to travel along the main county roads to find the way, and the inns where they could change horses and buy a meal. One day's riding covered some 60–70 kilometers. How were toilet needs managed? Finding rooms for the night? What about the supply of paper, ink, nibs—and a tabletop to write on? Where to find books to consult? And what did Linnaeus want to write, and what was he allowed to write in his diary?

He had encountered the most primitive conditions when traveling in Lapland. That was when he sought shelter in huts or overturned boats and made his bed on moss. Later, his bodily comforts improved. He preferred riding to coach travel, even as late as 1746, in Västergötland. Linnaeus enjoyed riding and recommended it as exercise. In Lidköping, he wrote enthusiastically: "Truly, if only each and every one knew of the uncommon utility for keeping healthy that riding offers in the summer, more would travel in that way in Sweden, and fewer folk would crowd into the spas. Seemingly they could with equal trust in the results ride as lie down by the spa water sources; as, at least, riding is a cure for sufferers of short breath, the beginnings of disease of the lungs, or the spleen, or inner obstructions, thus a treatment surpassing spas as well as medicaments."[4]

The public road network was what it was and, overall, perhaps not worse than in most countries at the time. Still, its reach was erratic. The stretch of road from Stockholm southward to Jönköping actually coincided with today's E4 motorway, but at first, Linnaeus went north, following the shores of Lake Mälaren. Military considerations meant that maps were often secret. Then, in 1743, Georg Biurman's "The Swedish Ulysses" (*Then Swänske Ulysses*) was published. It was a pioneering work that offered overviews of the country's roads, complete with tables of distances between all towns and cities. The regulations of inns dated back to the days of Queen Christina (she abdicated in 1664). The weather must have influenced the travelers' moods, choice of routes, and speed of travel—but how much? At times a lot, as in this passage from *Västgöta resa*: "The Fal countryside is ... much like Flanders. As the weather grew harsher, a

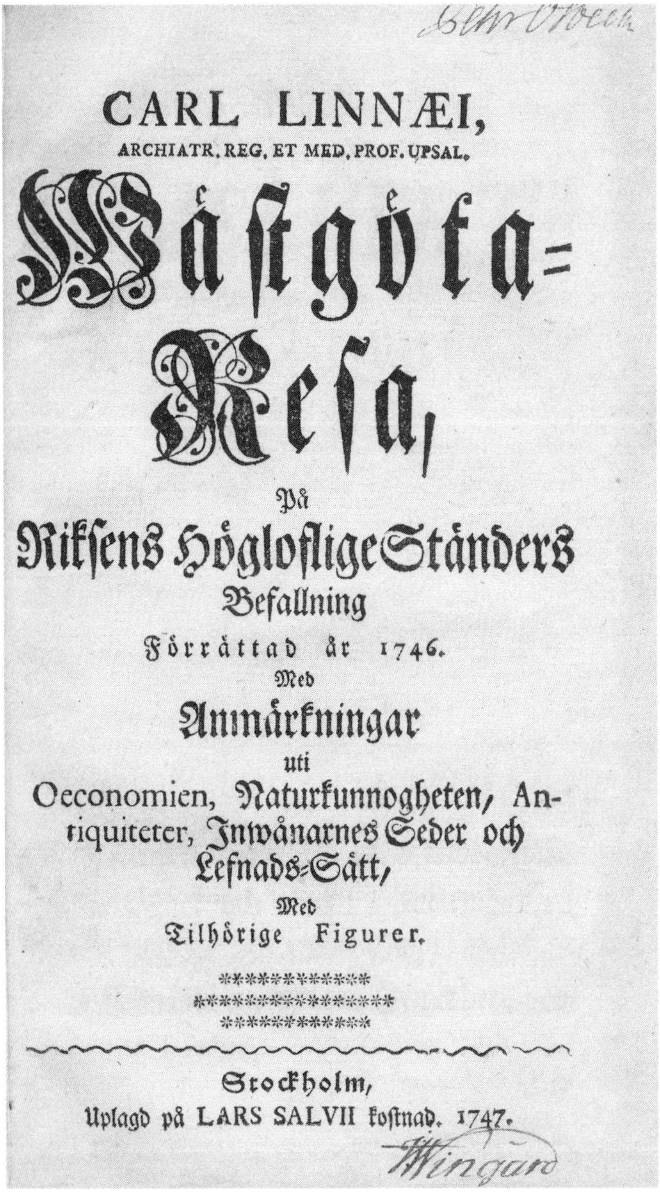

FIGURE 22. The title page of *Västgöta resa*, 1747. Note the *fractur* font as the text is intended for readers who did not know Latin. The text translates as "West Gothland Journey at the orders of the Nation's Honorable Estates. Executed in 1746. With remarks on economics, natural resources, antiquarian objects, the customs of the inhabitants and manners of living. With proper illustrations. Stockholm. Prepared for printing by Lars Salvius. 1747." At the top and bottom right, two previous owners have written their names on the title page: Pehr Osbeck and the archbishop, Johan af Wingård. Uppsala University Library.

strong headwind across the plains soon blew so hard your eyes wanted to fly out of your skull. The horse staggered, almost off the road."[5]

The journeys to Öland and Gotland in 1741 (the accounts were published first in 1745) went across the islands' limestone plateaus and sea stacks with their orchid flora. The governor of Stockholm, Count Rutger Fuchs, had issued the traveling passes referring to the parliamentary commission. Five more signatures by senior county officials would be required along the way. It must have seemed an awful fuss—but, then, there was a war on against Russia. The passports are dated according to the old calendar, eleven days early by our way of counting. The report was, more or less, Linnaeus's first published text in Swedish. He was 38 years old when it came out, four years after the actual journey. Six "young gentlemen"—their ages ranged from 14 to 29—came along as his assistants, each one with a specific task. All had signed the ten commandments or "laws" that were designed to keep the young men and their work in order. The fourth law stated that "the Journey should not be fragmented by gossip, chatter, songs, tall tales, jesting, rambunctious games, and vanity but all should be steadily and sensibly dealt with." The sixth ran: "No one should in the evening leave the *societé* in order to go to bed, take a meal, or spend time in any other way whatsoever until the protocol has been well acquitted and in the *actae* all has been entered what has passed and been observed so that each one will have part in everything."[6]

The work was dedicated to the new prince regent, Adolph Friedrich. In the first edition, the spelling is Germanic, the font German "Fraktur," and the style a mixture of grandeur and humility: "This modest journey was to be undertaken to study a few Baltic provinces with respect to their economy and natural history." The foreword sets out their commission in five main items: (1) investigate grasses and other plants fit to use as dyes; (2) investigate clays and other species of soils appropriate for porcelain, tobacco pipes; (3) record native healing herbs for the pharmacies; (4) record what otherwise belongs to our native country's *historiam naturalem*; and (5) keep a diary for the scrutiny of the commissioners.

What follows are some three hundred pages of usually brief entries introduced by a keyword. Judgmental passages are included: a town can be judged "a small spot" (*Södertälje*), "pretty" (*Nyköping*), or "without a distinctive setting" (*Linköping*). Notes were taken along the roads on plants and animals, what people were wearing and the homes they were living in, the land use and appearance of the landscape, with scientific as well as practical intentions. When they reach Kalmar, oaks impress them as being the most useful trees. But contrasts are noted, some of them very troubling.

One passage describes the sight of prisoners—"slaves," in fact—doing hard labor around the fortress before they are forced back into dark caves under the fortress walls; "their misery made the hair on our heads stand up."

On the island of Öland, just across the Kalmar strait, Linnaeus is amazed by the wealth of plants: "Here were the rarest plants, as were previously unheard of in Sweden and which, so as to observe them, I traveled in 1748 from Paris to Fontainebleau, where I saw them but once, never believing to be allowed to gaze at them once more." The meadows at Resmo make him exclaim: "The road passes through the most beautiful woods that could ever be seen, in beauty excelling all places in Sweden and so rivaling all in Europe."

Throughout, he often speaks of "we" and "us." It happened a few times that Linnaeus and his traveling companions were suspected of being Russian spies. At Ås, on 7 June: "The common folk gathered after the church service to confer about us, who we were and what errands we had. They believed us to be spies." They had to ask for a constable to protect them. At Böda, on 17 June, people were agitated and "with conviction held us for spies." Linnaeus switched to "I" to note that a large stone had been thrown and hit him "the outermost bony bump of my left ankle." The pain kept him awake at night: "I, who on many occasions helped those suffering from suchlike and other stresses, now have not the slightest cure with which to help myself." Now and then, he is shown the local misfits: at Sandby a hermaphrodite, at Ismanstorp "a daft child": "He was unhappily born, neither to his own satisfaction nor to the utility of *publicum* nor to praise his Creator."

Linnaeus was afraid of the sea, perhaps due to the episode of near-drowning in his youth. When about to embark on the ship to Gotland—"Gulland"—the storm was wild, and "hence we attended the service in Högby church" (what, not at other times?). Once out on the Baltic, the weather calmed so much "the ship strolled along." When, after many incidents, the group was ready to return to the mainland, Linnaeus wrote:

> At 6 o'clock in the morning, we stepped onboard the ship. Fearing for our lives, we came out of the harbor into a tumultuous sea. The friends on Gotland and in the city of Visby vanished from sight. Karl's Islands rose up. The northern winds began to whistle. The seas were raging. Our ship was cast hither and thither on roaring waves. Gotland vanished. The companions became seasick. The rigging began to give way. Despair filled our hearts and we commended ourselves into the hands of God.

It is difficult to do justice to the sheer variety of anecdotes and observations in Linnaeus's travel writing. On Öland and Gotland, Linnaeus liked listening to the crickets but doesn't record many conversations; people in Visby are said to be "good company." The group of travelers studied two hedgehogs, investigated their feeding habits, and apparently brought one of them along on a trip to Karl's Islands, where it "ate with rare appetite the many grasshoppers hopping in the meadows, also the May bugs that flew as dusk fell, and furthermore the many small snails with shells that clung to the grasses." Later, the hedgehog was picked as the iconic animal in Gotland's landscape. Reading the Gotland account, one might wonder at the absence of any mention of the medieval churches and the culture linked to them. The optimistic tone is striking but was probably characteristic of the whole expedition. Speaking about the hayseed used on Gotland, it was identified as a type of grass that suited the cattle and thrived even on meager, sandy slopes. "To my mind this observation is so great that I believe it to be sufficient to pay for all the cost of my journey to Gotland."

Linnaeus showed a persistent interest in what he called old wives' tales. They traveled to the "Blue Maiden" island (Blå Jungfrun) in the strait between Kalmar and Öland, an evil, magical place in folk belief—a place that "old wives' and fairy tales have dedicated to Pluto but not to Neptune. They say commonly that all witches must travel here (truly a quite troublesome journey) each Maundy Thursday." Linnaeus realized he was recording superstitions and said it was "more to amuse my readers than of any true utility." On the other hand, "the farmers' botany should not always be discounted, as the farmer has, at least in this countryside, his own names for almost all herbs. I took with me a decent farming man to walk in the meadows and there he knew many more herbs by name than I had predicted and his naming had mostly fine origins." Next, he included the names of some twenty-odd plants.

After listing examples of "tittle-tattle" at length—meaning all kinds of chatter, gossip, and rumor—he recommended the study of physics and natural history to all who planned to join the clergy. He reflected on the persistence of myth: "Nevertheless, it is curious to find with these and many such superstitions, how they have been preserved by the nation from the ancients long into heathen days. Some you receive again from the poets, soon before or after the time of Christ; still others have endured since pagan Sweden; some from Papist rule; some have been found artfully to explain what is presently practiced. I would declare it to be a fine argument that an investigator should assemble a sizable collection and attempt to show the origin of each and every tale." In other words: myths and beliefs from classical, pagan, and Catholic times should be systematically studied for their origins and authenticity.

"Botany for females" was a much more wide-ranging subject than one might think. During his journeys, Linnaeus took a serious interest in all female activities.[7] While traveling in Lapland, a local woman taught him medicinal uses of the poisonous monkshood (*Aconitum*). In Västergötland, he stopped by after the church service in Åmål to learn from "the old housemaids and other old women, who here owned more knowledge of herbs and household remedies than found in most other places." It must surely have been through his contacts in the Academy of Sciences that he heard of the noblewoman Eva Ekeblad de la Gardie, who described to the academy how to make flour and alcohol from potatoes, findings that led to her election as a member. The County Upland estate owner Charlotte Frölich wrote under the pseudonym Lätta Triewen on matters such as flax fiber harvesting and improved sowing machines—but "she wanted her name not to be disclosed so as not to enter common talk."

Västgöta resa (1746) was dedicated to a woman—the queen, Lovisa Ulrika, who at the time had given birth to a son, the future King Gustav III: "Such a Blessing for a Nation! Such a gain for the arts of learning!" The foreword introduced the report: "What during these journeys has been remarked with regard to antiquities, physics, economics, manufacturing, medicine, customs, and practices, I am now honored to humbly offer to my dear Homeland."

Plenty of remarkable findings do deserve attention; take Kinnekulle "inside which nature has demonstrated to us the anatomy of the Earth's crust.... I have investigated how, without ringing, pigs can be restrained from causing damage by rooting.... I have to be sure at times to interpose some jocular and harmless enterprises more to amuse my readers than of any true utility." However, it had been an uncomfortable journey "through daily and steady rain.... Traveling might have been still less bearable had not the learned philosophy student Eric Gustaf Liedbeck at his own expense offered me his pleasant company and also eased my labors by keeping a protocol and making clean copies thereof." Generally, then, a useful move. As the foreword assures us: "I have made numerous findings in natural history to intrigue and improve my learned readers; all remarks, each taken singly, I consider in my mind to be worthy and also each and every one would richly repay my efforts and *publici* cost"—a line of thought we have already come across.

The mass of information could easily become chaotic, but readability had been achieved by the chronological order, clear typography, good indexes, and, above all, a linguistically lucid text that frames the mixture of major and minor entries. The main stages in Västergötland took the group to Alingsås, Göteborg, and the coast of County Bohus; others included the "Shäfer" school, where sheep breeding was a main subject, and the Alströmer-driven textile manufacturing, notably in Alingsås. They came across a pipe factory that employed eighty workers, and went on to study the local geology painstakingly. "Lithogenesis (genesis of stones) may be a simple matter but much is still dark even after such few findings as were made at this time." After examining the different ground strata, Linnaeus was amazed at the Almighty's construction of our planet and concluded: "Thus, the stones speak when all other things remain silent." Socially, he noted that the problem of begging had been solved locally by employment in the clay pipe factory.

Knowledgeable readers might be surprised that Linnaeus visited Billingen without going to watch as the cranes danced on the shores of Lake Hornborga. For one thing, he arrived too late in the summer and, besides, there was no large-scale dancing until the water level in the lake had been reduced by nineteenth-century drainage schemes. However, he did see two places that matter in natural history: the inland hill landscape of Kinnekulle and the small city of Marstrand clustered around a fortress on the County Bohus coast. At Kinnekulle, his interest focused on the passage of time and theories of "decreasing ocean waters." In Marstrand, he marveled: "When I arrived at the fortress's side it was as if having been led into a new world. . . . Plentiful *naturalia* was offered up by the western seas, many so rare, indeed unheard of . . . that we by their sight were gripped with the greatest wonder. . . . We botanized on the bottom of the sea, as if in a new Sweden." Some fourteen pages followed, given over to variants of seaweed and similar observations—what the parliamentary commissioners made of all that detail is not known.

On 1 August, the group visited "the wise farm laborer," a man known as "Sven in Bragnum," whose collection of jars and boxes of medicaments filled his dark lodgings. He was in his early thirties, but his hair was already turning gray, and he had developed a prominent belly, presumably because his life was so sedentary. "My companions could not refrain from pretending diseases and consulted him accordingly. Liedbeck sought advice for his buzzing ears [tinnitus] and, among other things, said that the buzzing had moved from one ear to the other and hence asked if there were any communication between the ears. The man replied: "Yes, by three holes / yes indeed, by our lord's hard wood that is his cross / may you believe this." Magister Tengmark wanted to speak privately with Sven about an attack of venereal disease and, then, if hard liquor was permitted before the recommended medicine, as he was used to three shots a day. Not to worry, Sven said—three shots was neither here nor there. Linnaeus observed how the healer's ordinary patients were treated and found it admirable that he made no use of incantations and was "amenable about payment." However, he "lacked insight into anatomy, physiology, semiotics, dietetics, and botanics."

All in all, the visit seems to have been a rather unkind piece of social theater. The group showed greater respect for the wise woman "Ingeborg in Mjärhult" when they visited her on 7 August. Generally, these inspections were, of course, intended to bolster academic medicine and so control any competition to proper physicians from the so-called *empirici* and also from surgeons, who were thought highly suspect.

Linnaeus, dispatched as the envoy of Rational Practicality, also inspected schools, pharmacies, and hospitals. He still found time to describe many natural phenomena, and that despite the variable weather. He recorded observations on the plant preferences of cattle, a useful preparation for his dissertation *Pan Svecicus*. That the fact-finding mission paid so much attention to the bustling towns of Alingsås and Göteborg partly reflects the mercantilism of its taskmasters but is surely also a sign of Linnaeus's own ambivalence. When he wrote in the foreword about "jocular and harmless enterprises" to amuse his readers, it could be understood as expressing literary ambitions. Another example would be his sketch of the skua they called Elof: "This veritable Cossack who pursued others such as gulls until they must vomit. Their trait of easily throwing up has been given to them by the Creator so as they will sustain our Elof and his like."

Or it could be that Linnaeus, when promising a light touch, was actually returning to the tone of his famous Hamlet-style monologue when in the cemetery in Frändefors on 24 July. This is just a taste of that speech: "Should I remove some soil from the graves, then I would be taking away such elements which once constituted a human being and had been turned human by another human being; should I then take that soil to my cabbage patch and set in it cabbage plants, I would get heads of cabbage instead of human heads, but if I boiled these plant heads and served them to people, the elements will once more become part of human heads or indeed other components of human bodies; thus it is common for us to eat our dead and it suits us well."

It is interesting to compare the correspondences between the printed versions and the actual records of the three journeys, as dictated by Linnaeus in the evening to whoever was on protocol duty.[8] The original field notes have apparently not been saved, and the tidied-up versions were later edited before printing. The changes were not only corrections of language but sometimes also affected more sensitive matters. During a visit to the fortress in Marstrand, the condition of the garrison was described as poor. The men went hungry: "I have visited many nations in this world but never before seen serving soldiers afflicted by such misery." This judgment has been softened in the official text. In the entry on 8 July, he contemplated a night scene in words more effective than the original protocol: "Then night came and brought its darkness, the sky so overcast that on him no stars were seen though they were seen instead in their 100s where they lay scattered on either side of the road among the grasses; I refer to the great abundance of glowworms, under whose tail are 3 shining rings, although the last ring is but in 2 parts. All were females aglow with love so that their males were tempted to borrow their fire that shone but did not burn." In the printed text, the glowworms had to be given a longer introductory write-up. On the other hand, a brutal passage about how to catch a wolf had been deleted: once a female wolf is killed, the vulva should be cut out and used to smear the ground around the poisoned cadaver "as all male wolves will follow the trail to the poisoned meat, eat it, and die." Linnaeus asked himself: "Would perhaps she-wolves in a similar manner be tempted *per penem* from a wolf?"

The journey to Skåne was up for debate in a split parliament. On 1 April 1747, a proposal was presented, which was backed by the senior court official Carl Hårleman. Three estates assented, but the merchants did not, by twenty votes to eighteen, as the cost seemed too high. Still, it was accepted overall, but by then, Linnaeus had declined. He felt worn out and too exhausted for this new undertaking, even though, as he explained to the Academy of Sciences, he was eager to see Skåne "with attentive eyes" as it was the county "to prefer" above all others in Sweden. "But as I consider the long road and the trouble it heralds, I feel giddy. I will wear myself out before my time, with no one, least of all my children, among all others in Sweden to thank me for it . . . 400 thalers wasted on a carriage and for clothing another 400 thrs, in addition I would neglect seminars worth 800 ditto and a couple of *praesidia* [examinations?]." Besides, he would need an amanuensis. Because he had been appointed archiater—royal physician—Linnaeus had to pay more in taxes; also, his children were growing and needed a tutor.

He wanted to "live *commode*" in his study, he declared, and would be unable to travel the following year. However, he also displayed his negotiating skills. To extract the maximum income from every situation, he explained "he would not accept a professor's tariff of 3 thalers in silver but

FIGURE 23. Map showing Linnaeus's route through County Skåne drawn by Carl Bergqvist. From *Skånska resa*, 1751. Uppsala University Library.

wanted an archiater's fee of 4 thrs and thereto 4 horses." Once granted this, he agreed to travel as funding for an amanuensis was also part of the package. He would be "needed in the evenings to help him *tabulae* swiftly such as had been during the day recorded of findings and work conducted"— that is, to keep the travel diary.[9]

His Skåne journal was dedicated to the cabinet secretary Clas Ekeblad, owner of the manor house Stola in Västergötland—as Ekeblad rightfully should have been honored with a dedication in the *Västgöta resa*. Linnaeus's commission was issued on behalf of the king. It specified four sets of required records: (1) plantings and current numbers of walnut trees; (2) plantings of the Swedish variant of whitebeam, and where it could be introduced most easily; (3) any sources of gypsum—Linnaeus had so far searched for it in vain; and (4) flints, to avoid French imports. This brief list of requirements was to be set out in the travel report. This led to a new request from Linnaeus; as an archiater, he should be given a carriage and all expenses. It was granted.

Skåne struck him as a showplace for Sweden at its finest. Was there anywhere a natural scene more beautiful than this undulating landscape? South of the old town of Dalby, the land of Canaan spread out before him. This was how Sweden should be: cultivated fields flowing with produce and wealth. Unsurprisingly, the main intention behind the journey was to study agricultural practice and consider any improvements that might be useful, even in Skåne. Thus, the most sterile of soils would yield more if scattered with dried, ground-up clay found in the northeast corner, at Åsum, because the clay was rich in chalk. Once more, Linnaeus notes with satisfaction that "by this find I do not doubt this our journey would recover for *publicum* all its cost." This was, for unknown reasons, removed from the printed text. Other notes include the observation that pigs left to feed for themselves in beechwoods grew "so fat as be barely recognizable." In Lund, he visited the new botanical garden, expanded thanks to the work of the marvelous Hårleman. Also, no less than forty-five walnut trees were growing in Lund—a useful tree, the wood was the best for rifle stocks.

The *Skånska resa* grew into the largest of the Linnaean travelogues. It is also an account of the local nobility with their manor houses, castles, and tree-lined access roads. Linnaeus was traveling in a carriage, as a gentleman. He told of an odd, possibly symbolic episode, when he and Erik Oxenstierna were visiting an old friend, Esbjörn Reuterholm, and they were discussing dowsing rods and how they could be used to find minerals, not just water. Olof Söderberg, Linnaeus's secretary, thought it reasonable but his boss didn't. To prove his point, Linnaeus hid his purse, and the mineral diviners found nothing despite much searching. Much merry laughter, until Linnaeus went to find his buried thalers. Where was the purse? Foiled, he allowed Oxenstierna and Söderberg to use the rods in a different place. Yes! The rods twitched where his purse was buried. He didn't know what to believe.

The frogs were making themselves heard in Skåne: "In the marshes the frogs were playing, with incessant cackles . . . as if from geese or ducks. Thus, they must be of specific species and, from their language, I judged there to be twice removed from ours. . . . The bullfrogs were sounding off at Raflunda, as if bells in a distance calling to a gathering, which was quite pleasing to listen to."

The unwilling sailor Linnaeus was in luck when about to set out for Ven, an island off the northwest coast: "The strong storm prevented me today from sailing." He reflected on the passing of time as he stood on the promontory with the old fortress Kärnan, overlooking the Öresund strait: "It makes me giddy to compare the mountain here in Hälsingborg with its sand strata while I look back so far into past endless eras during which the sand flying in the wind could have built such large and tall objects."

An attempt to form a coherent view of Linnaeus the traveler has to take his many roles into account. Officially, he was *historicus* of nature and *oeconomus*, but at other times he operated as an ethnologist, a folklorist, and an ethnobiologist, always keen to chart old customs. He has had several Swedish successors (e.g., G. O. Hyltén-Cavallius, Sigfrid Svensson, K. Rob Wikman, and Bengt af Klintberg) and has indeed been called "the father of Swedish ethnology." However, he could be a contrarian in his contempt for folk superstitions and his ambition to offer "enlightenment," aspects of his thinking that influenced, for instance, his 1759 speech in the presence of the royal couple.

Yet again, much attention has been drawn to his beliefs in supernatural phenomena such as the "evil eye" and the *ignis fatuus*—the marshland lights known as will-o'-the-wisp, also believed to be souls of the dead. Linnaeus lingered over the supernatural in his *Nemesis divina* and speculated on topics such as doubles, premonitions, incantations, and remnants of papist dominance. A systematic study might well show Linnaeus to be one of the foremost recordkeepers of Swedish folklore of his time.

Prior to publication, all texts had to be submitted to a censor for scrutiny. Censorship was the responsibility of several authorities, including the diocesan administration, the decanal office of the university, the *collegium medicum*, and the *censor librorum*. Whoever it was, the censor had to be paid, whether a civil servant or not. This is an excerpt from the records kept by the *censor librorum*, Gustaf Benzelstierna: "Journey to Öland 1741, as of 15 May. Three sheets of paper under this *rubrique* were made available to me fr. Herr Kiesewetter together with a narrative said to have been composed by Prof. Linnaeus. As this was most carelessly written, in many places corrected and not dissimilar to a maculated text, I returned it with the suggestion that I could not be expected to revise such a hasty composition."[10] It had not been passed. Meanwhile, the Swedish language had entered into a process of standardization so that the censor increasingly functioned as a copyeditor. The total lack of interest in the quality of the writing is glaring, as is the length of time the text was left to gather dust: four years went by before this classic work was allowed into the hands of readers!

It was probably the action of the censor—also, perhaps, fending off a rival in the marketplace—that caused Linnaeus to "with mouth, hands, and all limbs" discourage Pehr Kalm from publishing his account of his 1742 journey in Bohuslän. It became available in print first in 1747. Kalm stated that "Professor Linnaeus tells me that, as for his newly printed journeys (öland/gotl) and he swears this by the Devil himself, he fears that come the next sitting of Parliament, once they have read the journals he will be held responsible and they will demand to know what he has done with the monies for traveling in 1742 as he has not presented them with more useful a work than these travel notes, in their eyes, dry and tasteless accounts of daily tasks."[11] Linnaeus would claim that his biggest mistake ever, one he would regret for the rest of his life, was to have the travel diaries from Gotland and Öland released in print. Still, he did it, and we happily forgive him.

"Herr Kiesewetter" was Linnaeus's publisher Gottfried Kiesewetter. While working on his later journeys, Linnaeus left him for Lars Salvius. Without the support of Salvius, the great travel diaries *Västgöta resa* and *Skånska resa* might well have been shelved, just like *Iter Lapponicum*. Mindful of past errors, Linnaeus notes that he had "a proper copy made of *Wästgöta resa*, and intend[ed] to send it with all dispatch to Secret Salvius, who would have it printed."[12] Half a year later, Kalm wrote that he "has now made a Register for Archiater Linnaei *Wästgöta resa*."[13] In 1768 Linnaeus seems to have been at work on a new edition of it: "Finally, I have now concluded *Westgötske resa*, a labor that has tired me greatly, as it has made me look up all names and 3 times read it through," he wrote to Lars Salvius on 8 November 1768. But his plan misfired.

At one stage, even *Skånska resa* got into trouble. Carl Hårleman objected to clearing land by slash-and-burn because it harmed timber production. His arguments impressed Linnaeus, who decided to delete his mention of land clearing and replaced it with a rather pointless passage about manure. Linnaeus instinctively shied away from polemics. To avoid quarrels, he wrote in Latin rather than Swedish: "I always dreaded fiery pamphlets as the horse the drover's rattle and find that the bluntest razors serve to remove the most."[14]

These comments make you wonder what would have happened if his classical works had never been published. Linnaeus himself did not list the Lapland journal among the volumes he intended to have published (in fact, it was first printed in 1811, in English translation). *Dala resa*, his account of traveling in County Dalarna, was also excluded, although it had been advertised in Holland. It only became available in book form in 1889. Of course, *Gotländska resa* was shelved because the censor had rejected it. Linnaeus was not keen to publish *Västgöta resa*, and he had not even wanted to undertake the journey through Skåne. If we hadn't been able to read these five volumes, our insight into eighteenth-century Sweden would have lacked perspective as well as detail and would have come across as gray and monotonous.

The travelogues were marketed on an upbeat note: "In our public sphere there appears already to be a yearning for many similar travel descriptions from locations in the wide expanses of our native land and by such means also to make new discoveries from all the 3 realms of Nature, through which this splendid Science could be excellently advanced so that many who in their current walks through life have remained ignorant of the best healing methods could learn already in childhood to revere the wisdom and never-ending goodness of the Creator." This is Linnaeus himself in full flow, presenting the benefits of the Västergötland journey. "Thus, it will be observed that our Native Country has been boundlessly provided with miraculous and useful things, were they only to be discovered. Would that we had such travel journals composed for every province in the entire Nation, certain to bring so far incredible advantages."

His *Skånska resa* ends on another note, also worth quoting: "Auctor has in this volume described a county which in regard to its natural advantages is among the very finest in our entire Nation. . . . . The exemplary diligence of the Auctor must be singled out for our gratitude as are his incomparable gifts which have enabled him to detect the most hidden workings and so apply them with gainful and advantageous effect to public life."[15]

Linnaeus is a camera that misses nothing, a one-man embodiment of the Official Records Office. In an ideal world, his reports would be correct in every detail; naturally enough, they are not. A contemporary critic Carl Hallenborg, a Skåne estate owner and a magistrate, noted 316 errors in *Skånska resa*—many minor ones that might have been due to Linnaeus not understanding the local dialect. It would explain spelling or naming mistakes, and possibly the misunderstood expressions—or were some deliberate because he disagreed with the person he had been talking to? An example: How long does a pair of wooden clogs last? Linnaeus's estimate was two years. Hallenborg begs to differ: "A man who wears them daily usually finds they have a life of just 6 or 7 months."[16]

The traveler Linnaeus examined the landscapes—Linnaeus-scapes—one after the other, displaying to the reader images of a nation of multiple opportunities. "Were all the provinces of Sweden to be studied in this manner, I envisage that one day we would surely see a Swedish Pliny be born, who could find excellence greater than the Romans in both remarkable and useful matters, so that our Nation will be able to boast of many hundred times more fine things than the fancy goods from foreign industry. . . . As for my task, it is sufficient that I have used the jack plane so that others after me may apply the smoothing plane with greater care to all."[17]

He had traveled widely and covered every Swedish county apart from Blekinge, Halland, and Jämtland; he even hurried through Finland at the end of the Lapland journey. On the road, he recorded the mosaic of cultivated fields and forest cover, the trees sometimes reaching the front doors of the cottages. He made notes of farming and forestry management, and the effects on plants of natural features, the calendar, and latitude. Pride of his country runs through his writing like the colors of the flag. He had encountered the poet Hedvig Charlotta Nordenflycht at Fullerö Castle, and she composed a poem celebrating his journey through Västergötland. It begins with the lines: "Now our Nation proudly boasts / Enlightenment and men of wit." Fredrik Hasselquist, Linnaeus's pupil, wrote about his vision of Sweden's natural beauty as exceeding India's: "With Linnaeus as a warden / our Sala soon excels Ganges's flower-garden."

What happened in the interval between his first and last journey—between 1732 and 1749? Linnaeus had moved on from the mythical landscape scenes he portrayed in *Lappland resa* to writing up useful records of local economic conditions, a change initiated in *Dalarna resa*; he, too, changed from the hearty young journeyman to the gentleman traveling by carriage. The emphasis on science and on personal comfort suited the mature man but also went with the spirit of the times. He journeyed in the summer, an obvious point; what he offered his readers was a sunny, flowering Sweden where Pan is playing his pipes and the Age of Liberty seems freer and happier than it actually was. For us Nordic folk, his narrative of "Summer Sweden" is a classic. But, of course, it rained on Linnaeus and his companions, too.

So, what has happened since, in the interval between his travels and our own time? His journals help us identify the changes. How many of the plant species recorded in 1749 have disappeared by now? How many new ones have been added? In what ways is land cultivated differently? One example: in Skåne, Linnaeus's 827 observations are replicated in 725 instances—that is, 88 percent can still be identified.[18] There are many questions about the extent to which landscapes have been restructured by agriculture, forestry, roadbuilding, waste disposal, and every kind of commercial activity. During the last half century, nature has been increasingly protected, so could it lead to a recovery of a Linnean Sweden?

Finally, a quotation from a hopeful disciple of Linnaeus's, who actually dedicated his verse to the favorite child of contemporary ambition, the entrepreneurial town Alingsås. The poet had more general things on his mind, though, and predicted: "Let it bring shame that once brought glory/ to cause a people's death in war / and praise what makes our nation increase/ well fed and well at peace."[19]

It is worth saying again: Linnaeus and his generation were ready to carry out a great mission to improve society. Sweden needed restoration, not rearming; its people needed feeding, not decimation. The Age of Liberty initiated a national journey. The next step out into the wider world was being planned; its consequences would be both tragic and adventurous.

CHAPTER THIRTEEN

A Language in Which Everything Matters

LINNAEUS ADMITTED IT HIMSELF: "It was true that Linnaei time did not allow him to cultivate his language; but thereto it is noteworthy that his *genie* was not in the slightest suited to languages, as he never learned English or French or German or the Lapp tongue, indeed not even Dutch, even though he had for 3 full years lived in Holland; nevertheless, he had in all situations done well enough."[1] Linnaeus was very much aware of his failings; he had apparently asked the professor of classics, Petrus Ekerman, to correct his inauguration speech.[2] As for his *Hortus Cliffortianus*, Linnaeus later called its style barbaric, coarse, and ill-tutored. He made similar self-critical apologies on several occasions, and most in a position to judge agreed that his Latin was a bit rough.[3] Add the misprints that he might have corrected if he had allowed himself time to proofread.

Even so, Linnaeus argued strongly in favor of Latin as the best language for science. In 1767 he pointed out to the university court that almost all scientific works, including the most important ones, were written in Latin, which made it the *lingua communis eruditorum*. It followed that someone unable to read and understand Latin couldn't be accepted as "a solid man of science" and, therefore, prospective students of scientific subjects must begin by studying the humanities. And, if so, dissertations must be presented and examined in Latin, for "should the Latin language be left out at *specimina academica*, a quack could push his way forward to *honores academica*."[4]

In the past, Linnaeus has been accused of writing "suburban Latin"—spoken of as "Svartbäckslatin" after a working-class area on the outskirts of Uppsala. He often broke with classical Latin usage; nowadays, the

[176]

emphasis is on the functionality of his language. People like Bäck, Rousseau, and Sauvages have made positive comments on his style. Linnaeus was well schooled in the works by classical poets and often quoted them, with a preference for Ovid. As a public speaker and writer, he sounded like a classicist, while in his letters, his language is half Roman, half Gothic. At one time, the letters exchanged between him and Bäck were written entirely in Latin, perhaps because they had agreed to keep the private contents secret.

He usually mixed Latin and Swedish freely, as we have already seen. The work of the Academy of Sciences raised the status of Swedish. Linnaeus's speech on "Curious Features of Insects" (1739) flew the flag for the academy's practical and patriotic stance, and the text of his travel journals also supported the new ethos. Generally, he wrote a great deal in Swedish; his style was studied and praised by Oscar Levertin, who recognized three distinct variants in hymns, fables, and travelogues. Thore M. Fries didn't agree: "From a stylistic point of view, it is, of course, impossible to rate his travel writing particularly highly." However, Henrik Schück, a professor of literary history, gave Linnaeus a high grade: "Since the days of Olavus Petrus, no one had written Swedish as it was spoken.... Linnaeus's style is vigorous, straightforward, and lucid, influenced by the Bible of Charles XII."[5] Linnaeus set himself standards for how to write: "In a simple manner, in short words that make the meaning clear." One commentator summarized his judgment: "[a language] in which everything matters."[6] Or, one might call it "anti-rhetorical."

Language can open doors, or shut them. Which language would Linnaeus choose for his 1759 speech in front of the royal family? King Adolf Fredrik and Queen Lovisa Ulrika were both Germans, and the queen spoke good French. But neither of them understood Latin, the usual language of choice for academics. Linnaeus decided on Swedish and delivered an enjoyable and splendidly rhetorical presentation. He had to ask for help with official correspondence in English or French; Johan Henrik Lidén undertook to pen a thank-you letter in French to Louis XV, who had sent the great botanist a collection of 130 different seeds and plants, gathered by the king's own hands in the gardens of Versailles.[7]

However, it is still a surprise to find that Linnaeus, who held a professorial chair in a Latin-dominated university, would write so much in Swedish—speeches, popular publications, lectures, and travel books. It also seems paradoxical that a man who comes across as lacking in sensitivity to poetry and art wrote reports from his journeys and on other subjects with the visual appreciation of an artist. Dalin and Holberg were the only contemporary Nordic writers he approved of. His everyday speech was inflected with Småland dialect, and "smålandisms" crept in here and there in his writing. His punctuation had its oddities, such as a preference for semicolons and an aversion to question marks—no matter, most of his questions were rhetorical anyway. Bäck, who had had many letters from Linnaeus, summarized his correspondent's writing in his speech *in memoriam*: "The rare and delightful manner of his letters in Swedish to supporters and friends cannot be imitated." Indeed, Linnaeus expressed a wish in a letter that the university would be "gentle and attentive to publishing in lucid and neat Swedish" and defended the native language as a cultural asset."[8]

He said in his foreword to *Öländska & gotländska resa*: "A language ornaments science as do clothes ornament the body, and he who cannot do honor to his clothes must allow them to honor him." The foreword to *Flora Lapponica* (1737) is full of metaphors involving clothes, while the foreword to *Västgöta resa* focuses on practical utility: "Our country wants for a Swedish Pliny, whose work would greatly exceed the Roman in curiosities and useful resources"—curiosities and useful resources summarizes perfectly the aims of his travels.

In his letters, the superlatives come thick and fast; he often feels giddy, and even more so, he "longs." "The blood was boiling in my veins and, on my head, each hair stood on end." He becomes ecstatic at the sight of the tea bush: "Never have I been as animated by any letter as by . . . live Tea Trees! Can this be possible? Is it truly a Tea Tree! I swear that if it is, I shall make the Herr Captain's name more everlasting than Alexander Magni." To Erik Laxman in Siberia, 1764: "I would fall readily to my knees, yes, kiss your feet, Herr Magister, should I with my entreaties have convinced Hr Magister to collect some seeds or insects for me."[9] Immediate and sensuous, he swoons and delights and gazes intently, he is curious and kneels to nature, wanting to understand it as well as use it. The curiosity is not just signifying his love of nature but his lucid, agile way with language.

The adverb "next" is repeated forty-odd times, again and again, in his lecture on "Curious Features among Insects" (1739)—it was, of course, a keyword for him because Linnaeus loved the narrative form as a version of the Baroque rhetorical concept of *amplificatio*. He was an orderly hoarder and one of the most devoted list-makers of all time, a squirrel-man—a *Homo collector*.

Linnaeus's use of verbal images, often religious, flourishes especially in his letters, which he wrote at speed. This comes from a 1763 letter to his siblings: "I have sent my disciples to every part of the world."[10] Classical motifs occur frequently, notably in *Iter Lapponicum*. As for his *Nemesis divina*, it could be described as a collection of moral fables on modern or historical topics. One example is the story of the seven wise men in *Cui bono*, with its thematic references to Holberg. He amuses himself with images of plants and animals: "What does the absurd Klein ramble on about? A debate with him is like racing a herd of calves."

He enjoys military images, often turned on their heads, as in "Flora calling up her beautiful army." The images make no reference to real fighting. Erotic images occur, and different traditions mingle.

We should allow ourselves a sample from Linnaeus, the poet:

> O mighty Venus / how great is not the power you wield over all living things on Earth / the most noble of mankind are still not set free from your rule / not David the righteous, Salomon the wise, Simson the strong, not even the Alexander of the great empire, none are free from your fetters / . . . / And how sternly do you not rule the wild animals / the cruelest carnivores, lions and tigers, you render them all as meek as calves / the strongest beasts, elephants and horses, you make crazed and dizzy / . . . / The timid fishes, hiding in the depths of the endless seas, you toss like sand onto the beaches / and the small worms frozen to death in the winter you reawaken in spring / yes, the very worms intended for the angler's hook you command each morning at dawn to rise from their dark underground caverns / indeed all that has life must obey you.[11]

It would indeed seem that Linnaeus's idea was that all must submit to the power of Venus, but some scholars urge caution. The poet Linnaeus tended to slip into the language of the proverb and the farmer's almanac. Of course, students of nature recognize the importance of experimentation in how they express themselves. August Strindberg, on several occasions, did speak of Linnaeus as the organizing spirit who ordered chaos and also infused the dull herbaria with vitality. He described Linnaeus as "a great poet who devoted himself to the natural sciences" in his collection of essays *Blomstermålningar och djurstycken* (Paintings of flowers and images of animals, 1888). In Martin Lamm's influential work *Upplysningstidens romantik 1* (Romance and the Enlightenment, vol. 1, 1918), Linnaeus's writings about nature are discussed in terms of their vividly "visual" style. Lamm contrasts the literary approaches to nature during different periods: freshness and joy in the Renaissance, darker tonalities during the Baroque. He identified something of the Rococo in Linnaeus's portraits of animals but also quoted approvingly from Henrik Schück:

Linnaeus was "a healthy, hearty, and cheerful child of nature." Also, his experience of nature was "primary"—that is, not in any sense literary: "In other words, that Linnaeus was epoque-making in his influence on Sweden's history of descriptive nature writing was not due to the modernity of his taste but his realistic, tangible understanding of nature."[12]

There are good reasons for objecting to Schück's judgments, as closer scrutiny has shown Linneaus to be a skilled rhetorician in control of a variety of stylistic devices—for instance, presenting himself as a plain, artless observer. Although his choice of words can be almost archaic at times, his syntax is relaxed and engaging. Lamm quoted the accomplished final passage in *Västgöta resa*—it demonstrates that Linnaeus was able to capture other seasons than the summer:

> All along the road from Fällingsbro, autumn constantly showed itself to our gaze. The forest still stood green before us but now appeared sterner than during the summer; rough land and meadow lay as green but without flowers now that all had been overcome, earlier by the grazing cows and later by the scythe. All over the fields yellow sheaves had been raised. Left once the straws had gone, the yellowing stubble was bestrewn with green weeds. The ditches stood high with water at the end of the wet summer and made yellow by the abundant burr marigolds. The road verges were covered with redshank, *Persicaria acri* [now: *P. maculosa*] whose spikes had begun to blush and droop. Farm laborers were everywhere hard at work. Some farmers were cutting down the grain with their scythes, while their womenfolk, quite white about their heads and arms, were binding it together and others carried it home in wagons, some were threshing, some crushing the clods of earth with wooden harrows, then the soil was smoothed with iron harrows and some sowed the winter rye, some raked the seeds into the soil, and some planed the field with rollers, while herds of children were singing and blowing their horns for the cattle grazing on the cut meadows, until the afternoon's gloomy winds began to blow and the bright sun to sink below the horizon as we arrived into the Garden in Uppsala.
>
> The land is alive, the field cultivated but finally at home.[13]

The young Linnaeus? Unknown artist, oil painting in the art collection of the Småland Museum, Växjö. See page 24.

The frontispiece of *Hortus Cliffortianus*, 1737, drawn by Jan Wandelaar. The intricate symbolism includes the revelation of Flora by Apollo/Linnaeus, standing on the dragon/hydra of idolatry in Hamburg—a forgery that Linnaeus had exposed. The central panel is flanked by the garden's fruit-bearing banana tree, and, to the left, by representatives of different parts of the world, all of which had supplied the garden with exotic specimens. In the lower right-hand corner, little putti point to the 100-degree thermometer—indicating that Linnaeus invented it. Uppsala University Library. See page 105.

Tea bush by Johan August Pfeiffer (1777–1842). Published in *Magasin för blomsterälskare och idkare av trädgårdsskötsel* (Magazine for lovers of flowers and practitioners of gardening). University Library, Lund. See page 144.

Linnaea borealis, introductory illustration in Johan Palmstruch's (1770–1811) *Svensk botanik* (A Swedish Botany), 1810. See page 14.

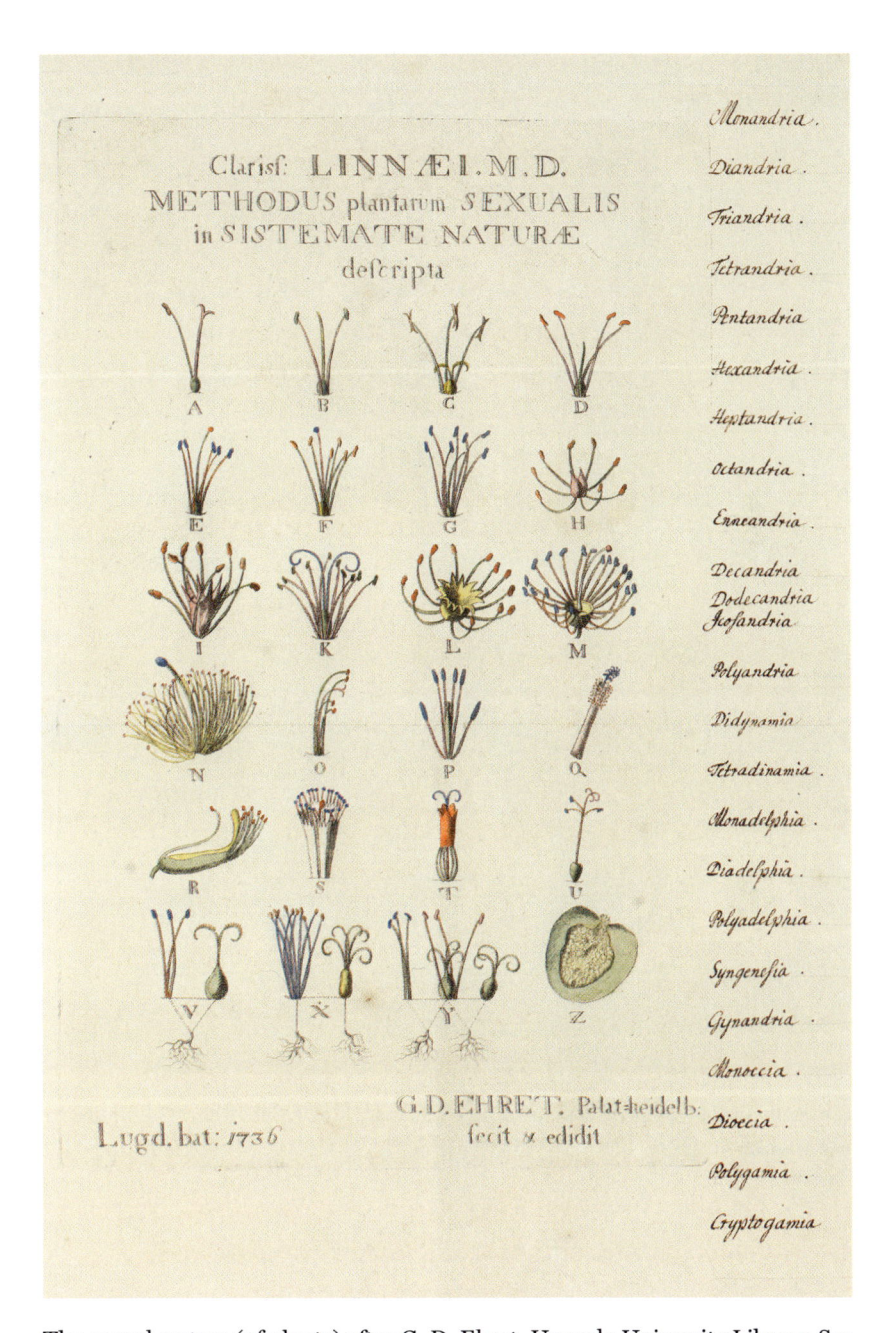

The sexual system (of plants) after G. D. Ehret. Uppsala University Library. See page 122.

Linnaeus's wedding portrait. Painting by Johan Henrik Scheffel, 1789. Photo by Michael Wallerstedt. The art collection of Uppsala University. See page 138.

Sara Lisa's wedding portrait. Painting by Johan Henrik Scheffel, 1789. Photo by Michael Wallerstedt. The art collection of Uppsala University. See page 139.

Bust of Linnaeus, hailed by Greco-Roman gods and goddesses. On the left: Asclepius with his serpent-entwined staff. Then, the winged Cupid, Flora, goddess of flowers, and Ceres, goddess of harvest. From Robert John Thornton, *The Temple of Flora*, 1806. Uppsala University Library. See page 10.

Uppsala city: Appearance from the south side. From *Draft of a description of Uppsala* (1773) by Johan Benedict Busser. Uppsala University Library. See page 150.

Prospectus for the Botanical Garden in Uppsala. From the dissertation *Hortus Upsaliensis*. Copper engraving by Johan Gustav Hallman (1726–1797). Uppsala University Library. See page 195.

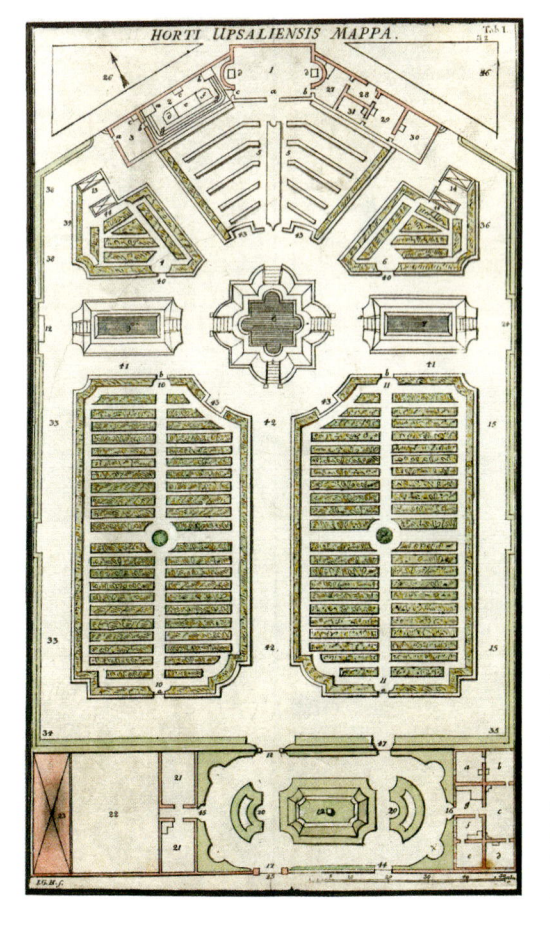

Map of the Botanical Garden in Uppsala. From the dissertation *Hortus Upsaliensis*. Copper engraving by Johan Gustav Hallman (1726–1797). Uppsala University Library. See page 196.

Wild strawberries from *Svensk botanik* [A Swedish Botany], 1810. Uppsala University Library. See page 302.

Linnaeus had his bedroom and study at Hammarby wallpapered with illustrations from the best botanical works of his time, among others by Ehret and Plumier. Photo by Mikael Gustafsson, TT News Agency. See page 321.

Portraits of three of Linnaeus's daughters hung on the walls of Linnaeus's Hammarby. Photo Mikael Gustafsson, TT New Agency. See page 322.

The wild-strawberry girl, Ulrika Charlotta Armfelt. Painting by Nils Schillmark (1745–1804). National Gallery, Helsinki. See page 360.

Linnaeus, aged 67. Painting by Per Krafft Jr., 1774. The art collection of Uppsala University. See page 378.

Carl von Linné Jr. Painting by Jonas Forsslund. Photo by Mikael Wallerstedt. The art collection of Uppsala University. See page 395.

Tahitian Omai, Sir Joseph Banks, Daniel Solander. Oil painting by W.J.S. Brown after William Perry. Royal Academy of Sciences. See page 397.

CHAPTER FOURTEEN

Flora et Fauna Svecica

THE TWO VOLUMES *Flora Svecica* (1745) and *Fauna Svecica* (1746) were among the outcomes of Linnaeus's provincial travels (extended editions were published in 1755 and 1761). There was, however, no corresponding print version of *Pluto Svecicus*—that is, an inventory of rocks and minerals. Local floras were already in existence: Olof Bromelius's *Chloris gothica*, Johan Linder's *Flora Wiksbergensis*, and Olof Celsius's *Flora uplandica*.

But Pehr Kalm testified to the lack of a proper flora suited to the needs of the studious young and urged that it should be printed "the sooner the better." Linnaeus agreed and decided to send his manuscript to Lars Salvius, printer to the Academy of Sciences. Kalm explained: "I then commenced to demonstrate to him how his hand was such that they could not be expected to find order in what he had written and surely the proofs would give him more trouble than he cared to think; it much brightened my day to find that Prof. Linnaeus had made such an unholy mess of the first sheets, to be about *Gramina*, that I am convinced that printer's lads would be sweating ere they had cast it all off for printing; it follows that Prof. Linnaeus was delighted when I undertook the task of writing out a clean copy." This was told in a letter to an appeal judge, Sten Carl Bielke, who was asked to keep secret the fact that Linnaeus needed help.[1]

Kalm continued: "Today, I walked hither, to Upsala, partly in order to tend to the foreign plants in the academic garden, partly also to hand over to Herr Prof. Linnaeus that which I have neatly rewritten for him from his *Flora suecana*, as it is currently being printed; I could not easily have been able to withdraw from helping him therein, and so I have written a clean copy of his *Flora Suecana*, with nothing remaining other than a few sheets of *Cryptogamia*, which I feel will be ready by the end of this week.

It seems that by midsummer, the printing of this *Flora* will be complete; it is printed at the shop of Salvius, who is in addition busy with printing the next volume, the *Zoologia regni Sueciae*, predicted to lie in readiness somewhat after midsummer."[2] It didn't happen quite so quickly: the year of printing for *Fauna Svecica* is 1746. These extracts reveal something about the early collaboration between Linnaeus and Salvius, but more about the role of the sensible, practical Kalm.

Later, Linnaeus wrote about his *Flora*, stressing its great scope and addressing his readers, as one might expect, from the patriotic high ground: "It was felt he had made it as clear as day how lacking in fairness are the judgments of Sweden by foreigners who insist that, here, nothing grows but pine, fir, and heather." During the decade before the second edition of the *Flora* was published, it had run to 500 pages and been extended by another 150 species. Useful information and economic advice were included everywhere—instructions directed to farmers and land managers as well as to young students.

Naturally, Linnaeus himself often dealt with matters related to *Flora Svecica*: "As it is inescapable that you, my Dear Sir [Lars Salvius], by next summer should have ready to present to us a new edition of this book as in its absence I would be obliged to remonstrate to higher authority." During his travels, Linnaeus had discovered so much, and collected so energetically, that the new edition might almost double in size. Hence it had grown "so unwieldy as to cause the student the greatest grief while taking it on herbations [plant collecting excursions]; no other means seem to answer other than that you, Sir, use a finer font." He rounds off his admonition in style: "In any circumstances, do not brush this aside, lest you should desire, dear Sir, to conserve the standing you have and to do yourself the best service, as well as myself and *publicum*." He uses the profitable sales of student textbooks as a prod to get Salvius moving. "The students regret that they cannot lay hands on the *materiam medicam*, on which I lecture. One bought a copy from another *studioso* for 65 thalers as he could not abide without it; for the sake of all that matters, send us with greatest dispatch 30 copies."[3]

That his *Flora* and *Fauna* had been written in Latin was seen to be problematic, and in 1749, Abraham Bäck made a proposal to the Academy of Sciences that it should support the translation into Swedish of both books and also of *Systema Naturae*, as this would provide useful reading for people like land managers, gardeners, and other tradespeople. It was proposed that the translations ought to be carried out in Linnaeus's home and under his supervision. Master Löfling, the tutor of his children, might be the right man. This is another expression of the utilitarian thinking that also characterized the academy's sponsorship of the provincial tours.

However, the members were put off by the translation costs, and the plan was shelved until the 1770s, when it was resurrected; it resulted in a translated edition, published by the Swedish Linné Society in 1986. It has grown steadily. The principles for any flora were established by Linnaeus in his *Philosophia Botanica* (1751): the work should include all spontaneously occurring plants in the chosen locality; these should be ordered systematically. The book should serve both as a textbook and as a tool for further investigations. It should preferably serve education as well as being an inventory and, finally, be useful by providing information in general and, especially, for pharmacists.

Collectors and correspondents in many parts of Sweden and Finland sent their findings to be added to the 1755 edition of *Flora Svecica*. Eventually, there were more than fifty of these, a measure of both the popularity of botany and of Linnaeus's central position. His old friend from the student days in Lund, Johan Leche, was the most devoted contributor. Linnaeus's own praise of the work was no exaggeration: "*Flora Suecica* is one of the most complete *florae* the world has ever seen."[4]

He felt as strongly about *Fauna Svecica*: "We have learned that the number of different animals found by *auctor* alone in one country seemed beyond the comprehension of many readers"; however, after thirty years of work, and with the support of many students "finally, we can boast that, as much as 30 years ago our *historia naturalis* was below many countries, as much has she by now risen above all others."[5]

As ever, Linnaeus kept working on new editions with interleaved additions. In 1787 Samuel Ödmann, a student who had become an ornithologist and entomologist, risked proposing a full revision of the work to the Academy of Sciences. Above all, he argued, references to humans should be excluded. His paper, which apparently caused a mild sensation, was sent to a committee for further consideration. The committee didn't take long to prepare its rejection. "Fauna" was a fairly recently introduced concept—Linnaeus had previously used the term "Pan'—but now Fauna and Flora made for a preferable combination of words.

A translated extract from the compendious foreword: "Why should it now be regarded as so important to devote oneself to zoology, the science in which we have since the days of Adam made so very little advance? We should, because the four-footed animals roam, concealed in the forests, the birds fly there, the fishes seek the depths, the amphibians their hiding-places, finally the very insects will avoid us hopping, running,

FIGURE 24. The frontispiece of Linnaeus's *Fauna Svecica*, drawn by Jean Eric Rehn. The many-breasted Diana bears witness to the rich Nordic fauna. Royal Library, Stockholm.

flying—indeed, to say it straight—every creature will flee to escape scrutiny and will not, like plants and stones, remain in fixed places."[6]

By 1746, after much hard work and using his many contacts, Linnaeus had been able to list 1,357 species belonging to the Swedish fauna; by 1761 another 1,000 had been added. He had been assisted by colleagues, clergy, and pupils acting as observers in every corner of the country, but also valued the folk knowledge about nature he had collected during his travels, as well as the work on *Flora* and *Fauna Svecica*. Evidence of his ethnobiological records come in examples such as this, from Åmål on 27 July 1746: "In the afternoon, we traveled into the countryside to learn of the botany of the valley from old maidservants and old wives, who in these parts were in possession of more knowledge of plants and household remedies than could with ease have been found in any other place."

Flora and *Fauna Svecica* have exerted a huge influence on regional and county-based collections of data on flora and fauna. Plants and animals have been registered nearly as carefully as the local human population—complete with names, places of residence, number of individuals, usefulness, and characteristics. There are direct lines of connection linking these works to modern species databases, and the great pleasure found in the rich flora and fauna of the nation.

CHAPTER FIFTEEN

Family Life 1

SCENES FROM A MARRIAGE

EVERYTHING HAD SURELY fallen into place; in 1744 or thereabouts, this is how Linnaeus saw his position: "Now Linnaeus had gained high repute, occupied a post to which he had been born, sufficient money, of which some had come his way through marriage, a Dear Wife, beautiful children, and an honored name." He and Rosén, once they had gotten their professorial appointments, swapped subjects. They agreed to share pathology and chemistry, but while Linnaeus took on botany, dietetics, and *materia medica* he rid himself of anatomy and physiology. The chancellor's approval is dated 21 January 1742.[1] After the death of Lars Roberg on 21 May 1742, Linnaeus was "paid in full," and in 1745, he went to Falun to receive his wife Sara Lisa's inheritance from her father—"although leaving for my mother-in-law the greatest share."[2] If, at this time, we were to place him within the scheme set out in *Metamorphosis humana*, he had reached the seventh stage of his life: "The heroic age, when man has reached the pinnacle of perfection or, if the strength of body and soul are combined, his summer solstice."

The prefect's house, tied to his role as "prefect" or curator of Uppsala University's botanical garden, is now the Linné Museum. The house had been built for Rudbeck in 1693 on a corner site (where Tovegatan meets Svartbäcksgatan) and had stone walls, as the risk of fire was on everyone's mind. Linnaeus, used to good living quarters in Holland, made certain changes. On 6 March 1742, he submitted his request to the court of the university, likening the miserable state of the house, garden, and orangery "more to an owl's nest or a robber's cave than a professor's home."[3] The university's chancellor, Carl Gyllenborg, issued the required permission, and

FIGURE 25. The Prefect's House, Linnaeus's home in Uppsala. Photo by Emma Schenson, 1864. Uppsala University Library.

the senior court architect, Carl Hårleman, made drawings appropriate to his client's status. When he had finished, the home had thirteen rooms: four on the ground floor, six on the first floor, and three in the attic. The ground floor had twelve windows, upstairs thirteen smaller ones, and there were four attic windows. The architectural detailing included blank windows, tall stoves clad in green tiles, dado-height paneling and wallpapers: good taste but no luxury. In 1743 the house was ready for the family to move in. The housewife ruled the ground floor and the householder the first floor; Linnaeus also used his quarters for informal seminars. Keeping their home warm was a problem, and they ran through a lot of logs, which served as a kind of currency. When the prefect's house was restored in 1935–1936, the finds included playing cards and a phrasebook from the Lapland journey.

Their first child, a boy, had been born in Falun, as we know, in January 1741. He was given his father's name and would, startlingly early, succeed him in various roles. We don't know how Linnaeus Jr. was brought up, but at the time it was generally true that at least in academe, the earlier the start, the better the likely outcome. His father discussed the matter with his French pen friend Sauvages, who, in a letter to Linnaeus

(20 January 1751), explained his pedagogic ideas for schooling his own baby son: "Once he reaches the age of four, we will only talk together in Latin in order that he becomes habituated to this language, without rows and beatings, just as did Montaigne." Young master Linnaeus struggled with Latin as well as with botany.

When, in 1763, his father summarized his qualifications in order to persuade the chancellor to let him take over the post of prefect, he pointed out that the boy "had grown up in the Academic Garden and there enjoyed the company of the plants since a tender age." He apparently had a tutor from time to time and was enrolled at the university at the age of nine, but was much later reported as a painfully hesitant Latinist unable to speak any other international language.[4]

In 1756, at the age of fifteen, he was ready to present himself for his academic dissertation: his thesis was full of observations but was not properly completed. He had composed it by himself and written in Swedish on the subject of the "Flower Clock." He reminisces about his father, who while in the orangery, had asked Rolander, his assistant, to run along to the library to get a book for him but, on the way past, have a look at the bed of *Arenaria rubra*. Rolander reported that the flower buds had not yet opened, neither on the way to the library nor when he returned minutes later. Soon afterward, Linnaeus walked past the bed, and all the plants were in full, glowing red flower.

Among Linnaeus's children, daughters dominated: first, Elisabeth Christina, known as Lisa Stina (1743–1782); then Sara Magdalena (1744); Lovisa (1749–1839); and Sara Christina, known as Sara Stina (1751–1835). The series was interrupted by a boy, Johannes (1754–1757), and brought to a close with another girl, Sophia (1757–1830).

Linnaeus didn't consider sending his daughters to school, perhaps basing his views on Rousseau's distrust of formal schooling. Otto Tullberg, a third-generation relative, insists that Carl "was extremely fond of his children" but that keeping the girls at home wouldn't have been Mrs. Linnaeus's preference. Sophia, her youngest, was sent off to school when her father was away and brought back home the moment he returned, a routine that was repeated for days on end.[5] Johannes grew distant from his father, but it seems to have been due to the boy's own attitudes. That none of the five children chose spouses from the academic world or the clergy might, of course, be coincidental but might also suggest a lack of suitable contacts and parental ambition in that direction.

During Rudbeck's work on *Campus Elysii*, a kind of home-based science education involved his whole family, but similar exercises don't seem to have taken place in Linnaeus's house. In his own copy of *Philosophia Botanica* (1751), a piece of paper with Sophia's full name in a childish hand has been glued to the page just after the end of the foreword. When she was little, Sophia came along when her father was lecturing and was allowed to stand between his knees. Sara Stina helped her father's friend, Mr. Bäck, when he was collecting butterfly specimens. Linnaeus's brother Samuel noted raised voices at home in Stenbrohult: "My daughters very much regret that they were not born into the other sex so that they could have gained from their uncle's knowledge, as studies would have been their choice."[6] We will hear from Sara Lisa on this subject later.

From the beginning, the union of Carl and Sara Lisa Linnaeus was seen as a marriage of convenience. Aspects of their arrangement might be compared to Darwin's thoughts on marriage: on two pieces of paper, he listed pros and cons under the heading *Marry* and *Not Marry*. "Marry" won on points, even though "Not Marry" included entries such as more "time for work" and "talk with clever men at the club."

Despite the romantic stories conjured up by his stay in Falun, Linnaeus and his fiancée can't have had many chances to be together. They might well have fallen in love, but Linnaeus's more than three-year stint abroad created a distance between them. In fact, they barely knew each other when they met before the altar and their lives were linked forever. Many accounts stress how different their interests were, as were their roles in the household. Sara Lisa must have had a busy, highly responsible life, managing the home in Svartbäcken where she was in charge of catering and all other services, not only for the family but also for often-troublesome student lodgers. Her work must have seemed thankless at times, and she had a sharp tongue.

A tangled story has emerged from their homelife, in some way analogous to the Greta Benzelia episode. It has so far been kept under wraps, but while the first part is a snapshot of the day-to-day life of a poor student, the later development illustrates an unknown aspect of Linnaeus's life. Additions to the list of characters are Elias Frondin, an auditor; his son Berge (Birger) and brother Erik; and Johan Gabriel Rothman, the son of Johan Rothman, Linnaeus's schoolteacher in Växjö.

Rothman Jr. had turned himself into a professional scandal monger, best known for his magazine *Philolaus Parrhesiastes* or "The Garrulous Dissenter" (1768), an anti-Hat political sheet full of "exquisite cruelties and such expression as would cause the ruffians down by the lock-gates to blush." Some of the contributors were learned men—*docentes philosophiae*—so it presumably can't have been just a matter of servants' gossip.

CHAPTER FIFTEEN

Young Rothman felt his existence was dire and wrote about his woes. Linnaeus had promised his old teacher to keep an eye on his son, indeed "to take the young man into his home in Upsala and care for him as were he a child of the house and also to make of him as great a *Botanicus* as he was himself." Such support was, however, denied him. Instead, he had been forced to earn a living by tutoring, a situation in which "he bore much severe ill-treatment." At Michaelmas 1743, he decided to move out once Linnaeus had offered to provide him with board and lodging in return for help with "certain small things." Rothman accepted the offer but then realized the lodgings were in Linnaeus's own home. "In Nov. 1743 he moved in with L, must pay bed and board himself, thereto candles, logs, etc. and furthermore endure all conceivable manners of chicanery, reproaches, and suspicions with regard to drinking, gambling, whoring, and every other vice. L *principe* was that a young man ought to be rather strictly supervised and put up with much hardship so as to make him a good citizen. The wife was as full of malice."

He went on to tell a story:

In Nov and Dec L. was in Stockholm, and then his wife had every evening the company of Mrs. Klingenstierna and Berge Frondin. They would be together til 3 or 4 in the mornings, and that without always being at the card table. K. would tickle F. often in the presence of R.

F. cried out, My gracious Auntie, leave me be, threw kisses and suchlike. Once she leaped up from the dining table, and that before the last dish had been served and went to him appearing exceedingly *paissonée* for him. Then Mrs. L came up to R about noontime, the clock said 11½, and when she had in a quite kindly manner assured herself of R's silence, gave him a Note to at once be carried to Fr, giving the reason of some unusual affair. R must so call on F who was at his meal to come outside and he blushed as he took the note that was so artfully pinned together that R did not believe himself able to undo it. Prof. F then went to his son, the maid told me, and strongly examined the boy, who with hesitation answered that R had desired to borrow a book and then showed himself unable to easily name that book when asked, and the father said that it could not be possible, and contrary to polite manners, to arrive at midday on such an errand.

One of L's maids told me that she three times before had been running across with such notes and that at one time the note dropped onto the street. Then R had at one time again such a commission and found the note in similar ways closed with a pin, and happened upon

Fr in his room, who with exclamations of joy received the note. Mrs. L then told him that the message was to arrange for Fr to make up the company for cards with herself and Mrs. Kl. Before he went to his sister, who otherwise kept him with her during her period of illness, L must after his return home have come to hold strong suspicions, since he had found her in a foul temper and so he frequently demanded of her to quit sickening for the Magister, etc. The bad blood between them grew a great deal. L often exclaimed that whoever married for money must bear with many offenses from his wife. She responded that he ought not to have married some lady who he knew could not stand him in any way. L asked, if so wherefore she had said yes. She then reminded him that he had addressed himself to her father who was so harsh to his children. R heard many other similar disputes. Then L attempted to win her favor with caresses although it would appear that by his *exès* in such a moment he made himself as *dégoutant* to her. *Enfin,* he in every way showed as little *conduit* as in all other things.

    At Christmas, he was about to arrange a festivity and commanded R to go around with all the invitations; though he said may 1,000 demons come for me rather than you should offer an invitation to Berge Fr and his brother. R promised obedience in this matter and when he came to prof Frondin, he was met by B Fr, who told him that the others were eating and that he would carry the message to the invited but seemed vexed that he was not among them. Braunersköld, who also ate there, stepped out and was exceeding grateful. B Fr then came out for the second time, still without being invited. R came home, was questioned by the Wife as to B F, had he been invited and was next ordered to without delay go there and do so or else there would be no festivity. Then R went to Solander, where L was, and submitted it to L that the younger Frondins should be invited but which L forbade with uttermost eagerness. Mrs. L nonetheless sent R off to invite them for 4 o'clock, undertaking all responsibility. L was in incomparable good humor on the day of the celebrations, commanded R as if a hired hand that he was to go with the housemaid to serve, clear away plates, offer drinks of wine, etc. By 4½ o'clock B F arrived and L then quickly lost his gameness, took himself off to the upstairs rooms to smoke tobacco with the older men but then ran up and down like a madman. L demanded from R if he had invited B F but R replied no and Mrs. L said it was surely accustomed for a Cavalier to be present where there had been grievances. L was inordinately troubled all evening and insolent to R and many others.

Pretty minor stuff, it would seem, but the rumor lived on for a long time and was topped up from time to time. The following quote comes from one of the chief sources of information about Linnaeus's private life, lines not referred to in the Linnaeus biography by Thore M. Fries: "At present, it has still not been forgotten in Uppsala how the Archiater's Lady Wife conducted amorous traffic with the current Librarian at the university library Herr Magister Birger Frondin and also with Herr Rolander. However, Herr Archiater was allegedly having his share in the company of his housemaid."

Not only but also . . . Linnaeus himself. Fries, self-appointed guardian of the precious Linnaean virtues, can't stop himself from also rescuing the honor of Nils Rosén on the issues of his squabbles with Mrs. Beckmann. The source has been annotated with several "sic!" Its publisher, Ewald Ährling, had described it as "a lot of housemaids' silly gossip and backstabbing," a critique Fries copied and concurred with.[7] So far, the source has remained untapped. It didn't take Linnaeus long to fire Rothman. He fled the country in the end, after working briefly as a ruthless party hack for the Caps.

Linnaeus reflected on the sexual drive and its transformations with age in *Metamorphosis humana*: a spark that develops into a fire, settles to glowing embers, becomes a dying flame, and, finally, ashes. At the seventh stage of life, the heroic one, men burn with desire for a true mate, at the eighth stage the flames rise in the domestic hearth and he longs for children, at the ninth the flame is fading as the children are brought up, at the tenth he works for his family, at the eleventh the old man rejoices at his grandchildren, and, in the end, only ashes remain. In the story just told, the flame is flickering uncertainly.

Grief and worries. Their second son, Johannes, was three years old when he died. This was said in a letter written by Linnaeus's student Daniel Solander: "Herr Arch. Linnaeus, his Wife, Mademoiselle, and Master Carl send their respects. By now they are much recovered but, a few weeks ago, Hr. Arch Linnnaei family was subjected to grief as their youngest son Johan Linnaeus, about 3 years old, died in the so-called Upsala fever aggravated toward the end by a powerful Crisis. Herr Archiater and Wife experienced rather strong anguish over this as they felt for the child almost too much love. He was not ill for long, no more than a week or so."[8]

Six months later, their fifth daughter, Sophia, was born but appeared dead: "This past Tuesday, my wife gave birth to a daughter after rather trying labor; the girl was born dead or appeared dead at birth however we breathed heavily into her and attacked with *insufflatoria medicina* so she began after ¼ hour to show a little life. At last, she drew breath as if for the last time and then at last she displayed more signs of life and by now seems to be quite well in herself. My wife is weak, however. God help her."[9]

The ancient Disting Market in Uppsala opened for trade in February 1754. Everything could be bought if you had the wherewithal, and the whole city was on the move. Sara Lisa was going along for some serious shopping, or so her husband said, as he warned his friend Bäck not to mention his trip to Stockholm.[10]

In 1758 malicious wags would have it that his daughters, garbed in corsets, were placed in the tenth edition of *Systema Naturae*: members of the species *Homo sapiens*, variant *Monstrosi*. It is a fact that the inventory after Linnaeus's death lists his 3 waistcoats and 3 pairs of trousers whereas his son's wardrobe contained 17 of the former and 13 of the latter. Linnaeus would express his skepticism about all excess but couldn't prevent his increasing status from manifesting itself in objects such as tiled tea tables and hundreds of linen napkins. For better or for worse—and for more luxury.

His household consisted, by my estimate, of the master and mistress of the house, their six children, a nanny, a housekeeper, two kitchen maids, two outdoor maids, a milking maid, and others—altogether, twenty-odd people, not including the shifting population of students and visitors.

CHAPTER SIXTEEN

In the Garden, at Herbations, among the Collections

"BOTANICAL GARDENS ARE like living libraries of plants," Linnaeus wrote once. There, *Materia Medica* is taught, but so is the ancient story about a mythical Paradise; proper ordering is practiced, as well as experimentation with acclimatization of plants, but it is also a fine place for promenading for pleasure and arranging displays to impress. Padua, Amsterdam, and London are homes to the oldest existing academic gardens. By the end of the eighteenth century, London added Kew Gardens to the older Chelsea Physics Garden, and the new institution became a center for introducing new finds. Pioneering gardens and vineyards flourished around Jan van Riebeeck's old fortress city at the Cape of Good Hope.

Then, there was Uppsala: "There is only one other garden to compare with it in human history, and that is the one in the Book of Genesis, which never existed outside words . . . the place where an intellectual seed landed, and is now grown to a tree that shadows the entire globe." The quote is taken from the essay "Seeing Nature Whole" by the English novelist John Fowles (*Harper's Magazine*, November 1979). In *Species Plantarum* (1753), Linnaeus listed the countries he had visited and the botanical gardens he had studied in Paris, Oxford, Chelsea, Hartecamp, Leyden, Utrecht, and Amsterdam. His sources, of course, included the herbaria he had examined, and the plants and observations sent to him by students and fellow botanists. All this learning had been channeled into books.

In Uppsala, the Academic Garden had been created by Olof Rudbeck in the 1650s and contained, among other features, a controlled irrigation system and a heated "pomerance house"—an orangery that included bitter orange trees and other tender fruit trees. This first garden lasted until the

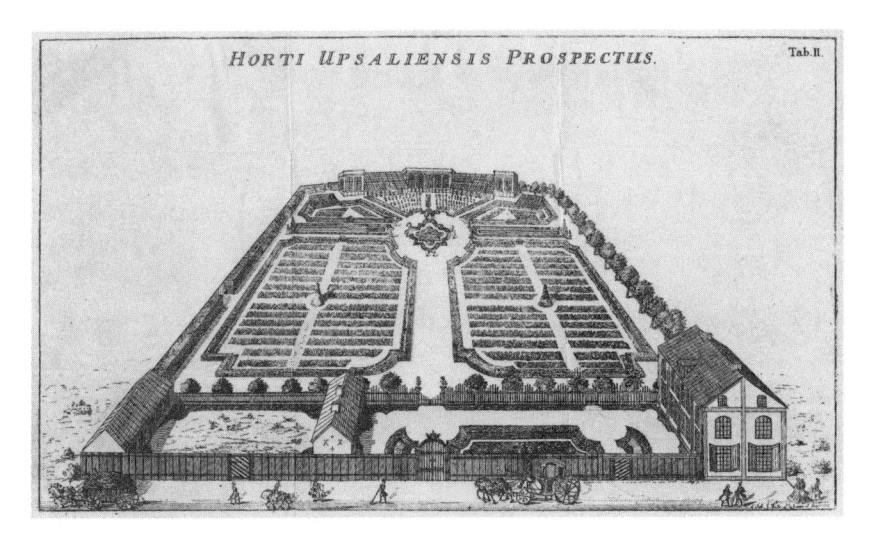

FIGURE 26. Prospectus for the Botanical Garden in Uppsala. From the dissertation *Hortus Upsaliensis*. Copper engraving by Johan Gustav Hallman (1726–1797). Uppsala University Library.

fire of 1702 but, although elements survived, Rudbeck had lost interest. He was an excellent botanist but went on to immerse himself in wide-ranging studies of languages. His new student Linnaeus was very interested, but bitterly disappointed when Nils Rosén took over the management in 1731.

The garden went into a decline lasting until 1741, when it was restructured by Linnaeus, Hårleman, and the gardener Dietrich Nietzel, who joined them from Holland. The new layout included a longitudinal axis along which were aligned ponds, greenhouses, and beds for annuals and perennials, ordered by the twenty-four classes in the sexual system of classification. As *Hortus Upsaliensis* (1745) points out: "No other Garden is arranged in a more orderly manner, none excels above his number of rare herbs and plants and none has propagated from seeds more specific forms; even though he is under a colder climate than all the botanical gardens of Europe." Rarities from the East and West Indies were grown in the hot and humid caldarium, including the "tree of Paradise," actually a banana plant, and exotic trees such as coconut and sagu palms, coffee trees, and shrubs.

There were a lot of problems with maintaining "small wet *hortus*"; the orangery was one case of challenging requirements for irrigation and heat. These two factors, as they affected plant growth, were of special interest to Linnaeus. In Holland, he had spent time socially with Daniel Fahrenheit, who published the scale and calibration points of his thermometer in 1724. Many others have claimed to be the first scientist to produce a

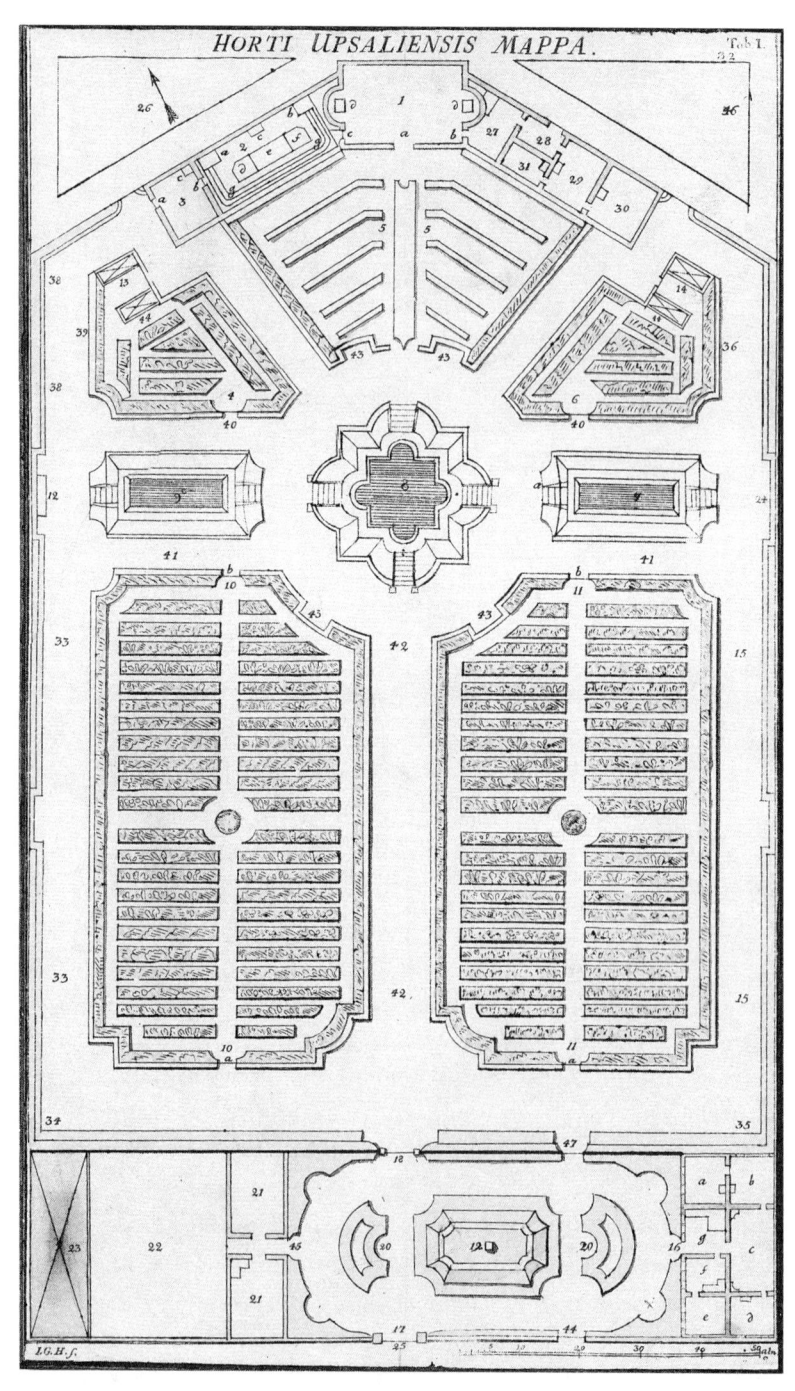

FIGURE 27. Map of the Botanical Garden in Uppsala. From the dissertation *Hortus Upsaliensis*. Copper engraving by Johan Gustav Hallman (1726–1797). Uppsala University Library.

standardized scale for temperature measurement. The Swede Johan Backman was one of them: in 1716, his brief paper on barometers proposed a 100-degree scale, with zero set for "a sharp winter" and "heavy heat" at 100 degrees. The frontispiece of Linnaeus's *Hortus Cliffortianus* (1737) includes a thermometer with a 100-degree mark, where 100° is probably set for the freezing point of water.

In 1741 Linnaeus ordered a thermometer with a 0° mark on the scale. Anders Celsius was the first to have manufactured one with a 100-point scale. It is not clear when the scale for "degrees Celsius" was completed, and some believe that it was Linnaeus's idea anyway. Internationally, this scale was known as Linnaeus's or "the Swedish" thermometer. In a letter to Berge Frondin from 1745, Linnaeus mentioned that Ekström, the instrument maker, had finished a thermometer intended for the orangery. "In this term I am to present my dissertation on the Academic Garden but cannot cite the degrees of warmth as I have no thermometer."

When it was delivered and tested, he stated that it was "signally well put together." In December 1745, the garden bought the thermometer from the Academy of Sciences, and in *Hortus Upsaliensis*, the temperature in the *caldarium* was expressed in degrees. It was the first time such a measurement had appeared in print. Does "100 degrees Linnaeus" have the right ring? Linnaeus was never timid about making claims to fame but never added the thermometer to his merits. It might well be that his master gardener, the highly skilled Dietrich Nietzel, should be honored for this innovation.[1] "Academic gardens offer numerous opportunities for young students to see and recognize plants of many and varied kinds, so that they will there learn all essential matters of *Botanique*, a science which provides the fundament of all private oeconomie and all *Materia Medica*. Upsala *Academie* can be argued as the only institution in our Native country to own such an academic garden, which had been founded and arranged at considerable expense, labor, and trouble."[2]

Linnaeus addresses the university court on issues of plants, cold frames, and manure from the academic stables. Traditionally, managing the garden was ranked as one of the costliest items in the university's budget. Applying for more money was a major component of the prefect's already heavy workload. Claes Annerstedt, a historian of the university, has declared: "The sacrifices made by the university in order to meet the Linnaei demands were indeed exceptional."[3]

It is true that Linnaeus produced a steady stream of new funding applications, but also that he was indefatigable in many other ways. He pleaded for planting skills to be added to the curriculum, conducted trials of new crops, and imported new ornamental plants—for instance, *Dicentra spectabilis* from Kamchatka. Whenever he received packages with plants, bulbs, and seeds, his unmistakable joy was communicated in his letters to Bäck: "I have no longer time to concern myself with illness. Flora is sweeping in, at the head of her entire beautiful army, so I must enroll with her yet continue to care for academic matters"; "My most dear Brother. Do not ever mention the theater at Drottningholm now as holidays continue and the garden is in its finest flowering"; "When M. Br has been to Drottningholm, travel to see our garden; I feel that he has merited such a sight for a few days. When my Brother has seen the wide world and its kings, come to meet a small printz in the realm of flora and observe if there is not much more for the eyes to see though perhaps not more for the mouth to consume. Let us walk about in our nightcoats behind the shut gate."[4]

The garden enchanted Linnaeus; it was his favorite child. "You surely know that I exist, body and soul, in the academic garden, that it is my Rhodos or rather my Elysium. Daily, buds open there to develop themselves into new delights; there I possess all the booty I might have wished for, whether from east or west, so much more precious than the tapestries of the Babylonians and the bowls and vases of the Chinese; the place where others learn from me and I learn for myself."[5] He could, for instance, take his pleasure in the exotic greenhouse specimens of sago cycad, cacao tree, and ginger plant. In 1744 he was sent a Musa or banana plant from Holland and took pride in it—as he did in the amaryllis, the lovely coffee bush, and the rice plant, the cluster of bamboo, the carpet of dragon's blood, the tamarind tree, the papyrus reeds, and so many others. Seeds and plants arrived ceaselessly, and his correspondence dealt mostly with this traffic, not least because he was keen to grow officinal herbs for the pharmacies. He paid particular, nervous attention to the tea bush. From his study in the prefect's house, Linnaeus was able to survey his domain and "control how diligently the workers carried out their required tasks, also that no mischief was done such as roughing up or seizing plants."

Being responsible for staff brought its own problems. The master gardener Dietrich Nietzel, borrowed from Clifford in Holland, was a great help at first. But he grew wayward, then fell for an adventurous Uppsala widow who eventually got tired of her conquest. Nietzel sought comfort in the bottle and died in 1756, killed by consumption. Linnaeus wrote about the practical consequences: "Nietzel's death robs me of many hours and his loss will not be replaced in my time."[6] He added that the academy's garden laborer "lies with a wealthy widow in the hope of marriage and consumes her good offerings. Falls so in love with some scullery maid, marries her; then he dies in misery, the widow shamed."[7]

His stop-gap solution was to recruit his son as a demonstrator and also to employ "a tired old man." A gardening apprentice called Lars Broberg had proved himself by going abroad to study and was appointed academy gardener in 1764, only to cause new problems: "Broberg steals *Adonis capensis*, takes them to Stockholm, after 2 years, takes a little one away. Not allowed to travel after taking my *bulbi*, so he quit. When I apprentice a lad to bring up as I would recommend, he takes on another behind my back, tells the other to go. When I plant *bulbos capenses* in my pot he lets me have only one and the others he kills or steals away."[8] Broberg apparently was not punished in court but was deemed morally unfit. Meanwhile, another gardener, a leaseholder at the Royal Gardens and Hop Plantations, had a fight with Linnaeus on the matter of acquiring dung from the university stables. The conflict, which reached the university chancellor's office, concerned a hundred wagonloads of manure, life-sustaining for Linnaeus's gardens. The case dragged on for a year, the documents and memoranda piling up to wagon-load proportions in the University Court and Royal Council.

The glorious garden existed to serve academic teaching and, to some extent, to be shown off to the public. Once inside the gate to the Academic Garden, the visitor encountered two monkeys restrained only by run leads fixed to bars in high-rise cages, where they could keep an eye on the traffic but in appropriate seclusion. A small menagerie was part of the garden. Linnaeus had been given an agouti—an unusual rodent—but the animal died suddenly and unexpectedly, causing him to write to the donor, Frans Bedoire: "Next my children I declare having taken nothing more to heart than this creature. Having in my charge agreeable Indian animals almost causes me fear and grief as their passing touches me so profoundly as that will surely shorten my own life; this one the most."[9]

Linnaeus's fascination for animals is obvious from his lecture notes and his animal portraits; he had a dog as a child and kept birds in his study as a student. He defended dogs against being used as vivisection subjects: "Dogs have over all times been martyred by the anatomists who have cut into them while still alive. Thus, they found the *vasa lactea* [lymph vessels] and Helmontius discovered in them the *circulatio sanguinis* [the circulation of the blood; he had forgotten that William Harvey, not Helmont, was the first to describe the circulation of the blood]. Anatomists have the habit of first slicing through *Nervum recurrentem* so as to stop the poor creatures from giving voice and scream." However, he didn't react against

FIGURE 28. *The racoon* or common raccoon (*Ursus lotor* L.), originally an illustration to its description in the 1747 *Acta*, Royal Academy of Sciences, now hanging in Hammarby.

the cruel old game of "beat the barrel to get the cat."[10] It is not known for certain whether Linnaeus had a cat but it seems likely, given his lecture that includes a knowing description of cats, in which he notes, among other characteristics, that they "smile with their tails."

What of other pet animals? If we accept the distinction that pets are allowed to enter the house while working animals are not, Linnaeus's raccoon was clearly a pet. A record from 1769 lists the following animals around the house: several monkeys, a raccoon, an agouti, and a guinea pig, and several birds including grouse. Exceptionally, one visit resulted in what reads almost like a reportage headlined "To keep in the house free or caged living wild and rare animals"; it is included in Eric Sefström's 1763 publication *Swenska Samlingar* (Swedish Collections). On 16 May 1760, Linnaeus showed the gardens to Sefström, who described seeing four species of monkey ("the capuchin monkey has a very frank, pure-hearted appearance"), three species of parrot, the guinea pig, a few turtle doves and peacock hens, and a goldfish. Explorers and other travelers, students and sponsors, including the royal couple, gave Linnaeus all manner of creatures. The museum collection of animals preserved in alcohol held donations from Gyllenborg, Lagerström, and "our most noble KING." The tour ended with the archiater's own specimens—his herbarium and collections of various insects and moths—and then his library. Linnaeus's talking parrot had grown famous in its own right. The parrot would call out when there was a knock on the door: "In ye come!" It announced its master's midday meal: "Herr Carl, it's twelve o'clock." Löfberg, the gardener, was greeted by the parrot's bossy: "Löfberg, blow your nose!" The peacocks were passed on to Mennander after causing "considerable grievance": they had eaten a barrel of grain and topped up with an entire batch of sweet, wheaten loaves. Still, "Beware of keeping them outside," Linnaeus instructed, "The foxes are crazed for them."[11]

According to Linnaeus, animals have a moral sense. To anthropomorphize animals is to make them comprehensible, and his attitude toward animals was similar to the folk beliefs and humor expressed in *Reinike Fuchs*, a collection of fables. He described the auk, or alcid, as "a learned gentleman in coat and waistcoat." He enjoyed the guenon monkeys: "Be so good as to recommend the monkey as I can without joking assure you there is none as amusing, as peculiar, and *differente* and especially as droll. A *comoedia*."[12]

Paying visits to collections was a must for people who traveled; the variety of terms used reflected the popularity of collecting: library, museum, chamber of rarities, laboratory, cabinet, and, for example, cabinets of *naturalia* and curiosities. Apart from books, the libraries often contained other objects on display such as portraits, busts, globes and scientific instruments, coins, gemstones, maps, objects from nature, and mounted dried plants in herbaria. Learned men had their own museums; Linnaeus liked signing off his letters as written "*in museo meo*," meaning "in my study." In Uppsala, he could investigate Joachim Burser's large herbarium, assembled during the first decades of the seventeenth century, and compare the range of plants with those on record in *Pinax*, the contemporary standard work by Caspar Bauhin. Linnaeus declared that "my collection of *naturalia* is to me like a book in which I read every day so it consequently cannot be held in as pristine, abundant, and refined a state as those less frequented."

Collecting was high fashion. A collection should, in a concentrated format, offer a view of the world, reflecting the work of the Creator, and to understand it properly, knowledge and assistants were required. Pharmacists became collectors—for instance, the Dutchman Albert Seba and the Stockholm-based Ziervogel—as did doctors like Worm in Denmark, Valentini in Italy, and Stobæus in Sweden. Royals and noblemen joined in: in Sweden, great collections were amassed by Queen Lovisa Ulrika and Field Marshal Carl Henrik Wrangel. Collaboration was to be wished for, but competitiveness drove the market, with seashells as one new-old form of currency. Certain types of objects acquired their own contact networks and trade routes. Generally speaking, the Swedish collections were modest apart from the royal ones. In a lecture, Linnaeus commented that collecting fever could become obsessional: "It is not fitting to every man's condition to construct for himself such a Cabinet, as when he gets a large lot together, he wants still more and can easily buy himself into poverty before he is satisfied."[13] He was himself unwilling to give away his duplicates and could be quite petty.

Maintaining and extending the university's collections were part of Linnaeus's professorial duties. In addition to his private collections, he was charged with looking after three more: the king's, the queen's, and Count Tessin's. In the Linnaean collections, the oldest datable object is an albino squirrel, caught in 1696 in the woods near Kungsör. It had been presented to King Karl XI, who had images made of it on three separate occasions. Lovisa Ulrika's collections were remarkably extensive; among the displayed items, one of the talking points was an alcohol-preserved royal fetus. She had also created the elegant style of the royal palace Drottningholm and showed off its fine library and art works. Between 1751 and 1754, Linnaeus spent nine weeks working on the royal collections at Ulriksdal, which included some 1,100 animals or animal parts in alcohol. More donations were arriving all the time.

Which objects were acceptable? A big-time cargo carrier, Magnus Lagerström, who was in charge of the East India transports, wrote to Linnaeus on 9 November 1748: "On that same occasion, I intended a *dens maxillaris* (tusk) of an Elephant but he is so large and heavy, it embarrassed me that I should ask a traveler for it, especially when I was not certain as to its welcome, hence in the event it is desired, delayed the errand until the next occasion." When Colonel Klinckowström sent a message about a sperm whale found beached near Strömstad in November 1749, adding a drawing of the beast and its teeth, Linnaeus thanked him politely but stressed that large bodies caused problems: "Such parts of an animal which are put in strong aquavit will be preserved; even every kind of jellyfish; large animals cannot be stored unless as a smaller one of the same shape can be found."[14] Drawings were the best way of recording animals too large for preservation. Plants were in general much easier to deal with, and insects even more so, which made them a very popular class of collectibles. Collecting was driven by notions of aesthetics and rarity, but sheer size imposed its limits.

A problem of language, apparently of a very delicate nature, turned up when cataloging the royal collections: How to name items in Swedish? Linnaeus wrote to Bäck: "What should *Amphibia* be called in Swedish? Not a critter in the *vermes* orders as these are invertebrates; would not *freak* be best suited, asks Prof. Ihre. Lord knows what I should do."[15] A few days later: "I am more than puzzled with the word *Amphibia*. It would be sweet to avoid the Latin for just this one *nomine classico*. Dear man, seek advice from Herr Sectr. Dahlin. No need in any sense to turn into Swedish the word *Amphibia* only to conceive of snakes and frogs, the ugly family. If there is no help coming then it must stay *Amphibier*." How was the matter resolved? The king made up his mind: *amfibier*. The problem had a royal resolution: the word became Swedish.

At this time, the idea of going on "herbation" was still something of a novelty in Uppsala. Herbations were botanical excursions or mobile seminars, involving all the participants in learning about nature. A few generations earlier, Olof Rudbeck the Elder had arranged herbations, as had Johann Steinmeyer, headmaster at the German School in Stockholm, who in the 1720s "commonly led his youths out into green places for purposes of recuperation but also taught them recognition of many herbs."[16] A decade later, the student Linnaeus led a group exploring the landscape; later, the navy medic Linnaeus looked for the plants growing at Stockholm's Hornstull and later still, now a professor, walked along paths around Uppsala. The trips grew more ambitious and varied, and became named by place: Herbatio Gottsundensis, Ultunensis, Hogensis, Danensis, Upsaliae antiquae, Waxalensis, Husbyensis, Jumkilensis. Linnaeus insisted in a dissertation (1753) that two conditions, above all, must be fulfilled to make current botanical study effective: botanical gardens and botanical excursions. It was recommended that clothing should be light and loose-fitting, and that apart from books, some pieces of equipment should be brought: loupe, botanical pins and a knife, pencil, vasculum, and paper. The season for herbations began with the first leaves in spring and ended when leaves dropped in autumn; a break was scheduled for the days of the dog star, the hottest part of summer. Every half hour, a demonstration would be held.

At most, an herbation was joined by 200–300 participants. Linnaeus describes one of his outdoor seminars in an autobiographical note: "Wednesdays and Saturdays, a couple of hundred *auditores* collected herbs and insects, carried out observations, shot birds, wrote protocols, and then, after having botanized from the morning at 7 o'clock until evening at 9 o'clock, they walked home, marching through the city with flowers on their hats, following their leader to the sound of drums and hunting horns on the way to the Garden. Several personages, also foreign such as gentlemen traveling from Stockholm would join Linnaei excursions." He ends on a proud note: "By now the Science has reached its pinnacle."[17]

The intention was, of course, to educate and, hopefully, add to the plant inventory and investigate new topics. The number of known local species did increase from Olof Celsius's figure of 470 to about 520. When the Swiss botanist Friedrich Ehrhart investigated the local flora (1776–79), he found another hundred species.

In later years, when the great leader's strength was waning, he was still memorable. One of his students reminisced:

> In my youth and while I was enrolled at the academy, I was happy to be able to follow him on botanical excursions. He walked bent a little forward and leaning on his cane, with which he from time to time pointed to certain herbs which he either picked himself or requested that one of those who followed him do so. After wandering about for a while, he would seat himself on the ground surrounded by his large following of students and lecture on the collected herbs, always seeking to encourage his audience to learn of and observe nature, practices which he considered one of the greatest purposes for the Creation of mankind. If it were not so (I heard him once say), then our Lord could have made the Earth of cheese and us the worms inside it."[18]

Another story, from about the same time, told by the doctor and family friend J. G. Acrel:

> From among his audience, he would select accredited civil servants, one who was an Auditor, for instance, and whose task was to make notes to his dictation whenever something new was found; one other, a Sheriff, was tasked with disciplining the troops, so that no disorderly incidents would occur. Still others were to be gunners and shoot birds and so forth. . . . Once the young men had thus enjoyed themselves in the open air from morning until evening, the return to the city began. The teacher would lead his troops and the youths follow behind their banner, marching after him to the sound of horns and drums all through the city to the botanical house, where a many-throated Vivat Linnaeus! brought to a close the pleasures of the day.[19]

J. A. Lindblom, archbishop-to-be at the time, notes in his autobiography:

> I continued my studies as described, but with the addition of a Collegium in Zoology and Botany under the famous Linné. Some small thoughtless *etourderie* caused me to become known among the many attending our botanical excursions and attracted the attention of our incomparable Teacher so I was on several occasions called to step

forward and in front of him examine herbs. I might well have advanced within this Science had not the Teacher, after only two excursions, fallen ill and been taken to his Hammarby estate, an event which caused the Collegium to be cancelled for the duration of the term. Nonetheless, it was gainful as I had come to know more of this Investigator of Nature, unmatched in the world, to hear his presentations of the Animal Kingdom and admire the acumen with which he, while conducting a simple demonstration, somehow revealed and exposed the secrets of nature to the light of day.[20]

This quotation comes from Samuel Ödmann's speech *in memoriam* of Pehr Osbeck. Both had participated in herbations:

Many other Sciences dreaded his Botanical wanderings. Youths hungry for learning rambled over the countryside around Upsala, where surely no patch of land exists with the slightest tuft of moss hidden from their scrutiny. During the breaks for rest, the finds were taken to the great Teacher for his judgment. He analyzed all, explained all. The glory of the Creator was constantly in his thoughts as nature's wonderous household was displayed. All was presented as planned and related. All placed in order and given purpose. The link that would seem a danger was shown to be essential for the whole. There he sat, the interpreter of nature, surrounded by his sons. All around were keen eyes and attentiveness.[21]

The astronomer Daniel Melanderhielm took a more hard-nosed view of these almost militaristic ventures:

An ever-augmenting army of botanici was being drafted. Himself the general, compagnies formed, captains and corporals appointed: Excercitie regulations announced, whereupon all began drifting about hither and thither over neighboring meadows and fields. The uniform consisted of small vests and tunics, the weaponry of insect nets and botanical pins. The trophies were wreaths of flowers and butterflies stuck to their hats with the pins.

Linnaeus charged the participants 1 specie ducat, which added up to boosting his annual income by some 400–500 of these gold coins. Melanderhielm and colleagues were simply envious; spending the night outside watching stars was less tempting. An exotic monkey interested the students much more than Newton's calculations, declared Klingenstierna, himself thin and impoverished.[22]

Academic elders found herbations a nuisance. It was tolerable that students enjoyed studying, but with food and music it looked like pleasure-seeking. These negative reactions may well have been because the herbations were privately arranged and, as we know, provided Linnaeus with an income. He had been warned as early as 1748 by Carl Hårleman: "Many of our best friends are as annoyed by the new style of clothing worn as by the new ways of living that will turn the minds of the young away from obedience and enterprise."[23] Hårleman's intervention in the herbations troubled Linnaeus and intensified his "crisis of nerves" in 1748–49. As he observed about himself: "The merry pleasantness and intense appetite for work were no longer felt."

CHAPTER SEVENTEEN

Ex Cathedra

THE YEAR'S QUOTA of lectures was advertised on a single folio sheet, pinned up on the university's notice board. Linnaeus felt that forty lectures per term was reasonable, but privately run collegiums were conducted in addition to the official timetable. Many sets of notes had been kept, most of them—about fifty—from his dietetics course. Of his zoology lectures, forty-odd sets are known, a number that might be doubled with a bit of effort. Some records are extensive, up to a thousand pages, others sketchier. Going through them, it is striking how efficient people were at note-taking, making clean copies afterward, and turning what was said into good Latin or Swedish.

In 1769, Linnaeus—by then ennobled to von Linné—proposed that professors aged fifty-eight and above should be allowed to lecture at home instead of enduring the cold in the Gustavianum, where the main lecture hall was located. That same year, Linné suggested that one lecture and one collegium per day was surely enough or else they would be "poor professors," unable to prepare properly.[1]

The official lectures were free, but students had to pay for private teaching sessions, which could be elementary or higher level. What mattered most: the books or the lectures? The latter were often criticized by people who argued that teachers would be better employed writing textbooks. Linnaeus was almost alone in defending lectures, although it didn't exclude wishing to have *Flora Svecica* adopted as a textbook. He also seems to have used the opportunity to manage several tasks at the same time: lecture, produce pharmaceuticals, and make a financial profit overall. He actually belonged to the small groups of academics who included their research findings in their lectures, an approach that the Romantics would later regard as of key importance.

Linnaeus was a very popular teacher: "Students who attend my official lectures daily invade and fill my garden; never before have there been so many, nor has any other professor ever had such a large audience. I lecture about such things as can be gainful to the state as well as for the well-being of the individual."[2] He dared to refer to social matters and allowed himself to be personal. He wanted to teach the young about what their native country could produce from all three realms of nature—"everything that feeds us and furnishes our houses, also how grass grows and how to cultivate land, industries like dying, manufacturing, home medicine, and pharmacy"—all in order to support "*oeconomiam privatam* in the parishes" and enable people so that they could themselves submit to the Academy of Sciences "such observations that seemed remarkable and describe them rightly."[3]

The number of "real students" can only be estimated. Some criteria have been suggested—for instance, that they should have defended a dissertation, enrolled in the Faculty of Medicine, or at least put their names down for lectures. When, in the summer of 1760, Linnaeus gave a dietetics lecture, 239 individuals were on the list and, of these, 183 have been identified. A total of 774 students have been found, the majority Swedish, but 22 were from Finland, 12 from Germany, 12 from Denmark, and 8 from Russia. Half of the Finnish contingent went on to become doctors. Linnaeus was confident of his power to attract students; when he was offered a professorial post at the university of Göttingen, he remarked: "Believe me, I would pull in half of the youth of Germany to study natural history at Göttingen."[4] Among his 435 identified Swedish students, some are known for their dramatic later lives. Or their oddity: the minister in Finnish Toholampi, Gustaf Cajanus, fathered 32 children in three marriages.

J. G. Acrel told of the lecturer von Linné in a later period:

> From the lecturer's chair he had an excellent manner of delivery, his own and unique to him although supported neither by an especially strong or fine voice nor by a more elegant pronunciation of the language (as his speech was affected by the Småland dialect), yet he never failed in the highest degree to charm his listeners. He knew how to give weight to his words by using short sentences in such a way as to convince all of what he was portraying. He who has heard him read over the introduction to *Systema Naturae*, over Deus, Homo, Creatio, Natura, et cetera, would be more deeply touched than by the most eloquent of sermons. Alongside this ability to persuade, he had the advantage of an incomparable memory and an orderliness of thought such that, on

the basis of small notes on a card, he could present a long speech or a lesson. Only rarely were his lectures written on anything larger than an oblong strip of paper held between his fingers so that he could with his thumb keep a mark at the last place where he concluded.[5]

Thus, the Linnaean tribe was created, which would populate Sweden for the foreseeable future.

What messages did Linnaeus teach from his lecture podium? Very properly, he began his lessons with indicating their purpose and applicability, as in this example from a lecture on minerals. The notes were taken in 1748:

> We see how the entire Globe of Earth is immersed in the Elements, namely the Fire, the Air, the Earth, and the Water. The Science of these is *physica*, of which it is not our intention to speak at this time, as she does not form the basis for our subject, from which much Enlightenment emanates, namely a science we call *historia naturalis*. This science encompasses all that the omniscient Creator has ever made, as in fact no things occur in all the visible world such that it is not *objectum historia naturalis*, that is, no object which is not stone, plant, or animal.... The Earth may well be cursed but still it bears all which will serve to our gain and utility although solely when we understand how to rightly employ it.... These are all matters which require insight into Natural History which here has been but little cultivated. And in this we are like the head lice that crawl caring for naught but their food. Without doubt this has caused us to be unable to investigate the great and awesome riches that Nature holds in Trust as they are so wide-reaching it took until we began to work *Systematice* on their use.... But, Gentlemen, this Science, as well as its matchless utility has for all estates its own great pleasure; and I would insist that each and every one ought at least to take some part therein, as anyone who remains ignorant in these matters will never advance in housekeeping. To that end, God saw to it that Paradise embraced every kind, and that Adam knew them and named them all according to their kind, species, and nature. Should we go through each estate we would observe the exceptional usefulness and indispensable necessity for us all and for each one in particular.[6]

This excerpt demonstrates Linnaeus's ability to switch seamlessly between Genesis and the present, between Paradise and contemporary Swedish society, Adam and himself, theology and utility—indeed, to show

how everything is connected in a vision of common sense. This is contextualized by another passage from the same year:

> On my *excursiones botanicae* I have in some aspects instructed you, Gentlemen, in praxis botanica, to know and name the commonly found herbs, use their proper names, and indicate their use; this, to bring about that you Gentlemen would acquire a taste for this Science. So, it is now my intention to go back and explicate to you, Gentlemen, the Fundamenta Botanica, and offer such principles within the science which can serve to distinguish them; as you will all have found in daily experience, a Medicus, indeed even a mere Quack, but one who has had some education in praxis medica, in the event that he heard or read something, will find it much easier to grasp or understand for himself theoretical truths and principles than would another unlearned one; if so, he will quickly be able to apply it to cases he has met in the past. As I have myself, so in order that you Gentlemen will better understand what I mean, I shall extract for you some examples."[7]

The different elements, theoretical and practical, must be complementary. Another noticeable feature is the formal address to his audience—"the Gentlemen" were predominantly teenagers.

In about 1748, he began a lecture on the animal kingdom by discussing the circulation of blood and lymph, and continued:

> That a human being is an animal cannot be denied by anyone, as she has a heart, hair, flesh, bone, and vessels through which are driven in different directions blood and other liquids;[8] she has all the *Sensus externos*, as we know from other animals, so she cannot be classed as a stone or a plant. If it follows that the human being is an animal, there must also be for her a place within the System. You will grasp at once that there is no place for her among the birds, as she has no feathers on her body. Nor among the Snakes as they have no hair, unlike the human. . . . Thus, she must by necessity belong to the first Class, which is among the animals with four feet and hair on their bodies and whose females give birth to living offspring who do not grow in eggs. And who the young suckle at their teats."[9]

The class that Linnaeus aims to describe was eventually given a name: *Mammalia*. In this passage, we observe his practice of defining by exclusion, an approach he also uses in other contexts.

CHAPTER EIGHTEEN

What Is More Precious than Life, More Pleasing than Health?

ERIK EURÉN, A student from County Värmland, wrote:

> Spring Term in 1747, I also studied Botany with Archiater Linnaeus, and collected a fine herbarium, with which I succeeded in impressing many that I was a great expert on herbs and could so amuse them with my elementary knowledge, which I however improved on gradually. I also attended Linnaei public lectures concerning illnesses and acquired a wide-ranging collegium in such matters. But the most excellent lessons I heard from him were on Diet or how a human being ought to live in order to stay in good health. Nothing can surely be of more use than to know such things. It must be so, as we often err against the principles, if only through ignorance, and also given that one error alone has the power to cause us, before our time, to become sickly.[1]

Linnaeus would later recommend Eurén to the headship of natural history at the gymnasium in Karlstad in Värmland, and his ex-student got the post. Another pupil, the medical student Sven A. Hedin, wrote in the same vein: "If He spoke about the rules that govern diet, He often allowed his students to burst into loud laughter by painting for us the foolish whimsies of fashion, and would speak in a light and pleasant manner, with amusing anecdotes, thus instilling the most useful learning about the care of health and its preservation."[2]

In the winter, Linnaeus lectured in medicine, with a preference for dietetics. The venue was Gustavianum, and the lecture course was on the timetable from 1742 to 1772. He was due to return to teach on the same theme in 1775–76, but it came to nothing as he was by then recovering from a stroke. His thoughts on the subject were collected in the bulging manuscript of an opus he called *Lachesis Naturalis* (Lachesis, one of the three Fates in Greek mythology, was in control of the human life span). He wrote: "The work treats the topic of natural diet from such understanding as has been found in narratives, case histories, investigations, knowledge of different peoples, travels, and studies of physiology, therapeutics, physics, and zoology so that customs and demonstrations are joined by observations. A *philosophia humana* which is Know Thyself." This title was probably penned in the 1740s, but the text is constantly elaborated with new examples and opinions. The published version is selective but can be extended by following up in the seminar notes.[3]

Health mattered above all to Linnaeus: "Good day, Gentlemen! Do you feel well? How are you? Would that you kept in sound health, as it will gladden me much. Stay well! Such are our greetings. What is the highest *finis*? Wealth perhaps, coins aplenty, goods? Or high office, dignity, honorary titles? Would it be wisdom, learning, understanding? No, health has the highest rank, all other things are as nothing to it."

On the subject of health care:

> I prophesy this: the day will arrive when you are ready, after much effort and trouble, to enjoy the rewards of your life's hard labors. But then a guest will come to knock, indeed hammer on your door: Sickness, the servant of Death. The doctor will be sent for, the apothecary called. Then appeals will be made to another doctor; were he too unable to help, no succor can be had. It could be that you live far from doctor and pharmacy. It could be that the doctor cannot come because he is hindered by attending to wealthier patients. It could be that the medic who calls is a quack; maybe you do not yet know of such people, but then all that is yours would be in the hands of a skipper who knew neither chart nor compass nor rudder.[4]

Like other doctors at the time, Linnaeus tacked between the different medical schools of thought. At heart, he was a follower of Hippocrates, convinced that nature is the best healer, and so unwilling to cut into the human body. Anatomy, the basis of modern medicine, seems not to have interested him much, and anyway the deal with Rosén made the subject part of his colleague's territory. Attention to the environment is central to the Hippocratic approach; factors crucial to health were how and where you lived, what you ate, and the rhythm of your life. They were also the subject matter of dietetics, which in its totality had become the star turn of Linnaeus's medical teaching. His huge range of reading about nature and its interactive systems paid dividends in this area, too.

The overarching concept was the Greek *diatia*, the simple life. This, and bloodletting to maintain the body's internal balance, constitute the basis for good health. In the seventeenth century, dietetics was highly regarded by "iatrochemists," who understood physiology in terms of chemical reactions, and included Friedrich Hoffmann; Herman Boerhaave, the great teacher; and the Swedes Benedictus Olai and Johannes Cesnecopherus, who wrote *Regimen* (1613); and Anders Sparrman-Palmcron, who was the most influential. His *Sundhetsspegel* (Mirror of Health, 1642) was a well-informed and widely read pamphlet.

Linnaeus insists that the ideals of exercise and moderation in all things are never wrong, while bloodletting or "feces medicines" (animal and human excrement used in medicines) should only rarely be recommended. Everyone should learn to live the right way. He uses a proverbial turn of phrase: "Live like a wise man and die like a simple one, said the cat to the bat." A range of cultural influences combined in his mind: classical, Christian, Gothic-nationalistic, an investigative approach to medicine and natural history as well as personally observed case histories. The Hufeland fixation on "the art of a long life" became fashionable toward the end of the century, but when Linnaeus was instructing his young gentlemen, the theme was not at all prominent. On the other hand, he imparted quite a lot of kitchen knowledge.

Linnaeus saw dietetics as including air (climate, in our terms), sleep and rest, movement and exercise, food and drink, and excretions and their control. Sense organs should be at peace with the mind. Linnaeus's expansive manuscripts on the subject have only been selectively published. Much remains to be sifted through and will eventually be hauled into the light, especially what he wrote on food and drink; what he has to say about luxury is also of interest. Linnaeus made references to some useless fantasists but also to the oracles of the time such as Santorio, whose book he kissed. It is possible to discern a continuity through his forty years of work, but the material must be asked the right questions about changed perceptions, and revised experiential and ideological concepts.

Moderation matters. This is one of the rules in *Diaeta*: "You eat to build strength not burden your stomach." If you eat too much, your stomach will lose its elasticity and tone. "In Stockholm some years prior to this, I observed as I was in the eating houses that some men bought for themselves each as much bread, soup, meat, etc. as would serve 2 but were nonetheless as skinny as skeletons." It was important to drink with your food: "I know of some who have not taken much drink with their food as they thought it harmful and all have been incommoded by obstruction but as soon as they began to drink sufficiently, quickly found relief." This is what animals do, Linnaeus averred, and would approvingly quote the proverb: "Eat like a cat but drink like a hound so you will live long and stay sound."

You are meant to eat and drink in moderation—advice that Linnaeus will also have expressed in terms of maintaining the body's "humoral balance." Food and drink supply the "warmth in our blood." Well-fed animals can survive in the fiercest cold but on the other hand: "I have seen in the winters so many little birds frozen to death, of which all have been quite empty in their stomachs." It sounds as if Linnaeus had dissected the birds, which is, of course, possible. Variation, he urged, is a good thing: many and diverse dishes stimulate us to eat more. Eating more does not necessarily make you fatter, and to be fat does not at the same time make you "fleshier"—that is, "more muscular" in modern terminology. It is important to eat right but also to be on the move in the right way. Moderation is the watchword. He was keen to stress that drinking is not harmful.

However, cooking might do harm. "Is it then the case that man's stomach is made of copper and animals have stomachs of flesh, indeed is this the distinction between us?" Linnaeus asks rhetorically, referring to the human habit of eating cooked food while still hot. "Hereof we see that all who eat their food hot look as if already dead, due to the stomach being too slack to digest properly. Still, it is believed that the human cannot digest raw meat although her stomach is so constructed and made by nature in the likeness of the dog's." As for people known to eat raw meat, Linnaeus adds: "Nonetheless, the Samoyedic are said to do this. Many wild men also, as it is the natural way. And the elders, the patriarchs, did not grow old by taking no care and the Lapp is healthy by the same reason, that is, by not eating hot food." His next argument certainly sounds persuasive: "When I can cook a stomach in a pot until it comes apart, why could I not ruin him with heat while in the body?" Hot drinks are just as bad: "All tea and coffee are harmful as such substances are emollients and soften the stomach."

What should people eat? What have we been created to consume? Our teeth are mixed-purpose, some to suit a herbivore, others a carnivorous predator: "That mankind is in the same class as the Apes can be seen from the face, teeth, stomachs, hands, feet, etc." Apes eat meat and fruit, although he notes that some people were anxious about eating fruit, especially in the North, where they never ripen properly—or else have gone bad. But this is mistaken, he declared: "I have always eaten fruit in the greatest quantity without aggravations." However, his conclusion was contradictory: because "the more foolish animals eat plants, the more cunning and perfected eat meat," it follows that the most perfect and intelligent species of all creatures, human beings, must eat meat.[5]

With the exception of predators, people should stick to eating such animals as are closest to us in the classification system—the worst, and wholly rejected, option is to eat insects: "*Insecta* ought not be eaten, as all are jagged inside us. Of the same kind are crayfish, lobsters, and crabs and suchlike, as they are as clearly insects as the lion is an animal." And oysters and seashells are the same: "I know of myself never to have eaten any of these without reactions." Despite reports of insects used as foodstuff, bread made from dried and ground locusts, and the like, Linnaeus remained doubtful. Oddly enough, he commented that "so far, no *materia mineralis* is eaten"; odder still, shells of crustaceans and mollusks are classed as minerals in *Systema Naturae*. Linnaeus came up against matters of principle in a long passage on worms in cheese. It was accepted that, as these worms have been feeding on cheese, they were perfectly edible, but what was not known is that "the *vermiculi* wake up in the stomach." Besides: a fox who eats nothing but lamb still remains a fox, and "we do not expect that a man of wealth who has eaten nothing other than the finest cuts of meat, sugar, sweetmeats, wines, etc. would taste more pleasantly than an ordinary man." Linnaeus finds these speculations quite amusing. Ask your gut what to eat and listen to what it tells you: "French chevaliers will eat small songbirds, entrails, and bones, but a gnat's dropping can cause nausea." Next, Linnaeus quoted the Bible: "Ye blind guides, which strain at a gnat, and swallow a camel."

Among the four-legged animals, herbivores are good to eat but meat-eating ones are bad. Sometimes, his turns of phrase take on magical overtones, as for instance when he describes freshly slaughtered meat plunged into the cooking pot: it has "a strange and abominable taste, what gives me gut torments as it should be cooled and not be eaten with its soul." The soul isn't tasty at all! Linnaeus has also written that God in his "ceremonial laws" (the Old Testament rules for what to eat and not to eat) intended to instruct in what was healthy food for mankind; no need to look beyond this for any wider meaning.

He is very keen on the discussion of the naturalness or otherwise of eating meat. It focuses on the human dentition and, hence, where to place humans in the taxonomic order of living creatures. A related question concerns how to understand the present in the context of classical ideals. At first, Linnaeus seemed to approve of vegetarian principles but later came out in favor of being omnivorous. It should be stressed here that vegetarianism was not an unknown concept. Examples of practitioners include Anna Ovena Höijer, the lady poet of Värmdö, and Christopher Polhem; the latter has explained that milk and fruit are our natural foods and that only dogs, our guards, should be allowed meat. Around 1754, Linnaeus's pupil Mennander wrote in defense of animals, and similar thoughts were expressed by the lacto-vegetarian and instrument maker Johan Backman.

Why did Linnaeus dislike mushrooms? Why didn't he believe in potatoes? What was his problem with crayfish? He wouldn't consider eating horse meat either. "Mushrooms are better suited as food for flies than for people," he stated in *Flora Lapponica*. "The fungi are deadly . . . and I shudder to think of the known fungus with its unknown poison," he exclaimed in his 1740 lecture in the House of Nobility. He described mushrooms as an "errant horde" in a frequently quoted phrase from *Deliciae Naturae* (1772). As for potatoes, Linnaeus had never "dared recommend them as much as others of my fellow Swedes. Those who eat them *in loco natali* [unclear what he meant] became as those suffering with leprosy." Furthermore: "To plant potatoes in good soil is easy but to make Farmers, yes, even their swine, eat them will surely be more difficult."[6]

Drinking was another favorite subject. All prepared drinks were inferior to fresh water. "No animals will drink our drinks; Virginians and Lapps drink but the *pura* water. The latter are as agile as line-dancers." And: "I never felt better than in the mountains where I could do nothing but drink the pura water, water running over clay." Drinking water was guaranteed to protect against "stones, gout and podagra, dropsy, and innumerable other afflictions." Nor did water lead to constipation, one of the scare-story topics of the time: "We know that frogs and lizards do not crawl inside us but should we drink water containing their spawn, it quickens in the stomach and will make us fall terribly ill with a repulsive and miserable sickness."

Beer and wine form a large part of this subject. Rule number one is that they harm you much more when taken hot. "Beer comes in many sorts; the longer boiled and the livelier, the better it keeps. . . . To our good fortune it will act as its own purgative"; he means that we will quickly get rid of it. "When I drink beer or wort quite fresh, I often burp with sour heartburn, proof that she becomes vinegar in my stomach." Oh dear! Wine is special: "A divine liquid, a heavenly nectar but must still be drunk with care.... From its drinking to excess will come podagra, gout, [kidney] stones." Wine would be much healthier if "wine waiters" did not treat it in different ways "so they will get it sold."

Linnaeus composed a drastic description of the progressive decay of the alcoholic. "Whoever begins to help his appetite with strong liquor, who takes a dram for the cuckoo, or so that the cold will not hurt him and then warms himself with another hearty one, or who takes a drink to get the cabbage down so it will not cause *colique*, or, indeed, by taking ever more aquavit, attempts to drive away sorrow, worry, intrigues, aches, other grievous things—such a person will acquire a new passion called aquavit sickness, and it is one which nothing in the whole pharmacy will help against, so he will in the end have a body that is half-dead, trembling, and hectic. In the end it will succumb to severe dropsy."[7]

When *Lachesis Naturalis* was published, it provided new, fiery material for the teetotalers' propaganda. Their campaign also engaged academics, and in a small collection of essays printed in 1907, the publisher A. O. Lindfors wrote an introductory chapter on "Linnaeus and Sobriety," which is full of distressing examples of the evils of drink. He also discussed "Linnaeus the popular educator," a figure in line with Luther, Grundtvig, and Ruskin, who, far from being confined in the past, still is a relevant, warning voice. Linnaeus had observed during his *Skånska resa* that getting drunk on aquavit too frequently seemed to worsen the miserable lives of the poor. Perhaps it was true that some two hundred years later, Linnaeus might have saved some people from a sad fate; hence, it is ironic that, nowadays, we can buy "Linné aquavit." However, for him, alcohol was above all a medical issue while Samuel Ödmann, for instance, saw "excessive drinking as highly damaging to diet, manners, church services, and all society."

Linnaeus reflected on the subject of tobacco smoking: "It is strange indeed that an evil-smelling herb that tastes bad, is poisonous, and makes the user's mouth stink should be so well-liked in all corners of the world." His attitude toward the habitual use of tobacco, and consumption of coffee, tea, and chocolate was in principle negative. These were all substances that had not been known to antiquity and were therefore not recommended in classical medicine. Furthermore, they were prepared as warm or hot drinks, a habit he considered dangerous in itself. In terms of the economy of the state, all these habits should be rejected, and besides, they were of the exploitative upper-class habits and never practiced by natural people like the Sami (he was wrong there). Linnaeus confessed on 17 March 1772: "Tobacco I have smoked often and perhaps too much, and

FIGURE 29. Kitchen interior, from *A House-Keeping Primer for Young Females* by Cajsa Warg, 1755. Uppsala University Library.

from that time [his first intimation of a stroke] onward I have withheld on that matter. I was earlier constrained to use tobacco against my toothache, which has plagued me from my early years in this world and by which usage I rid myself of him."

"Habit is, as it were, second nature": the quote from Cicero had become one of Linnaeus's key themes, and he says in *Lachesis Naturalis*: "Habits are worse than Satan's temptations. Lord help whoever acquired a bad habit." He cited an example: "When the traveler first arrives in Stockholm or Amsterdam in the heat of summer, he is suffocated by the privy smells; stay for 14 days, and he senses none of it."

Linnaeus issued a harsh judgment of his own time: "For how long has the world not existed without cooks, brewers, teahouses, coffeehouses, aquavit distillers.... Given how many 1000s whom fouled wine has struck down by fatal *lithargyrio* [sleeping sickness], such draughts should be forbidden on pain of execution." He also quoted a passage from Petrus Lagerlöf's *Spanmålsskötsel* (On the care of harvested grain), which stated that the Swedish authorities used to forbid wine imports because "it was observed how men thereof became like women, so softened and lacking in strength as not to endure labor and warfare"—sounding a Gothic note. Linnaeus clearly supported the notion that the Swedes were in an unpolluted state until the middle of the seventeenth century: "During Queen Christinae reign begins the Swedish pride in knowing of the Gallis."[8] The modern Swede began to eat like the English, drink like the Germans, dress like the French, build like the Italians, smoke like the Dutch, and down hard liquor like the Russians.

The popularity of Linnaeus's lessons in dietetics was not lessened by the lecturer's habit of featuring the students' own problems in his lectures. He might warn them: "Those who study lack *motus musculorum*, causing the fibers therein to become enfeebled and also the lung, a frail tissue, to grow lax and pneumonia follows." On the temptation to drink wine and spirits, he reminded them that "little harms less"; and on encouraging exercise, "to be mobile in moderation is to live freely but too much is slavery." Moderate application was also right when learning: "Studying makes wise, too much makes you mad, not wise." It was, of course, a little absurd to lecture on food to young men who would probably never have to cook a meal in their lives. The paradox was resolved by his explanation that he taught general precepts, not recipes. His predecessor Lars Roberg had said: "A youth ought to be in the kitchen and watch the womenfolk as they clean out their fish, scald and pluck their chickens, bleed the lambs and calves." The genre of the cookbook was established at this time: 1755 saw the publication of Cajsa Warg's best seller *A Helper with the House-keeping for Young Females*. Bengt Bergius's great work *Tal om läckerheter* (On Delicacies, 1780) shows how interested he and his contemporaries were in the joys of a good table. As for Linnaeus, he admits that he left all housekeeping in his wife's hands.[9]

His lectures on dietetics are invaluable because, in them, we can—almost—hear his voice. We also hear the spirit of his age in his discussions of what is luxury or misery, what is the relationship between medicine and morality, and also medicine as a social mission. These lectures are part of a long, classical tradition, but adapted by Linnaeus to his audience. Many hundreds were listening, but the effect of his teaching outside the lecture theater is difficult to assess, as is the extent to which the teacher lived as he taught.

Linnaeus the medical man is at the forefront in his expositions of mistakes in contemporary medicine; what he saw as obstacles to good practice was set out in twenty-one items, in a brief dissertation entitled *Obstacula Medicinae* (Hinders to the art of healing, 1752): (1) practicing has become dull routine; (2) ever-changing hypotheses; (3) poor diagnoses of morbidity; (4) poor knowledge of pharmacy; (5) the activities of quacks; (6) doctors being too cautious in prescribing medicines; (7) doctors prescribing too small doses; (8) pharmacists being ignorant of botany; (9) doctors being similarly ignorant; (10) use of combined medications; (11) use of mixtures of incompatible medicines; (12) ignorance of the natural classification of plants; (13) insufficient use of folk remedies; (14) insufficient educational travel outside Europe; (15) failure to study botanical treatises; (16) careless preparation of medicines; (17) neglected plant cultivation; (18) doctors and pharmacists being ignorant of medicines from other countries; (19) general ignorance of exotic plant species; (20) dull routine dominating pharmacies; (21) insufficient care taken in the collection and storage of plants.

As can be easily seen from the list, many items overlap and many of them concern botany. Linnaeus often wrote in order to attack the practitioners he called "half-educated" and the outright quacks. He defended academic medicine as it was taught but also argued in favor of plant-based medication. Such drugs were "simple" but often powerful, and their strengths can be understood from the natural classification of the plants. However, his emphasis is, in particular, on the primacy of theoretical knowledge. It will be clear from later chapters that his ambition was to clear away all these "obstacles" and give medicine a new start.

Linnaeus was a populist medical educator. His writing about observations made during his Swedish travels is a case in point, as are his almanac essays, which are modeled on his lectures on dietetics. He wrote a dozen or more essays aimed at a wide readership: two of them were directed to seafarers, others dealt with fevers and the use of fir and pine products in healing. Yet others described the usefulness of knowing about the reproduction of plants and the effects of tea, coffee, beer, and aquavit. The latter pamphlets were reprinted many times, as they were seen as particularly useful. All this contributed a great deal to the public health information of his time and included many pieces of advice that have been noted by cultural historians: the many uses of pine sap and chopped pine needles; the observation that fir wood is an excellent material for making fiddles; and the placement of small fir branches on the head is a tried and tested cure for headaches. Generally, pieces of information were lined up without any linguistic finesse. The Creator is frequently referred to in all matters related to the intricacy of reproduction in plants, and that it depends on the sexes of the plants is not treated as titillating—it makes good sense, especially as "it is easily seen that nature commonly uses like means and causes as to reach like outcomes."

Even though this chapter has in the main discussed the course on dietetics and mainly what was said about food and drink, Linnaeus was a medical man who wrote on a wide range of topics. Works such as *Genera morborum* and *Materia Medica* were very successful. Still, dietetics was Linnaeus's major medical specialty and, besides, an area of knowledge that expanded with the passing years. He published several dozen works in this area although the exact number is dependent on how the contents are classified.

Ordering the multiplicity of illnesses was an important element in his academic work. Placing a particular condition in its class would indicate some of its significant characteristics and also help doctors establish that they were talking about the same problem. Linnaeus corresponded over a long time with the professor of medicine at Montpellier François Boissier de Sauvages de Lacroix. Their collaboration was valuable to them both.[10] In the academic year 1741–42, Linnaeus lectured for the first time on the systematic classification of illnesses; the last of these lectures were given in 1774–75. Pehr Osbeck's lecture notes from 1746–47 show that Linnaeus had placed illness in nine classes, given here in the original Latin: the illnesses, *morbi*, are given a dominant characteristic—(1) *critici*, (2) *phlogistici*, (3) *doloriticii*, (4) *mentales*, (5) *privatii*, (6) *spastici*, (7) *deformans*, (8) *evacuatorii*, and (9) *chirurgia*.

At this point, it should be emphasized that, after his years as a doctor in Stockholm, Linnaeus did no clinical work; this was in Rosén's professorial territory. As professor of theoretical medicine, Linnaeus seems to have had no practical contact with *Nosocomium upsaliense*—Uppsala Hospital. He was not keen on anatomy, as we know, but did formulate medicines and took a great interest in the pharmacopeia and *materia medica*. He and Abraham Bäck collaborated on the first Swedish national compilation of illnesses and appropriate medicines, the *Pharmacopeia Svecica* (1775). It catalogs some 380 substances, with possible combinations producing hundreds of "composite medications" and listing also many hundreds of useful plant-based preparations. Among the illnesses treated, hypochondria and hysteria were seen as especially common and also treatable with a range of medications.

In addition to this major work, Linnaeus wrote dozens of shorter dissertations on medical topics. His contributions, in volume and influence, deserve more space than they have been given here.

CHAPTER NINETEEN

Academic Amusements

YOUR THESIS "DISPUTATION" was the high point of your university studies, the ultimate academic goal. Disputations could be *pro exercitio* or *pro gradu*—for the purpose of practice or for a degree.[1] Pehr Kalm never sat examinations but was appointed to a chair at the university in Åbo. This fact points to the relative weight of factors such as regulations, finance, praxis, and your relationship with the professor as well as indicating the importance of how the respondent expressed himself in writing: all in all, it was an intricate system. The dissertation came with "paratexts," and the dedications were especially important: you praised whoever had funded you, your teacher and your parents, and should add some more or less felicitous verses in celebration of friends and fellow students.

The disputation should advance the respondent's career, but it did not always work that way. If presided over by the actual author, which was quite common, he could hardly fail you outright. But the whole performance was aimed at showing off the respondent's ability to defend his—or, often, the *praeses*'s—thesis. Even so, it wasn't critical: Carl Wänman's dissertation, written by himself and on the subject of illnesses in seafarers (*Sjömännens sjukdomar*, 1768), was set out in paragraphs of just two and half pages, and so poorly defended in terrible Latin that the faculty decided to fail it. At that point, royalty intervened because Wänman had brought the queen specimens for her *naturalia* collections from his travels in the East Indies. So, the faculty passed the dissertation, and Dr. Wänman went on to enjoy quite a successful clinical career in Finland.[2]

Linnaeus defended the role of disputations as tests of knowledge but was also critical as can be seen in a 1760s essay entitled *Academia*: "A confrontation that seems it might end with murder, is abruptly concluded with compliments and congratulations. Following custom, *praeses* steps down from the cathedra as the victor. The performance ends in festivities."[3] It could well be that some present-day colleagues nod in recognition at this.

The promotion—the moment when the successful respondent shows off his claim to a place on the heights of Parnassus—was a costly exercise. You had to pay both to participate and to throw the expected party afterward. A note in the protocols of the medical faculty records that music was lacking in 1743, but on the other hand, the playing in 1768 was excellent. Linnaeus declared that the promotions should be accompanied by "less bong"—true, that the prince (Gustaf?) would often be present, but "for one *promovendus*, all the academy must be on the move, the bells rung, the guns fired, and music more costly than the entire set piece be played, while lessons cease as all *professores* perforce must be present when an elaborate oration is presented, for which programs must be printed, and *cursors* made to parade in full uniform accompanied by guards." A brisk summary of academic pomp, although we know little more about the proceedings.[4]

Supervising dissertations and presiding at disputations were important elements in Linnaeus's academic teaching. His dissertations, some 186 in total, make up a significant portion of his written work. It might seem a very large number, but his successor C. P. Thunberg presided over 294, though in the main they were repetitive works of cataloging. Linnaeus saw his academic papers reach a wide, international audience after publishing them as *Amoenitates academicae* (Academic Amusements) in six, or possibly eight or ten, volumes. Internationally, the work has long been misunderstood because the authorship has been attributed to various respondents.

The disputations were indeed high points but surprisingly few descriptions remain. Attribution has become a never-ending game for historians, but the rule of thumb is that the *praeses*—the chairman of the proceedings—is the author if the respondent is described only as "defendit." Otherwise, he would be explicitly indicated as the "auctor," or at least the text should identify his contributions. Linnaeus's dissertations seem to have been the work of the respondent in only ten or so cases, although some might have been responsible for the Latin translation or the illustrations.[5] The respondent would refer to his supervisor in terms such as "Linnaeus has demonstrated" or "our famous *praeses*."

It is relevant that the presiding professor charged for his presence, which boosted the disputation business to industrial proportions. Petrus Ekerman, professor of rhetoric and eloquence, who had grown wealthy after presiding at 516 disputations, launched a special offer: "For 200 thalers, my dear chap, I'll pen your dissertation myself."

Linnaeus clearly thought of the dissertations he had supervised as his own: after all, they were included under his name in the series *Amoenitates academicae*. In his case, such presentations were not just practice runs, or meant to be of strictly local interest. Sometimes, Linnaeus's continued interest in the topic was indicated by later annotations. In each case, the dissertation was a research paper by an observant *praeses* with seminal ideas. J. G. Acrel wrote this appreciation of Linnaeus's authorship, which has been widely quoted: "All disputations he wrote *dictando*, in part in Swedish and other parts in Latin, which the respondent had to tidy up and make orderly. Although he was himself not much concerned with the Latin, he would surely express his pleasure at good writing and also express the opposite. To write a presentation for the disputation would therefore take not much longer than three hours, as they were but lectures on the topic from which the respondent took notes."[6]

The dissertations had print runs from a few hundred to around five hundred copies. By 1778 the number of printed dissertations at Uppsala University had reached 7,450; the counts in Lund and Åbo were lower but still considerable. The language of first choice was Latin but Hebrew and Greek were acceptable; Swedish was later permitted in certain subjects such as topography, economics, and natural sciences. Linnaeus would present his theses in Latin but a handful were translated into German, French, and English. They were regarded as of international interest, and amendments indicated where the Latin texts collected in "Academic Amusements" had needed correction.[7]

Attempts have been made to sort Linnaeus's dissertations by subject—such as botany, zoology, and so on—but precise subdivision turned out to be difficult. Page numbers varied between 8 and 50 pages; the average was 25, which makes the estimated total 4,300 pages. The first five years of production demonstrate the wide range of Linnaeus's interests. In 1743 the first paper is on dwarf birches; he explained his choice of subject in the introduction as intending to treat a native Swedish subject according to all the rules of academe though this patriotic line of thought was abandoned in the 1744 treatise on the fig. Few realize that Linnaeus used observations from his travels on the islands of Gotland and Öland for a later dissertation on coral reefs—just as few are aware that Darwin was the first to explain the creation of coral reefs. Linnaeus's 1745 dissertation on the academic garden *Hortus Upsaliensis* is important because of all the information it contains on the garden's history, plant population, and its greenhouses. *Sponsalia plantarum* has become a classic; it is a reworked version of his publication of some fifteen years earlier on sexual reproduction in plants. The title page has an equally classical illustration of the analogies between an egg and a seed and also of wind pollination of the woodland plant dog's mercury. Under the motto *Amor unit plantas*, it was for a long time the logo printed on the Swedish 100 kronor banknote. His 1748 paper on the tapeworm was the first zoological thesis in a long series. *Oeconomia naturae* from 1749 is a key work on chains of cause and effect in nature. In the same year, his favorite pupil, Pehr Löfling, contributed *Gemmae arborum* on the subject of tree buds and further speculation about principles of reproduction in plants.

Another dissertation published in 1749 was truly forward-looking: *Pan Svecicus* deals with animal feed. What do cows, horses, sheep, and pigs eat and what do they reject? He was familiar with the subject from home and based his discussion on team project work of a kind that has become common in our time. A group of Linnaean pupils collected many, varied observations and organized them in long tables. Linnaeus wrote that all this "builds the foundations for a wholly new science in Natural History, as it shows how all Botanists each in his locality can institute such inquiries and so judge whether grazing is good or evil."[8] Arguably, a very significant aspect of this work is that lack of space had forced all the contributors to use the binomial nomenclature, a few years before the publication of *Species Plantarum*. It would become the first in a small group of monographs about domestic animals: the dog, reindeer, sheep, and guinea pig.

More dissertations followed: in *Semina muscorum detecta* (1750) he claimed to have identified the "seeds" of mosses, which made the cryptogams analogous to the observably seed-bearing phanerogams. *Plantae hybridae* (1751) treats the strange phenomenon of asymmetric "sports" in plant populations (peloric plants).

By now, we have reached number 33 of Linnaeus's dissertations, with a large number left to consider. He acted as the supervisor for about a quarter of these. In 1752 he wrote a study of native edible plants—quite a long list. The subject had practical aspects, and in 1757, a year of major crop failures, his text was translated into Swedish. In 1752 possibly driven by winter temperatures as low as −31°C, he published a paper confirming that cold weather was not always conducive to good health. Also from 1752, *Noctiluca marina*—the Latin name of sea sparkle—discussed bioluminescent organisms in general; that same year, Linnaeus presided at a dissertation on *Rhabarbarum*—another fast turnaround.

FIGURE 30. Guinea pig. Illustration in the 1754 dissertation *De mure indico* by Linnaeus's respondent J. J. Nauman.

His study of the guinea pig, *De mure indico* (1754), deserves a mention not only because of the charming illustration but also because the respondent, Johan Justus Nauman, had an unusual and sad fate ahead of him. Nauman, a pharmacist's son, had joined his county "nation" at the university but was eventually thrown out, probably due to alcoholism. He became a tramp, and his life ended in 1778, when he was just thirty years old.

CHAPTER TWENTY

Appetite for Work, Weariness, Communication

LINNAEUS'S DAYS AND nights followed nature's timetable, as was proper. His autobiography claims that he slept between 10 p.m. and 3 a.m. in the summer, and for twice as long in the winter—between 9 p.m. and 7 a.m. He claimed to work "around the clock" and his exceptional productively is proof enough. "When Linnaeus had an idea, he always wanted to write it down. Among the children, however much they played and shouted, he could pull down the lid of the bureau and settle down to write. He would let nothing disturb him and did the same when visitors were in the house."[1] As he said himself: "I lecture 5 hours daily, at 8 a.m. with Danes, at 10 in *publice*, at 11 and 12 with Russians, and at 12, *privatim* with Swedes. Checking the proofs of the second edition of *Fauna Svecica*, he added an explanation to "why I write to you at 2 o'clock at night."[2]

In 1761 he wrote to his pen friend Nicholas Jacquin: "While my colleagues are free daily to take pleasure in their daily lives, I spend night and day on the exploration of a field of knowledge so wide that thousands of hours would not suffice to bring the work to a close, not to mention the fact that every day I must give myself time to correspond with other learned men—the entire routine makes me grow old before my time." To another friend, he penned this note, demonstrating the full span of his talent for feeling hard done by: "All this has crushed my spirit. Hereafter, I shall sleep until 8 o'clock."[3]

While clearly not a shining example of moderation, he recommended sensible restraint to all his students: "I have known quite a large number of gifted men, but few have also been of sound body. Their habit has been lay in bed by 11 in the evening or 12, and rise at 3, to assure them of time to become learned which they have achieved. However, while running to catch that shadow, they dropped the meat." Linnaeus's favored source of such sensible admonitions was Bernard Ramazzini's *De morbis artificium diatriba* (1700, 1713), a classical work of clinical medicine. As his son remembered him in 1778: "My dear departed father would do no work unless in a reasonably good mood. He rose early in the mornings and when he woke, he would light a fire and soon settle down to write. However, should he perceive the slightest torpor, he walked away from his work and rested from it while he smoked a pipe, or else he would throw himself on the bed. He had the gift of falling asleep straightaway and snoring, after which within a quarter of an hour he would be once more cheerful and go back to his work. In this manner he would continue many times during the day, until the early evening at 4 o'clock when he liked to have company in order to dissipate that which had during the day occupied his thoughts."[4]

One aspect of his working life is so self-evident that it is easily forgotten: good light was a necessary condition for the time-consuming work of plant examination. Daylight was precious, and to keep one's eyesight intact, it mattered to get as much light as possible from the windows. Otherwise, the student depended on candles, either smelly tallow ones or the more expensive wax candles—all, of course, entailed a fire risk. Joseph Banks wrote about his deceased friend Daniel Solander: "That part of the day, which is the brightest, he devoted to Botany though his companionable spirit would never allow him to also spend his evening [studying] in the *Museo*."[5] All we can do is wonder how Linnaeus could possibly have completed all his writing despite the lack of good light. Even the wax candles were inefficient, flickering and evil-smelling, eyeglasses provided only poor focus, and street lighting was nearly nonexistent. The Enlightenment was simply—dark.[6]

Another cause of wonder is Linnaeus's capacity to handle intellectually the ever-increasing flow of information that began with the discovery of the New World. Already as a student, he had been working on comparative dichotomic tables, and also used diagrams and long lists of different nature specimens. Such paper-based systems later developed into a card index, organized alphabetically, which gave unlimited room for additions. This approach became widespread by the end of the century and revolutionized data management.

The cards measured 7.5 × 13 cm and were marked with the name of the genus.[7] His backup system helps explain his well-documented ease as a lecturer, using only a brief note to aid his memory; Acrel has confirmed this. A more traditional way of storing data was to make marginal annotations in his personal copies of his printed works, as for instance in *Systema*

Naturae.[8] These updates came in handy when new editions were to be published.

Success early in a career brings continued demands for more of same. By 1748 Linnaeus's appetite for work had become complicated by recurring depression. "A malignant year," he noted in his autobiography.[9] He was referring to the criticism of the herbations, and in particular the unofficial personal accusations against him, bandied about by the *juris* professor who was probably offended by the fun and games. Then, Hårleman sent Linnaeus a letter "which practically struck me dead and killed sleep for 2 months." The letter, dated 28 July, stressed the general outrage: "Many of our best Friends are mightily upset by the donning of such clothes and also the new manners which divert the minds of youths from all proper obedience and behavior."[10] Two months later, Linnaeus wrote to Bäck: "It was a good thing that you, dear Brother, alerted me from my sleep; I admit that I now and since a long time have loved quietness and peace; nor have I any longer the spirit to sail outward if I can remain in the harbor."

Next, Hårleman made a more focused attack, this time against the passage about clearing land by slash-and-burn in Linnaeus's *Skånska resa*: "The French have a saying that no man is so deaf as he who will not listen; such is also the way with the son [he meant Linnaeus] who is deaf as he appears to continue to defend his old Weaknesses due to the same having been instilled in him in Smålen [*sic*] inter alia some fine authority which in no way honors him, however allow me now to see the page of the return from Småland before I altogether decline to be a loving father!"[11] Linnaeus backed down, but not until 1750–51. By then Engelbert Hallenius, one of his best friends, had objected to Linnaeus's work *De curiositate naturali*, and, in a very personal attack, it was prohibited from publication abroad.[12] "These 3 crushed Linnaei spirit so that he threw his pen away and declared that he would from now on work only moderately." Indeed, a malignant year, as a fourth censure hit him: in *Ouvrage de Penelope*, La Mettrie ironized about Linnaeus's mission to systematize.

Worse still, his father died. Should his state of mind not be diagnosed as stress-induced burnout? Could it not be that one component, his arthritis, hindered his work on the sixth edition of *Systema Naturae* as well as other publications? In any case, after his first decade as a traveling, organizing professor, Linnaeus was fatigued.

Then, in 1752, he received a knighthood in the recently created Order of the Polar Star (Nordstjärneorden).

"Students, bailiffs demanding debt payments, slaps, fights, the Frisendorfer family and old Norrelii estate, stories of royal Physicians,

Librarians, the Celsius family, clerical synods, markets, orations, teaching programs, christening festivities, such matters we act out at our comedic meetings. We hold forth on such things and forget much of what needed our urgent attention. Thus, the world plays on. We do not miss time before she has already gone on her way." Linnaeus gave a snapshot of the memos on the vice-chancellor's desk and went on to philosophize in his letter to Bäck dated 16 January 1750.[13] Again, a year later: "I have now for all of 10 years given my body no peace, neither day nor night. No carrier's mare has been as sorely battered as I have done it to myself. I have at last taken note of the vanity of it all and reckon it has done nothing but bring me harm, bitterness, grief.... After a month of *exercitie* I find myself somewhat improved, seeing my folly and my previous illness."[14] He sometimes felt lonely, as he admitted in his letter to his siblings and wider family at New Year's 1768: "At Christmastime when it is practice to lay a table for the cruelest of wild creatures, it is given to You all to speak together and joke in innocent trust of each other." He continued: "My wife and family can travel to her family in Fahlun for New Year's which is barely 200 miles away; I stay here alone as I have no strength to travel 500 miles to mine." It might be lack of energy, or lack of will, but whatever the reason, work kept piling up.

The periods of depression still recurred. In 1756 he wrote to Bäck: "Now only the heavy years remain, when step by step you approach death and all joy is fading." Again, in 1758, also to Bäck: "I have no more time to write today, the weary hand flags. I am a child born to misery and had I but had a rope and English courage, I would long since have hung myself.... I fear my wife is once more with child; I am old, gray, emaciated and besides the house is already full of children; who will see to it that they are fed? It was my day of misfortune when I took the professorial chair; had I only held on to *auream praxin* [here, "lucrative practice"] I would surely have provided very well for mine."[15]

Suicide was believed to be a grave sin but did occur, and had a Swedish defender in Johan Robeck, a citizen of Kalmar in Småland. Robeck had set out his case in *De morte voluntaria* (1736) and later followed up his plea by killing himself. Linnaeus's acolyte Artedis died in a manner that could be interpreted as voluntary, and there is also this opaque note in the *Lachesis* manuscript (translated from the Latin): "Fear of Death. Those who are heavy of heart rescind this inclination. I at four hours; to die at one's own hand."[16] Does the time refer to the dreaded Hour of the Wolf?

Knowledge is generated by communication and spread through being circulated. Knowledge must travel, adapt, and grow in the new surroundings. Researchers have every reason to seek new contacts and communicate in as many directions as possible. Linnaeus would not have identified these activities with "networking," now the subject of so much theoretical interest. The spider—if the concept refers to its nets—had an emblematic context of "touch" among the five senses (the other four are represented by the lynx [sight], boar [smell], ape [taste], and hare [hearing]). If one were to pick one Swede as *the* contemporary specialist on nature's own networkers, it would have to be the civil servant Carl Clerck, a good friend of Linnaeus's and author of *Aranei Suecici* (1757), a work that laid the foundation for the international nomenclature used to classify spiders. As Linnaeus put it: "This study, which I today have the pleasure of receiving from you, Sir, is one of the most beautiful, if not the most beautiful work ever published in Sweden."[17]

Communication networks come in many shapes: within families, between friends in different places, and in collaborations within and between universities, industries, and other institutions. Academic, social, and industrial networks often interlink more or less informally. Linnaeus was a member of the Academy of Sciences, a teacher and dispatcher of traveling students, and a correspondent. His networks were growing, linking him also to the world outside the universities, as well as making him a member of a large assortment of European academies.

Contact networks can be visualized by circles and arrows in diagrams. One of these has been based on the archives of Linnaeus's rival Albrecht von Haller: his existing letter count tops 17,000! Among other things, this demonstrates just how far north and hence how isolated Scandinavia was—Uppsala was just about the only place included. On the other hand, unlike Linnaeus, Haller had no American correspondents.[18]

The popular image of Linnaeus with his butterfly net and vasculum, stalking insects and flowers across lush meadows, has little bearing on the man bent over his desk as he had to, managing his correspondence, writing prescriptions and memoranda, issuing certificates, and taking on the work of the university court. Writing letters in particular must have taken up much of his time: about 5,000 have been archived, but his total has been estimated to between 8,000 and 10,000 letters. They vary in length, and the apparatus of polite address can occupy a lot of space: repeats of "Knight and Archiater Carl Linnaeus" soon fill a sheet of paper. Letters to Linnaeus have been fairly systematically kept, unlike his own letters to correspondents. Notable exceptions include his letters to his friend Bäck, while Bäck's letters to him are missing. The estate let Bäck have them back in order to prepare his speech *in memoriam* of Linnaeus. It is also very sad that the letters from another important correspondent are missing: Pehr Wilhelm Wargentin was the secretary of the Academy of Sciences.

The most inexplicable of the gaps in the collection is the almost total absence of letters from the family. One might ask, for instance, what happened to the letters exchanged between the engaged couple during Linnaeus's years in Holland? Could it be that there was no room for Sara Lisa's letters in the bag he packed for the return journey? But Linnaeus had saved and brought back letters from Gronovius, and also retained letters to Stobæus. Comparatively little remains from his time abroad, partly explained by the 1827 fire in Åbo when Linnaeus's letters to Johan Browallius went up in smoke. The records also contain many so-called minus-letters, missives that were mentioned but cannot be found. In the majority of letters, Latin and Swedish were used just about equally (about 2,500 exist in each language); there are a couple of hundred in English—these are significant— about 150 in French, 100 in German, and just 24 in Dutch.[19]

Linnaeus's roughly 600 correspondents were scattered all over Europe; his proud lists in *Vita* testify to this. He must have drawn up some kind of system to keep track of them. The rate at which the network widens can be quite easily worked out: back in 1735, five people wrote to him and he wrote to just three; in 1736, the numbers had grown to twenty and nine, respectively. A decade later, he addressed sixteen individuals and wrote frequently to Bäck. In 1756 he received thirty foreign and Swedish letters, and by 1776 he had thirty-four regular, mostly foreign correspondents. On 25 February 1774, the Abbé Duvernoy in Montpellier dared to write to the very busy Swedish professor, who replied on 6 May, saying that, had he had ten hands, he could not have found time to write to everyone who wrote to him; but he did answer all the same.

The idea of publishing the foreign correspondence occurred to Linnaeus as early as 1769. He informed the potential publisher, Gjörwell, at length about them, "a larger bundle than can easily be lifted," and worried that they would make tiresome reading for people unversed in natural phenomena. Another concern was that the many tributes would cause his readers to conclude that the publication was motived by vanity. In print, they would fill some fifty volumes and to copy them all out would be like "forced labor." He asked for time to consider the project.[20]

In the 1770s, the Swiss botanist Albrecht von Haller published letters from Linnaeus without asking leave. J. E. Smith's translations into English of a selection of letters by Linnaeus and other leading botanists came out in two volumes in 1821. Various minor selections were also published, and larger ones followed later in the nineteenth century, above all the schoolteacher Ewald Ährling's collection of Linnaeus's letters to Swedish correspondents. Ährling also completed key studies of bibliography, and it was partly through this preparatory work that Thore M. Fries could begin on the definitive publication, announced at the Linnaeus celebrations in 1907. The plan was for two series of volumes of Swedish and international correspondence, respectively, both intended to be complete. Even early on, the completeness was rather compromised by arbitrary decisions: one letter was rejected as "utterly incomprehensible," probably an "harcèlerie" (harassment). A letter from Linnaeus's unpleasant brother-in-law Carl Fredrik Bergcrantz was also dismissed: "of no conceivable interest."[21] After a brisk start, the pace slackened and, with about a third of the job done, the work came to a halt in 1941. The task was restarted in the 1990s, aided by modern data handling. Fairly soon, some 250 years after its inception, it should be completed.

The conditions of professional correspondence are becoming clearer. Penning the text tended to be quickly done, driven by a lack of paper and time, and awareness of the demands of the postal service. By New Year 1731, he apologized to Stobæus for sending "such an ill composed and confusedly written letter without having first a clean copy made. . . . Hope now however that my youthful faults will be excused." Praised for a letter, Linnaeus responded: "I am thankful to my Br. whom it has pleased to flatter me for my letter to Mrs. Hårleman; surely it cannot be right as, may the Lord punish me for it, I wrote it out free hand, without concept and so speedily as I write to all, and in a room full of people."

Well-turned letters of introduction helped oil the hinges of seemingly closed doors. Linnaeus was on his way to England when Gronovius wrote to Philip Miller: "Dear sir, I don't doubt you have heard that the King of Sweden and the University of Upsala have sent a Gentleman to make observations there. He is here in town [London]."[22] His reputation was boosted by having traveled in Lapland, been painted in full Sami dress, and published *Flora Lapponica*.

All this was news.

Donations and gifts oiled the operation of the network. To keep his sponsors happy, Linnaeus asked Bäck to collect six bound volumes of *Fauna Svecica* and "hand these to the chancellor [Gyllenborg?], Tessin, Horleman [*sic*], Colonel Palmstierna, Count Ekeblad, and Baron Höpken. If you would be so good as to deliver these with your own hand when the occasion allows and you, Sir, happen to pass by; I can think of no one who I believe would do me this favor with a lighter mind and honesty."[23] At that time, these six men were Linnaeus's most important supporters and backers.

Distant friends and colleagues were remembered by their faces in portraits, which covered at least one of the walls of the grand salon in the prefect's house and were listed in *Hortus Upsaliensis* (1745). Among them were likenesses of Vaillant, Boerhaave, and Tournefort, and the total amounted to more than twenty images. Some were missing, or worse: for instance, Linnaeus's spiky colleague from the years in Holland, Albrecht von Haller, was exceedingly irritated to learn that his portrait had been hung upside down.

The naming of plants was one means for botanists to strengthen contacts with colleagues. A tall, exotic evergreen with splendid flowers was given the genus name Dillenia, a fitting tribute to Johann Jacob Dillenius, who cut a splendid figure among the botanists. Gronovia is a genus of climbers, named after a man with an exceptional reputation as a collector, and it was the famous Gronovius who decided that a small flower found in Lapland, an inconspicuous and overlooked herb with a brief flowering period, should be called Linnea after Linnaeus as "he is like it." In *Species Plantarum*, plants given genus names based on Swedish notables included Frankenia, Tillandsia, Celsia, Rudbeckia, Bromelia, Browallia, Artedia, Baeckia, Solandra, Hasselquistia, and Forsskålea. Some quite shameless proposals came up, as when Francis Masson resisted having a Massonia given Thunberg's name and insisted he wanted this from "the Father of Botany" (1775). Tracing ironies and insults in the names was quite popular. Siegesbeckia is a disagreeable type of thistle, Bufonia is a variant on the Latin word for toad, and the three members of genus Browallia are said to correspond to the developmental decline of Johan Browallius, according to Linnaeus's German visitor Johann Beckmann: *humilis* (humble), *elata* (elevated), and *alienata* (alienated). Beckman specified further: the plant species corresponded, in descending order, to Browallius's way of looking up at whoever he was speaking to, the way his mood could swing from humility to arrogance, and the way his arguments in the row about falling water tables had became "almost foolish." Thore M. Fries, keen to save the reputations of his heroes, contradicts this anecdote with chronological calculations and also refers to Browallius's premature death. Fries believed that the anecdote might well have been "concocted," or else Linnaeus "cannot escape rightful criticism for a much too intrusive, even improper jest at the memory of a deceased friend."[24]

His most valuable colleagues were his traveling students, trained to be amateurs in the true sense of the word: "lovers." His disciples were meant to love their teacher and their school and subject. Eventually, they became members of the teacher's network. They acted as Linnaeus's eyes

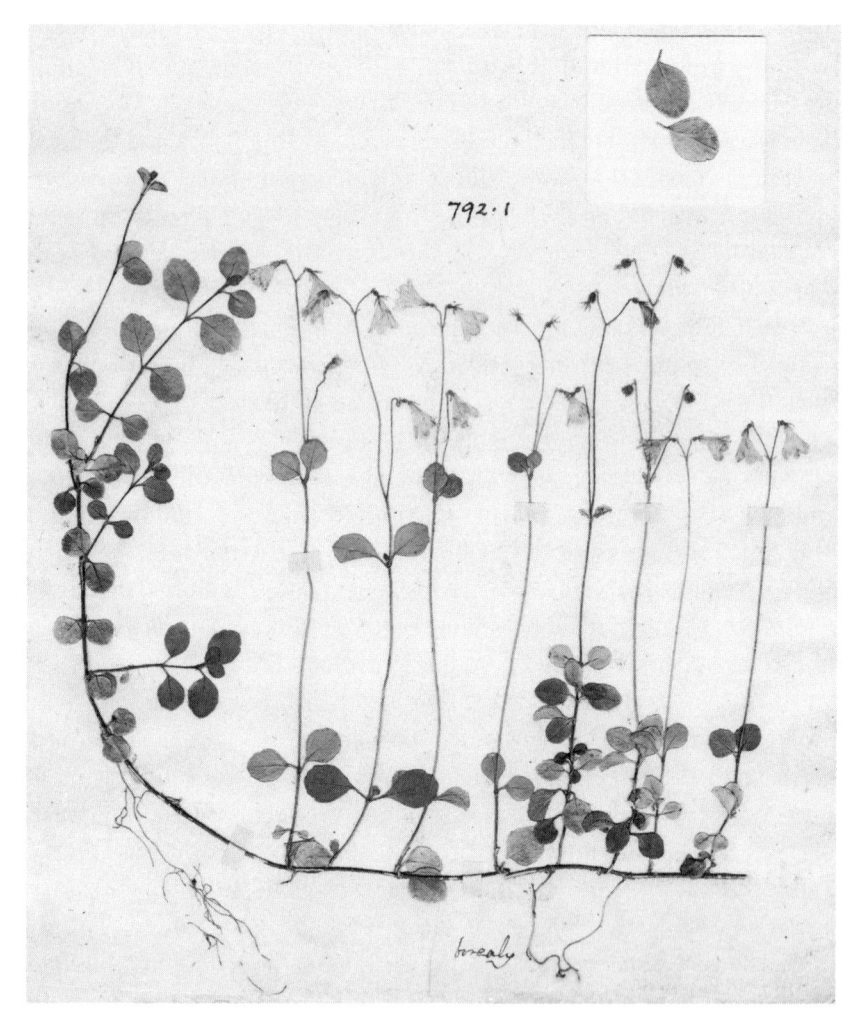

FIGURE 31. *Linnaea borealis. The Linnean Herbarium.* Linnean Society of London.

in distant places, supported his work for almost half a century, and formed between them a relatively coordinated network. They shared a language and came to regard themselves as a group, often expressing their happiness at belonging. Johan Otto Hagström, a doctor in Linköping, wrote to Linnaeus on New Year's Day 1766: "Among all lovers of Natural History, you feel as one of a dear tribe and without concern about Nation.... Thus, I love a Torbern Bergman, a Modeer, a Rolander, a Solander, men whom I have not yet seen.... Therefore, I count those as of my closest family in connection with the Science. It gives me more than delight to learn that the whole world over, work is being done on Insects. You, Sir, and Archiater, have in these men planted a light."[25] In a letter from 1753, Linnaeus himself speculates on what the group can do to deal with noisy opponents: "I have been of the habit to comfort myself with the comical anecdote of the Bitch who happened to find a nest where she could give birth to her pups and when a horde of dogs came running to drive her out of it, asked them to let her stay for so long that her little pups had grown large enough for her to take them into the forest. So, they [the dogs] left and then finally they came back by which time the pups had grown and had teeth in their mouths so she told them to go to it, as it was time to settle the matter of the nest."[26] But could it be that his disciples/pups counted a Judas among them?

It is tempting to apply the term "school of research" to the generation of Linnaean followers. They were not a homogeneous squad, and their future fates would differ markedly, depending on where and when they traveled. Many died in the field and have been regarded, to varying extents, as victims of their teacher's ambitions. Seen in those terms, he—or, arguably, his science—carries a heavy burden of responsibility. It is true that Linnaeus pressured his pupils to travel without informing them about the risks. However, it is worth keeping in mind that high mortality rates did not stop other young men from joining the East India Company.

CHAPTER TWENTY-ONE

When Linnaeus Wrote, Salvius Printed, and Tessin Bought the Books

LINNAEUS WROTE SOME seventy books—some were short, some voluminous, and some came out in many editions. His innumerable essays had been published, as had the 186 dissertations he had presided over, almost all of them written by himself. His contacts among printers and illustrators were significant, as were his views on their work and fee structures. He cared about the appreciation of his work by scholars, as well as reviewers and the wider readership. The title of this chapter suggests that, despite inevitable problems, his publications were successful. How did he make sure that all worked out well in the end?

In the world of publishing, Sweden was a small country. Its book industry had long been in the hands of immigrant printers, book traders, and publishers. Most of them came from Germany, including men with names such as Russworm, Lochner, Grefing, and Kiesewetter (until quite recently, Swedes used a version of the German word *Buchladen* to mean "bookstore"). The Swede Lars Salvius was an upstart who had married into the job but eventually became the country's leading publisher. He was an adept marketing man, with excellent contacts among politicians and senior civil servants. In 1752 the Printing Trade Society was formed, and its first catalog (1753) listed the output from 16 printing workshops based in Sweden. The list covered 68 pages and, of these, 28 (40 percent) dealt with works from Salvius's production line. The Royal Printers, run by Peter Momma, were close runners-up. In Stockholm, the book trade had concentrated in the Old Town, with addresses in Stora Nygatan,

Västerlånggatan, and Stora Gråmunkegränd—streets lined with printers and bookbinders, bookstores and a lending library (it later moved to Norrbro). Salvius and Momma, who came from a Dutch family, both worked on internationalizing the book trade.[1]

One recurring issue was whether to print in the native language or aim at an international readership. The Academy of Sciences made the patriotic and financially less demanding decision to publish in Swedish. Sweden had to be cultivated as a marketplace, and the books had to be adapted accordingly. In the academy, Linnaeus spoke warmly in favor of Lars Salvius, insisting that he "uses more beautiful fonts and prints more accurately than any other; also, he is Swedish and lives in Sweden, so he is the first native printer so far able to debit Swedish books with foreigners and so earn foreign thlrs. You, Gentlemen, are Swedish and ought to serve Swedish people." A concise, political statement into which Linnaeus managed to squeeze five patriotic references. He continued: "This Grefwing is a foreigner, as we know. His credentials should be taken from him as his printing is so *vitieux*—indeed, so corrupt that no observation is printed without many faults in it. I know that myself as I had many times corrected *errata* in *Öland resa* and yet they are there in print, as the firm without doubt cared nothing for them." Linnaeus rounded off: "May you behave like Swedish men, not only say so in your speeches, but act accordingly."[2]

Scientific texts should reach an international readership, and in the eighteenth century that meant being produced in Latin. For typographical purposes, Latin was printed in the Antiqua font—a "Classical" style—while Swedish, a Germanic language, was still usually printed in Fraktur, one of the older "Gothic" typefaces. Fraktur was the font of choice if the intention was to reach a wide audience by selling cheap books in large print-runs. The Academy of Sciences chose Fraktur for the first year of its annual volume of *Proceedings* but then changed to Antiqua. Linnaeus's *Öländska och gotländska resa* was printed in Fraktur by Kiesewetter, but the later *Skånska resa* went to Lars Salvius, who had it set in Antiqua. The Linnaean journeys were printed in a hybrid typographical format because of the mixture of predominantly Swedish text and descriptions of species in Latin. Later, the Antiqua typefaces became widely used, also in literature aimed to interest a bourgeois readership. The font issue is only one example of the important midcentury modernizations affecting the book market (the Fraktura styles, however, were still used, especially for religious texts, well into the twentieth century).

Other considerations in book production concerned stylistic matters and whether a single volume was preferable to two (first chosen for *Species*

Plantarum). The Linnaean book was typographically challenging, with indentations, italics, and scientific terminology, all of which demanded a high level of printing skills. Salvius's firm was responsible for some of the most complex, high-status publications of the time: the *Proceedings* of the Academy of Sciences, Dahlin's history of Sweden, and Linnaeus's *Flora* and *Fauna Svecica*. Salvius knew how to handle the authorities and sent them a steady flow of submissions. After a few years, he had reorganized the way Swedish books were printed and sold, often using patriotic arguments such as the risks to the national balance of trade unless Swedish industries were favored. The Hats, the party forming the government, were in thrall to mercantilist ideas and wanted to support exports in particular. Salvius profited more than many other entrepreneurs from the 15 percent export credits on offer from the beginning of 1757. He had recruited as clients the finest names in Swedish academic and literary circles, partly because his publishing house offered its writers decent fees. His weekly magazine *Lärda tidningar* (Learned News, 1745–73) served a range of purposes—as a general forum for news and comment, an advertising sheet for his own publications, and a semiofficial information channel for the Academy of Sciences. Salvius also ran an international network of commissioners with bases in Paris, Antwerp, and Leipzig, important for international sales.[3]

From 1745 Salvius published Linnaeus's works; thus began a period during which, as it was said, "Linnaeus wrote, Salvius printed, and Tessin bought [the books]." Linnaeus wrote to Bäck in 1756 pleading that "whatever you do, see to it that the great Tessin is kept on in his position. This, as the sciences will last for as long he lasts because they were once aroused to life by him." More about Tessin to come.

In his letters to Salvius, Linnaeus is friendly and businesslike, and aware of the work-related stress of their professional roles: "It is well understood that you, Sir, do not hurry me with the text of my Wästgöta travels. I have awful amounts to do but hope in a short time to get something more read." On 14 January 1747, Salvius wrote to say he had taken delivery at a payment of 3,000 thalers of Dutch paper currency for Linnaeus's forthcoming book but had no manuscript to print. If only it were to be in his hands soon, *Hortus* could be ready by April—"which I doubt would have been the case, were this task to be set up in Holland." Many years later, on 30 July 1765, Linnaeus ended a letter to Salvius by saying, "I wish you the best of luck with your work, however would you, Sir, as the eminent Director, consider dispatching the work, as soon as it comes out, to London, Montpellier, and Italy, which will surely make the proceeds sufficient answer." "The work" was probably *Systema Naturae* in its large twelfth edition.[4]

Linnaeus was continuously preoccupied with plans for new books or new, extended editions of old ones. In 1765 he felt that *Flora Svecica* must be given a new edition: "You have, Herr Directeur, already grown old and rich, but that a revised version is needed cannot be denied. I am prepared every day so with me there will be no lag."[5] The third volume of *Systema Naturae* was the subject of his note to Salvius on 15 January 1768: "My old Benefactor, Dear man, do not scold me. If M. Benefactor had watched me as this past week ended, I assert that M. B himself would have dragged me from my work with force. The days are short. And without light I cannot with the best will rightly examine the stones. I press on more than my strength will allow."

These are two recurring themes, health and those two constantly hurrying each other on. As the years passed, the tone of their exchanges grew more and more jovial, the seriousness and warmth of their good relationship finding expression in the genre of the New Year's letter: "When I think back on my most and best times, nothing has progressed in this world more evenly and pulled along the work of this work more than you, Sir, by now over the last 30 years. I thank you for all your friendship, all your forbearance, all the care you have taken. We have both soon spun our spindles full; we will see how others go forth to follow us. I shall, until I am in the grave, with every *estime* revere Your company and trust in this world."[6] Linnaeus, however, has suggested a rather more down-to-earth take on Salvius's importance to him; as he said in a letter to Bäck dated 12 March 1773: "It is an exceeding harmful thing that D. Salvius should die before me; I wished to have him first publish both *Materia Medica* and *Species*. Now it may well be no one will dare do it."[7] Salvius's printing presses and workshop were sold to a J. G. Lange, who probably had a German background.

Illustrations mattered hugely to the kinds of books Linnaeus authored. For many years, he could only dream of the splendor of his and Dionysius Ehret's *Hortus Cliffortianus*, composed and printed in Holland. That gold standard was within reach in a few of his later publications: *Museum regis Adolphi Friderici* (1754) and *Museum Tessinianum* (1753). In Sweden, such skilled work was mostly done by the competent Carl Bergquist and the elegant Jean Eric Rehn under the aegis of the Academy of Sciences. Linnaeus both needed and desired fine illustrations: they were essential for proper descriptive work as well as status symbols. The conditions demanded that he be compensated by making his technical language as clear and precisely detailed as possible. A bonus for his readers was that the lack of illustrations made the books substantially cheaper to buy. Linnaeus was aware of the situation: "Auctor fears that, by this condition, botany will one day go under. Were one to buy, for example [a long list followed], such a library would demand expenditure of a few thousand copper plates."[8] What really counted in specific cases becomes clear from his enthusiasm for Clerck's great work on Swedish spiders. Generally, natural history as a science had to remain verbally expressed knowledge, apart from pedagogic diagrams which to be effective had to be simple.

For natural historians, being able to draw was one of the more desirable talents. In Sweden, many of the leading ones were good illustrators—like Lars Roberg, Olof Rudbeck, Urban Hiärne, Magnus von Bromell, Charles De Geer, Carl Clerck, and Erik Acharius. Opinions of Linnaeus's ability varied: the entomologist Felix Bryk held him in high regard—"His drawings are often done with a sure hand and elegant lines"—while Wilfrid Blunt, an art historian and Linnaeus's biographer, wrote that while Matisse spent a lifetime trying to draw like a five-year-old, Linnaeus managed to do it effortlessly. Linnaeus seems to have agreed with Blunt: "I understand less of how to draw and drawings as I have not learned the art."[9] However, it should be pointed out that Linnaeus knew very well how to do simple diagrammatic drawings; his early notebooks were crammed with sketches, tabulated observations, and diagrams. For instance, his Lapland records are full of drawings of plants and ethnographic objects, often with a functional slant. It is possible to see this as sheer happiness at portraying what you see, a joy that died away later, even though his thinking remained fundamentally visual. Generally, why do children stop drawing things?

It is unmistakable that Linnaeus found tables and diagrams essential for grasping systems and studying their functions, and was notably keen on layout and placement of texts, which mattered greatly to understanding. He liked having frontispieces to his books and enjoyed designing them, as he did for *Flora Lapponica* and probably *Hortus Cliffortianus*. The absence of images other than the linnaea (!) in *Species Plantarum* was dictated by the potential bulk of an illustrated version, as was also the case with *Flora Svecica*. He pleaded for illustrations to be limited to four safe, fundamental characteristics: number, shape, position and proportion. In *Genera Plantarum*, a famous passage defines why he would say no to genus but yes to species illustrations, a rule that was applied to most subsequent dissertations: "I cannot recommend representations to determine genera, indeed I reject them even as I admit their value to little boys; I also admit that they might be of some use to the unschooled. Before the letters were made known to us mortals, before speech, all must be expressed in images. But, as soon as letters had been invented, we had a safer way to convey ideas. . . . Hence, we should attempt to express in words all the particular features as clearly—if not more clearly still—than

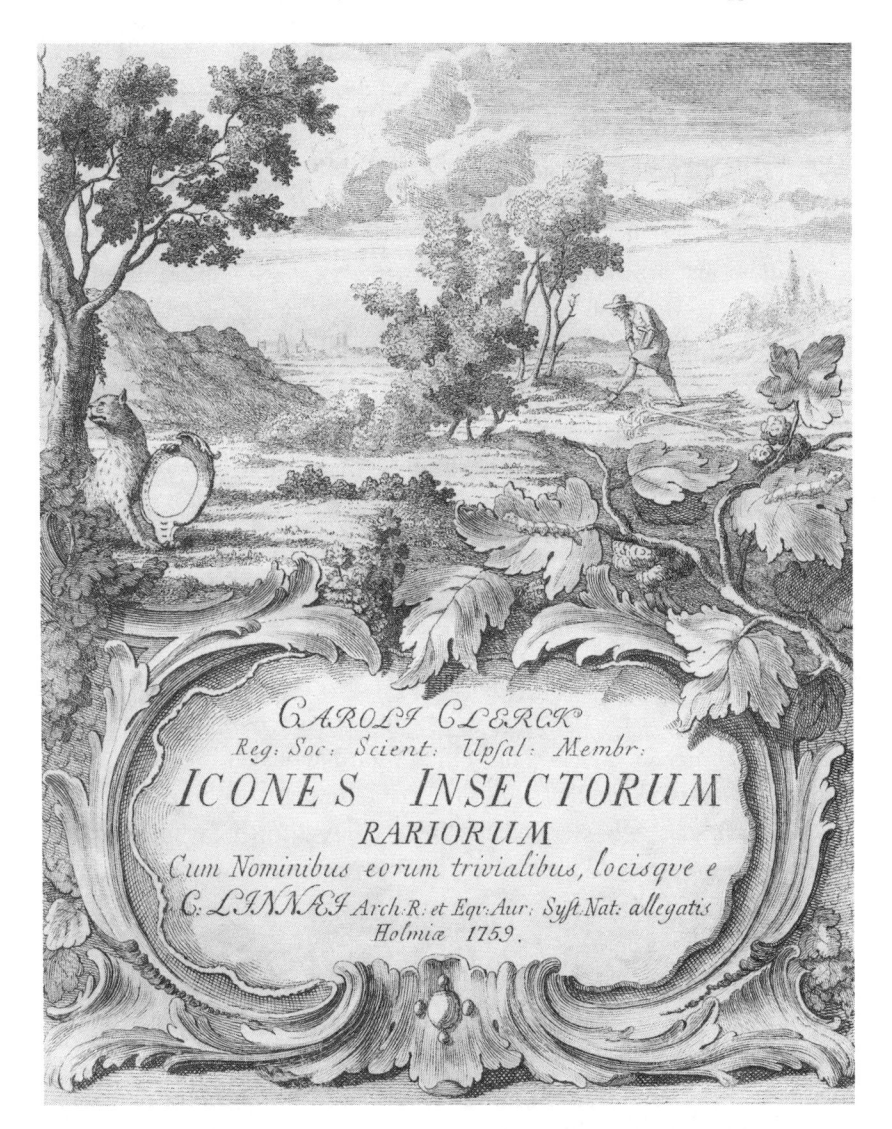

FIGURE 32. The motto of the Academy of Sciences—an old man digging—shows the utility of etymology; the lynx, to indicate that you need good sight to study insects. A few silkworm larvae are shown on the leaf of the mulberry tree. The title page of Carl Clerck's *Icones insectorum rariorum*, 1759.

others do in their fine drawings." From such thoughts emerged the purest form of Linnaean plant analyses, intended to rely on the letters of the alphabet linked to the organs and structures of the plant.[10] Linnaeus created a simple word-and-number system as the descriptive tool of natural history, but all the same, he would have liked very much to see some of his works join the club of expensively produced books, as once with *Hortus*

Cliffortianus. The means were not available, and he lamented: "It is miserable these days for us here in Upsala, that we have no one here who can provide us with one single drawn figure. Hr. Hallman went away on a journey suddenly and unexpectedly just as I had planned the following day to use his services." His complaints did not stop; even in 1763, he could write: "Imagine that for 27 years . . . never to have been able to raise one fine and honest drawing." Is his inability to organize a handsome edition of one of his works a source of real grief? "If only during my life in the profession one such draftsman had seen to it that the discoveries I have made and described were well drawn. Now my time is almost gone," he sighed in 1774. An old man by then, Linnaeus was over the moon to learn from the orientalist Jacob Jonas Björnståhl that the Princess of Karlsruhe was planning a great illustrated Linnaean work for just that year.

It was true that the Swedish lack of illustrated works was symptomatic of a relative paucity of both funds and artists, but it is also relevant, and worth emphasizing, that Linnaeus's handbooks were cheap and easily distributed. The negative aspects of production actually helped make natural history widely accessible and popular.

Fees and print runs were genre- and author-dependent, changed over time, and were paid in money or in kind—for instance, in free copies. It is difficult to compare what was on offer in different circumstances and then transfer payments to modern terms; that exercise is further complicated by the dedication system that entailed a kind of begging. Lavish praise sometimes paid off; in the reign of Gustav III, men of wit and learning had a chance to acquire a pension. Salvius had this to say on the subject: "It could well happen that from time to time that a decent enough *auctor* makes himself known, but does not require a payment for some minor piece that he has handed in to be printed; however, and on the contrary, most of the *auctores* have begun to raise their *proemia pro labore* to *alterum tantum* [fees have doubled]. They go by the rate of exchange in so far as when a ducat is valued at 18 thrs, they were pleased enough to receive 3 copper plates in notes for the sheet of paper but when he costs 36 they demand either 6 plates or the ducat *in natura*."[11] It sounds complicated and must have been.

In Sweden, Salvius initiated a system of decent fees, but there were large variations between pay scales. The Uppsala theologian Nils Wallerius received 12 thalers per printed sheet for his *Psychologia empirica*, but Linnaeus got 18 thalers per sheet for *Flora Svecica*. Linnaeus could demand advance payments and recruit people to help with the writing: "Many thanks for the ducats but the detailed description will arrive by the next mail, as I cannot get my amanuensis to join me here in the country."[12]

He was, as always, in a hurry, and the sense of urgency grew more pressing toward the end of his life: "My days run past speedily and what I have to do must be done soon." He continued: "We are already old and I believe of the same age; we have both exhausted ourselves. We have followed each other through time. With the distinction that you, Herr *Directeur* have been *Fortunae* and I but the poor *Palladis filius* [son of Knowledge]."[13] Though one might ask whether Linnaeus, too, had benefited from *Fortuna*.

To make sure of a good mention, Linnaeus wrote the reviews of his books in his publisher's magazine. The self-appraisals make a great source of comments about the scope and development of his activities, and of how he regarded his work. In a printed volume, these appreciative reviews cover almost four hundred folio pages and occasionally contain information not included in the books. Contrary to what one might have expected, the wall-to-wall praise that dominated in his early career became more subdued once he was an established authority. As he wrote, on the 1748 publication of the sixth edition *Systema Naturae*: "Herr Archiater has suited this work well to lectures and the education of youngsters at the Academies, but it would nonetheless be useful if the learned author's many activities would allow him, for one, to summarize it as a brief compendium for Schools and Gymnasia and, for another, to edit the work in its entirety with all its species, the former being intended for the young and the latter for their teachers." On the publication of the twelfth edition in 1766, he wrote: "The work as a whole is aimed so that each animal and plant should show his own name to the Reader and the name should be conducive to the Reader finding all the noteworthy *auctores* who have written about the same; so that by such guidance each and every one may be able to study in the Book of Nature. . . . This Edition leads us to congratulate not only the Author but also our Swedish Nation and all those everywhere in the world who are curious and hungry for knowledge."[14]

His work had, in the main, a target readership among students and botanists-to-be—people who needed handbooks such as *Flora Svecica*. His role as a populist educator was very significant and his influence reached far into the future. The spirit of his account was to turn botany into a democratic science, appealing to both state authorities and small-time farmers. The idea that botany was the people's science became reflected in the flower symbols of Romanticism—the *Blaue Blume* (blue flower) image. The wide appeal of botany attracted women as well, at first as practical knowledge of use in their work around the house and home. Botany and zoology were among the first areas of academic study open to women. Now and then, the world outside was told of learned females:

"Mrs. Burmannus has one among the prettiest collections of *papilions* from every part of the world." A year or so later, the marchioness of Rockingham was said, after the death of her husband, to have "developed a fondness for the *Botanique*, such that she reads all Linnaeus's writings in Latin and thereafter knows how to examine plants."[15] Linnaeus was also informed in 1764 that Lady Monson understood botany very well and had a beautiful collection of plants and insects. When the after-dinner toast to the king was announced, she would always drink to Linnaeus first "as he is the king of the realms of Nature." It was suggested that naming a plant after her would be a good idea. Linnaeus apparently followed the advice and wrote to her to express his highest regards.

Select bands of ladies joined Linnaeus's herbations; in 1747 the group included Madams Schönström, Feman, Landtberg, and Linnaea, and a Mademoiselle (?) Elvira or Elvia.[16] The editors of Linnaeus's *Bref och Skrifwelser* (Letters and Writings) apparently didn't consider the five letters from Anna Christina Lagerberg worthy of inclusion, even though she wrote on subjects he was keen to hear about, like the worms causing havoc in harvested and stored grain, and cures for callouses and corns. Once, she even enclosed a piece of bone.[17] Ulla Sparre, who was married to Tessin and shared with him an outstanding library of books on natural history, also took an interest in Linnaeus—though he remarked: "It may well be that the countess Tessin is serious about botanizing; were it the case she would surely be *unique* in that science among her sex."[18] Elsa Beata Wrede also told him that she was an enthusiast. There is, however, no known contact between Linnaeus and Eva de la Gardie, the first woman elected to the Academy of Sciences. On the other hand, he wrote flatteringly to the countess Catharina Charlotte de la Gardie: "But as God is my witness, I was astounded to find what I had not seen to that day: while customarily, Nature hands its gifts in such a way as to offer one eminence in beauty, one in wisdom, one in gentleness, one in yet another quality, now he has given all to one person and so I could not but be struck dumb."[19]

In his dissertation *Auctores botanici* (1759), Linnaeus devoted a special section to women botanists. Four of them, among 250 men, had "made themselves deserving of Botany": he named Hildegard of Bingen, Sibylla Merian, Elisabeth Blackwell, and Jane Colden.[20] Later, in *Deliciae Naturae* (1772), new women were added. He argued that, once you have become skilled, it is as quick "to read Nature as any other book; yes, this is true of women at least such as Lady Monsson in London, Anna Blackburne in Oxford, and Miss Colden in New York." The three were the most distinguished "women in botany" at the time, but eventually others would join them.

CHAPTER TWENTY-TWO

Linnaeus, "the Sexualist"

LINNAEUS'S LIVELY INTEREST in sex is a characteristic often cited in attempts to portray the man. People of course refer to the sexual system he introduced into biology and the moral outrage it is believed to have caused. The academic community has tended to accept that his contemporaries were scandalized; the notion that female readers found it especially challenging has been repeated but actually seems to be a later construct. One might note that when Linnaeus's student Johan Haartman had completed the translation of the "sexual system," the published volume was dedicated to a woman, the already mentioned Ulla Sparre. True, the classes were termed "regiments," and a politer tone introduced into the author's sex-based descriptions of pollinations, a process he would aim to make more acceptable by using poetic imagery of weddings between classical gods and goddesses (what if the analogy had been boys and girls having it off after a dance!).

In his oration at his old friend's funeral, Abraham Bäck spoke of Linnaeus's "light-spirited and playful method of showing herbs in flower as if males and females paired in their wedding bed, engagements which tempted even the beautiful sex to study botany." Perhaps the poetic possibilities of the approach were thought particularly appealing. In any case, no one in Sweden, male or female, seemed to have felt any moral reservations about the summery love play of stamens and pistils in meadows buzzing with bumble bees.

What were the reactions abroad? Linnaeus claimed: "My Books were among those forbidden to be imported by the Church State, except now Prof-Bot. Maratta has been through his dissertation and Minasius joined the Profession in Rome, who shall now *in publice* read my *Systema sexuale* by the agency of Cardinal de Zelades." As far as we know, Professsors

Maratti and Minasius, the latter a linguist in Naples, had no connections with the Vatican, and no work by Linnaeus has been found in the index of forbidden books. The church apparently did not object to Linnaeus's system—but some Roman botanists did.

In 1778 in Lichfield, Erasmus Darwin was pondering the right English words to use when turning the ideas in *Systema sexuale* into a poem: should he use "males" or "masters" or perhaps "lords" or even "cuckoldoms." The poem, discretely published abroad and given the title *The Love of Plants* (1789), was a success and included in Darwin's more scientifically intended work *The Botanic Garden* (1791). Generally, the English readership took the whole idea in stride, even though it was later suggested that the system "scandalized many," and some felt that the vocabulary should be toned down "for the sake of the ladies."[1]

The Duchess of Portland, who had an outstanding collection of *naturalia* and specialized in shells, embraced the Linnaean nomenclature and system with enthusiasm, regardless of any alleged impropriety and eventual mistakes. Both aspects, however, had offended the botanist Emanuel Mendez da Costa: "Science should be chaste and delicate. Ribaldry at times has been passed for wit, but Linnaeus alone passes it for terms of science, as His merit in this part of natural history is, in my opinion, much debased thereby."[2] The views on the matter held by another great shell collector, Queen Lovisa Ulrika, are unknown. It might be that men took offense more readily.

Around the beginning of the twentieth century, a historian of medicine in Finland argued that "Linnaeus in fact used coarse expressions and had given free rein to his imagination." As proof, he offered this erotica from *Hortus Cliffortianus*, for the genus *Gratiola* (hedge hyssop): "Oestro venereo agitata foemina stigmate hiat rapacis instar Draconis, nil nisi masculinum pulverem affectans, at satiata rictum claudit, deflorit, foecunda fructum fert." He felt this kind of thing should not be translated, so we will not do it here, just in case.

The self-taught British botanist William T. Stearn described the more dramatic aspects of Linnaeus's system of sexual reproduction in plants as "a gift to Freud" and called Linnaeus "a botanical Peeping Tom." Given his charting of the private activities of plants in their "bridal chambers," why not the Dr. Kinsey of flowers? Bill Bryson's best-selling *A Short History of Nearly Everything* discusses how Linnaeus created analogies between mollusks and women's external genitalia, and quoted names used in *Praeludia*, for example, *Clitoria*, *Fornicata*, and *Vulva*—thrilling and shocking then and, apparently, now.[3]

What governs nature and hence mankind? A note by Linnaeus from 1730 suggests a pansexual view of history. He saw an analogy between the fall of man and the sexual awakening at puberty: "What were then the forbidden tree and its apples, of which Eve took one, and felt it tasted good, other than *poma Adami* [the word *testiculi* was crossed out], although I will not dispute or take the part here [the words "defend or deny" crossed out], although they soon hid themselves." Further: "In the middle of the Paradise Garden—*in medio corpore Adami*; the snake was treacherous—penis; *fons aspectu pulcher—testiculi*; delicious to eat—*usus titillans*; were then ashamed—*a coitu tristitia*; saw that they were naked—*pudor*; would die—*qui procrearent*.... But as soon as Eve took *poma Adami*, her thoughts became different, and she liked the taste so well she forgot the other things [delight at flowers and birdsong]."⁴

Diaeta Naturalis, a work from Linnaeus's youth, often links observations to the sex drive: "It is true that the first law is *crescite et multiplicamini* [increase and multiply], and is indescribably habituated into all bodies. So, I need not describe it as in my mind I am convinced that all feel as I do. Morality is often of no help." Later, on the similarities with the drives of animals: "Love is for all animals the greatest *oblectamentum* [pleasure]. As we observe excellent men who are enlightened with fine wit, we find they were all horny. True, lust makes one stupid. Love, however, has been the cause that we and the world, yes, all who live, are alive. Love's *abstinentia* for 100 years would ruin the world. Hence the pagans thought her [Venus] a mighty goddess." But he noted differences between human beings and animals: "You rarely see *marem* [the male mate] bite or beat his female, *ut canis, taurus*, etc. But the man often so to his wife." He reflects: "Why do people kiss, why do men touch the breasts." And who was the "I" in this quote? "When I lie with someone ugly, I feel anxious but with someone lovely, it feels that I have done well."⁵

The better class of contemporary garden was not only decorated with ponds and topiary but also with statues and busts. Linnaeus had spotted a lead copy of the Venus de Medici (first century BC copy) in the Piper Garden in Stockholm, and was keen to acquire it. His bid was accepted and the statue placed in the Academy Garden; a drawing from 1773 shows a man admiring the beautiful goddess. The weekly *Posten* (it lasted for two years, 1768 and 1769), published by Anders Berch the Younger, commented sharply (Berch Sr. and Linnaeus were not on the best of terms): "I cannot here refrain from my duty to name one location in which such a representation, an unclothed figure of Venus in full size, is on display for all to see: The Royal Academic Garden in Upsala; it surely presents an offensive prospect to the young students. The garden holds no other statues; thus, one catches all eyes. Why not replace Veneris with a Priapus? He was the god of gardeners to whom sacrifices were made." For reasons of propriety, the statue was later moved and promptly disappeared. A new copy was put in its place in 2009.[6]

Human sexuality is the subject of a short text that has been ascribed to Linnaeus, albeit on slightly shaky grounds: *About the way to become together* (*Om sättet att tillhopa gå*) was first published in 1969; the publication was based on a recently discovered set of notes. There is evidence that he had written something of that sort. J. G. Wallerius writes in his autobiography: "Her Roy. Highness [Ulrika Eleonora] had heard that Linnaeus brought into the House of Nobility a band of youths from every kind of estate, held lectures on among other matters *de Sexus femini genitalibus*, which news did not please the Queen."[7] Then, another similar set of handwritten notes has been found, both stating as the author "The Archiater and Knight Carl von Linné," using the title and ennobled name conferred in 1757. Since both sets had been copied, the conundrum just grew more tangled. The texts read like Linnaean lectures and in their frankness about sex seem close to what was said in the dietetics course, even though it had no exactly corresponding content. A similar lecture on sexual biology and mores—but more tightly formulated, shorter, and older in style—has been ascribed to Olof Rudbeck the Younger. It, in turn, can be shown to have derived from the medieval best seller *De secretis mulierum*. The theme of sex is as old as the hills . . . or older.

The subject matter in *About the way to become together* is treated positively and linked to the contemporary theories of reproduction but also to what is nowadays known as a "physiognomic hypothesis": "It would seem that no one can create himself otherwise than he is, just as the face displays tempers such as chastity or lasciviousness, or a good or evil mind." Among later propositions along the same line of thought: "If in mind cruel, proud, and lacking in compassion, her glittering eyes may well attract a boy to climb on top of her, and she will be more likely thin than fat, and such girls are always willing for *coitum*." The easygoing style fits with a nonacademic audience. The notes deal with the process of coitus and how it changes the body, continue on a respectable note with a section on "the conception of a child and its birth into this World," and end with "signs that a woman is violated," which doesn't mean "raped" but that she has "lost her virginity," become pregnant, and is in that sense troubled or ailing. The tone affirms sensuality in both sexes and insists that denying one's lust is harmful: "Women's state responds to that of men, as in them is also awakened an almost similar fermentation and stirring of the blood." The differences in erotic drive are more related to age than gender. The explanation is presumably that Linnaeus has biological perspective on this theme while, for example, the erotic folklore takes a social view.

A long-established treatise on Linnaeus's imagery states: "[His] interest in Nature's love-life is very well documented in his writings, from early youth and well into his 50s. After his youngest child and best-loved daughter Sophia was born in 1757, the erotic elements disappear from Linnaeus's language which overall becomes poorer in images and less poetic." This observer is the linguist Jöran Sahlgren, who also remarked: "Although Linnaeus in his own life had had proof of the power of Veneris, he never succumbed to his passions, as can be deduced from selected extracts from his writings and from the evidence of history. The enraged Venus never drives him out ere the sun has risen. Linnaeus built an altar to love at the domestic hearth, and for those who raised glorifying statues to Venus in city squares, he held no sympathy."[8] Well, I never!

We know that he did raise a statue of Venus, though tastefully, in a garden.

In the manuscript of *Lachesis Naturalis* there are plenty of relevant passages, although only a scant selection of mostly Latin extracts was included in the printed volume. Linnaeus quoted from William Harvey's writing on the subject and also from those of Johan Hoorn, the father of Swedish obstetrics. The topics vary, first from one animal class to the next, then on to the artificiality of laced corsets, and the symptoms of coitus. There are ethnographic comments along the lines of "*Intoxication* always weakens *venerem*," Hottentots have only one testicle, and so forth. The array of themes is rather different from *About the way to become together*, but, overall, the contents are very similar. Many of the notes in *Lachesis Naturalis* make fascinating reading, but none comes across as particularly personal—except for the pervasive interest in erotic matters.

This perception was, more or less, what triggered the exiled entomologist and Linnaeus expert Felix Bryk to start a debate with his brief pamphlet *Linné als sexualist*, printed in fifty copies in 1951 and dedicated to the king. The Swedish Linnaean Society was not pleased: "A curious piece of writing," it commented in its annals. The magazine *Bokvännen* (Friend of Books) exclaimed: "Linné as an erotic obsessive! Such blasphemy! As if this were a value judgment, a denigrating insult."[9] Bryk covered all the angles: plant sexuality and obscenely named mollusks, the mother's role as wet nurse in *Nutrex noverca*, ascribed by Bryk to an Oedipus complex, and then on to a string of confused examples from the lectures on *Systema morborum*. Bryk had let it be known that he was planning a bigger study of the subject, but nothing came of it.[10]

What else could one do with this thesis? Distort it, perhaps. Suggest that Linnaeus's great output "was rooted in EROS who determined all that was central to him in his life"? The pamphlet could actually have been entitled *Bryk als sexualist*, especially when we recall that his own publications include the notorious *Neger-Eros* (Negro Eros, 1928).

CHAPTER TWENTY-THREE

Curiosity-Driven Research

LINNAEUS INSISTED THAT the study of nature should be driven by *curiositas*; he always wanted to be curious and to investigate curiosity.[1] To him, being curious entailed the very essence of studying nature because the word also has a sense of rapture. The dissertation *De curiositate naturali* (1748) described how, from birth, the human being instantly shows that she has functioning sense organs, which she will eventually learn to use to the fullest: "While she is only a small child, she will learn to recognize first people, then animals, followed by plants, and then the simplest kinds of soil." The outcomes of curiosity can be rank ordered, from the single observation to the generalized conclusion. Where we start out from in life also affects the interpretations we make of observed phenomena: "That which we learn through our senses would be different had we fallen to Earth from the Moon. At first, he would conceive of our Earth to be like his Moon but with time the systems would become clearer and so would the terrible devastation that is brought about not only by animals but by mankind. Without the slightest mercy do those persecute the weakest creatures. But as we learn to inspect more closely the wonders fashioned by the Creator, we are astounded."

Time and time again, Linnaeus wrote about the sensual pleasure or rapture he experienced in studying nature. We must not forget to turn our eyes upward, to see all its wonderful manifestations—as urged in the quotation printed on Sweden's old hundred kronor note: *Omnia mirari, etiam tritissima*—wonder at everything, even the most commonplace. *Miror* is another of his keywords, used variously to mean: "I am amazed, full of wonder, enraptured, vertiginous." *Seeing* had been enhanced as developments in lens technology improved the microscope and the telescope, although Linnaeus was never entirely comfortable with the new tools and preferred to rely on the biological eye. He ended the dissertation with the conclusion that "to observe nature is to experience before your time the marvels of Heaven." In Salvius's *Learned News*, Linnaeus stated challengingly: "Is not he who is *curieus*, and attentive as he scrutinizes the work of the Creator, the one who is truly called, unlike he who despises such careful examination and walks away from the Creator's handiwork?" Who indeed is the best interpreter of the Creation?

The Faculty of Theology at Uppsala issued a warning about *De curiositate naturali*, more precisely because of its misuse of the words from the Bible: it was felt that the author had made too free with the words in Ecclesiastes as he translated them into Latin. Expressions had been used that were not compatible with what was understood to be sound theology. The faculty sent one of its members along to carry out essential corrections at the disputation itself. They chose Engelbert Halenius, actually a good friend of Linnaeus as well as the man who had permitted the dissertation to be presented. After this event, Linnaeus never trusted any members of the clergy again, as we know, and singled them out as one of the reasons for his exhaustion and mental crisis.

Linnaeus never tired of preaching that the study of nature was in the service of God. The concept of a scientific theology has not gone away nor have the arguments about how to complement theology with a philosophy of the natural sciences. In a 1750 letter to Linnaeus, von Höpken referred to another dissertation and assured its author: "I have once more read the disputation by you Herr Archiater *de Oeconomia naturae*, indeed I often read it one time after another and always with the same pleasure, marvel, and enlightenment. Greater proof would not be needed to persuade the godless of the Creator and a more beautiful treatment of the subject no minister would have the wit to produce. I am much moved by it."[2]

This was how Linnaeus saw mankind's role, and how he understood human nature, manifest above all in his binominal classification of man: *Homo sapiens*. It is our allocated definition, as we lead the system of animals in *Systema Naturae* (1758), a breakthrough in the process of recognition of what might be termed a naturalistic view of humanity. You become a *Homo sapiens* by being curious, and mankind might well have been named *Homo curiosus*. In a draft, Linnaeus emphasized that the study of nature should be done "without prejudice"—a phrase that might suggest that he had accepted Francis Bacon's ideas about the human mind and the ease with which it was misled by "Idols." Radical doubts had also been formulated by Descartes.[3] But what we pick up on here is more like the strains of esoteric alchemical teaching, supposedly secret but still not exceptional in any form: some are enabled to see far into the obscure mysteries of nature, "to peep into God's council" as Linnaeus said.

Still, curiosity is not all. Aspects of usefulness—*utilitas*—seemed at times to rank more highly still in the works of Linnaeus. Utilitarianism was a prevailing ideology of his period, and the word *utility* has already turned up many times in this account. The ideal was to be generally useful and serve societal welfare by opening and closing the correct doors—for one thing, in national terms, it is so annoying when the fat fish swim away to benefit another country. The dissertation *Cui bono?* (1750) was designed to show off the wide applications of "usefulness," and the text was translated into Swedish (published in 1753), which indicates that it was seen as important to a nationwide readership.[4] The snappy title, which translates as "who benefits?" but could be taken to mean "for what use?" was often understood to be ironic and dismissive. One of the reviews took that line: "People who quite often entertain us with this much worn-out Question 'For what end is it useful?' are in general and most notably Blunderbusses who have no right judgment and a drowsy and rarely cultivated understanding. She [the question] is posited by poor minds and less-educated brains, yes, by such who have failed to advance even as far as their common sense would allow." Linnaeus's traveling students had experienced the public's inability to comprehend what they were trying to do and, indeed, so had *preases* himself—for instance, when the Sami failed to grasp his interest in investigating the reindeer flies.

"We are created to offer our Creator praise and devotion. It cannot be completed by less than our coming to know Him through His work, either by revelations or in admiration of nature and the wider Creation." Utility should be linked to gratitude, and Linnaeus cataloged the many ways in which his fellow Swedish citizens would find it only too easy to discover examples of the Creator's favors; "too easy" because seeing nature purely in terms of use can be trivial. But when famine struck in 1756, Linnaeus responded by compiling a critically useful list of emergency foodstuff. It was of technical interest because he used the binominal nomenclature, but of national importance because of the insights it offered the king into the misery endured by the people: "when so many poor citizens are threatened by the terrible countrywide plague of hunger, which causes not only them to wane but, even more dreadfully, to hear the wails of their small children as they suffer privation and terror of death for want of food."[5]

Curiosity and utility often interacted, but Linnaeus also wanted to guard the pure science of botany—the wish to know about plants in their own right and establish its essential difference from the old herbal compendia in which systematic observation mingled with notions of use. He stated emphatically in *Fundamenta Botanica*: "Botany is the science of knowledge about plants"—plain and unadorned, shorn of all references to utility and the like. This also applied to the study of the marvels to be found within even the smallest creatures.

CHAPTER TWENTY-FOUR

Nature and Culture

IT IS AN OLD PREJUDICE that in some idyllic past, man lived in harmony with nature; of course, people lived *with* nature as well as *of* it and, often, *against* it. According to the First Book of Moses—Genesis 1:26–27 (Let [man] have dominion over . . . all the Earth)—mankind is in control of the animals. The Stoics thought intention and order can be found in nature, while the church—and Descartes—believed in a more or less sharp division between man and nature. Technical and scientific advances have increased man's capacity to rule nature; what has that power brought, what is (still) nature? Besides, what is culture?

It is relatively easy to identify the texts that are relevant to the intellectual ground covered by Linnaeus; at the time, it was not considered obvious to see nature in terms of its beauty. The Latin *natura* was used to correspond to the Swedish "Creation," and by definition God's work was beautiful. For the eighteenth-century Swede, useful was often understood as synonymous with beautiful. When Linnaeus viewed the strait between his country and Denmark from the fortress in Helsingborg, he noted the city's good location by the sea but regretted the trade imbalance—"otherwise, this city has one of the most excellent aspects." Beauty was perceived in what man had ordered and, in general, he was not enamored with the wilderness. True, he had discovered the beauty of the high mountains, but was that "new world" in the distant North beautiful, sublime, or terrifying?

Culture is a word that only rarely turns up in Linnaeus's writing, and when it does, the reference is to gardens. Linnaeus thought cities were in the main unhealthy; Stockholm and Uppsala stank. He appreciated centers of industry and trade such as Alingsås, Göteborg, and Helsingborg and gave them over-generous grades for utility. His journeys through Swedish counties are, above all, aimed at examining the economy and can make dull reading. Linnaeus enjoyed the leafy meadows on Öland and Gotland because, he said, they were like well-tended parks; the juniper bushes looked like the cypress trees of classical landscapes. But, now and then, a special moment moved him into new aesthetic territory: in 1746, he climbed the flat-topped hill Ålleberg in Västergötland:

> At the very top of the mountain I perceived a fine spectacle as I looked down onto the large and smooth fields spread below, lit by the sun's rays but the clouds were like large black stains somehow cast forth on the grassland and fled away as they would otherwise do across the sky, caused by the shadows cast by the fleeting massed clouds as they moved past the sun. Surely every child has seen the shadows of clouds flying over the Earth but still will never have seen anything to measure up to this dispersal, all fragments but dark as night.

Earlier, no one else had seen and described the cloud shadows chasing over summer meadows in this way.

Are systematic and ecological approaches inherently opposed to each other? Are activities like collecting and fieldwork in any way contrary to studies of plant geography and sociology? The Linnean papers *Oratio de incremento* (1744), *Oeconomia naturae* (1749), and *Politia naturae* (1760) discuss interconnectivity and interdependencies in nature, notably involving climate, on variables such as distribution patterns. A natural cycle of transformation, defined as "create-protect-destroy" was elucidated in the dissertation *Oeconomia naturae*, which nowadays reads simultaneously as modern and anchored in older ideas. In discussions of cyclic processes, Linnaeus applied the concept of housekeeping—*oeconomia*—an idea that lasted until the early twentieth century and Ernst Haeckel's "ecology."[1]

The transformation cycle in *Oeconomia naturae* can be summarized as "one individual's death is the other's bread." He began by describing the origin of the mineral realm: the stones are conserved through air and water but gradually destroyed; the destruction of one kind always serves to generate another. As always when discussing minerals, he kept it brief but was all the more expansive in his treatments of the animal and plant kingdoms. "The destruction of animals often happens in the manner of the weaker serving as food for the stronger; yes, each animal has its own enemy to combat. . . . The predators clean up." Finally, he describes how all this must be made to serve human welfare.[2]

Politia naturae examines the work of the Great Housekeeper: "Nothing is created in vain. Were it that all sparrows in our country had succumbed, all that we have planted would be preyed upon by the insects. Had we not kept Cats, all would be consumed by Rats. Had *Libellulae* (dragonflies) been created as large as lions, what would have been the fate of men and other animals? The smaller the animal, the larger is the excess numbers of that species; and the more long-lived, the more limited is their generation of offspring. The hawk has annually no more than four eggs but a bee has more than 40,000 and so forth."[3]

Richard C. Stauffer, who studied both Charles Darwin's *Origin of Species* and Linnaeus's *Oeconomia naturae*, singled out the Linnaean work as a major source of inspiration for Darwin alongside the geologist Charles Lyell's *Principles of Geology*; he also referred to *Politia naturae*. Stauffer argued that Linnaeus's thinking on ecological topics was "crude but meaningful" and cited several of the themes that also appeared in Lyell's works and were later included in Darwin's development of the theory of evolution.[4] Darwin's apparent rejection of the Linnaean version of the origin of the species is paradoxically contradicted by his acceptance of the concept of nature's economy.

Linnaeus did not see the order in nature in terms of harmonious unity. *Politia naturae* describes nature as a state of all-out continuous warfare, and his writing is suffused with a strange emotional chill: "The bloodbaths and everyone's war against everyone else may at first appear as very horrible. However, these are not matters that should shake us especially. It is true to say that animals have lives which soon pass and flee away as if in a dream. Each one who is born must also die." In his thinking, we detect not only scientific theology in the idea that everything has been created in order to serve another but also notions of Providence, that all is designed for the best. He spotted interdependencies and symbioses everywhere—human beings were not excluded: the decaying human body is but one link in the Great Natural Cycle, and human practices help spread natural life-forms. "That there are of late more often apple trees growing along country roads is an outcome of lads who have eaten fruit and must stop along the way home to carry out their natural functions."[5]

All is connected. From *Curious features among insects*:

All that which the almighty Creator instituted on our Earth Globe is of such a wonderous order that not a single being exists who does not need the help of another that he might live. The Globe itself with its stones, ore, and gravel is naturally nourished and born from the elements. Plants, trees, herbs, grass, and mosses grow of the soil and the animals, finally, of the plants. In the end all are transformed into their first substances. Soil becomes food for the plant, the plant for the worm, the worm for the bird, and the bird often for the predator. In the end the predator is eaten by the bird of prey, the bird of prey by the worm, the worm by the plant, the plant by the soil. Indeed, man who turns all into his own use, will often be food for the predator, the bird of prey, the worm, and the soil. Thus, all goes around.

It is possible to play with the concept "nature," attempt to give it any number of meanings, and still miss some of them. Nature could be understood in the economic or aesthetic sense, or touristic, or, of course, Linnaean. It is easy to spot the contradictions in the latter, and if a connection is sought with the Linnaean traditionalists, the results would still be varied; even in these circles, what was meant by nature in, say, 1907 differed from how Linnaeus himself saw it.

In *Diaeta Naturalis*, he noted the contrasts between city and countryside, but the differences he listed became less pronounced later:

> In the cities, the stink of open sewers and other uncleanliness is so bad that the air many breathe is therefore not good. In the countryside, the air is sensed as full of such greenery as nurtures the eye, wherefore the Creator has made all of Earth green; here play flowers of every kind and in various *coleurs* that excite and give pleasure to all mankind. The trees swing their leaves and make a pleasing murmur, the birds join their songs into the glorious chorus, all of *regnum vegetable* give off a lovely odor. The insects dance about through the air and settle here or there as if a *pochade* [jewel or bead?], yes, the marks of the unfathomable Creator are seen wherever you turn. He would surely be made of stone who did not delight in such a scene.[6]

As we have already noted, Linnaeus was not using "nature" here and spoke instead of "creation" or "scenery." Terms of approval tend to be polite, as in "delightful prospect" and "charming spot," and are often used to describe combinations of nature and culture: cultivated landscapes, peaceful rural settings, and pretty views.

Linnaeus hardly ever discusses hunting. At the time, and until the law changed in 1789, hunting big game was a privilege of the nobility and forbidden to everyone else. To what extent did Linnaeus feel compassion for animals? He wrote in a late note: "When I was in Orsa a cucu flew outside the window and sang; I was keen to have him [sing?] and the pastor sent out a man who shot *accurat*, but the bird flew over a tree. The man followed, and he flew back and began there to pick at his feathers, but then

FIGURE 33. Martin Bernigeroth's portrait of Linnaeus in *Philosophia botanica*, ca. 1751. Royal Library, Stockholm.

the pastor swore he would not get away; the manservant came along, he flew to the next tree and there was shot. I asked the pastor what he knew of it, the answer was, it was wished for and so [I?] felt nauseous." He was in County Dalarna and didn't seem to have cared how the cuckoo was killed at the time but pondered over the episode twenty, perhaps thirty years later. What, precisely, was bothering him? The morality? This account is followed by a note on his dreams about the murder of Caesar.[7]

He was enchanted by birdsong: "In Skåne, I often walked about all night to listen to the nightingale." In Dalarna: "When I was in Lycksele, I heard at night at eleven o'clock a bird sing in such a melancholy tone

FIGURE 34. Linnaeus in his normal clothes, captured by Jean Eric Rehn, 1747. Uppsala University Library.

that I could do no other than go outside and seek him out." It was apparently a small, gray bird. In the spring of 1755, when Linnaeus was in Skänninge in Östergötland, he took pleasure in the voices of many birds. These moments have been taken from lectures on the potential utility of music, given much later.

Interest in the weather and in the night are often part of a sense of closeness to nature. In *Somnus plantarum* (1755), we meet Linnaeus wandering about outside in his nightshirt to investigate how plants sleep: they are hard to recognize then, not necessarily because it is dark but because it is a time when they change shape and appearance. Plants lack nerves and the animal type of sensory apparatus, and so do not sleep in our meaning of the word but in some analogous sense. Linnaeus was also fascinated by phosphorescent nighttime phenomena: his dissertation on sea sparkle (*Noctiluca marina*) listed bioluminescent organisms: foxfire (a fungus), glowworms, the minerals *lapis bolonienses* (from Bologna; a baryte) and glittering forms of feldspar. He did not investigate this topic himself, but his interest is linked to the elusive possibilities of natural electricity.

To describe nature, he sometimes used military metaphors, at other times showing a preference for mythology. Passages about nature were sometimes decorated with classical fantasies such as fluttering butterflies named after Trojan soldiers.[8] References to mythology were plentiful but can seem random: we have come across the Furies and Andromeda, others included naiads and named figures such as Nemesis, Lachesis, Adonis, Asclepius, Ceres, and Pluto. Much of this was drawn from his favorite classical writers—Ovid, Horace, and Vigil. Some readers might associate this with the Swedish balladeer Bellman, but Linnaeus has made no mention of the great contemporary poet, who also used imagery with creatures such as naiads and nymphs and, of course, Venus, the queen of love. Linnaeus's writing was never pastoral, though. He came closest to the poet's Bacchanalian fantasies in *Inebrantia*, a dissertation on intoxicants, in which he told a story of the participants in a wild drinking party turning into animals.

CHAPTER TWENTY-FIVE

Entrepreneur and Economist

A FACT THAT SHOULD BE RESTATED: Linnaeus had been born in a poor country. When he grew up, everything had value and was saved. When the harvests failed, people turned down the pace of life to a minimum, or else faced death. When the long years of the Great Northern War came to an end, hordes of the starving, homeless, wounded, and sick were drifting along the roads, and many who survived never had a decent life. Linnaeus told a story from the late 1720s of how twenty-two impoverished persons died after eating a toxic pea soup, offered to them by the well-meaning Mrs. Rudbeck. The Sweden we encounter in the history of the first decades of the eighteenth century is very far from the opulent image the country acquired during the splendors of the rule of Gustav III. Some of the early utilitarian texts make us smile—or recoil, as when Lorens Rothof argued in his *Hushållskatekes* (Household Catechism) that everything is worth saving, including human urine. Testing innovative solutions to problems was essential, but so was "housekeeping."[1]

People from Småland are allegedly tight with money, and Linnaeus was true to form. He was also hostile to "abroad," especially before his stay in Holland. Right or wrong, anything foreign was by definition suspect. In the early days, he used the word *gaudy* and meant "grotesque" as applied to imports. He was exceedingly irritated by the French lifestyle, understood to be all about dressing à la mode, young dandies with cavalier manners and unhealthy, overdecorated and costly foods. Added to the straightforward economic reasons for being concerned was a deep-rooted resentment toward papistry and the metropolis of Paris. His views on the topic of women's fashions were even more vigorous and often rather misogynist. "Foreign" also came to include countries outside of Europe. Later, Linnaeus's tone grew milder, perhaps because by then he had been successful in Europe.

Linnaeus proves on many occasions that he had an alert social conscience. Traveling in County Dalarna, he took note of the parish records, which showed that immeasurable numbers of people of all ages had died of hunger between 1696 and 1707. He recounted the observation that "those who had fallen from this cause and about to die, would if given a very little milk, come to and show some signs of life." For want of anything better, people were driven to cook up a watery brew of white or reindeer moss, eaten with water and milk.[2] He missed nothing and would stop to reflect on what he saw: the Falu mines, the harvests the farmers brought in, the starvation diets. His 1756 letter to the king on the subject of the poor harvests and hungry people foraging in the forests included his suggestions for useful foods, which have already been mentioned but deserve quoting again: "When so many poor citizens are threatened by this terrible countrywide plague that is hunger, which causes not only them to wane but what is even more dreadful, to hear the wailing of their small children as they suffer deprivation and the terror of death for want of food and not being able to help them."

It had struck him how important it was to create an inventory of what nature could offer. By then, he had published twice on this subject: *Flora oeconomica* (1748) and *Plantae esculentae patriae* (1752). Both existed in Swedish translations, which Linnaeus would now like to see widely distributed. Raising the nation's level of knowledge about plants would produce helpful results: "He who knows the herbs will never have to lose his life should a poor harvest strike the land again." During an epidemic of cattle sickness, Linnaeus generously handed out food and coins to the hungry. His friend Bäck told of how Linnaeus could not bear to pass a starving person without offering alms.[3]

The country was divided into two camps in so many different ways: poor-rich, country-city, north-south, Hats-Caps, mercantilists-scientists. Another divide had opened up on the issue of how to deal with economic policy: by native Swedish means or colonial exploits? It has been claimed that Linnaeus had little time for foreign trade and actually thought of it as surplus to requirements. He even assumed some sort of divine ordinance in geographical differences: in the South, there was a need for more abundant vegetation to compensate for their infection-laden air and all the crocodiles and snakes. His take on the Swedish economy was optimistic to the point of naiveté, especially in view of the country's loss of Great Power status, but was also grounded in his conviction that something must be done to lift people out of hunger and misery. It is true that his remedies were, on the whole, quite unrealistic. They were, however, expressions of his social conscience. Besides, he enjoyed telling the higher-ups what the nation required.[4]

This mid-eighteenth-century period is commonly thought of as rife with cantankerous and cynical political posturing, but this view has recently been modified.[5] Luxury was fairly consistently seen as foreign and improper, in contrast to plain, Swedish manliness. For the upper classes, a reasonable amount of luxury goods was acceptable but not for the bulk of the population or, at least, less and less so as you descended the social ladder. However, gainful things brought from abroad, notably botanical acquisitions, were celebrated. As an economist, Linnaeus wanted to improve the contemporary conditions and, above all, eliminate poverty and starvation. He joined the ongoing debate about luxury but argued for a radical attitude. He wanted to rid the nation of all luxury goods—his lectures on dietetics were to a large extent concerned with that goal. His vision of a Swedish utopia was not so much a place of material welfare, as a country where lost knowledge had been recovered and so created a nation of paradisial health and beauty.

In other words, he had distanced himself, for example, from Adam Smith, who sought progressive future developments. "The utility of plenty," von Höpken's defining phrase for how consumption would make the country flourish, was most likely to appeal especially to the comfortably well-off, but even they held on to the expectation that the plain, home-grown products would suffice for Sweden. This is Carl Gustaf Tessin, in his 1764 account of his horticultural experiments: "This year, I have planted in my garden lingonberries and blueberries with the intention to find out if these vegetables can be grown to greater height and perfection."[6]

The first principle was to exploit the native country. A sizable labor force was needed to do this. One shaming aspect of the country was its small population. Småland, for instance, could surely support twice as many inhabitants and "nourish itself twice as many if not more," if the moorland could be used for agriculture. Linnaeus believed in, or at least argued in favor of, the cultivation of Sweden's northern highlands. In 1742 he wrote to the Academy of Sciences: "It would be an improvement of the Lapp economy when a few forest-living Lapps on some area with many ash trees, would first begin to farm elks, and if the country could later become used to more of this."[7] Linnaeus believed, it turned out, that elks would be excellent sledge-runners but poor replacements for oxen. This notion wouldn't go away, even though it sounds unreal. It was just one of many attempts to adjust the economy to our given northerly conditions.

In 1732 Linnaeus took an interest in pearl fishing for the same reason, and began to test a growing site in the local Fyris River. He drilled small holes in the shells and introduced tiny chalk fragments to stimulate pearl formation. The experiment met with success. A secret government committee discussed his method, and in 1761 Linnaeus sold his patent on pearl cultivation to the Chamber of Trade with the proviso that he would appoint his successor himself. The parliament of 1760-62 debated new attempts to improve the Swedish trade balance by selling pearls. Linnaeus insisted throughout that, for the sake of national welfare, the method must be kept secret. He was by then thinking of the likely personal rewards. The original proposition was reduced by half, and the merchant Peter Bagge was initiated into the secret, but Linnaeus was content. He had received enough to pay off the loan he had taken out to buy the Hammarby estate near Uppsala. How Bagge did from the deal isn't known. The secret committee had the executive responsibility, and the parliamentary groups accepted what was proposed, with the exception of the merchants, who moved that "this proposition should rest."[8]

Axel Reuterholm noted in his diary from 1738-39 that "of today I remember little more other than that Linnaeus dined here.... I heard an absurd observation, namely that all the *Plantae alpinae* or mountain herbs, when they are taken down from the hills and one wishes to cultivate them in gardens, would not thrive in the soil beds on the ground but must be planted on high shelves and up on the fences or on roofs in order that they should do well." An absurd idea? We recognize it easily from various places in Linnaeus's writing, for instance *Oration de increment telluric habitable*; foreign plants could be acclimatized if they originally grew at the same latitude and in the same climate. Linnaeus was awarded a medal for his proposal that mountain slopes in Europe should be cultivated: "To the enhancement of our mountains it would be no small contribution if *publicum* would initiate at first a small garden high up." A catalog of plants to be tested was to be compiled—for instance, the cedars of Lebanon were surely likely to grow on the mountainsides. Saffron should also be tried, even though Linnaeus warned his students that saffron in large quantities can cause laughing sickness (he mentioned the case of a girl he had come across in Falun in 1749).[9]

In the 1740s, the notion of acclimatization became one of the triggers for dispatching his pupils on international journeys. The plan to send Pehr Kalm to China via Siberia is significant in this context: he would search at "Swedish latitudes" in order to collect a varied set of useful plant and animal specimens for introduction in his home country and so, hopefully, make it richer. Linnaeus weighed Siberia and North America against each other and in the end thought America a better option. Growing tea in Sweden was one of the top priorities.

Kalm's taskmaster, and Linnaeus's correspondent and colleague in the Academy of Sciences, Sten Carl Bielke, also ran a much-admired experimental garden at his Löfsta estate in Uppland. He had good contacts in Russia and joined forces with Linnaeus to propagandize for the exceptionally hardy Siberian buckwheat. Many shared the enthusiasm for geographical extrapolation, which could become extreme—at one point, the import of buffaloes was considered. An often-described experiment involved planting mulberry trees in Lund, with the express intention of importing silkworms and starting to produce homegrown silk! Linnaeus felt confident and commented that it had indeed been proven that the white mulberry tree transplants easily to our climate, so it would obviously follow that "these useful critters both could and should be reproducing in our dear native country."[10] The academic discussion grew lively for the project was both expensive and laborious. The sponsors remained hopeful for another few decades, but the only result was a collection of small and exceedingly expensive pieces of cloth.

In his 1752 dissertation on rhubarb, Linnaeus spoke warmly about this plant, which was not only edible but a source of medicines. Rhubarb extracts were sought for their laxative effect and generally "purifying" actions, appealing in a period when constipation was a common affliction. Expectations grew and others picked up on the idea; Linnaeus used his Russian network to get hold of the desirable plants. A Linnaeus student, Pehr Osbeck, brought rhubarb plants back from China in 1752 but managed to pick the wrong species.

One example of the craziness of this line of horticultural research is the case of Linnaeus and the bananas. In 1736 he published a paper on how a banana plant flowered and set fruit, which became a horticultural sensation. In it, he also speculated about what kind of vegetable he was dealing with—since everything is be found in the Bible, perhaps it was a Dudaim or a Phoenix? He had discovered quite a few clues to suggest it was actually the Tree of Knowledge because, when the fruit is cut, the surface bears a cross. A crucial argument, as it were, because although he had shown that all life forms must be paired in order to reproduce, the Tree of Knowledge was a single specimen—and, behold, it is a fact that the banana has hermaphroditic flowers! It followed that Adam and Eve were not using fig leaves to hide their nakedness, but banana leaves. Consequently, in *Systema*

Naturae the banana was given the name *Musa paradisiaca*. Linnaeus triumphed in 1755 when the banana plants in Uppsala set fruit. The harvest was presented at the royal court, and it was possibly Queen Lovisa Ulrika herself who was offered a feast of a few small, green bananas.[11]

An embarrassing failure, as told by Anders Ehrström in 1764:

> Among other vegetables, he prescribed from America a species of cactus on which the coschenill louse lays its eggs from where its larva comes out, and this critter yearly produces the color given the name coschionell [carmine]. It was well done and the plant full of larvae came to Upsala but, as bad luck would have it, the Old Boy himself was not at home. The head gardener, who received the plant, considered that the crawlies were vermin that must be cleansed away, which he went about with the greatest of care, so that no single larva remained. As soon as Linnaeus had come home, he asked for the arrival of the prescribed cactus and if there were any worms on it?—Yes Sir, the gardener said, it was full of worms but you must not concern yourself, Sir, because I have cleansed the plant from all of them.—Hell's bells, man, no one will thank you for your troubles. Your hard work has caused much harm, Linnaeus told him.[12]

There were other madcap attempts of this sort, even though Linnaeus understood well enough that all life forms engage in intricate interactions, and he sometimes questioned the right of people to try to control nature. However, the century created an entire investigative industry aimed at taming exotic plants in the name of *cameralism*—the administrative theory that favored a strong centralized state—as part of a general drive to also rule over nature. Eighteenth-century man was rarely as optimistic as when taking on the challenge of developing new crops. Similar trials of biological acclimatization were set up all over the world, and the ideological foundation for botanical exploration was the intent to colonialize or at least prepare for trade expansion. Arguably, the entire plan, judging by trials such as Linnaeus's pearl cultivation and the dispatch of student-collectors, was misguided.

It is easy to smile at these ideas, but it is also true that Sweden needed to improve its means of production across the board. Linnaeus and his contemporaries had no proof that the planet's resources were limited; dig deeply enough and your spade will hit hidden treasure. Linnaeus, for one, believed in taking every opportunity—"or else, the blame will come to us in time for not being capable of cultivating."

Theological presentations ended with *Soli Dei Gloria* or SDG for short, meaning "Glory to God alone." Many other texts used *Soli Patriae Gloria* (SPG)—"Glory to the Fatherland alone." Even though Sweden's most eminent scientist was in an ecstatic mood, much of the glory consisted of a banana or two in the botanical garden, a frost-stricken tea bush, and a few pieces of silk.

Linnaeus, his top students, the East India Company, the Royal Academy of Sciences, Uppsala University, and many other individuals and institutions stood there in the end, perplexed to find themselves with their beards caught in the letterbox.

CHAPTER TWENTY-SIX

To Describe the World

NO ONE KNEW SWEDEN better than Linnaeus—but what of "the world"? Where was it, anyway? Which geography was Linnaeus meant to describe, and how would he go about developing his understanding of it? Some of his elementary knowledge might have come from a couple of school textbooks, by Johann Hübner (1726) and Albert Gieses (1728), respectively. From school, Linnaeus would remember definitions and concepts such as continents, nations, regions, languages, zones, the tropics of Cancer and Capricorn, climate and winds. The early teaching of geography would have followed the Bible and placed Jerusalem at the center of the world but would also have included classical ideas from Herodotus, Pliny, and Strabo and Gothic ones from Rudbeck. During his years in Holland, Linnaeus would have felt, in different ways, the winds of change from the distant sea but wrote relatively little about any direct contacts with the outside world. However, he lists numerous imported plants in *Hortus Cliffortianus*, and its famous frontispiece displays representatives from the four continents. Curiously enough, he did not discuss these samples of living geography in any detail nor about how his outlook had widened as he left the meager plots of Småland for the huge surfaces of the oceans. Later in life, as he visited nobility and royalty—Count Tessin, the queen—he could consult handsome volumes describing life in distant parts of the world. Above all, though, he had by then learned much from studying the many works of natural history collected in his own library. He seems not to have owned a world atlas but had acquired an Åkerman-made globe from the Cosmographical Society's workshop in Uppsala. The society had been inaugurated in 1758 and was symptomatic of the growing nationwide interest in geography.

The chemist Torbern Bergman, on behalf of the Cosmographical Society, published his *Physisk beskrifning öfwer jordklotet* (Physical description of the Earth, 1766), an impressively modern account for its time. Bergman's books, the description of the world from 1772 by the lector at Strängnäs academy and later Bishop Stephan Insulin, and Åkerman's own small book *Atlas juvenilis* were all found in Linnaeus's library.[1]

Strangely enough, this aspect has not been researched by any of the scholars interested in Linnaeus. When *Diaeta Naturalis* (1733) is inspected more systematically, it turns out to be full of observations of folk customs in foreign countries, suggesting a fairly active research effort. Here, for example, are comments on marriage customs: "In Congo, it seems impossible to be content with one wife. *Virginiani, Siberiani* take as many wives as they are empowered to feed. Tartaric *chamanus* recorders may sleep with whoever they so wish, wives and maidens, without her allowed to say no."[2] Or this, an expression of popular psychology: "In the days of Queen Christina began the first of Sweden's pride through making acquaintance with France." In this passage, he discusses some of the results of his reading about clothing, nationalities, and comparative medicine. He does not admit to any widening of his horizons during his years in Holland, but it must have changed his understanding of seafaring and trade, of spices and strange peoples from far away. Also, Holland was a center for cartographic work. That countries were interconnected became obvious, if it hadn't been before, when everyone learned about Lisbon after its destruction in the earthquake of 1755. The activities of the East India Company also contributed, and Linnaeus was delighted when he learned that the Academy of Sciences in 1746 had been granted permission for a yearly scientific mission abroad and would send one man along with the company's ships. He congratulated the academy and himself, too, as someone who has been allowed to take part in such a fine feast; [I] also have been given a new lust for life yet a few years more to watch as our nation will stand out brilliantly among all the *curieuxe* kinds of peoples; we, who once could hardly tell the difference between fir and larch would now teach the foreigners to count the eggs in the [freshwater] polyps."[3] (The last comment referred to the new discovery by Abraham Trembley.)

In his botanical lectures, Linnaeus divided the world into five climatic zones: southern (Ethiopia to southern Africa), eastern (Siberia to Syria), western (Canada to Virginia), northern (Lapland to Paris), and Mediterranean (around the Mediterranean Sea).[4] Paradise was the fixed, central point and placed either in the Persian river delta, or on Cyprus or Ceylon (noting its mountain Pico d'Adam), but most often on the Cape of Good Hope: "None of the world's regions known produces such peculiar and different products of Nature as does Caput bonae spei, as I have for all time wished for someone knowledgeable about nature could be sent there, a task for which first now a much-desired occasion has been granted to us."[5] (He had Anders Sparrman in mind.)

What led to the global reach of Linnaeus's explorations of nature? One of the precedents was Rudbeck's worldwide *Campus Elysii*. No boundaries are indicated in *Systema Naturae*, but on the other hand, no programmatic statements were made in the 1730 edition about the necessity to go global. The opportunities opened up with the establishment of the Swedish East India Company. However, the question should probably have been formulated in these terms: ultimately, the task God had given him from the start included all of Creation.

The first volume of the year's *Proceedings* of the Academy of Sciences contains an early outline of the research program for the students' travels, and has several geographical notes under the heading "Findings on the places of plants." It discusses the observed climatic conditions and how soil types, the air, the weather at sea, and many other factors affect plant distribution. He was well aware that great botanists had been traveling widely; in *Philosophia Botanica* (1751), he named Scheutzer investigating the Alps and Tournefort the Levant, Alpinus's visit to Egypt, Marcgrave's and Piso's to Brazil, and Plumier and Gronovius traveling in America.[6] He modestly omitted any mention of his own journeys in Lapland. Still, the section about exploration abroad seems a little haphazard. The incitement to study new territories must be the capacity of experiencing wonder at all things, even the most ordinary ones: the important goal should be to describe nature more exactly than anyone else had ever done. The explorer's records should include the distances traveled, the geography of the area and its physical characteristics, contain notes on its botany and mineralogy, describe rare animals and their appearance, food preferences, and habits. Notes should also be taken on the local economics, its architecture, agri- and horticulture, as well as on specific customs and traditions at weddings and burials, superstitions, and preferred antiques. The reward for this labor is a widening knowledge of plants, animals, and minerals, and their influence on and usefulness to mankind. In the end, it was all a matter of God's plan for His Creation, which is everywhere present to those who bother to look. This is how he put it in a letter from 1752: "Should [Mårten] Kaehler not arrive in Cap. B. Spei [Cape of Good Hope], he will surely arrive in Guinea or whichever other coast. The World is wide and everywhere teeming with the wonders of the Creation."[7]

For his own part, Linnaeus was very much of two minds about going abroad, at least when it concerned traveling for pleasure. As he reflected in

1750: "Happy is the person who is content to stay in his house all his life, in the air he knows, and does not have to take journeys hither and thither. However, if the journeys of some are gainful to themselves and their native country, they must also have learned something unless the proverb should be true about them, namely 'the sow went to Rome and returned still a sow.'... Such has it been since the dawn of all sciences and, in that way, the sciences have arrived from the East in Europa.... The other element needed for journeys is that they should be to such places of which little or nothing has been discovered by us, and which should possess many of the wonders of nature and the arts. Indian journeys are very necessary, as of them we have become aware of how, there, nature itself brings forth all that belongs to the basic needs of the inhabitants."[8]

A general framework for so-called *apodemic* (the Greek word for being on a journey) literature, to which Linnaeus made several contributions and, especially, the dissertation *Instructio perigrinatoris* (1759), had been set: the traveler should ideally be between twenty-five and thirty-five years old, have a very good knowledge of natural history, and be an able draftsman. He should beware of a too easygoing lifestyle; avoid running up debts and getting involved in fraud, political or religious; watch his diet; and abhor gambling. In his descriptive work, he should aim to stay so true to nature that the reader feels present and observing the phenomena himself. He should not trust his memory as it is in the main treacherous, but should immediately—if it could be done discretely—write down all he has observed in his notebook and not go to find a bed to sleep in until he has edited everything and has an orderly record. Among the goals of the journey, increased knowledge seems to dominate and colonial prospects are not mentioned. The set of instructions state that the model travelogues are the previously published Linnaean journeys, including his own.

The dissertation *Cui bono* (1752) should be added to this group of works; by way of an introduction, it narrates the fable of the seven wise men of Greece, who ask Jupiter to be allowed a three-day trip to the moon. Once back home, they have little to show or tell because they had simply wasted their time on "the pretty women of the neighborhoods" instead of investigating or finding answers to the question "What is it useful for?" Linnaeus was indeed no friend of relaxed tourism.

These journeys of exploration often lasted for several years and should be regarded as qualifications for an academic career. It is questionable whether the travels of the acolytes should be seen as "projects" in the modern sense of being integral to a research program spanning many years.

However, despite all the gaps, they had aims set by a plan or at least a vision for the future. Religion contributed to the vision, and the words of Ecclesiastes 104:24 were often quoted: "O Lord, how manifold are thy works! In wisdom hast thou made them all: the earth is full of thy riches." The *plenitudo* of the Creation, its diversity, made it endlessly rich, at least in principle. The other gravitational point was the utility in worldly terms. To describe and list the global plenitude of nature, firm principles had to be applied—rules of terminology, language, and taxonomic systems—and these were compiled by specially trained natural historians, sent abroad at the behest of Linnaeus, who would receive the resulting records and have them published under the general title *Systema Naturae*. This is the essence of the Linnaean methodology or, by analogy with "Newtonian science," "Linnean science" or "Linneanism." The scope of the task was such that it required a fixed, precise language, a formulary, which the acolytes must have mastered after due teaching and laying on of hands. It was, of course, an impossibly optimistic project, with a conclusion that still eludes science. The extent of nature was clearly thought of as somehow manageable; Linnaeus himself figured that nature's three realms would contain some fifteen thousand species.

The exploration project took shape during the 1740s and '50s, and during that period Linnaeus directed it. He inspired the idea and formulated the scientific rationale for using the traveling as a means to systematic knowledge. It is possible to posit other reasons: career advancement and monetary gain, adventurousness, the hope to escape from studenthood or become known to posterity; perhaps meet famous people, create a work of art, or collect precious things. To get ahead of the competition is a strong motive as is to advance from student to investigator (true, the concept of "doing research" was not current, but the lure of a high academic title—medicus, doctor, magister—certainly was). Making your name known was a tempting reward, as was the chance of gaining a permanent post at a university. Pehr Kalm's qualifications for his later chair were simply his journeys. Jacob Jonas Björnståhl was appointed to professor in Lund while still traveling (he never came home). His successor Matthias Norberg gained his post in a similar way; his appointment came through while he was in Constantinople. Adam Afzelius was off to till Sierra Leone for idealistic reasons but, it has been said, career ones as well (he did become a professor). Linnaeus himself had based his career more on the journey in Lapland than his disputation in Harderwijk. The same can be said about his successor Carl Peter Thunberg, whose merits were linked to his experiences in South Africa and Japan.

People traveled experimentally at first—testing the water. Pehr Kalm went to Karelia, Roslagen, County Bohus, and considered in turn Iceland, the Holy Land, the Cape province, Russia then China and Persia, then China, Japan, and America. The journey to America was finally supported by the Academy of Sciences, and Kalm could set out. The official deliberations were often surrounded by wearying uncertainty; once, after a positive decision about a proposal to explore Russia, Linnaeus was seen to dance and hop about in his chamber and said, "I am sure it was well 10 times, that this journey would be so much better than one to Caput Bonae spei, or one to the north part of America." Kalm, apparently always such a safe pair of hands, had been swinging between resignation and fear: "Would an American journey come to something or stay buried in a midden?" He read travelogues: "My keenness for it cools as I learn how unlucky travelers for no reason had been carved into pieces by the Americans, especially the Iroquois, and were also caught up in big adventures; but I read hardly 3 lines on from such stories before I feel hot with desire to see where the rare and useful plants are growing, which would benefit both cattle and folk; not to speak of all other curiosities."[9]

Linnaeus was the entrepreneur who kept the business afloat: he formulated the reasons for selecting a place and the travelers, sought financial support and worked through the reports of explorations. Often, funding depended on a patchwork of investors: on 27 February 1750, Linnaeus informed Hasselquist, who was in Smyrna at the time, that the Academy of Sciences had turned down a request for more money; supporting Kalm had emptied the kitty. But Linnaeus completed a donation jigsaw, starting with 100 thalers from his own purse. He had persuaded the archbishop to part with another 100; Council Secretary Carleson donated 200; Rosen, Bäck, and Schützer 100 each; Ankarcrona 200; Salvius 50; while Billing and Grillarna both gave 150 and Broman 300.[10] By 20 April, 18 sponsors had provided 2,400 thalers, later increased to 3,000.[11]

In 1747 Linnaeus spoke warmly about the same man in an application to the Academy of Sciences on Hasselquist's behalf: "You will be aware, dear Sirs, how the very greatest of *patres Theologi* strive to discover the true and right version of the Holiest of Writs; their greatest concern has been *animalibus et plantis Biblicis*. How impossible it is for them at present to clarify this matter, for no one has yet investigated what is growing and alive in the Holy Land. Soon every nation in the world will be studied by traveling *Botanicis* but, to this day, no *Botanicus* has visited the *terra sancta*."[12]

The journeys of Linnaeus's apostles could fill another book; there are quite a few already.[13] A new kind of geography emerged as they traveled, starting from nodes established in distant places. There were highly competitive natural history institutions in London, Paris, and Berlin but also in Königsberg, Göttingen, and Tranquebar. Now Uppsala had joined the club. The scattered distribution of rivals but also of collaborative centers set a pattern still relevant to our ongoing globalization. Quoting from a recent overview: "Swedish literature on Linnaeus has shifted its focus from individuals to, increasingly, ideas, theories, methods, and institutions in science history, recently also looking at broader historical developments."[14] Combining some or all of these approaches is surely possible.

The journeys have variously been judged to be outright failures, cynical forerunners to colonialism, thrilling adventures, expansions of cultural horizons, and sacrificial—too great a sacrifice—on the altar of science. Linnaeus's enterprise dispatched around twenty young men and, of these, a third didn't return (Tärnström, Löfling, Hasselquist, Forsskål, Falck, Berlin, and Bartsch). Evaluation of their fates and apportioning blame varied over time. Ewald Ährling, to whom Linnaeus is always noble, believed the master was grieving for the loss of Löfling, Tärnström, and Hasselquist. Others—for example, Otto Fagerstedt and Sverker Sörlin—found it cynical to send young men into such dangers and cited the Tärnström case: pressurized by both Linnaeus and the Academy of Sciences, the East India Company allowed him onboard the Calmar as a ship's chaplain in 1746. He had been given plenty of instructions (Linnaeus's list began with tea bushes and mulberry tree seeds). Somewhere off the coast of Vietnam, the ship got becalmed, with no winds for six months, and Tärnström fell ill and died. The grief and anger of his widow was directed at Linnaeus, who wrote of her awkwardly: "Lord help me with the widow who cries out for a conservation to support her, as I am married and unable to act. We have no priest in the Academy. How to find something for the poor widow: she cries to heaven about me who advised her husband." The letter to the academy is dated 29 April 1749, the same day Linnaeus left for a much more sedate journey to Skåne: "I would dearly like to honor Tarnström with an extract from his travel journal now that he has turned to dust, but when God will give us time for this I cannot tell."[15]

His friend Bäck offers us quite a different picture when it comes to Hasselquist, Löfling, and Forsskål: "His sorrow was like that of a tenderhearted father stricken with longing for his only son, after the message had arrived." And, later: "Nor was he slow in having statues of honor raised." Again: "How much had he not longed to hold Thunberg in his arms once the traveler was back home"; he died just before.[16] Ever since Bartsch's demise, Linnaeus must have been aware of the risks. In his letter

FIGURE 35. An East India trader by a quiet lake. Drawing by Jan Brandes (1748–1808), who had previously lived in County Halland and on Java. As an older man, he became a pastor in Tuna near Vimmerby. Rijksmuseum Amsterdam.

to the academy concerning the journey made by Hasselquist, Linnaeus wrote: "Should he still be alive for his homecoming, his name will rate more highly than that of any other Swedish man."

The news of Hasselquist's death in Smyrna led him to write to the academy: "I can never again hope to find a disciple as able to do these journeys as no one is surely going to take such a risk with his life." Elsewhere, he made similar, more or less emotional comments about lost acolytes.

CHAPTER TWENTY-SEVEN

Nature's Order 2

IN THE 1740S, Linnaeus completed his provincial journeys; his great systematic works, which laid the foundation of his fame, were creations of the 1750s: *Species Plantarum* (1753) and the tenth edition of *Systema Naturae* (1758–1759). What had or had not happened during the thirty years that steadily saw new editions of *Systema Naturae*? *Systema* is the Arc in which all Creation will be penned in an orderly manner for the greater good of mankind. The first edition ran to twelve pages (Leyden, 1735); the fourth edition had grown to 108 pages (Paris, 1744), and the sixth to twice as many again (Stockholm, 1748). The tenth edition was 1,384 pages long (Stockholm, 1759). When the work reached its twelfth edition (Stockholm, 1766–68), it had swelled to 2,300 pages.

This is how Linnaeus presented the recently published sixth edition in a letter to Tessin dated 9 September 1748: "For this little work, *Systema Naturae*, as it is here under Your Excellency's most distinguished name, I have for my entire lifespan together compiled and now for the sixth time winnowed through, so that it might include a greater share of the Almighty's masterpieces than any other which has seen the light of this day." Even so, much more would be added in the tenth edition, published in two solid volumes (an animal volume of almost 800 pages and a plant volume of 550 pages) and printed by Salvius. The third volume, about minerals, was not printed. Arguably, the work signifies, or should signify, Linnaeus's breakthrough as a zoologist. The author himself regarded the tenth edition as "a new book in many ways. The first tome of Animals serves to show that Linnaeus has become more prominent in zoology than he has in any other part although *Botanici* recognize him as master." In this book then, zoology has overtaken botany.

FIGURE 36. Title page of *Systema naturae*, 1740, showing nature as a well-kept garden. Royal Library, Stockholm.

Now, he launched the category "Suckling animals" or Mammalia "as they alone and no others among the animals have teats; whereby it also makes clear that man, as the First, must take his place at the head of this herd." Whales now end up in their right order in the system. Polyps, recently discovered "like a new province," now joined the zoophytes. The presentation is framed to show what has been observed within the royal collection cabinets, and those belonging to Count Tessin, the Gentleman of the Court De Geer, and the Academy of Sciences, and "the wonderfully strange and many animals brought by His Disciples" as well as, of course, observations drawn from Linnaeus's own wide-ranging correspondence. He invited his friend Bäck to have a look at the proofs and judge "what labor I have done to determine so many things with *differentiis* and *synonymis* from so many and confused authors."[1] Tessin awarded Linnaeus a gold medal for that volume.

Throughout *Systema Naturae* and *Species Plantarum*, he refers to three principles designed to show the foundation of nature but also to give it new meaning: all living beings come from an egg; reason tells us that all living things have, since the beginning of time, been created by a pair of parents of opposite sexes; nature never leapfrogs its own rules. From these principles, the natural historian can derive concepts such as continuity, hierarchy, variability, and constancy. The dictum that "nature does not jump"—*natura non facit saltum*—is found in *Philosophia Botanica* (1751) as part of an explanation of how to seek "the natural method."[2] Linnaeus provides a fragment of the natural method and its application and then, at the end of the paragraph, returns to nature's abhorrence of discontinuity. This must, of course, be put into context: he constructed the rules and chose examples of a new or renewed scientific approach to nature.

Overarching statements of this kind often have an ill-defined status. When Linnaeus wrote "[In our time] we can now count as many species as were created in the beginning,"[3] his claim seemed to be backed by the narrative in Genesis, even though it says nothing explicit about constancy.[4] However, it was assumed that this was a plausible view of how the created world would function. Above all: What would be the point of the labor to systematize nature if its mode of operation kept changing all the time? Linnaeus did not explain any further. His devices and principles should be taken together as the axiomatic, unchanging foundation for the contemporary activity in the field—hypotheses that did not require testing. Similar maxims state that "every living creature gives birth to its own kind" and that "like breeds like." Yet another (not one of Linnaeus's this time) asserts that "nature abhors a vacuum," which rules out any explanations involving "a jump." Accounts of Linnaeus's theories usually omit any ideas about the dimension of "natural time," but he realized that understanding the role of time was a necessity. At this stage, Linnaeus was captivated by the idea that creation might still be continuing within the God-given framework of classes and orders, although with a flow of additions at the levels of genera and species. Nature would gradually become "filled" through hybridizations and so become less and less prone to have gaps or apparent "jumps." He had arrived at a new insight—namely, that nature grows toward completeness, as it were.

It is perfectly possible to look for links between his axioms and the social ideals of his time. In the main, we now prefer to see the eighteenth century as a period of relatively peaceful human enterprise with an inherent dislike of abrupt changes of direction—that is, until 1789. Linnaeus had been supporting the reformist policies of the Hats, but politics had only a marginal appeal to him. He was far too anchored in classical teaching on virtue, with its emphasis on balance and restraint. However, he was interested in the contemporary ideas about development and changing conditions, and wanted to elucidate distinctions as observed in nature. Nature's fuzziness fascinated him, and he saw it, for instance, in the transitions between plants and animals, and between monkey and man. The trouble was, of course, that this made any notion of a simple hierarchy harder to maintain. Nature and human society alike were like complex webs, and the metaphor Linnaeus used was, as already discussed, the map, while other metaphors went for the tree or the net.[5]

Linnaeus concentrated on the important boundaries between nature's three realms and the orders within them. As we have noted, these three categories and their basic definitions had been established by Aristotle: stones grow; plants are alive and grow; and animals are alive, grow, and are sensate. There were transitions between these realms: the questions posed by zoophytes or animal-plants, and lithophytes or stone-plants, including the true nature of corals, filled a great deal of his correspondence with John Ellis, an English linen merchant and author of the pioneering work *An Essay on the Corallines* (1755). Ellis had raised crucial questions about reproduction, freedom to move, and sensibility in general, based on the characteristics of organisms such as the mimosa, Venus flytrap, and polyp. On the matter of zoophytes, Linnaeus decided "thus, their larvae are plants, their flowers, animals, and such will it remain."[6] Dualism in nature, quite simply.

The difference between "the chain of nature" and "the ladder of nature" is that the former stresses connectivity, the latter distinction. One might have assumed that the Linnaean system with its sharp definitions would come down on the side of the "ladder," but his thinking was not so straightforward. His fascination with continuity was growing, not least because of the popular interest in it and the growth of evidence through discoveries all around the world. It was nearly impossible to fend off the narratives about humanoid apes, and about corals and other ambivalent creatures of the sea.

These two constructs of nature drove the debate between Linnaeus and the French philosopher Georges-Louis Leclerc Comte de Buffon. It has come to be seen as a classic stage in natural history, both because the two men had genuinely different views, and because the arguments they used were central to the development of ideas about nature in the eighteenth century. Linnaeus and Buffon were both born in 1707 but into very different social and financial circumstances. Buffon, whose enormous work *Histoire naturelle* was the only true rival to *Systema Naturae*, moved quickly to publish his criticisms of Linnaeus's definition of species—an unnecessary human artifice, he thought. Their debate did not deal in petty details—grand issues were at stake: essentialism and Aristotelian teleology, the reproduction of plants, the concept of species, nominalism or realism, the denial of the original Creation as the guarantee that no new species would be generated—in fact, the accounts of Paradise and Genesis. Besides, the French scientists, and Buffon's circle in particular, objected to the Linnaean decision to give the sexual organs a crucial role in plant life, while rating all other features as secondary.

The essence of the conflict between the two men—never fought in the open field—was the opposition between Linnaeus's distinct lines of development, with their sharp boundaries and well-marked junctions, and Buffon's vision of "almost imperceptible degrees of change" along the chain of nature. Linnaeus was said to have taken a more realistic stance as he, in the spirit of Plato, posited the species and the genus as realities, not just human inventions. "But this is metaphysics," exclaimed Buffon. Linnaeus, who didn't read French, had encountered his criticism at the time of the publication of the French translation of *Systema Naturae* (1744). Insofar as Buffon's philosopher colleagues expressed any views, they amounted to finding that the Swedish work was "not-nature"—unnatural. This criticism could, of course, have been made in the opposite direction. Any signs that Buffon might have changed his mind to move closer to Linnaeus's views have not been discovered so far.[7]

The maxim that "nature doesn't jump" had Linnaeus's authority behind it and remained widely accepted in the nineteenth century. Several key passages in Darwin's writings refer to it in the context of continuity of change and development in nature following the excess of variant organisms being born. Linnaeus is a spiritual presence in the early version of *Origin of Species* (1841): "As natural selection acts solely by accumulating slight variations, it is natural to ask if new organs might not emerge in a species or smaller groups within a class. That this is a rare occurrence is indicated by the old expression and rule of natural history: *Natura non facit saltum*."[8]

But, at the same time, the extended "chronological chain" was gaining in popularity as a model of nature. And Linnaeus, more than is usually recognized, continued to adapt the framework of nature to the new findings, work to which he actively contributed. Buffon and Linnaeus disagreed, but their areas of disagreement kept shifting. Arguably, Linnaeus began to change his mind in 1742, when his student Magnus Ziöberg discovered a strange plant while on an excursion to Roslagen archipelago off the Uppland coast: an ordinary specimen of toadflax with a deviant flower. The baffled Linnaeus at first took the plant to be an artifact, a flower fraud as bizarre as the hydra in Hamburg. It turned out to be the real thing and was given the name *peloria*. In the dissertation *De Peloria* (1744), Linnaeus explained that this was no monstrous aberration but a cross between two types—a hybrid. From this followed a "wonderful conclusion": new species can be generated in the world of plants. This revelation revolutionized contemporary botany, just as much as his use of precisely described fructification parts for classification. It is one of the classic moments in the history of biology. In 1999 a *Nature* article ascribed the innovation to a mutation in the *Lcyc* gene.[9]

Linnaeus set out the hybrid idea in the dissertation *Plantae hybridae* (1751) and followed up with comments in other essays and papers. He was able to list some sixty-odd cases, grouped by whether the hybrid was formed between members of the same family or different ones—but included a number of "unproven" examples. He was intrigued enough to attempt his own trials of artificial pollination and actually succeeded in creating an authentic hybrid. He also drew on instances of animal hybridizations, not only the crossing of donkeys and horses but also the crossbred pups born to his own poodle.

The freakish outcome of a presumed love match between a hen and a rabbit was strangely enough included in many serious debates. Linnaeus's interest in abnormal couplings with offspring such as de Réaumur's "hairy chicken" was related to his hypothesis that reproduction requires mutuality. Once more, nature turned out to be full of contradictory phenomena. Sentences like "We count today as many species as have been created in the Beginning" had to be explained away as well-intended extras. Anyway, the notion was removed from the later editions of *Systema Naturae*.

Linnaeus's rapid conversion to acceptance of hybridization might seem surprising. To be fair, it should be pointed out that Linnaeus wasn't the only one who was easy to persuade. Buffon, who was skeptical about de Réaumur's claim, also thought it unlikely that Linnaeus could be right to make dogs, foxes, and wolves members of the same family, as they had such different relationships with human beings. Be that as it may, cross-breeding in the animal kingdom gave new impetus to investigations of plant reproduction. Were hybrids fruitful or genetic blind alleys? How to construe the process of reproduction in the context created by the new observation of the behaviors of the freshwater polyps? In 1751 Linnaeus wrote to Bäck: "My battle nowadays, could I only be bothered, is to create new plants. I have found *Plantas hybridas* more generally than *Animalia hybrida* and quite a number of them. I believe myself about to open the door to a copious chamber of Nature, though she does not open without creaking."[10]

A new set of discoveries within groups now known as *Polyzoa* and *Coelenterata* drove the intense interest in nature's interconnectivity: corals turned out to have an "animal nature," and the freshwater hydras reproduced like plants. These observations would cause no end of trouble for scholars of natural history and philosophy. The discussions about the corals had been going on since at least the 1725 publication of a paper by Marsigli illustrating its "flowers." Two years later, the physician and committed naturalist Jean-André Peyssonel presented findings to show that corals were animals, a statement met with disbelief at first. The mineralogists were unwilling to let go of these structures, seemingly so close to stones, and the young Linnaeus placed them among the plants when he wrote about them in *Pluto svecicus*. Bernard de Jussieu's observations, announced in 1742, seemed decisive: corals were animals. Linnaeus had been thinking along similar lines while on his journey to Öland and Gotland, and presented it in his dissertation *Corallia baltica* (1745)—but hesitantly. Surely, to suggest that these small tenants could build comparatively enormous new homes would be like proposing that gall-forming insect larvae could build entire oak trees. Bäck countered this with an argument as simple as it was compelling: How come the same wormlike creatures are found in the same corals? By the time of the sixth edition (1748), the matter was settled: corals, branching like plants but with stony-hard exteriors, are truly animals.

The polyps played an important role in the overlapping of kingdoms. Abraham Trembley described all his artful experimental studies; notably, he had sliced a polyp into 30–40 pieces and found that, over 24 hours, they developed into as many complete creatures. Linnaeus picked up quickly on the news and worked it into his lectures and arguments. Once more, Bäck made a key contribution: he was the author of the final presentation for publication in the 1746 *Proceedings* of the Academy of Sciences. Linnaeus—who thought the name polyp (many-footed) was absurd and "introduced by ignoramuses"—decided that the proper name was to be *Hydra* and that they were to be placed in the order of zoophytes.

Zoophytes, literally "animal-plants," were useful entrants into the chain model of nature, and joined, together with the lithophytes or "stone-plants," the lowest order of animals, *Vermes*. The explanation comes in a brief dissertation *Animalia composita* (1759): plants and animals may be thought of as distinct because the latter move around freely and the former are stuck in one place by their roots, which provide the organism with nutrients. However, at the boundaries between nature's realms, you find mobile plants like the freely swimming algae absorbing nutrients without having roots, as well as animals stuck in one place. Different organisms move in and out of the borderlands, where the least developed animals merge into the group of cryptogams. The belief that only animals are sensate—the unique distinction of animal life—is challenged by the mimosa and by the telegraph or dancing plant, which "rotates" without being touched. Reports from Florida described the moving parts of the flytrap (*Dionea muscipula*), the first flesh-eating plants to be known as such.[11]

To naturalists at least, the 1740s was the decade of the tapeworm and the freshwater polyp. Examining just the *Proceedings* of the Academy of Sciences during the years of Linnaeus's life, these animals were the subject of scholarly papers by academics such as Carl Magnus Blom, Roland Martin, Lars Montin, Otto Friedrich Müller, Lorens Odhelius, Nils Rosén von Rosenstein, Herman Spöring, and probably others. Linnaeus's unusually lengthy study of the tape worm—*Taenia* (1748)—can be taken as his response to, as well as an extension of, the research on polyps. The respondent, together with a few students, had also contributed material from the city's street dogs and helped with the dissections. Linnaeus considered the tape worm to be "of polyp nature" possessing "the life principle in each segment." As each segment was capable of independent existence, it seemed likely that the creature's brain was distributed along its length and that the entire worm should be seen as a continuous chain of polyps, each one of them equipped with a mouth and a digestive tract. It reproduced by shoots "like the quick root" (couch grass) and thus must be classified as a zoophyte. "Were we then to note how all living animals generate themselves, we might well find it would take place such as it does in the plants or in the Tape worm, where one twig or segment after the other protrudes its own bud or seed from which another will grow out, so all living things that come of Plants and Animals, indeed of Mankind, were nothing but a chain that reaches from the first moment of Creation to the very end of the world. Verily, there is nothing new under the Sun." The phenomenon expanded into the explanation of the lives of all animals and plants.

What Linnaeus seems to be saying here is that human beings are in some ways analogous to tape worms. When, in 1758, he tried to comfort his mother-in-law, who had been worrying about her rather wayward son Pehr and his imminent wedding, Linnaeus applied his ideas without much finesse:

> May I develop this more broadly and use an example, namely that of a tape worm. Each part of the worm has its own life. Were these segments to be separated, each one would continue to live and grow bigger. Which part would then be the greater than another worm? The last segment gives birth to its offspring which will in turn do the same and so onward. There is nothing different from other animals more than each part sometimes falls away from the other. But were the umbilical cord to remain intact always, the children would be strung together each to the other, like a tapeworm's, and then it would show all the more clearly that they were as one. It follows that young Brother Petrus and the other children are nothing else than their Mistress Mother herself and dear departed father-in-law.[12]

To make sure his point got across, Linnaeus drew a picture of his unsavory worm, complete with its different generations. A note added to the letter reports that his wife refused to let him send it, and his stunned reaction. He wasn't just playing a joke on his relatives: in his 1754 *Naturaliesamlingar* (Collections of *Naturalia*), he risked telling the royal couple that all living animals are generated in the same manner as plants and tape worms, including human beings.

Whole series of theories on reproduction were launched in the eighteenth century. Already before 1700, the physician Charles Drelincourt had managed to dredge up from his predecessors' articles a total of 262 thoroughly false hypotheses; nothing could be more certain—as someone with a sense of irony remarked—than that Drelincourt's own would be the 263rd. The number would rise steeply over the following years. Linnaeus's proposition was the so-called marrow-bark, and he seemed to believe in it firmly. It was not very popular but pops up here and there in Linnaeus's writing, reaching its completion in the tenth edition of *Systema Naturae*. Its background in the history of ideas is complicated, but something analogous can be traced back to Plato. It was presented for the first time in the dissertation *Gemmae arborum* (1749). Löfling, the respondent and

[288] CHAPTER TWENTY-SEVEN

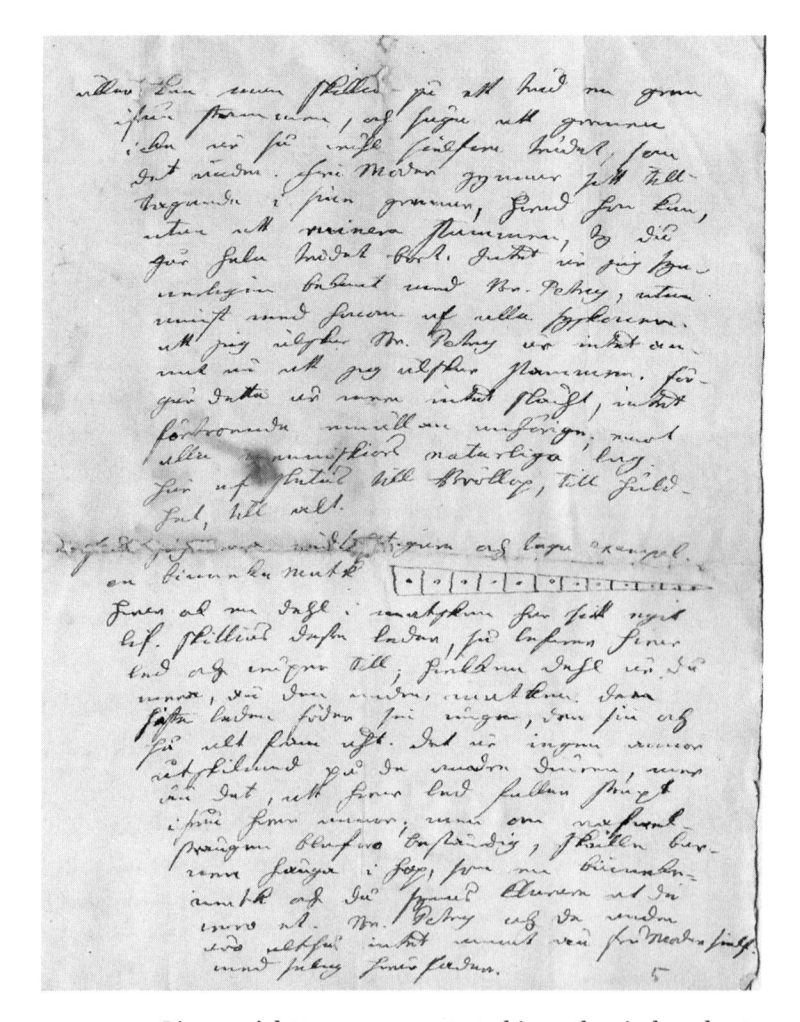

FIGURE 37. Linnaeus's letter—never sent—to his mother-in-law about continuity of life in the human species, based on evidence concerning the reproductive mechanism of the tapeworm. Uppsala University Library.

also Linnaeus's secretary, had contrary to custom written most of the text himself and also provided material for it. The thesis is that a plant grows like a polyp and reproduces with eggs (seeds). The bud should therefore not be regarded as a new plant but a continuation of its parent, while the seed is a quite new, reproduced individual. That the marrow grows from the roots upward proves that reproduction can continue without any need for new creation. In *Philosophia Botanica*, Linnaeus moved the arguments along by using a perhaps rather tortuous analogy to animal anatomy, by including the brain, spinal cord, and peripheral nerves in the construct "marrow."[13] In plants, the marrow is wrapped in bark but strives to erupt and succeeds

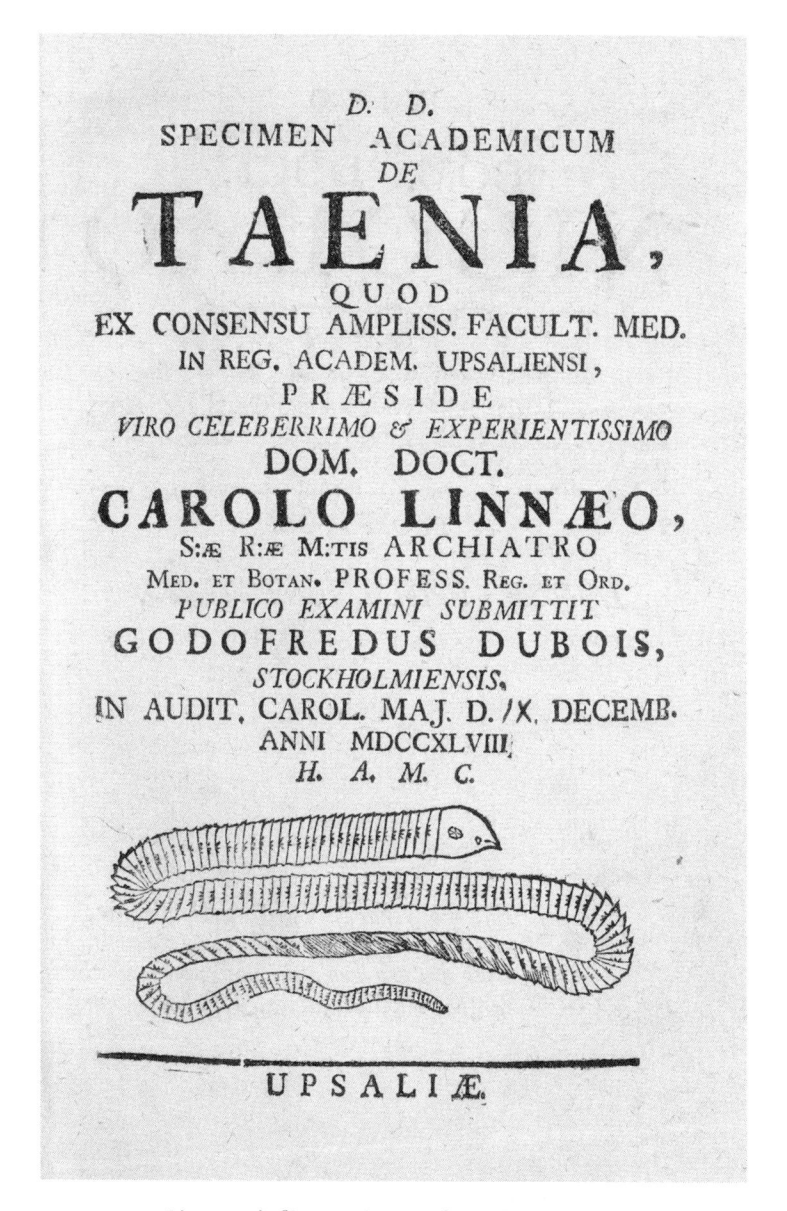

FIGURE 38. Linnaeus's dissertation on the tapeworm, *Specimen academicum de Taenia*, 1748. Royal Library, Stockholm.

where the resistance is least: at the tip where it can unfold into a flower. The task of the bark-father's wood is to protect and nourish: thus, the bark is masculine and becomes the pistil, and the enclosed marrow is feminine.

The dissertation *Metamorphoses plantarum* (1755) should have a brief mention in this context. It begins with a declaration that all is subject to change. The strange transformations of insects entail "undressing": one outfit after another falls off until the creature is finally naked. Many plants undergo a similar set of changes. Like the metamorphoses of insects, their "undressings" do not lead to transformations but, as the outer parts of the flower are peeled off, the inner plant is exposed. The marrow continues as the pistil and is concentrated in the seed. This vision had poetic potential; for one thing, the butterfly was seen as the flower of the insect—a flying flower. The idea of the *prolepsis* of plants was an outlying consequence of the periodicity of the leaf buds. Linnaeus's assertion that a bud contains five generations is plucked out of the air. His love of the number five will be seen again.

Briefly, how far did Linnaeus want to take his hybridization idea? We have an account by his pupil Claes Alströmer, who moved on to the wider perspective on development after first discussing the peloria and instances of cross-breeding in the plant and animal kingdoms. He considers various indeterminate families, fits in cattle between rodents and "nags," then places cartilaginous fish, like skates and sharks, between snakes and fish, and tortoises between "crawlers" (Mollusca) and corals. Crosses between orders as well as families seemed plausible to Alströmer and Linnaeus. "Let us now postulate, as is quite likely although it cannot be proven to completeness, that since the time of the Creation of all things, the animals were in no more numerous than the present natural families."[14]

Linnaeus created order in nature by establishing a pattern for a fixed terminology. He rejected fuzzy definitions based on color, size, usefulness, and overall appearance in more or less precise illustrations, and instead used the number of stamens and pistils as a definitive measure. This offered an opportunity to replace wordy and subjective descriptions with concise classifications that are more compact, and hence cheaper. The botanist's task was to make inventories by sorting specimens using a summary description to determine the classification. Nature was structured on five distinct levels; plant specimens can be grouped and regrouped into larger and larger "boxes," or categories. The series, in ascending order, begins with the variety—often unnecessary—then species, genera, family, order, and class. The final, overarching category is the plant kingdom.

Linnaeus's science is encoded in his nomenclature. In *Philosophia Botanica*, he formulated its essence: "*Nomina si nescis, perit et cognitio rerum*" (if the names are not known, understanding is lost with them).[15] Although this perception is often ascribed to Linnaeus, it was first recorded in *Etymologiae* by Isidor of Seville, whose life ended in 636. Linnaeus's central contribution to knowledge was his epoque-making introduction into biology of the double name convention—the binominal nomenclature. Indeed, it is often claimed to be a lasting insight. In the example *Linnaea borealis*, the first name is that of the genus, the second of the species. This convention was applied to 5,900 members of the plant kingdom in the first edition of *Species Plantarum* (1753), and to 4,400 members of the animal kingdom in the tenth edition of *Systema Naturae* (1758). The expert William T. Stearn proposed—and others agree with him—that the two-name approach was practically a byproduct that emerged by chance out of the main task he had set for himself, which was essentially to name families and species. Stearn traces the origin of a more systematic approach to *Pan Svecicus* (1749), possibly intended to be a space saver but not to replace the diagnostic method of polynomial classifications. Linnaeus wrote about himself and his system: "*Trivialia nomina* were not heard of before; but he appended such to all plants. It was the same as placing the pendulum in the clock; through him doing this, botany sprang back to life."[16]

At a stroke, names became easier to remember, and the amount of space on each page increased. Linnaeus's contribution was actually fivefold: (1) improved overview of the system; (2) terminological precision increased; (3) because Latin was consistently employed, the binominal structure was internationally comprehensible; (4) the approach applied universally; and (5) at the time, *Species Plantarum* included all known species.

Species Plantarum is more than a catalog of names: it is a critical overview of a field of study. The author's goal was to provide a flora for the entire world, no less. To achieve this, he had to examine herbaria, and those of his disciples were critically important. He began this task in 1751; Löfling's and Hasselquist's collections were significant but Kalm's provided the largest number of specimens: 89 out of their total 190.

Species Plantarum became a major part of his life's work. Adoption of the economical two-name classification system grew during the second half of the eighteenth century, and the task Linnaeus had set for himself proved essential. The list came to include twelve thousand plants and animals with the added "L" at the end. That capital letter gives Linnaeus the most distinctive posthumous symbol in contemporary biology. His contribution has been described as a poetic act of heroism, and as a mass-christening. The match of name to organism could be startling, as when Trojan warriors lent their names to butterflies. People sometimes objected to novel names that changed what had become common usage. Some offended current taste, others needed interpretation, like *Homo sapiens*. That one is undeniably worth understanding—we will soon attempt it.

Nonetheless, this essentially technical reform has proven to be Linnaeus's most permanent contribution. In 2014 he won the Wikipedia title "person with most links to his/her entry" as "L" is found in all binominal terms.

We can follow his work on the naming project in, for instance, his 1 May letter to Bäck in 1750: "I now begin to strike *Species Plantarum* utterly from my mind. . . . I would wish to leave it for an *inventarium* once I am gone. Then the world to come should see that I could have completed it had I only had the time and inclination. But am I meant to work myself to my death? Is it meant that I shall never see or taste the world? What do I gain with this? But such cunning only dawns on you toward the end." When Löfling became his assistant, things moved along more easily. On 28 June 1752, he wrote: "Ever since I had been to Stockholm, I am jolly, able to work, and quick with it; you see, gentlemen, what powers you exert. Before, I was downhearted, melancholy, unable to do anything."[17]

It is worth emphasizing once more: the great significance of the Linnaean reforms was that his names were applied universally—that professionals could communicate in the same language. True, one outcome was that familiar local names, often vivid and magical, faded from popular memory. The name reform entailed a disenchantment with nature, in Max Weber's sense of the unidirectional tendency of modernity.[18]

As part of his creating thousands of binominal names, Linnaeus introduced new terminology, notably "fauna," to match "flora"; it is derived from the word *Fan*, in turn a form of the old Swedish word *fänad* meaning cattle or, sometimes, just "animal." "Flora" was not his invention: in Sweden it was used, for example, by Olaus Bromelius in his *Chloris gothica* (1694), and by Johan Linder in his *Flora Wiksbergensis* (1716). In this earlier usage, it was normally applied to descriptions of the plants found in a particular territory or region. Fauna didn't become widely accepted at once; it seems to have been more frequently used only in the 1780s, and there were rival terms. Interestingly, Linnaeus chose the feminine form for his freshly minted terms—except *Pluto* for the mineral kingdom. The concept *ammalia* (animals that suckle their young) was used as a replacement for *Quadrupedia* (four-footed) in the tenth edition of *Systema Naturae* and was sometimes translated as "animals with teats." It led the science historian Londa Schiebinger to tell Linnaeus off for his breast fixation.

And, yes, the name begs the question: Why was it picked? Milk secretion is after all just one possible animal characteristic and, besides, one that applies only to about half of any species—the females—and then only during the relatively brief period of lactation. Why not select hairiness as a significant feature and call the group *pilosa*, the hairy ones? The breast may be seen as a sign that women are closer to nature than men, so could that have been the reason? Linnaeus had actually argued strongly in favor of women nursing their babies in *Nutrix noverca* (1752). Schiebinger believes that the names he coined show nature becoming infiltrated by the norms of the "European middle class," and not for the first time either.

The first order of the class of *Quadrupedia* had been named *Anthropomorpha* but was renamed *Primates*, often translated as "the first" in lectures. In the beginning, Linnaeus called the first species of the first order the *Homo diurnus* and the second *Homo nocturnus*; the humans of the day and the night, respectively, were then renamed *Homo sapiens* and *Homo troglodytes*. The "troglodyte" category had been criticized for several reasons. It had flickered past in Mennander's lecture notes and is one of the expressions indicating that the old and the new zoology were not as different from each other as Linnaeus wanted people to believe. The background to the species name was to be found in a travel book by Nils Mattson Kiöping from 1665: it mentions observations in the Far Indian Island chains of creatures called *cuckolacks* and likened them to *Homo sapiens florensis*. Empirically, the account had little to show for itself, and gives the impression that the species was more of a construction, fashioned to fit a link into the chains of beings—a "missing link."

Which means that we have reached the point where we can scrutinize Linnaeus's perspectives on our own species, which he had named.

CHAPTER TWENTY-EIGHT

Homo sapiens

IN MODERN HISTORY, it was Linnaeus who drove the ordering of humans into the animal kingdom. According to Mennander, in the lectures in 1733, *Homo* ended *Sectio 6a, Fructivora*, and was noted for the features "tail absent" and "in the common, 32 teeth." He also recorded, "Here it is surely a matter of one species, but the white Europeans, brown Asiatics, black Africans, and red Indians are variations. On *Amboina* people have been seen who are as snowy white as chalk, they are called *Cucorlacro*. They whine like children and come out only at night and that is to steal, which is why people kill them and call them in their language 'insects.' In the genus, *Homo*'s first neighbors belong to the ape species *Simia*, subdivided into apes with and without tails." [1]

Linnaeus took the apes seriously, much more so than the interest he would take in any other single group of animals except the human one. Early on, he seemed very aware of the similarities between the primates, but he constantly added more layers of evidence. In his dissertation *Anthropomorpha* (1760), he lined up major examples of organisms on or close to the boundary between the families *Homo* and *Simia*. He had observed apes in the *vivaria* at Clifford and in Uppsala, and made plenty of space in his lectures for lore about apes—still, he had never seen a chimpanzee or a gorilla himself. He listened avidly to travelers' tales and alternated between being clearheaded and too trusting. He reported, more or less faithfully, testimonials as to the high intelligence of apes, clearly looking for links—that is, for creatures who would reflect the abilities of humans and close the gaps in the chain connecting us and them. Linnaeus pointed out that some apes are less hairy than humans, walk upright, and use their hands as we do—the only difference is their lack of speech. Hence it is difficult to be precise about what is so unique about mankind.

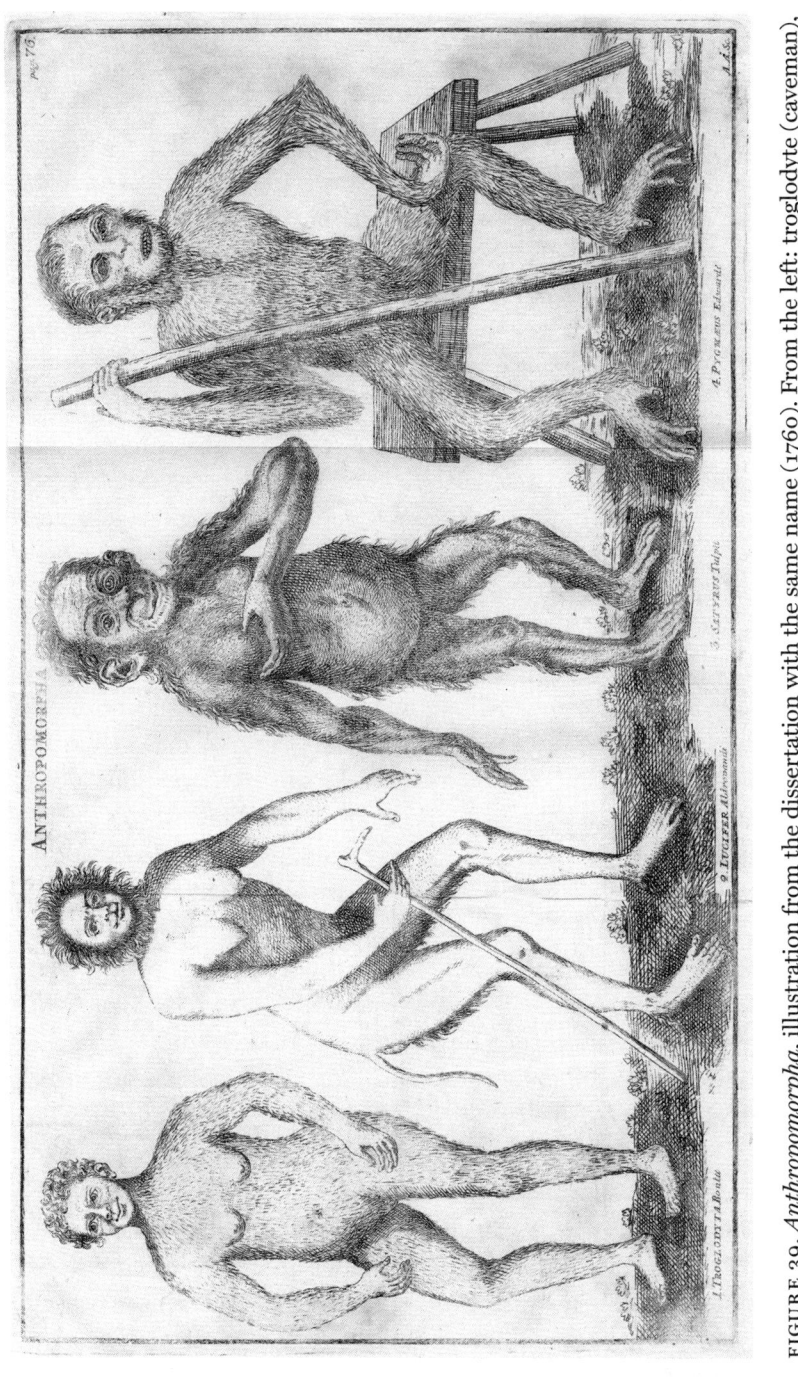

FIGURE 39. *Anthropomorpha*, illustration from the dissertation with the same name (1760). From the left: troglodyte (caveman), Lucifer or Homo caudatus (tailed man), Satyrus (chimpanzee) and Pygmaeus (pigmy). Drawing by Christian Emanuel Hoppius. Engraving by Andreas Åkerman. Uppsala university library.

Starting with the first edition of *Systema Naturae*, he divided the human population by continents: *Europaeus, Asiaticus, Afer*, and *Americanus*; four, as opposed to our seven, as nowadays South and North America are separate continents and, in addition, we count Oceania and the Antarctic. He ordered his named peoples in an apparently hierarchical order but broke with expectations by placing *Americanus* first. He meant the native populations and not the European migrants, which is another expression of his primitivism. We have seen this preference before in his travelogue *Iter Lapponicum* and in his criticisms of Europeans in general and the decadent French in particular. He characterized the species variants according to the classical teaching on the temperaments. Some of the normative rules are inherently strongly negative and lead to unreflecting condemnations of various people, including Europeans but usually not the Sami. He was not very well read on the subject of Native Americans and usually modeled his ideas on the Sami: "The Virginian drank nothing but water until the arrival of the Christians. The Lapp drinks the *pura* water, knows of no other drink, is as agile as a line dancer. . . . The Lapp has rather few household possessions [and just like] the Virginian he builds his house around a pole. Kettle or bowl or drinking vessel is on the house floors, the mussel gives bowls." The tent is the home for free nomadic peoples.

Has anyone ever seen a "yellow" human being? Galileo's argument that "colors are mere names" and that an object's perceived color is an inessential quality, meant that natural historians increasingly came to see describing by color as a mistake. Linnaeus set out to eliminate this source of error. Color is changeable and has no defining value; in *Philosophia Botanica*, Linnaeus explained: "Color is wonderfully variable and hence of no value for definitions."[2] His colleagues may well have been influenced by the current printing techniques which made using color so expensive that botanists preferred black-and-white illustrations. But as the eighteenth century drew to a close, use of color as a characteristic was revived.

Linnaeus's view of mankind was no different from how he saw other organisms. He wrote this in *Critica Botanica* (1736):

> The Almighty created man as described in the Holy Book, but if small characteristics were to be enough there would be today thousands of human species. This is true as true there are persons with white, red, black, and gray hair, their faces with some white, slightly red, brown, black, and the nose may be straight, short, curved, blunt, or like an eagle's beak, there are giants, dwarves, fat and thin, straight-backed and bent, ridden with gout and lame people. But which sensible persons

would call them different species? Let us therefore find secure characteristics and expose the false that seduce us. I declare that size, place, time, color, smell, taste, utility, sex, monstrosity, hairiness, span of life, quantity, etc. are not singularities with which to characterize a species.

This set of variations could be studied at the level of the nation: the first edition of *Fauna Svecica* (1746) begins with an account of the kinds of people who live in Sweden: either Goths, Finns, Lapps, or various mixtures of them. In the second edition, printed in 1761, the first species is named: *Homo sapiens*. The order, however, is not *Primates* but *Magnates*—evidently, Linnaeus had not yet made up his mind. His definition of *Homo sapiens* as *naturae regnorum tyrannus* (tyrant of the realm of nature) was eye-catching. The word *tyrant* should be understood in its classical sense of a ruler who rebelled in order to take power, and whose rule may be more or less tyrannical. The rank order of the native human varieties—now different from *Systema Naturae*—places at the top "Swedish people mixed with other nations," followed by Goths, Finns, and Lapps. They are described by body build, color of hair and irises, but without value judgments. This was the last time the Swedes themselves were included in the Swedish fauna.

Later, Linnaeus expanded his model of the four main human variants by adding a fifth category of *Monstrosi*, often misunderstood as "the monsters." He wanted to illustrate what climate, artifice, and other environmental factors can do to people. Individuals belonging to the group were not "naturally" made that way, nor did the fact of their inclusion always have a negative slant or classify them as freaks. It is on an observational basis that Linnaeus included the Hottentot who had only one testicle, and the Patagonian hunters; he might easily have added, say, fashion-mad women using wasp-waisted corsets for the right hourglass silhouette. The concept was multidimensional, but the fundamental idea was to have a category for people who were malformed, or hybrids, and also for the creatures of classical fables. Mostly, they have been found elsewhere: "*Monstreux* people one rarely sees here, or never, even less the paralyzed." The word *monster* is derived from the Latin verb "to show." The thought was, generally but also specifically in Linnaeus's case, that a monster demonstrates something out of the ordinary, something that has happened or soon will. The laws of climate and habit apply in all cases, but to varying degrees, something that this fifth category makes clearer. The features characterizing the Lapps have their origin in their lifestyle and the climate in their homeland, and are not due to any inherent type; a Lapp is a person who "by his customs, foods, clothing, and language is obviously of the same people as other

Sami." In this way, the group *Monstrosi* does not specify any consistent features, but all should be regarded as more or less changeable.

A necessary remark concerning the name *Sapiens* that Linnaeus gave our species. He did not intend *Sapiens* to be seen as an absolute quality of human beings but as a virtue we can acquire through good upbringing and natural education. It should be understood in the sense of "wise" rather than "sensible"—as something we can become rather than a quality we are born with. That it is a conditional descriptor can also be concluded from his contrapuntal use of the term *Homo ferus*, the wild man. If we do not cultivate ourselves by learning, we become like wild animals and cannot aspire to be a *Homo sapiens—ferus* (wild) is the other option for us.

The theme of *Homines feri*, "the wild men," forms an important part of Linnaeus's lectures but is also a characteristic strand in the philosophy of the eighteenth century: What does our species amount to if isolated from society, forgotten in nature, and brought up by animals? He was not talking about societies existing in natural environments but the solitary individual or, by all means, twins like Romulus and Remus who were fostered by a female wolf. Above all, he referred to the reports of people abandoned by their kind, of children left to be cared for by wild animals. They learned to eat raw foods, ran around on all fours, couldn't speak, and were frightened of humans. Linnaeus collected lists of such cases. The subject has many ramifications and has been keeping psychologists busy for generations, including the present. The wild boys and girls fascinated Linnaeus especially as counterphenomena, also critical to the definition of *Homo sapiens*: What is a human being without society, what can she be called unless she looks upward from the feral state?

Sapientia, wisdom, is a quality that must be conquered and, worse still, if we don't succeed, we are not *Homines sapientes*. Like the Renaissance philosophers—Pico della Mirandola for instance—Linnaeus believed that mankind was facing an existential choice and, in his case, we could choose insight into our debt of gratitude for all that our existence offers. *Homines feri* is interpolated between *sapiens* and the genus *Simia*, the members of which can actually be *sapiens*, too: a *Simia sapiens* was found in his posthumous notes! This quote is taken from his dissertation *Meniskans cousiner* (The cousins of human beings), a preliminary to *Anthropomorpha* (1760): "Many are probably of the opinion that between the human being and the ape the difference is as great as between night and day; but if you were to align a highly enlightened prime minister in Europe with a Hottentot as are found in *cap de bonne esperence*, you would hardly believe that they were of the same extraction."

The path to wisdom is indicated by *Nosce te ipsum*—Know thyself! *Systema Naturae* begins with this instruction. Linnaeus was criticized by Johann Theodor Klein because he told his reader to find the answers by staring at himself in the mirror. Linnaeus added a footnote to the sixth edition (1748) to explain what he meant: *Homo sapiens* has the capacity to understand him/herself—but we don't like the challenge. *In terms of physiology*, you should realize that you are a frail piece of machinery; *in terms of dietetics*, that if you have learned how to feel healthy and calm, you will be happy; *in terms of pathology*, that you are nothing but an inflated bubble until you burst, hanging from a hair in a fleeting moment of passing time; *in terms of nature*, you, who are the miracle of nature and the prince of animals, are but a laughing, weeping animal; *in terms of politics*, that you, instead of doing what is right, you follow what is evidently wrongheaded and, even as a newborn, you soon cover yourself in the clothing of habit; *in terms of morals*, you are, under your unrefined mask but a simple-minded being, rude, dependent on others, vainglorious, wasteful, and quarrelsome; *in terms of theology*, knowing yourself means realizing that you are the utmost aim of Creation, the Almighty's masterpiece on this planet Earth.

He struck no overtly Christian notes. This question-and-answer motif is present throughout Linnaeus's thought and is intended to situate himself as the original version of *Homo sapiens*.

Linnaeus celebrated the human mind, as in this passage from his presentation on "Curious Features of Insects":

> We have not been granted strength like the Elephant, but good wits to learn how to tame even the strongest. We have not been given the front feet for digging through the soil like the Mole, but reflection taught us how to bore through rock. We have not fins and rumps shaped like the Fish, but by the use of our brain we have learned to swim to both Indias. We have not been granted wings like the Birds, but calculation has taught us how to take the bird down from the Skies. We have not been given eyes as sharp as the Lynx, but thoughtfulness has shown us how through a tube we might see spots on the planets and through a microscope observe the blood vessels of the louse.[3]

Various aspects of Linnaeus's writings have been added to the history of racism. Labels such as "racism" and "racist" are frequently misleading when applied to circumstances and people in the past; such judgments should surely be adapted to the time and the available knowledge. For instance, Linnaeus defended the Sami against the rumor that they were sorcerers—or "poison-blenders," an accusation made against them by Johan Linder but which neither Linnaeus nor Rudbeck accepted.[4] Generally speaking, Linnaeus's vision of the Sami contrasts, on the one hand, to the city folk and courtiers and, on the other, to the Hottentots, the untutored wild men. The Sami leads a life in the wild but in an enlightened way, while the Hottentot is the lowest of the low. Linnaeus's position is ambiguous because the animalistic is normally good in his eyes; the message in *Diaeta Naturalis* can be summarized as "the more like an animal, the better the human being."

Then, how to evaluate Linnaeus's social views? He created a hierarchical model for ordering humanity, and the different types were described complete with value judgments; in all the rankings, the African always scored least well. Without prejudice, he classified humans with the animals and consistently used the concept "variety" rather than "race." Variety is in principle influenced by environmental factors, and in his hierarchal lists—never quite the same—the highest-ranking were normally the Indian rather than the European people. Finally, Linnaeus was always unequivocally positive about the Sami.

There are plenty of other counterarguments. Linnaeus never classified humans according to a polygenetic model—that is, he did not posit a Creation involving Adam and Eve in one place, while the rest of mankind arose from somewhere else. He was not into measuring skulls. He never wrote or said anything that could be interpreted as anti-Semitic. Perhaps most importantly, he did not attempt to sort people by intelligence, simply because *sapiens* is linked to *sapientia*, a trait that isn't unique to the European but characterizes all mankind, if only they would follow the condition of wisdom, which is to know oneself.

Another example of how Linnaeus would argue, in this case about the differences between man and woman: "I would never have been persuaded that man and woman were one species among us, *structura vestimenti externa* is very unlike what applies to the whole body, where God has clearly created all alike excepting the birth parts. The Lapp is quite another way. The animals are another way. God himself has made *differentiam sexualem* in the beard which is now a sign of shame and contraband although by God's Creation but which mankind is not content with. Hark, now the vessel of clay speaks to its maker; why did you make me thus?"

But the idea that sexual identities are nothing but social constructs would probably have been a step too far for Linnaeus.

PART III

The Old Linnaeus

1758–1778

FIGURE 40. Wild strawberries from *Svensk botanik* [A Swedish Botany], 1810. Uppsala University Library.

A Bowl of Wild Strawberries

No one, not even doctors, can escape the infirmities of old age, and taking the waters at resorts was a favored remedy. Linnaeus never tried it, even though he wrote a dissertation on the practices at spas: *Levnadsordning vid en surbrunn* (1761). He had found another cure: "This entire time I have been here to recover my strength in the countryside, rest and eat wild strawberries, whereof I have been pleased daily." This is why the emblem of this third and last section of his biography is the wild strawberry. He is said to have consumed something like two and a half liters daily. Still, it is a problematic curative, as it is, of course, available only in the summer.

Linnaeus wrote a dissertation about this (atypical) berry: *Fraga vesca* (1772). He believed eating them was a good way of cleaning one's teeth, and unbeatable when it came to purify the blood of the poisons causing podagra (gout in the big toe) and other arthritic conditions. Eat as much as you can get hold of every third to fourth day, he advised, an insight that came to him after an attack of gout that struck him down in the summer of 1750: "At the time, he felt near to death's door." Someone brought him a bowl of wild strawberries and he fell asleep after eating them. When he woke up, he asked for more and finished more or less a basin of berries, then went back to sleep. The trouble recurred the following summer, and he medicated the same way, and repeated the procedure when podagra afflicted him, now less severely, in the summer of 1752, and again for a fourth and fifth time. Thanks to the wild strawberry, he lived for almost twenty years without any more symptoms. Many followed Linnaeus in this, and one outcome was that the price of wild strawberries inflated almost six times.

CHAPTER TWENTY-NINE

Honors

LINNAEUS HAD REACHED a professional pinnacle. Anders Johan von Höpken, a member of the upper parliamentary chamber and university chancellor, wrote to him on 7 October 1762: "Were it not for the presence among us of you, Herr Archiater, the Swedish would be counted among such nations as are still under a barbarian darkness." From the same letter: "I wonder how soon the Revered Archiater would expound to us his remarkably beautiful observations on the medulla and the formation of the Earth as were told to me in Upsala. Your thoughts are so elevated that I find myself taking an inordinate pleasure in them." A last quote, this time from a letter dated 4 January 1776: "Apart from Newton I know of no one other than Herr Archiater who has during his lifetime enjoyed such widespread praise such as all strive for but is given to few to win before their death."[1] Linnaeus commented in *Metamorphosis humana* (1767): "But when a tree has grown to its tallest, it must fall."

Linnaeus combined such breadth and multiplicity in his work and so many contradictions that he seemed to have many personalities. He had been christened Carl after the then king, Charles XII; from childhood onward, he would in turn respond to Little *Botanicus*, Prince of Botany or King of Flowers—however, not King of Animals, or of Lions, or anything of that sort. He preferred *Princeps botanicorum* and wanted the title on his gravestone. In a letter from 1746, Albrecht von Haller had called him, in reference to *Fauna Svecica* and not without a measure of irony, The Second Adam.[2] Arguably, he was more than that; Adam's task was to name the animals—plants were not on his agenda. In 1792 Linnaeus's biographer Dietrich Stoever coined the phrase "Deus creavit, Linnaeus disposuit"—God created, Linné ordered. It had been a successful teamwork, as Bäck had already noted in 1744: "It gives me heartfelt pleasure that the Orangery stands in such glory; but it ought to stand under this inscription: *Deo & Linnaeo*."[3] Adam or Linnaeus, featured as the giver of names, dominates the frontispiece of the 1760 edition of *Systema Naturae*. It is relevant that his botanical garden was known as the Paradise. As he provided ever more aspirational schemes, he might well have been called the lawmaker of science, the second Moses, or perhaps the Moses of Nature. Linnaeus was a charismatic leader, and the students who joined his innermost circle became known, of course, as his disciples or apostles. Later, other celebrated professors would also gather around them such bands of selected followers.

So far, the frame of reference has been Christian, but the Baroque mania for mythologizing drew heavily on classical tales. In the frontispiece of *Hortus Cliffortianus*, Linnaeus is portrayed as the naked Apollo; when he was offered membership in an academy, Linnaeus was given the epithet Pliny of the North, though he was displeased with the geographical limitation. As early as 1737, when he was elected to the *Academia Naturae Curiosorum*, Linnaeus became *Dioscorides secundus*—the members gave each other names drawn from antiquity. During his Lapland travels, he played at being Perseus and, in one of his autobiographies, with allocating military ranks to himself and his fellow botanists. He grants himself the top rank of General in *Florae officiari*, followed in descending order by Major General Bernard de Jussieu; the Colonels Haller, Gronovius, van Royen, and Gesner; then Lieutenant Colonels, Majors, Captains, First and Second Lieutenants, and the Master of Rumor-spreading (Heister); followed by Sargeant (Siegesbeck)—a total of thirty-two names. Including Linnaeus, four were Swedish: Kalm was a major, Olof Celsius and Johan Leche captains.[4] In Tessin's fairytale, Linnaeus is the stoat or ermine: "His [the ermine's] lust was for seeking out Books that told him of Nature's fostering in grasses, bushes, trees, flowers, and seeds as all belonged to its housekeeping."[5] Jacob Wallenberg summarized Linnaeus's power over all realms of nature: "His Majesty of the Kingdom of Herbs, Duke of the Crocodiles and Mermaids, etc. as well as Master of all four-footed animals, feathered creatures, and insects."[6]

Then, in 1762, Linnaeus was knighted and changed his surname to von Linné. In his homeland, it would become the significant name above all others—even though it was his for just the last fifteen years of his life. It was said that "only one *Linnaeus* was as great as *Linnaeus*, and he could not have become greater in European eyes, certainly not because he was now called *von Linné*."[7] His own reaction was to strike a humble pose: "*Linnaeus* or *Linné* make no difference to me; the one is *dialectus latinus*, the other *suecus*."[8] The French ambassador, charged in 1765 with providing a French magazine with news from Sweden, asked: "As Linnaeus was the only name of a Swedish scholar of whom I had heard before arriving in Sweden, a name that is known in France and also in many other places—why then abandon a moniker so well known, famous, and respected in all of Europe?"[9] He was right: Linnaeus has become an internationally known concept while Linné is familiar only in Sweden. In modern times, catalogs abroad often omit his first name as Linnaeus is sufficient. He is not so much a person as a name-giver. From 1760 on, Linné often dropped the "von"—it is not clear why he preferred simply Carl Linné.

The official titles were added to all this naming: he was a *medicus*, doctor, archiater, knight, professor, rector, inspector, and member of learned societies (they liked presenting his name on the title pages of their various publications). Once he had bought the country estates Hammarby and Säfja, he became not only a farm manager but one who also had an obligation to provide a certain number of tenant-soldiers in wartime. Then there was the game of pseudonyms: when he joined the Academy of Sciences' competition to find the best ways of fighting infestations on fruit trees, his entry was submitted under the name "C. N. Nelin. N. Minist."—meaning something like "Carl Nilsson Linné, Minister in the name of Nature." Torbern Bergman won the prize.

Different roles come to the fore in portraiture: Linnaeus as a Sami, as a Roman, and as a naive nature lover in a flower meadow, but also on banknotes as a supporter of the state. It mattered to him, of course, that he was a husband, father, and—eventually—grandfather. There were the pet names and the jocular ones; abroad, his observational skills earned him the attribute Lynx or lynx-eyed; in modern times, we amused ourselves with Chaos von Linné, Carlsson Linné, Carl from Linné.

How is identity—a self-image or persona—actually shaped? What was truly relevant: being a scientist, a senior academic, a scholar? Which virtues are emphasized in particular? Linnaeus himself liked to use the metaphor *Theatrum Mundi* or "the great theater of the world" in which we all have roles to play. In that context, there are "regions to the back and the front" to pay attention to, according to the sociologist Erving Goffman: one man shows off in private, another in public. Linnaeus is also a brand, used to name hotels, cafes, streets, and a mountain on the moon. His linnea flower is a symbol and a logotype.

It goes without saying that all these titles, all these different roles, brought power. His American student Adam Kuhn reported to Linnaeus from London in 1764: "Here, it is customary at once to drink to the health of the King but she [Lady Monson] drinks always first to you as you are the King of all the realms of Nature."

In the Theater of the World, leading roles are allocated in many different ways.

CHAPTER THIRTY

Among Students and among Senior Academics

TIME PASSED. THE world was changing, and Sweden, too. The Hats had gone to war against Russia with nothing like the desired result. The Office of State Census collected statistics about Swedish society, the king's attempt to expand the royal influence failed, but the coup by Gustav III in 1772 succeeded and centralized power replaced party politics. In between these dates, there were the forgotten Pomeranian potato war and the truly glorious legislation in 1766 that supported freedom of expression. For some time, the country was happy—but it was a treacherous mood; by 1773 severe crop failures were followed by widespread famine.

After eight years away from Uppsala, an anonymous writer visited the city around 1764 and was taken by how much it had changed in such a short time: "I came of an evening, was lodged in a new and pretty, wall-papered chamber. The clock struck 7. I waited to hear the incommodious hollering of the so-called porridge-clock or the old cracked Castle bells. Time passed. The clock struck 9. Then I heard the ringing out of clear tones from the bell and soon enough, the call of the fire warden.... I waited to hear the next call to lower all lights." Uppsala was showing a new face to the world: "I walked along one street after the other. The church looked new, with neat towers, spires inlaid with copper. *Academia Gustaviana* and the library, now gloriously improved. The Court building very fair. I could no longer recognize the street by the Mill. The Hospital improved. The hospital church had somehow moved elsewhere. The Cathedral bridge, new with a splendid stone span." Hårleman was probably just about to start clearing in and around the castle. But it didn't take long before Uppsala was on fire again: it began to burn on 30 April 1766.

"Our Lord was merciful and protected me this time," Linnaeus wrote, but he took the precaution of moving all his belongings to a barn beyond the boundaries. "All was in the greatest confusion"; a third of the city went up in flames.[1]

As ever, the university was a city within the city, with hierarchies of its own. Linnaeus was a professor, dean of the medical faculty, rector (a senior but short-term, rotating post), and active in the county-based student colleges or "nations." By origin, he was a member of Småland's nation and served as its inspector or link-man between the nation and the *consistorium*—the university court. He was the prefect (chief curator) of the academic garden and also secretary of the Society of Sciences, once the source of funds for his travels in Lapland. It seems he didn't do that job all that conscientiously.

The students at the university were allocated to a nation by home county, and the membership meant that they had access to study space and social contacts. Because the sociability could get out of hand, the university had decided to appoint an official—an inspector—to oversee the nations' activities. Nowadays, the nations have their own buildings but, in the mid-eighteenth century, students met in some agreed place in town, or in the inspector's home. In 1743, Småland's nation elected a successor to their current inspector, Anders Celsius, and Linnaeus won by 31 votes against 15 for Petrus Ekerman and 9 for Johan Ihre. Many wrote later about the paternal care Linnaeus offered his fellow Smålanders, who gathered around him either in his garden, or at Hammarby; there were as many as 116 get-togethers during his inspectorate, which lasted for all of three decades! Serious trouble surfaced only once, during the complicated election in 1770 of a new curator or head of nation, which forced the members to hold meetings three times over eight days. During Linnaeus's inspectorate, some 308 students came and went at the nation, of which 91 chose the church and 22 school teaching; 81 studied law, 16 medicine, 9 joined the Trade and Customs Office, and 8 became academics or worked in the university administration. From what we can find out, it is clear that Linnaeus was a busy and engaged inspector. It was particularly true of his late period of service, which has been described as a golden era for the "small Småland." Its students were the only ones invited to eat the food served after Linnaeus's funeral.

Linnaeus insisted that his post at the nation should include seminars on current topics followed by something to eat. While he was in charge, the orations and disputations took on a more Swedish-patriotic character, connecting the subjects to topographical studies. Events in the royal family were duly noted as were many funerals of colleagues (Linnaeus spoke at the graveside of Andreas Neander). The inspector was also meant to admonish and punish misbehaving students, keep an eye on anyone straying from the teachings of the church, make sure everyone paid their dues and was aware of the fire regulations. Of course, he would warn against too much drinking in pubs or "basements"; when crowds of fellows were regulars at Flodberg's Basement but failed to turn up at lectures, Linnaeus was in despair. Now and then, he apparently gave terrifying sermons against "pub-crawling." The record from the meeting of the nation's board on 22 March 1770 noted that "Hr Archiater uttered that he would see it better if they would quench their thirst in their own chambers." In 1747 he involved himself in a quarrel about which newspapers the nation should subscribe to, and in 1770 he addressed the students in writing to instruct them always to speak in Latin at the nation's disputations as there had been a complaint from the diocesan chapter at Växjö, their home cathedral city. He also had to control institutional bullying at the nations, or overlook milder forms, depending.

Inevitably, the students were often rowdy. One example of an intervention by Linnaeus was when, in 1747, members of Småland and Västgöta nations fought in the cathedral about seats in the student gallery. That time, he decided to ask his old adversary Rosén to write the incident off. Rosén did, but Halenius, the inspector of Västgöta nation, seems to have borne a grudge and ungratefully attacked Linnaeus about a dissertation a year or so later.

As teacher, archiater, and inspector, also the rector at a later stage, he was respected. What happened at a bash on the night of 30 April 1760 is a case in point. A bunch of students had been partying on the slope of Castle Hill, playing music and firing off guns and fireworks in traditional celebration of the end of winter. The city guard turned up and warned them off: anyone still around at half past nine would be arrested. Quite a few of the rioters surrounded the guard, taunting him for being a "tatty tramp" and marched him off to the square, where the unhappy man was dumped into the "sausage-pot"—the student arrest cell. By the time Linnaeus's carriage was about to pass the crowd, the driver didn't dare to continue, especially when pistols were fired between the frightened horses' heads. But at that point, a couple of students stepped forward, grabbed the bridles, and escorted the carriage across the square to the academic gardens. It took place "with no row" because the rector was liked and respected.

During Linnaeus's time in the dean's office at the medical faculty (1741–76), a total of 284 medical students were enrolled. In 1749 it was calculated that it would take the novice five years to qualify. Linnaeus's courses included botanical garden–based demonstrations every summer, one year of *materia medica* and medical semiotics, then a crucial course in signs and treatments of illnesses, followed by one and a half years each of dietetics and natural history. Meanwhile, Rosén taught anatomy, pharmacology, physiology, pathology, and praxis. Medical studies also entailed an introductory course in "philosophy for medics" run by the faculty of philosophy. For various reasons, the years to qualification were usually extended by a few more terms.

"*Consistorium* was not his pleasure nor appointed task as he had been tasked with other matters of a more engrossing nature than such to be dealt with there." Still, as shown by his biographer Thore M. Fries, Linnaeus was a dutiful attender at meetings. His name is found in most of the recorded protocols and agendas of the faculty for the years 1742–77. From his installation in 1741 to the middle of 1776, he was present at no less than 1,902 meetings—Fries counted them—an average of one every week! This work for the university court didn't include the responsibilities that came from holding, from time to time, the rotating posts as rector, pro-rector, dean, and pro-dean. The consistorium met in the Carolingian Academy until 1755 and then in its new, dedicated building. The agendas listed items about courses, examinations, staff appointments, and student issues such as their clothing and manners. The court also handled allocations of grants and the wider university finances, election of the chancellor, defense of the professors' privileges, and cases to be examined and judged according to the academic jurisdiction. Research is still to be done on aspects such as who was in what camp and the increasingly critical role of politics.

In a political context, Linnaeus has been described as the "archiater of the Hats." The Hat party members were, as we have noted, critics of Arvid Horn's peace-oriented diplomacy, which was backed by the Cap party. Linnaeus's allegiance would seem confirmed by his friendship with Tessin, Höpken, and Hårleman, but all the same, he claimed in a letter to Bäck (May or June 1756) that everyone in Stockholm called him a Cap.[2] Especially in his younger years, his policy ideas made him come across as a Cap party supporter, but his close relationships with the official elite probably made him change sides. Fries seems to have decided to tone down all references to politics: "In those matters, we can be quite brief as he in reality was preoccupied himself hardly at all with them." This dismissal is presumably based on a widely held belief that party politics during this period, the Age of Liberty, are a particularly dark aspect of Sweden's eighteenth-century history. Linnaeus did not subscribe to a newspaper but read the papers in his nation: Småland took *Stockholmske Post-Tidningar* and *En ärlig Swensk* (An Honest Swede). He mostly sympathized with the Hats but deplored the infighting and "the moral depravity which those on either side are guilty of." He wanted to avoid becoming involved in the quarrelsome squabbling in the capital, and preferred the maxim "He is the happiest who lives in the shade." Linnaeus seems to have aroused himself from moderate political apathy just once. It was at the revolution stage-managed by the king, Gustav III: "Everyone is singing *jubilaeum* for his Majesty, I the loudest," he wrote to Bäck on 4 September 1772. Linnaeus was loyal to the king—yet another "good" quality mentioned by Fries.[3]

Naturally, he complained. He wrote to Bäck in February 1768: "Today early, we had notice of Hr Julinschiöld's *cessio bonorum*, which meant that we were sitting in *Consistorio* all day long.... The entire time we are made not to think of anything other than sitting over in the *Consistorio*, having been dragged from our own tasks."[4] And again: "Melander, *theologiae* professor *Upsaliensis*, drives the proceedings always with great vigor within *consistorio academiae*. One day, we had been at that table until after 6 o'clock in the evening with some vexing dispute, Melander having been the strongest actor on behalf of a most unjust cause, when at 6 o'clock his head twisted toward his back. Fell down beneath the table, was carried home; never more sees a healthy day. All went home, beating their breasts and saying the Lord pays heed to all our intrigues." Indeed, he seems to have included himself in "our intrigues." The event was presumably in 1745, the year of Eric Melander's death.

An anonymous pamphlet that made the rounds at the 1747 election of the chancellor offers an overview of the professorial establishment in Uppsala; it was called *Sagan om Herrgåln* (The Tale of the Manor). The setting is a royal manor house farm, and the king's horses and oxen are reviewed and sometimes get an individual description. It has an obvious political aim: the animals are the professors, with the horses as the Hats and the oxen as the Caps. Among the horses, Olof Celsius is Big Blackie, and Engelbert Hallenius is Old Piebald, an elderly but still hard-working horse. Li'l-Foal, Linnaeus, "is small but trots strongly, plows deep and harrows well, has a jolly neigh, forages perhaps not so much but likes well enough some dry slope or meadow hays with clovers and other flowers within. That horse is used by the Master for long journeys." A few more horses pass before the oxen are paraded: they include Big Britches, who is Rosén, and Arseblare, who is Ihre, "an odd ox, always smooth and slippery." The horses leaped and the oxen quarreled until a cat sneaked into the Estate; the point of the story matters less than the flippant tone. Note, for instance, the hints at Linnaeus's short stature and the poor grades given to Ihre, the Cap.

Linnaeus was much engaged in education. The chancellor of justice Carl Gustaf Löwenhielm referred to this in his 1746 memorandum to the parliamentary estates on the matter of how natural history might be used to boost the economy generally, and farm management in particular. He emphasized that no student at the university should be awarded a humanities degree if he were lacking in such knowledge. His message concluded with an expression of hope that Linnaeus would soon have trained a cadre of teachers to staff the *gymnasia* or grammar schools. The memo was submitted for consultation to the university and also to the notorious committee on education, the first commission on education policy in Sweden—but by no means the last. The clear aim is to provide professional training for state employees and leave science research to the Academy of Sciences. The examination system should be adjusted accordingly.

What was Linnaeus's response? He "favored being at liberty to choose a course of study, the right to be free of bureaucratic strong-arming and to enjoy learning, in line with the academic ideals of the nineteenth century, as was his defense of the lecture as a teaching method." Did taking this stand signify anything about Linnaeus's preferred political party? Was he no longer a Hat? Rather, the simple answer is that he backed the students and prioritized their interests ahead of political calculations.[5]

His opinions of education are brimming over in a long letter written in 1749, intended for Bäck but never sent:

> From Upsala *academie* I believe we have offered up more learned men than from any other place and academie in Europe.... Let us then take a look at the last 20 years and how incomparably the inclination in the country has risen for the sciences.... But what might the cause have been? Maybe pressures from the authorities? No, not so.... Our academie is now much changed in such ways that I hardly know if it is still the same as once or if it is now transformed into an illustrious gymnasium as new laws have stated that no one has permission to travel for a few leagues outside the city without instantly set about counting *studiosos*, noting absentees, and not offer anyone such means to travel abroad or else be charged to pay a fine of a thousand thalers.... It follows that it becomes ever harder to respire but instead one must keep the bowstring tensed all year long. What formerly took place with pleasure, I fear will now be learned by more enforcement.... A nag who with daily beatings is forced to walk onward will degrade

immeasurably from the high-spirited horse she once was. . . . However, do not believe that I in any way plead my own cause nor that I am in a mood to obstruct the authorities; no, my higher authority I will faithfully and humbly obey for as long as one warm drop of blood courses through my veins.

He continued in this vein and, reading on, it becomes clear that if Linnaeus had not had a family to think of, he would long ago have accepted the offer of a chair at the university of Oxford.[6] The teaching alone was demanding, as he revealed in the same letter: "The gentlemen in Stockholm have always said that the Upsala professors do no real work as they apparently have not understood such posts to be about other than to read at times and at other times sit about for a morning as an Assessor. Such people believe that learning is shaken out of the sleeve as if from a *cornu copiae* [horn of plenty]. . . . A professor, who in the morning has spoken *publice* for a whole hour, finds he becomes tired from it so that he until noon cannot arouse himself to do his own work; the afternoon passes in like manner with private [lectures]. Saturday in *consistorio* for all of the forenoon and perhaps on the Wednesday in *minori*"—a minor meeting of the university court.[7] So, Linnaeus complained but affirmed his loyalty to the authorities. In the same letter, he also got around to regretting that no academic would be joining the commission of education, which he thought would be seriously damaging. All of which is still very recognizable, across the centuries.

Being a member of the consistorium entailed, as we have noted, having to enact the judicial function of the academic establishment, which was in force until the mid-nineteenth century.

In 1767 the housemaid Stina Carlsdotter was charged with having drowned her newborn baby. The father, a farmhand called Jöran Sundgren, was assumed to be aware of this and charged with complicity. Both were servants in the house of *juris* professor Olof Rabenius, and hence the case was to be heard and sentenced by the consistorium. Stina had confessed and the law determined her sentence: execution by decapitation and the body to be burned in public. Jöran's case was less clear-cut, so he was sentenced to birching and a fine. Linnaeus belonged to the group who argued that his punishment should be just a fine "as the experience testifies that the delinquent is not straightforwardly improved [by being whipped] but also is depraved so completely he will have no shame or modesty to further hold him back." When the case was examined in a higher court, this was accepted. Overall, though, the process wasn't made any less terrible: Stina's execution was delayed by almost half a year because of the quarrel between university and city about who was supposed to pay.

Generally, the law of the country was followed—not only by Linnaeus but also by other leading scientists on the judges' bench, including Anders Berch, Torbern Bergman, and Daniel Melanderhjelm. Linnaeus did not refer to Stina's fate in his letters but returns to it when discussing a series of female child-murderers punished by execution in *Nemesis divina*.[8] It was to be the first decapitation in Uppsala in almost thirty years, though it would not be long before the next execution of a housemaid.[9]

Linnaeus managed to fit in three periods as university rector. It sounds ambitious but was less heavy an undertaking in his day compared with current demands. The post rotated every term between the professors, and it was quite common for the same man to take it on several times. Linnaeus was rector first in 1750 and then again in 1759. When he retired from his third rectorship in December 1772, he ended on the most beautiful note, speaking to his most dear subject, the delights of nature, *Deliciae Naturae*. A still more definitive endpoint was reached in December 1777, when young Carl Linné was installed as his father's successor. Then—rest, at last?

CHAPTER THIRTY-ONE

Family Life 2

HAMMARBY

THERE WAS A WINTER-LINNAEUS and a summer-Linnaeus. Then there was a commoner and later an ennobled Linné, a courtier and at-home variety, an Uppsala and a Hammarby Linnaeus.

This was the era of Rousseau, and finding a home in the countryside was a goal for those who had the means. Linnaeus and many of his colleagues bought country farming estates, which also ensured the family's daily bread. As for himself, he explained the escape to the countryside in terms of health, morals, and finances. In March 1763, he proudly told his relatives in Småland of his purchase: "I now take pleasure in traveling on Sundays to my country manor house to rid myself of the stench of the city."[1] Lecturing in 1767, he stated that in order to enjoy a long life one should live at some height but not on a slope, unless one happened to be born in such a place: "I am myself an example of this point. For as long as I lived on the ground floor of our house, all in it passed in poor health. Always, *Terra natalis* ought to be preferred." Fresh air was as important: "On this matter, too, I have an *exemplum domesticum* inasmuch as, if I were not to go to my country house in the spring I would be shattered, and were we to go there while in ill-health we would be well the next day." Implicit in all this was that the prefect's house in Svartbäcken was in an unhealthy part of town.

In the autumn of 1758, the year after being granted nobility by the king, he bought a manor house, which was an aristocratic privilege. His acquisition was Hammarby in Danmark parish near Uppsala, where he would live for long periods. He informed Bäck: "Now, with heartfelt greetings to my Brother, I must confess that I am unfit for life. I have always dreaded debt as much I do snakes." He had purchased "a small seat farm" for 40,000 thalers and had borrowed half that sum. He had done it, he said, for the sake of his small children. The price included a working farm called Säfja, also in Danmark parish. A tenant farmer managed Säfja, providing the owner with a regular rent income. The following year, Linnaeus bought the neighboring farm Edeby from Johan Ihre and covered the cost by asking the university to lend him the full 40, 000. He wrote to his mother-in-law, the widow Moraea at Sveden: "These are the first times when no sensible persons ought to keep money hidden away but should at once spend it on properties, whatever the prices." Three years after the property transactions, he had paid off his debts. His obligations as a member of the landed gentry included spending on cavalry men's cottages and, in wartime, providing the householders with a horse, weaponry, and other equipment. There actually was a war on, though thankfully not on Swedish soil but in Pomerania, which dragged on for almost five years of sporadic and inglorious fighting. The four soldiers Linnaeus had to sustain bore names that related to the farms their livings were attached to: Säfbom and Säfgren lived on Säfja land, and Hammar and Hammarblom on Hammarby.

The manor house at Hammarby was rebuilt at considerable expense, and completed in 1762. When the visitor arrives on the grounds, paradise islands come to mind; one can sense the presence of Adam on the lower slopes of the hill or "hammar" that forms a backdrop to the site and gave the house its name. The previous owner, Anders Schönberg, was the head of the royal hunt and had been living in a more modest house, now part of the west wing of the main building. The Linnaeus family were quartered there at first, but in 1762 "Linné builds at Hammarby, as he notices himself to grow feeble, so that the children may have a house." He repeats this: "I build for my widow and my fatherless children." The grounds were turned into parkland; he lived in the manner expected of an estate owner and an ennobled personage. Hammarby is a costly demonstration of how Linnaeus aspired to play the role of von Linné. The rooms on the first floor were quite splendidly decorated, including gilt wallpapers. Famously, he had his bedroom walls covered with prints of illustrations from Georg Ehret's *Plantae selectae*—many had been colored by hand. The Latin motto above the bedroom door—*Innocue vivito, numen adest*—is probably best translated as "Live harmlessly, the divine is near"; it is taken from Ovid's *Ars amatoria*, not from any gospel. The front room was hung with botanical illustrations of West and East Indian plants, drawn by Charles Plumier.[2] The portraits of celebrated friends—Tessin, Höpken, and

FIGURE 41. Hammarby, etching on copper by Fredrik Akrell. In Linnaeus's *Anteckningar om sig sjelf* (Notes about himself), 1823. Royal Library, Stockholm.

Schefferus—were reminders of the owner's status. Portraits of Linnaeus's parents were added to the Hammarby collections much later, when the estate had become state property.

Hammarby offered security when the city was burning. The German traveler Johann Beckmann kept a diary and described carefully what happened: "*Der 30. April [1766] war für Upsala ein erschrecklicher Tag*" (April 30 was a terrible day for Uppsala). The fire began at half past ten in the morning at Vaksala Square in a wooden house belonging to a leather trader. It was very windy, the hats blew off people's heads, and the fire spread rapidly. Beckmann had never seen such incompetent firefighting: no one seemed to have grasped that nearby wooden houses must be demolished, and no one carried fire axes. Soon enough, adjacent streets were burning, and once the wind died down around five in the afternoon, the devastation was widespread. By then, Linnaeus had had his herbarium and library moved out of harm's way, and Beckmann had packed his things and put them in the yard. Drums were beaten to call the population. Everyone was forbidden to smoke tobacco for 48 hours. Once the fires had been extinguished, the damage was counted; 87 ruined houses and yards. The Linnaeus family traveled to their country house, where they stayed for the entire summer. Linnaeus thought a third of the city had been reduced to ashes.³

Linnaeus thrived at Hammarby. "A wealthy farmer I might have wished to be, had I only been used to it from childhood, but rather than that, were it not so cold, to be a wealthy mountain Lapp. Now, I wish to own a small manor on an island in a good position so that I, without troubling myself too much, can find enough to eat. And, in a not too large *compagnie*, while the time away in conversation with some singular and well-meaning friends, so that I may look after my body and enjoy being in the studio of nature." Hammarby was a working farm, but Linnaeus did not become a farmer: the land was rented out to two tenant farmers. After Linnaeus's death, the 1778 inventory included 14 cows, 2 yearling calves, 7 pigs, 18 sheep, 8 turkeys, 20 hens, 7 geese, 2 ancient brown carriage horses, and a discarded cavalry horse. These numbers increased once the very competent Sara Lisa had taken charge at Hammarby. What a lot of cackling, neighing, braying, and baaing! The horses were small, their height barely above 12 hands. Cows, with or without horns, were also smaller then than now. Linnaeus's cows had pretty names: Summerrose, Lovely, Maidensweetie, Starrose, Lily, Flowery, Hammarrose, Goldie.⁴

The farmer's life was ruled by the weather. The summer of 1764 was the third one in a row that was dominated by long periods of drought and poor harvests. Conditions in 1771 were also awful, but for different reasons: "Never in my life has the farming man in these parts suffered such misery as this year. The spring wheat grew poorly. The rye came up quite thin but it is growing now. The Lord alone knows about the barley. It trials you to watch the sowings that will feed you but are not ready to get under roof. What will the farming man live on for the rest of the year? Where can he buy seed? Wherefrom can he take to pay expenses and tax? Wherefrom do we take to pay wages? Still no day without rain. Hay is now 1/3 part against last year."⁵

Today, little has been changed on the estate; the visitor meets farming practices adapted to conditions set by nature and culture. A visit is warmly recommended: Linnaeus's Hammarby has no comparison except Darwin's Down—and the other way around.

Foreign students were offered lodgings at Hammarby and shared in family life. Johan Christian Fabricius, Adam Kuhn, and Johan Zoëga spent two full years there, in 1763 and 1764. They lived some two miles away from the main building and rose early—at four in the morning during the summer. Linnaeus would arrive at six when breakfast was served. Afterward, he gave lectures, usually until ten o'clock. The students returned in

FIGURE 42. Linnaeus had his bedroom and study at Hammarby wallpapered with illustrations from the best botanical works of his time, among others by Ehret and Plumier. Photo by Mikael Gustafsson, TT News Agency.

FIGURE 43. Portraits of three of Linnaeus's daughters hung on the walls of Linnaeus's Hammarby. Photo Mikael Gustafsson, TT New Agency.

the afternoon to see him in the garden. In the evening, there were card games with the mistress of the house, who liked *tressette*—the point of it is to acquire three sevens. That more than a million officially stamped decks of playing cards were manufactured in Sweden testifies to the popularity of such games. Beckmann confirms that Linnaeus enjoyed joining in.

This was how he wanted to manage his life: "I now live steadily in the country during the summer, have 5 students with me in the neighboring farm and read to them twice a day for 3 *à* 4 hours at a time; eat wild strawberries; botanize and work on my *Systema Naturae*, now to be printed again and finally."[6]

According to Fabricius, fun and games were often part of life: "On Sundays, the whole Linné family was usually with us and, from time to time, we would call on a farmer to join us and bring his instrument, a so-called *nyckelharpa* that looked like a violin. We would dance in the barn and so there was great liveliness. While it is true to say our balls were not elegant, the company not numerous, the music below critique and nothing but minuet and *polska*, it is also true that we had a very good time. The old fellow smoked a pipe with Zoëga meantime, looked on, danced sometimes but only rarely, would do the *polska* in which he excelled over all us younger ones. He much liked it when he saw us growing very jolly, yes even noisy, as he otherwise feared we did not enjoy ourselves. I will never forget these days, all these hours, and feel happy whenever I think of them."[7] Fabricius said that Linnaeus almost always looked at ease and content at Hammarby. He described his teacher in a short, red robe and green fur hat, with pipe in hand, chatting very merrily and pleasantly, and laughing heartily.[8]

A note about the music: there is still a "music machine"—a hurdy-gurdy—at Hammarby, which is in the estate inventory but without a note on when it became a house possession. The recorded music is hard to identify, but it has been claimed that its roll plays a version of the popular melody *La Folia*.

Life at home has been described like this: "I walked over to the Archiater v. Linné's but did not find him at home and had instead to endure the very driest and most flavorless *entretien* with his old crotchety wife as none of the daughters was at home"—the words are Gustaf Adolf Reuterholm's, writing around 1773. He returned on another occasion: "He was at home himself, his wife and daughter being out visiting. The old boy sat talking about past experiments as he had conducted them during his Skåne journey, at the farm of my dear departed Herr Father. When I had been sitting there for a full hour, discussing with the old man, I went over to see young Gyllenborg."[9]

Fabricius described Sara Lisa Linnaeus: "She was a large woman, strong, despotic, selfish, and utterly lacking in culture. She often drove the joy away from our get-togethers. As she was quite unable to share in our talk, she also did not care for our company."[10] Sara Lisa may not have charmed the visitors to her home, but her position was far from easy. There were many and demanding guests, but she was outside the circle and had other interests.

What was served at Linnaeus's table? "In his life, Linnaeus was very dietetic, however he occasionally deviated therefrom. He condemned coffee, condemned aquavit, and advised much restraint with wine-drinking. But when my countryman Herr Bäckman held his disputation *De fundamentis ornithologiae*, I observed Linnaeus at the usual disputation feast consuming a little of most things. He commented: "It was quite well said by the old Romans that a man should have a *gula docta* [learned gullet]. Such is also mine: I can put everything in my mouth but I can also desist." This quote comes from the 1764 memory notes by the Finnish chaplain Anders Ehrström.[11]

Regrettably for us, Linnaeus rarely spoke of his own fare in his teaching of dietetics. "I am accustomed to plain home cooking and from it I feel the best," he insisted once. Long-distance visitors such as Johann Beckmann were invited to supper, but no one kept any records. A number of ill-tempered comments that concern Linnaeus's wife might suggest that the meals were not exactly sumptuous; she was, of course, the person who decided the menus. It is possible that she owned a copy of Mistress Warg's cookbook, but it reveals little: the recipes of the time list sauerkraut, small or large herrings, steamed puddings, mutton, carp, pigs' trotters, baked apples, cheesecake, dumplings, porridge, turnips, and beer. Many of these items were eaten everywhere in the country until quite recently, even though we by now have started to regard many of them as positively immoral, almost as bad as eating roasted songbirds, or as just plain unappetizing. In a quick autobiographical note, Linnaeus described his contacts with food: "I was brought up in poverty, so could eat with contentment the simplest dishes; with Clifford lived excellently well; *hinc redux in Patriam* in Stockholm, forever invited to dine with magnates. When next, a professor in Uppsala, I ate just a little of a few things, could not consume much but when so many dishes arrived, ate much more than usual, felt dried out, and became *exsucus* [exhausted]."[12]

Speaking of the good life and the good wine—this snapshot of Linnaeus comes from a 1773 letter to Johan Olof Kalmeter: "Yesterday the yacht with the wine *bouteilles* came. The Wine was reasonably good but in motion and sharp to taste; I do not know if its motion might have been brought about on the ship or yacht as it was quite tepid when it arrived. Being stirred onboard ship, it has the first degree of acidity and then it is dangerous for me who easily get *migraine* (I suffer from ache in half my head) and then I dare not drink it. Might be the work of the strong thunderstorm that passed a few days earlier started the work?"[13] Linnaeus knew his wine ...

Adam Afzelius visited Hammarby in the summer of 1768 and was greeted by a friendly host: "'You're so kind as to come to see me,' he said. 'Yes, but you're surely well tired and thirsty to have walked thus far in the warm weather; you will be needing a drink of beer!' He asked at once for this to be brought, but it took time as the housewife appeared not to care for treating wandering students to beer. In an angry mood, he went to remind her of it and soon returned quite upset and then said, as if to himself, 'It is difficult if one is not obeyed in one's own house.' So, the beer came, Linné brightened."[14]

Flowers are usually emphasized as Linnaeus's great love, but he also had a warm affection for animals even though they were no longer seen as moral exemplars, as in *Diaeta Naturalis*. They had belled a stoat at Hammarby and its approach scared the rats away; "She's a fine one," Linnaeus said.[15] Their parrot spoke with a Småland accent and would call out "Step inside!" and Pompe the dog, who had become used to strolling with Linnaeus to Danmark parish church on Sundays, would walk there on his own if his master was not around. After stopping briefly at the stone where Linnaeus usually smoked a pipe, Pompe went straight to the Hammarby pew and lay down. When Linnaeus had had his fill of talk about God, he would leave the church, and Pompe followed his example after a brisk "woof." This anecdote is told to many smiling but often doubtful children, but the matter of church dogs is an old one; Laurentius Petri noted it in his Church Order of 1571. Linnaeus liked the crickets in the bedroom at Hammarby singing him to sleep, and "only when they became too numerous could he be persuaded to with poison eliminate some of them," a descendant informs us.[16]

Another nocturne for grasshoppers, in Adam Afzelius's notes from around 1770:

> Mistress Linnea was a sharp lady, who ran an earnest regiment and often gave her housemaids strict marching orders; hence, was not a person to cross, as they say. Linnaeus, then, was a mild-mannered and restful man, although brisk and lively, and he loved peace and quiet.

Tired from his day's work and perhaps of hearing his wife's rowing, he used to go early to bed and sleep away a little of it. Then, when the wife had bustled about and withdrawn from her day of challenges, so that everything was silent in the house, he would get out of bed and sit down to work all night toward the morning when he would again go to have a rest and sleep for as long as he believed himself to need. But while during the night he was sitting up alone, he very much liked to hear from another living creature near him. From a baker who lived in the lane opposite Linnaeus's home, he therefore acquired for himself a cricket whom he had given lodgings in the tiled stove in his study.[17]

Let us finish this visit to Hammarby on an evening in 1762, when Elisabeth Christina spent a little time in the bower. "When I, this past summer, was at my Father's, Herr Archiater Linnaei country house Hammarby, which is situated some six miles from Upsala, Indian Cress [*Tropaeolum majus*] had been planted to fashion the bower, and I noticed that evening as I rested for a while in the bower that the flowers of Indian Cress flashed strongly so that it struck me as strange."

Her father encouraged her to make her botanical note public; it was noticed by Goethe and her name was attached to it, but she does not appear in any other documentation of the plant.[18]

CHAPTER THIRTY-TWO

Friends and Enemies

LINNAEUS HAD SEVERAL friends who were both aristocratic and cultured: Carl Gustaf Tessin, Sten Carl Bielke, Carl Hårleman, Anders Johan von Höpken, Charles De Geer, and Lovisa Ulrika. When he was younger, learned men had been his mentors, but now the status of his patrons came from much higher up in the hierarchy. Membership in the Academy of Sciences meant having colleagues from every class; all were knowledgeable and some wielded power. Another circle of friends had been formed among the young Prince Gustav's tutors: Linnaeus, Tessin, Dalin, and Klingenstierna.

Linnaeus had Tessin to thank for both being allowed the title "Archiater" in 1747 and being awarded the Order of the North Star in 1753, but in a way, Tessin was also a pupil. He wrote to his teacher in May 1757: "I await, indeed long for Herr Professor Linnaeus. I need to be taught by him. My sluggish, elderly mind requires fire and warmth. Professor Linnaeus is like a spirited and tireless lord lieutenant, who would rather resolve a matter twice than allow delays in its resolution." Anticipating a visit to make a start on curating Tessin's collection of minerals, Linnaeus wrote, also in 1757: "As I come closer to the end of my life's span, I have earned a body exhausted before its time, which cannot be restored. That I have lived in Sweden in the time when Your Exc. has been holding the reins has been to my great benefit." In his 1770 New Year's letter, Linnaeus wrote: "My heart will burn with gratitude for as long as warm blood still courses through my veins, and I shall follow Your Excellence in forthcoming events as if they were my own, and although I cannot anymore, I will when ready to be placed in my grave, still be sensible of the sincerest and profoundest thankfulness, such as a child might ever feel toward its fondest father."[1]

Their relationship, essentially that of a patron and client, or of father and son, grew more familiar, as expressed in the address "brother" or even "my dear brother." Letters to Carl Fredrik Mennander might in five or six lines contain as many "my brother / dear brother." Despite contacts often being limited to letter writing, friends established a form of family connection that could be firmed up by becoming godparent to the other's child. "Nothing could be dearer to me than to see for a few days my Distinguished Brother at my small country house some 5 miles from Upsala and converse together once more before we *valedict*"—did he mean "from each other" or "from this life"? The letter was written in 1766.[2] Invitations to Mennander recur but never seem to have materialized in a visit; yet again, in 1769, he elaborated: "Had I been allowed to receive my old friend here, in this castle of the air that I have built, I would have had double pleasure thereby, as it is here that I keep my herbs, my *conchiliae* [seashells], my insects, my corals, and my stones; in a word, all my reality. I weep over the absence of My Distinguished Brother."[3]

Linnaeus's closest correspondent and dearest friend was Abraham Bäck (1713–1795), so often referred to here. Bäck grew up in Söderhamn, studied in Uppsala, traveled abroad for four years—first to Holland, then on to England and London, France and Paris. While in Paris, he supervised the printing of the fourth edition of *Systema Naturae*, and his care in seeing it correctly produced must have made him a trusted colleague. On his travels, he reported on the sensational polyp story and contributed findings of his own.

Bäck ran a large medical practice and became a crucial operator in the construction of Sweden's medical care system. He collaborated with Linnaeus on a new pharmacopoeia: the 1775 *Pharmacopoea svecica*. Linnaeus was his patient and their families met socially. Over five hundred letters from Linnaeus have been preserved but, sadly, Bäck's letters are missing except for one series from his time abroad. Most of them were commandeered by Bäck, when he was preparing his speech *in memoriam* for Linnaeus.

Bäck was Linnaeus's most trusted confidant and helpful colleague; personal as well as medical and professional matters were discussed between them. They passed on news, and contacts at court—even though Linnaeus claimed to prefer avoiding them. They kept in touch for the best part of forty years. Initially, they addressed each other along the lines of "Most Noble and Widely Famed Herr Professor" and "Excellent and Most Skilled Herr Assessor," but by 1748, Bäck had become "My old and honest Friend and Brother." The tone between them, judging by Linnaeus's letters, was frank and warm, and their friendship never became laddish or gossipy. It continued functioning as before even when, at least in the 1740s, Bäck was engaged in intensive correspondence with Nils Rosén concerning aristocratic patients whose care they shared.

They exchanged mutual favors; Linnaeus's caring support for Bäck's son Carl Abraham—given the combined names of the two friends—was matched by Bäck's for young Carl Linnaeus Jr. It took some fifty-nine letters, but then, being a godparent brings important obligations. Their relationship was probably not free of conflicts and disappointments; for one thing, Bäck stuck with being an organizer and practitioner and never became the scholarly investigator Linnaeus had hoped to work at his side.[4] Both Carl and Sara Lisa Linnaeus enjoyed the company of the Bäck family. When Sara Lisa became very ill with the Uppsala fever around New Year's Day 1755, Linnaeus wrote heartbreaking letters to his friend: "I become torn apart by her wailing and sighs, it has been almost a month. . . . I have not ceased weeping; even now, every other night my little wife's life is under threat." His friend tries to help with medical advice.[5]

For Linnaeus, the Bäcks were a kind of second or reserve family; worries were shared. Linnaeus wrote to Bäck, full of concern about his wife Anna Charlotta: "God help your Wife, our Sister; pitiable females, they are in harm's way and must so often wear bloodied chemises. May the Lord God sustain her. See now to it above all that my Br. finds a decent and clean wet nurse so m. Br. will not have to change the child and it may suckle for the full 2 years; see to it the wet nurse is not too old."[6] When, in 1767, Anna Charlotta died just thirty years of age, Linnaeus wrote to try to comfort his friend: "Thank now the Good God that My Br. in his finest years has owned such a pleasing wife. Keep in mind that, now, nothing but the dregs are left in life's pitcher, keep in mind that the carriage will run away downward further all the faster. Look back on the living that has passed, is it not like a dream? Thus, what follows will seem the same, and in the end, all will be like a dream." Linnaeus ends the letter in distress— he cannot offer any real comfort: "But this is to ill-use my Brother, to wash the Moor for too long; I cannot let my dear Br. go from my thoughts, as I am alive."[7] Such intimations of frailty, of *vanitas*, were not uncommon and belonged to a mixed religious and Baroque inheritance, suitably Christianized. Psalms was a favored source of quotations: "Behold, thou hast made my days as a handbreadth; and mine age is as nothing before thee: verily every man at his best state is altogether vanity. Surely every man walketh in a vain shew; surely they are disquieted in vain; he heapeth up riches, and knoweth not who shall gather them." He sent a similar letter of comfort to Eva, Lady Memsen, and ended it with Linnaeus "settling in my corner to relive my heart in floods of tears."[8]

Bäck's home was in Stockholm, and the distance to the capital could at times overwhelm Linnaeus: "Now I have begun to long to see Dr. Br. but why will My Br. never more meet with me in Upsala? I am sleeping in Upsala and weary much with days of leisure as I once wearied with work. In good company, I will travel to spend one, at most three, days to Stockholm even though I have not the least thing to do there. But to journey to Stockholm and not find M. D. Br. there would be like going out to sea without getting a fish."[9] Linnaeus didn't always like to stay the night with Bäck, because he wanted to protect his friend from having to invite his guest to a good dinner; it was sufficient "if I may come in the evening to converse." One invitation had been offered to both Sara Lisa and Carl, but Linnaeus had persuaded his wife, who had spotted the invitation letter, that she shouldn't go: "My wife and I always find room to spare in a bed, but of an evening, we prefer to speak on our own"—no Sara Lisa, then, but why not invite Wargentin, the next best Brother?[10] Then again, how expensive everything has become. Linnaeus complained in 1773 to his friend Wargentin: "When traveling to Stockholm, as I will now, many things are necessary such as a carriage, horses, servants *et sexcenta alia*. When I journeyed all over Lapland, I went alone. One visit to Stockholm now costs me more than exploring all of Lapland."[11] A nightcap changes the persona of a great man like Linnaeus into someone ordinary; true, the terminology is a little opaque. His biographer wrote: "At home in his own house, he mostly went about dressed in a short nightshirt and a velvet cap." The tricky part was to get the correct cap on his head during Stockholm visits. He wrote to Bäck: "Both the nightcap and the furry cap I sent back the following morning with Dr. Wahlbom," and another time "If I have forgotten my nightcap in the bed in My D. Brother's house, I will surely have it returned sometime."[12]

It was not uncommon to place one's offspring in a friend's house to give the children as varied an upbringing as possible. Eleven-year-old Carl Abraham Bäck went to stay in Uppsala in 1771, and told of his experiences in a letter home: "Sunday 8 June, I bought a few little things, watched riding at a manor house called Eklund [Eklundshof?] and then Cousin Carl as he fenced bravely at Porath's. Listened to a disputation under Uncle v. Linné. Taken in a carriage, brought pharmacy things. Afternoon, to Hammarby. . . . Afternoon, first to the post and promenaded a little. At home, wrote one page of an *explication* of Voltaire, in the garden with the archiater." The next day, a visit to Gustavianum to see the Augsburg art cabinet and the Silver Bible. Dance practice. More parsing of Voltaire—truly, an ambitious education. On the Friday: "Explication and interpretation of Cicero. Garden and the animals with little Hinric trotting after me. Voltaire in the afternoon, then again to the dancing master. The post was taken for me and Mrs. Linnaeus went through the various herbs and caught *phalaenes* [night moths]." What an unusual sight—Sara Lisa with a butterfly net! He also listened to the story of the journey in Lapland while going for a walk, went to countryside herbation sites such as Gottsunda and Vålsätra, celebrated Midsummer's Day on an island in Lake Mälaren, at the Baroque castle Skokloster—then more Voltaire, botanical names, Cicero . . . which of Voltaire's works did they read, and who taught him? Surely not Linnaeus? An idyllic time—but their godson died at the age of fifteen.[13]

A quick check has revealed that Linnaeus's friends generally didn't live in Uppsala, and that he seems not to have been in close contact with his relatives. A rare mention of family from 1763: "I now can take my pleasure to travel on Sundays to my country house and so rid myself of the odors of the city; when I am there, I reflect 100s of times on my absent siblings and wish that I could have them visiting once." Of course, friends and acquaintances in Uppsala are not going to be well documented in correspondence, so the possibility of local friends can't be excluded. After the fighting over professorial posts, Rosén and Linnaeus seem to have become reconciled. Rosén took charge of Linnaeus's treatment when he was ill with pleurisy in the spring of 1764, and he "thereafter felt an incredible friendship for Rosén."[14] Linnaeus and Wallerius remained on unfriendly terms, but he got on well enough with Ihre and saw his neighbor Klingenstierna socially—at least, their wives played cards together. And there were always his students, who had flown out into the wide world.

Inevitably, there were tensions; the jealousy of colleagues was reflected even in their entries in books of patronage. Around 1780, the Uppsala student C. J. Knös wrote: "At the time when Upsala at one and the same time was illuminated by a Linné and an Ihre, the Jalousie in each toward the other ruled their spirits. Therefore, when some student or otherwise another person presented his book of patronage to one of them, Linné would always write: *Famam extendere factis, hoc virtutis opus*; but the Ihren always wrote just the opposite: *Non magna sunt, quae tument*" (The work of virtue is to extend reputations with acts *versus* It is not the great who inflate their name). Born within a couple of months of each other, Ihre and Linnaeus were the most famous academics at the university and political rivals as well. They were competitive, and that drove an ongoing duel. Such things happen . . . even in academic circles.

Who were the anti-Linnaeans, the young lions who roared impatiently in the brushwood? Torbern Bergman, Peter Jonas Bergius, Daniel Solander—anyone else? In 1757 Linnaeus wrote to Wargentin: "Three of my pupils, whom I have favored with every service in the world and brought up on my lap, have joined forces to everywhere diminish me." The physiologist Roland Martin and the medics David Schultz von Schultzenheim and the Bergius brothers made up the group of treacherous students, just as Rosén and Wallerius had done in the past. Another colleague, Samuel Aurivillius, was making a fuss about salaries and private advantages; clearly wanting to needle Linnaeus, who charged a fee for private lectures, Aurivillius remarked that "I have not yet noted for my part a gradual accumulation of wealth so that I therefore sought to profit from my lectures."[15] However, according to Thore M. Fries, the battle with Aurivillius in 1758 was "the only conflict with a colleague with which he had to deal during his long period as a professor at the university."[16] His relations with Peter Jonas and Bengt Bergius, both collectors on a grand scale, grew tense from time to time. Linnaeus could never cultivate *Linnaea borealis*, the twinflower, in the botanical garden, but to his huge irritation, on 24 July 1774, Bergius succeeded. This was the reason why "he that spring fell ill during a lecture and said it was a species of stroke although he seemed as he was before."[17]

Holding positions of power, such as being a censor on behalf of the Academy of Sciences, brought with it causes for hostility. The schism with Jonas Bergius was allegedly triggered by Linnaeus's report on the soya bean.[18] It sounds harmless and commentators have suggested that Bergius overreacted, but the situation was complicated by their competition for objects to add to their respective collections. There were many letters agonizing about the interpretation of Linnaeus's reports on various subjects. "Rather fine" was his stock judgment, but he sometimes let rip, as in this dismissal of a Finnish chaplain's conclusions on tree dieback and how it affected cattle: "I have hardly ever seen such a feebleminded work in print. No simple farmer could write or think more poorly."[19] Wargentin, with his many connections in useful circles, often had to mediate and moderate. On 13 November 1764, he wrote to Haller in French: "that you are displeased with M. Linnaeus... is hardly surprising. One must realize his eminence to forgive him his caprices. The entire world holds him in high regard. But hardly anyone loves him; not here either." Wargentin added: "This is said between the two of us."[20]

The interaction with the royal court was a special case; the royal house had supported Linnaeus for decades. Lovisa Ulrika, the wife of Adolf Fredrik who became king in 1751, had been interested in Linnaeus because she was an avid collector of *naturalia* and he was an international name in the business. In 1746 the queen visited Uppsala: "I ingratiated myself with Linnaeus by admiring his plants and insects and my audacity to respond in Swedish to the speeches." Later, Linnaeus was requested to curate the royal collections. The queen wrote to brother Fredrik the Great in 1752: "I amuse myself by ordering them [butterflies and shells] in the company of a professor from Uppsala, who is a leading connoisseur and scientist. . . . He is very entertaining, a great wit, although one has to overlook his manners which have something of the peasant about them." She wrote about Linnaeus to her mother: "He is a most amusing man, who has all the witticism in the world but lacks its style and so, for both reasons, constantly enlivens my days. In the evenings, he has been tasked to go for walks with the king and hardly a day passes without him finding a way to delight and humor us all."[21] It became a Linnaean family tale that the queen had asked for one of the daughters to be presented in court but Linnaeus had sternly refused.

In Hammarby, not only the spirit but even the air was different from Drottningholm. Linnaeus told his students that he once had to sleep in an unheated chamber at Drottningholm. He wrapped himself in a quilt at night, but after staying there for a month, he began to suffer from "a horrible toothache": "He crept about like a worm. Finally, it came to him that the very air was infected. The Arch. and Knight took himself off to China [the garden folly], was so cleared of the torment."[22] He pursued the theme in a letter to Bäck, in the summer of 1753: "I shiver when I hear Drottningholm mentioned as I cannot get hence when I want to. A poor prisoner for long, now in Drottningholm—a sheep upon an island"—here, Linnaeus alluded to the hymn by Erik XIV, the unhappy sixteenth-century king.[23] He much preferred Hammarby, but did as he was bidden by his two royal clients. He recognized well enough the value of royal goodwill, was of two minds about the luxury, flattery, and power, and also driven by the wish to complete a grand opus, something to stun rivals abroad in the international race to acquire the best collections.

But the work on the royal collections was floundering, and he wrote to Bäck on 11 December 1753: "His Roy. Highness's cabinets printed catalogs goes so slowly I fall asleep over it. Never have I been subject to suchlike. And His Majesty is so quick to want more. . . . Dear man, submit on my behalf to her R. Maj. that, once the Queen's cabinets are to be in print, I should be given free hands to take on and arrange the matter with the printer."[24] After the failure of Queen Lovisa Ulrika's 1756 coup to reinstate the king's political power, the situation changed somewhat. When Linnaeus, in his capacity as rector of the university in Uppsala, gave a speech to the royal couple,

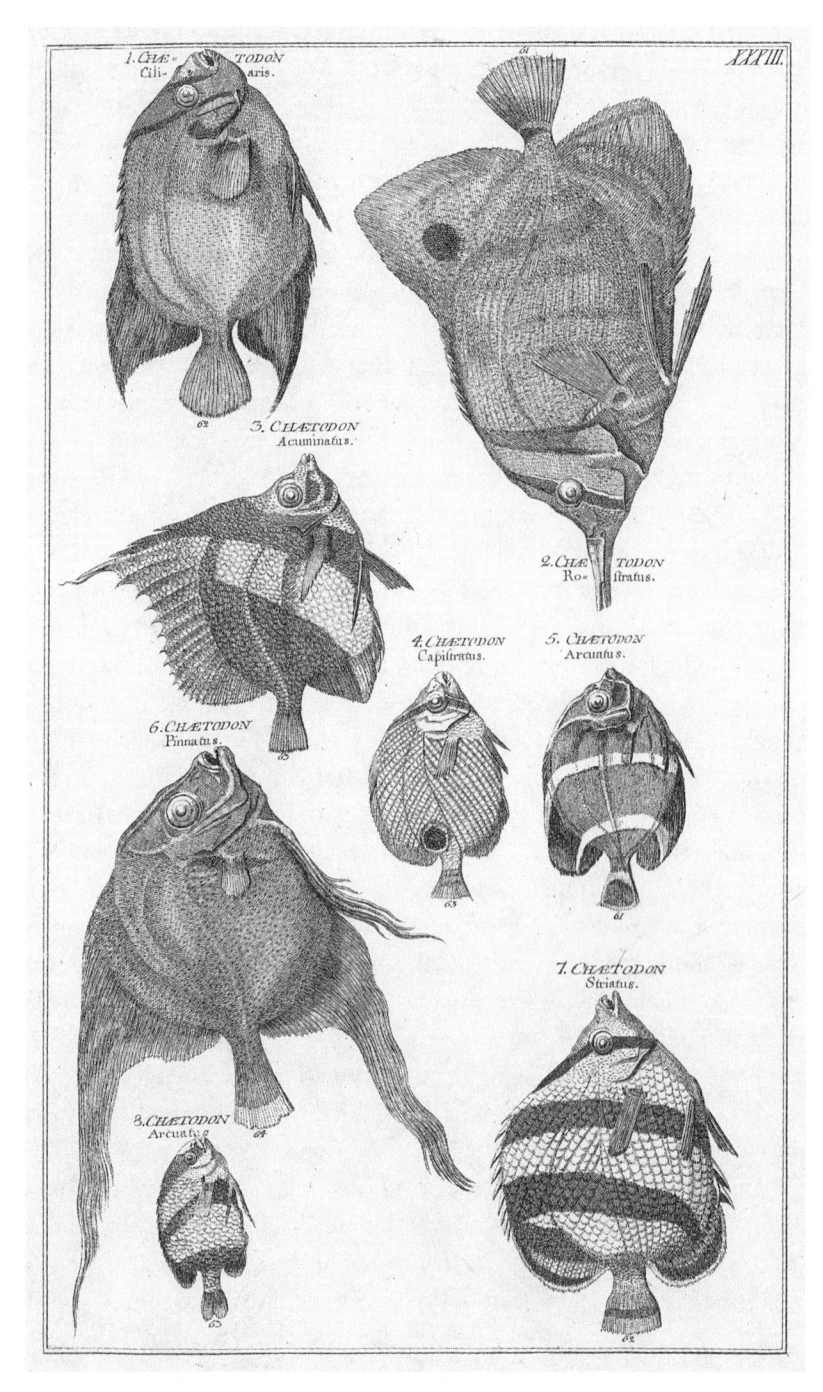

FIGURE 44. Fish. From *Museum Adolphi Friderici*, 1754. Uppsala University Library.

he painted a glowing vision of Science—"The sciences are thus shining a light that illuminates people who wander in darkness"—and then, in the last third of his oration, turned directly to the king and queen, employing a familiar metaphor: "Our Sciences receive their vigor as if from the Sun when a gracious Authority considers its most dedicated practitioners with more benign eyes." The final words looked toward the future: "Later, the world coming after us will surely be envious of our good fortune; should we then make ourselves unworthy and forget the benevolent attention of our most gracious Royal house?" What followed was an opening up of the relationship between the royals and Linnaeus: "He then was offered the pleasure of daily conversing with such a great and excellent Queen and the gentle King. He would have to become a courtier, he who had never intended it." But despite moaning to Bäck, he was naturally flattered.

The setting was brilliant, French and Swedish vocabularies mingled, and powder whirled in the air. The library was rearranged in a wing, together with the coin cabinet and the collections of minerals and assorted biological specimens. He would give little lessons to people at the court; games were played. A reluctant Linnaeus was made to play blind man's bluff, so, against all the rules, he peeked and managed to catch the queen. After that, he didn't have to join in anymore. Prince Gustav (later, Gustav III) asked Linnaeus to "tell me about animals" and so he spoke of "midges, and the birds who eat them, and swans and pelicans and sea lions." The crown prince told his mother that Linnaeus had replied *"avec sa vivacité ordinaire"* to a question about a monkey, saying "for God's sake, it must be given to him. I will send my muff so the farmhand can put the poor little animal inside it. Otherwise, he might freeze to death." In June 1769, the prince visited Hammarby and, as the newly crowned king, Gustav sent Linnaeus nature specimens from Surinam for Christmas 1774. The king visited Uppsala in August 1775 specifically to see Linnaeus. His adjutant, Carl Johan Ehrensvärd, noted in his diary: "The old boy is much decayed, Linnaeus seems to fade with his science. He has no successor." The last time the king and the scientist met was probably at Drottningholm in 1776.

Linnaeus enjoyed some aspects of life at court—the monkeys, for instance:

How do my young Mistress and Master hold themselves at court? I believe the old fellow to be an uncommon court buff if only he is not teased or heckled as he cannot take a joke. Old fellows do not take kindly to harassment. But his wife is prickly and grumbles, however not meaning it badly; for she is now in a blessed state; I venture to speculate

FIGURE 45. White-throated capuchin monkey. From *Museum Adolphi Friderici*, 1754. Uppsala University Library.

she will deliver first when spring comes. Then, she must be given a warm room; though she is herself not as persnickety as the other ladies at court, her children when first arrived are quite tender, if they get a chill, a cold in the nose will follow and hang on for a long time; which should not be wondered at as they are quite naked, just as are ours are and also descendants of a southern hot extraction, so very far from the home of the Lapps. I have asked myself for what reason the Old Fellow, each time he caresses his beloved since she was blessed, grows malevolent and seems to want somehow to assail her; that is something he never did when the female was being flighty. As soon as the female has taught the offspring to move about and climb, the son will keep closer to his father than to her and the father admonishes the mother whenever she chastises her child, as she will do not too rarely. So, see to it that their winter quarters are such as that they can lie together and warm each other, as they will then enjoy life more, just as we do, but better tie them a little away from each other so that the old fellow does not rob her of all food; without her getting away from him she would die of hunger.[25]

Could it not be that Linnaeus added something of himself to the portrait of the "old fellow"? Perhaps it is the aging couple in Uppsala and at Hammarby, who are reflected in his narrative—or it could have been the royal couple in their castle. Then again, he might have just been pondering over the eternal mysteries of living with a partner.

CHAPTER THIRTY-THREE

Problems

LINNAEUS, AT THE age of twenty-three, had already conceptualized his entire set of future aims; much later in life, he wrote to Tessin: "At the age of between 20 and 30 years, we believe ourselves to know and understand everything; at least I was never again as convinced of my learning as when I was 24." That was in 1731, and the process of formulating his botanical taxonomy was underway. But much remained to be done, with new and old problems to be solved.

The Feasibility of the Project

As late as 1753, Linnaeus could write in *Species Plantarum*: "The number of plants on the whole of this Earth I have calculated with fair certainty to be much smaller than is commonly believed, namely it reaches at most 10,000." By then, the true size of that number was, however, dawning on him and his colleagues. Already in 1748, the complexity of what we now call "biodiversity" had challenged Pehr Kalm on landing in America: "I arrived to a new world. Wherever I looked there were plants I had never before seen. I shudder with dread at this revelation of such a large part of natural history being unknown to us." In *Oeconomia naturae* (1749), Linnaeus estimated the number of species in the different classes: plants 10,000; worms 2,000; insects 10,000; amphibians 300; fish 2,000; tetrapods 200—adding up to a total of about 26,500 living species on Earth. His early biographer J. E. Smith wrote, "We dare to predict that, while *Systema Naturae* was but the first attempt of its kind, it will doubtlessly also be the last. Natural history is by now too extensive a subject for any one individual to take over his leading position as the universal naturalist."[1]

A school textbook in physics from 1779 explained: "Already described species are rising in number to 7,180, and it is quite likely that even now as many unknown species will be found." When Thunberg had Botanicum built in Uppsala, the herbariums were planned to have room for 15,000 species. It didn't take long for window embrasures and all other free surfaces to fill with herbarium pages, all demonstrating Thunberg's dedication to, as well as success in, the collection race. His long series of dissertations on *Nova plantarum genera* (fifteen parts) and on *Museum naturalium academiae upsaliensis* (all twenty-nine parts) provide evidence of Thunberg's capacity for hard work. Seemingly, the questions driving the search for an all-embracing system had given way to specialization and detailed recording of species characteristics, methods that in later times have been termed "descriptionalism."

Scientists were finding it ever more difficult to unite the concepts of the perfection of God's Creation and science's search for completeness and truth. Can nature be contained in tabulations? When the travelers return, how much room must be set aside for their discoveries? The number of species might well leap from ten thousand to a thousand times larger, for how to define a species? Some saw it as a greater glory to find as many as possible, others wanted to limit the number; botanists grouped themselves into "splitters or lumpers." If new species were allowed to emerge while others vanished, any attempted overview of nature's variety was becoming steadily more impossibly complex. *Systema Naturae* appeared to be forever a work in progress—would it ever be completed?

No one could fail to notice that the framework had to expand, despite growing numbers of diligent researchers and the publication of large works. In *Tabula affinitatum animalium* (1783), Johan Hermann calculated the number of possible variants that would result from combining ten variables from two species of beetles: it was 10, 172, 640. Several other such calculations from that period produced equally dizzying conclusions and raised the question of whether every species had to be fitted into a system. The insect records in particular overflowed any preexisting order; so, entomology liberated itself and became a specialty in its own right. Instead of referring to the closure of a circle it is more appropriate to say that it was broken; that is, the old idea of the encyclopedia—which means "entire circle of knowledge"—faded, dreamlike.[2]

All this entailed specialization and created a need for new ways of publicizing science: beginners' textbooks, connoisseurs' splendidly illustrated works, collectors' handbooks, specialist journals, and for those able to cope with a whole field, bibliographies. In our jargon: information overload. The supplement, which Linnaeus called Mantissa, was an essential new form of publication. He couldn't yet face another edition of *Systema Naturae*, but he did produce two mantissas.

The Invisible World

The ever-increasing number of identified species was only part of the story; the number of families and classes also grew, and there was even a possibility of a new kingdom. For taxonomists committed to an encompassing format, it must all have been very unsettling. Linnaeus's dissertation *Mundus invisibilis* (The Invisible World, 1767) has been admired for his innovative thinking but has also been evaluated differently. He had largely ignored the microscopists for a long time, but he had become intrigued: they were penetrating into the darkest rooms of nature and opening doors to reveal marvelous things—"doors, as if unlocked by Baron Otto von Münchhausen." His work suggested that the black deposits on cereals were not "soot" but tiny living creatures, which he had described in his great work on housekeeping. A similar powdery material had been discovered in fungi such as puffballs and milk caps—indeed, even molds were made up of some kinds of seeds. On being immersed in tepid water they developed into worms, clearly visible under the microscope. The wormlike creatures would form a finely meshed tissue that swelled and eventually turned into the fungi that produced the powder in the first place! This process, Linnaeus declared, posited several questions: Shouldn't the fungi in that case be included in the animal kingdom? Or should they join a new kingdom of nature, a *Regnum neutrum* or *chaoticum*, in which for instance the polyps would also belong?[3]

Mundus invisibilis was really a commentary on the final pages in the animal section of *Systema Naturae*; it discussed the most extreme organisms of the zoophyte class that included organisms with partially recognizable names: *Hydra*, *Pennatula*, *Taenia*, *Volvox*, *Furia*, and *Chaos*. Down there, we are in the deepest basement of life, expressed in forms between animals and plants—a repository of life's secrets. Linnaeus stressed the role of the microscopists and their wonderful observations. Münchhausen's "discovery"—known to Linnaeus, who corresponded with this German land steward and botanist—was surely about to make world history as it established a new kingdom in nature. He consulted other experts and initiated the Swedish queen by sending her a demonstration of this astonishing phenomenon. Johan Carl Wilcke, a skilled marine biologist, was asked to confirm Münchhausen's findings. He failed, saying that he had examined the material in maximum magnification but had not seen the slightest sign of life. Presumably, the wrong kind of soot and, anyway, too old.

Linnaeus speculated on the theme of live carriers of infection, or bacteria, in his 1757 dissertation entitled *Exanthemata viva* (Live Eruptions), but far from being a new idea, it was relatively common, with predecessors such as the old belief in panspermia, the criticisms of ancient hypotheses of generation, and the discoveries of microscopists about small-scale life forms. Linnaeus took these notions as building blocks for a new construct, a species of live molecules, or of free marrow and similar notions. His speculations affected his reinterpretation of human conception; in a 1772 lecture, he asked the testing question "Could one not claim that human beings were fungi?" Just as fungal spores "ferment" and develop, so surely the spermatozoa? If so, these "molecules" are by definition animals; they should be classified as such. Notably, Münchhausen assumed that fermentation drove development, and Linnaeus added decay by rotting—both processes were central to the alchemists.

By now, many thought it worth questioning the traditional tripartite subdivision of nature and what was happening to it. Thomas Martyn, professor of botany at Cambridge, apparently said that, at times, Linnaeus seemed quite mad: "Our poor kingdom of plants will soon be crushed into atoms by, on one hand, the animals and, on the other, the fossils." New observations about the Australian fauna eventually led to a new subdivision into monotremes, marsupials, and placentals. Linnaeus did more than just skim the reports from James Cook's explorations when they began to come in, at first from Banks and Solander. He found that they caused problems. He heard about them via his friend John Ellis and then from Joseph Banks himself, who passed on the Hawkesworth account. Banks had withdrawn from Cook's second voyage and been replaced by Johann and George Forster, so he had to ask J. R. Forster. Forster sent Linnaeus a quite thorough lot of descriptive material, but it arrived too late for him to make use of it. The findings from the first voyage included observations of the kangaroo, and Linnaeus used his new method of cataloging. He created a card index for the animal: 4 molars; a sheep's size; a hare's voice. His son then entered the information into his set of marginal notes for the new edition of *Systema Naturae*. However, lack of time meant that the discovery of a new continent's flora and fauna didn't leave any deeper imprints.[4]

The Dream of the Natural System

The classification of plants was deliberately based on "artificial" principles, and the hunt for the basis of a "natural system" continued for the rest of Linnaeus's life. He had made his first suggestion of such a system already in *Classes Plantarum* and then again in *Philosophia Botanica*: "It is therefore called *Methodus naturalis* when the herbs are arranged in the order in which they would naturally occur together. This order is difficult to establish, as it is true that we do not have all plants known to us. Therefore, only very few are able to take this matter forward and even a great *botanicus* such as Hr. Arch. Linnaeus must admit to being unable to make any great *decouverts*."[5] However, he had given some thought to the issue: "Should an *ordinum naturalium* be considered, one should once more attend to: 1. Which are the herbs that are closest together; 2. Characteristics by which they are distinct; 3. The effects or *vires* of their taste and odor." The third point is worthy of special attention. Still, the natural system ideas were normally ignored and spoken of only to the most favored—fee-paying—foreign students.

Linnaeus came up with a map model as an alternative to the chain of nature, or the ranked scale. It stressed continuity as well as hierarchy: all plants develop mutual affinities with those around them, comparable to an area of land on a geographical map. The idea had been visualized in an illustration that was shown during lectures on the subject. The notes and images were published posthumously in 1792 by his pupil P. D. Giseke. The map shows the orders portrayed as larger or smaller islands in numbers that vary from one version to the next. The relationships are proportional to their stated affinity, and the transitional families were inscribed, although Linnaeus allegedly told Giseke that he would never reveal them. The straits between the islands seem to deny the concept of continuity but should be taken as the white patches on maps before all geographical areas had been explored. When this idea is discussed, it is often pointed out that the famous adage "nature abhors jumps" had been crossed out in his private copy of *Philosophia Botanica*. It is a moot point because the phrase is still there in the final passage; it could just be a case of eliminating a repetition.[6]

"Such jealous folk and good-for-nothings," Linnaeus wrote to Bäck in 1764 on the subject of Michel Adanson's natural families of organisms: "But I have no quarrel with Adanson. . . . as his *Methodus naturalis* is the most unnatural of all. . . . None of his classes become natural, only a mingling of everything. One must suppose that God did the one thing before he did 2, 2 before 4, that he first did one species of each genus, then that He mixed different genera so as to produce still more species."[7]

It is notable that Linnaeus never refers to a natural system for ordering animals. Did he think that that system was already in place or that the task was impossible? Linnaeus would never see his hoped-for dream of a natural system become reality. Meanwhile, a new biological genre had emerged: a dictionary of natural history, arranged in alphabetical order and apparently providing a possible compendium of all new knowledge.[8] The idea attracted customers and tempted the publishers. Linnaeus had noted the trend and accepted when, through Carl Gustaf Tessin, he was offered the task of compiling a *Dictionnaire portatif d'histoire naturelle* by the publisher Jean Marie Bruyset in Lyon. The project got underway in 1757 but folded a couple of years later, as the work became more and more unmanageable. The reason might have been a self-imposed goal of completeness or intimations of just how meaningless it was to abandon principles and lose a systematic and truthful approach—even though it was, of course, practical with its portability and alphabetical order.

Questions of Species and Their Development

Lack of clarity affected not only the definition of a species but also the possibility that it was not a constant entity. Contrary to what many believe, past thinking was much influenced by transformations and metamorphoses, crossings between species and ideas about degeneration. By formulating his straightforward laws with their set boundaries, Linnaeus had actually given natural history a more modern basis.

In his later work, the species was relegated from its earlier function as a cornerstone in the system to a more occasional role. Instead, the genus was emphasized, often because it was simply too difficult to decide whether you were dealing with a species or a variety. It became more practical to concentrate on the groupings at the next higher level of the system. It was also important to define the distinctions between different categories— class, order, family, genus, and, within the genus, species and variety.

In his own copy of *Philosophia Botanica*, Linnaeus had crossed out the phrase *nulla species nova*. He combined his ideas about hybridization and the marrow/bark interaction in the 1762 dissertation *Fundamentum fructificationis*, and then, in 1764, set them out in numbered items in the sixth edition of *Genera Plantarum*: 1. From the beginning, the Creator of the universe dressed the marrow of plants in bark or differently constituted principles, from which emerged so many differently shaped individuals as there are orders. 2. The Almighty mingled such orders of plants between them so that there emerged in turn as many *genera* from the orders as there are plants. 3. Nature itself mingled its genera from which emerged so many species of the same genus as there are today. 4. Random mixing of these species took place from which emerged as many varieties as here and there have been observed. 5. These processes follow the laws of the Creator, going from the simpler to the more complex, the laws of nature bringing forth hybrids; the human laws are based on observations *a posteriori*.

Linnaeus didn't publish any corresponding overview of how he saw the system operating in the animal kingdom but, in his lectures, his explanations indicate that he saw the same system being applicable: "God began from the simpler [and moved] to the more complex. He created an animal, a plant, let them multiply on the land. He reshaped the animals in so many classes as we have today, whereof affinities have emerged between different families and genera such as camels, deer, and oxen. These are then mingled between them [in nature] whereof the red deer, the reindeer, and the elk have emerged."[9] Such were the lines of thought that led Haller's spokesman, the Swiss J. G. Zimmermann, to accuse Linnaeus of atheism. In 1778 Linné Jr. responded to Bäck as follows:

> Never was my Dear Departed Father an atheist, and he could not endure those who voiced such notions. His collection in Nemesis surely bears witness of his thoughts about God and so do many passages in his works, especially in the introduction to The System, where he spoke of his belief that the species *animalium* and *plantarum*, also genera, were of the present time but the *ordines naturales* were acts of the Creator. Had these latter not existed then none of the former could have come into being. We regard it as the very most unlikely propositions that any such should take place later, namely Reaumur's mixing of rabbit and hen, an event that magister Acrell claims to have seen in our times in France. That mingling between plants takes place and so produces new species would seem to convince many *a posteriori* in the cases of many herbs and especially Capenses, as they are exposed to the most wind. There is where we have genera with the most numerous species.... I claim that my D. Father conceived of the world thus that when the globe existed at first, one small point on it was bare whereupon the Creation began; as the earth was formed herbs were created. How have enough room for everything? Afterward the animals were created such as were able to live of eating plants, and then the animals were created who

lived from these other animals. Nothing was sufficiently completed in days but in many 100s of years: it had to be because had Our Lord at the same time created *animalia phytivora* as he did the plants, and the same day *animalia carnivora* as He did *animalia phytivora*, then the Creation would have been ruined. Therefore had Our Lord first among animals and plants created *ordines* and then also *sensim*, then He would have with His almighty power without tampering with His laws easily have made them mingle and from that the genera would be brought forth, and later left them to the laws he had given Nature and implanted in every growing thing that *sensim* would mingle and produce species; how far this would reach may well not be for mankind to set out to investigate; it would be daring enough to wish to reason concerning what had taken place.[10]

Linné Jr. composed a summary of his father's thinking from 1760 that, overall, reads correctly on all critical points; it testifies to the free communication between father and son. "Proof" such as Reaumur's rabbit-hen chimera is on the same level as the creatures listed in the *Paradoxa* group of *Systema Naturae* (1735), examples contrary to the laws of nature. God the Creator was still required, but the old models turned out to suffocate life forms when applied to nature as observed. Instead, the Creator "mingled" nature's orders. And so forth . . .

The Golden Years Came to an End

Natural history was about to become unfashionable. It is possible to follow the natural sciences as they lose their economic and materialistic applications, their ideological function, and their audience. The newspaper *Posten* printed on 20 May 1769 a pretend letter from a Nature Investigator addressed to a Mr. Oeconomus: the writer reminisces about when, in his youth, he dressed in loose-fitting shirts and joined a "troupe of grass-finders" led by the "very well-known" Herr von Linné. So many enjoyable times were to be had in the meadows, and no one dared "be crabby" if the land was trampled to bits as it would single the complainer out as an enemy of the natural history, at the time a breach of the highest law of academe. To the point: "So tell me dear friend, how is it now with the Natural History and *Oeconomics*, which at that time I was told to be one and the same Science? He who couldn't classify herbs and tell the difference between a mouse and an elephant "based on their teeth" was no *oeconomus*: "I have now learned that the great, most Honorable Sir has himself acquired sizable Country Properties and I am totally persuaded what is the culture there must be the very model for all *Oeconomie*."[11]

During the Gustavian era, wit and literary learning became more highly valued than science. In the poem "Porträtterne" (The Portraits), Anna Maria Lenngren satirizes the "old school": the countess boasts of her ancestor "the widely-traveled president / who knew flies' names in Latin and in Greek / and once the Academy's thanks did seek / gifting an earthworm from the Orient." As early as 1750, Linnaeus himself realized what was in the air. In his 1768 letter about the educational system, he addressed the university court, the *consistorium*, on the subject: "Since 1750, I have sensed how the sciences have more and more declined and still continue waning in our country. Even at that time, a new epoch had begun for the schools of learning: the academies were being remolded from academies to *Gymnasia illustria*"—he followed up with a long, critical analysis of this troubling development.[12] On another tack, it is also possible to pick up on a sense of the approaching romantic nature philosophy.

He also had to confront his own aging and the dulling of his senses. As for that "invisible world," Linnaeus could no longer see clearly without aids, and he also had to start supporting his other senses. He was no microscopist and the genius of his eyes was clouding over. A display case in the Linné Museum contains his pocket watch, snuffbox, ear tube, eyeglasses, and alarm bell to alert the servants—equipment linked to his five senses and their continued function. You might say that growing older was forcing him to improve and change his sensory perceptions, but he was, of course, also conscious of having to follow developments in his scientific specialty. In 1758 John Ellis gave him one of the famous Cuff microscopes, but it seems to have mostly been gathering dust. When Linnaeus went on his travels, he brought a small loupe. The naturalist Anders Jahan Retzius recorded Linnaeus admitting that he hadn't used his compound microscope except once, when he tried to repeat Münchhausen's results. Ellis was skeptical of both Münchhausen's and Linnaeus's claims that they were about to force open the door to the new realm of nature.[13]

CHAPTER THIRTY-FOUR

A New Synthesis?

WHICH ROUTE WOULD his work follow now? Could a new synthesis be discerned beyond the current problematic state of his science, was there any new challenge to fire his enthusiasm so late in life? When science historians expound on his life's work, the emphasis is usually placed on the young, optimistic Linnaeus, who created the taxonomic base of biology and persuaded the world to apply the rules in his *Systema Naturae* to the entire diversity of nature. Why should his 1766 essay *Clavis Medicinae Duplex*, just thirty-odd pages long, deserve the amount of attention it will be given here? It looks like a collection of pharmacological tables with an added map of the body, charting the physiological responses to different events. It could be described as a summary of a traditional form of medicine with roots sunk deep into the past, and drawing heavily on natural history. However, aspects of it make it into Linnaeus's medical bequest, a will to be read in the future, its author insisted, because only then would it be comprehensible to its readers.

Clavis is in many ways Linnaeus's most defiant piece of writing and probably the hardest to interpret. It has arguably failed completely because it demands a great deal of close reading and so has rarely or ever been given even the benefit of the doubt. The starting point for the discourse in *Clavis* is a property peculiar to each plant called its "power" or "virtue," a concept that can also be found elsewhere in Linnaeus's writings. The introduction, as well as the rest of the text, is the outcome of merging the biological taxonomist's ideas with those of the speculative medic. Its fundamental claims are that human beings consist of marrow and bark, and life is a manifestation of electricity. They are followed up by a mixture of systematic ordering of illnesses, physiological observations, and pharmacy. *Clavis* also provides keys to the mind of Linnaeus the man, who

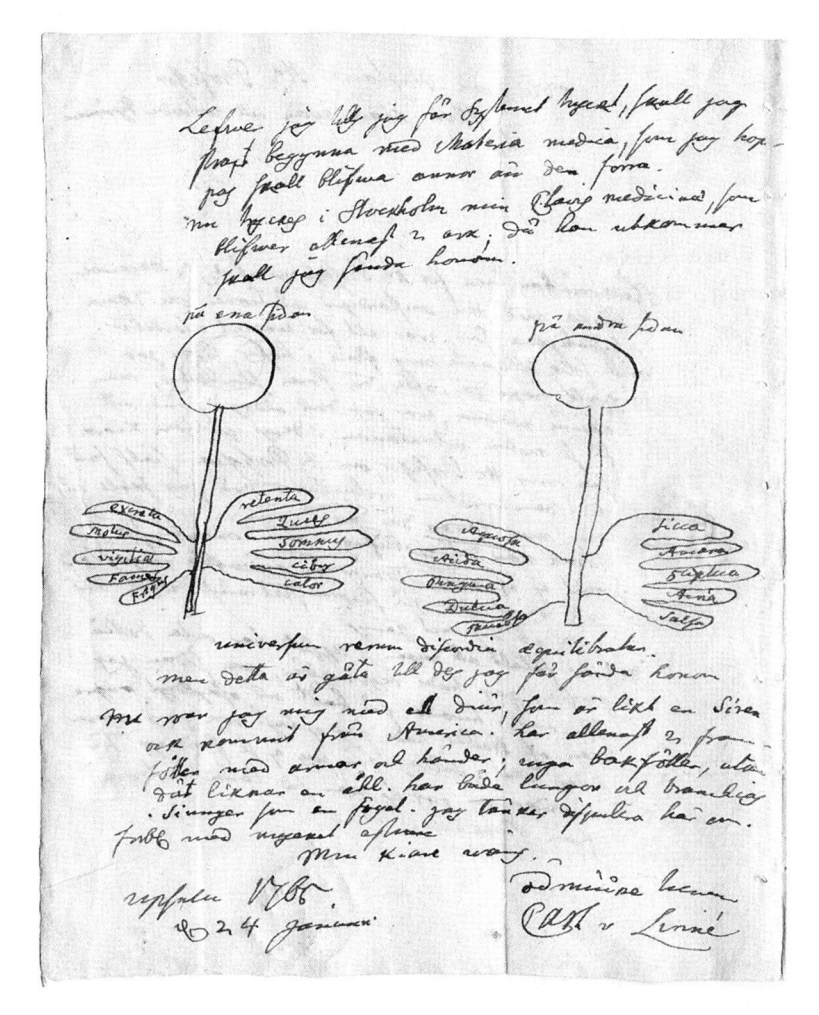

FIGURE 46. Linnaeus's description of what was to become the frontispiece to *Clavis medicinae duplex*, in a letter to J. A. Murray from 1766. Uppsala University Library.

intended it to be a masterful survey of his medical thinking. He had been working on the tables in the book for ten years and "so written that each word has been weighed.... We are thus able with good reason to assure its essential significance in every part." Indeed, he later characterized the booklet as "the finest jewel in medicine."[1] In other words, it may be short, but it is also a highly aspirational work. While the format would have fitted into his series of dissertations, his ambitions for it were much greater.

The mid-1760s were hectic years for Linnaeus, when his time was above all taken up with completing his crowning work of natural history, the massive twelfth edition of *Systema Naturae*. Actually, the contents of the opening pages of *Clavis* and of the introduction to *Systema Naturae* are similar in many ways. This was a mainly solitary labor because there were so few people around with whom he could share his thoughts on medical subjects: his students were too green and his colleagues too gray. But his old friend Abraham Bäck was there, and Linnaeus wrote to him on New Year's Day 1766: "This autumn I have been amusing myself with elaborating *clavem medicinae*. There are 2 keys to the sacred temple of medicine. One of these I am certain to have found once more and I believe to have the second one right. Methinks my Dear Brother will smile at this but were m. D. Br. to see him, I venture to think m. D. Br. would judge more gently." While he knew that his exalted expectations from his work may seem impossible, the verdict should be judicial. It may sound a little surprising that Linnaeus had been "amusing himself," because *Clavis* at first comes across as a dry and erratic set of lists. However, the many pages of annotations, all in tiny handwriting, bear witness to the mass of work that lay behind the finished text.

One helpful approach to *Clavis* is to consider the language imagery. Linnaeus had used the "key" metaphor already in his first presentation of the sexual classification system in *Systema Naturae* (1735). It was meaningful for someone who had grown up in a world of Christian worship and knew the icon of Saint Peter holding the key to the Kingdom of Heaven, and the symbol of two crossed keys. Symbolic keys were also closer at hand: the key to the university was integrated into the insignia of its rector, and the key to the city into the mayor's. In alchemical and hermetic literature, only the serious adepts were given in their care the key that gave access to the innermost secrets of nature. The idea of a key recurs in the many "universal languages" devised during the Baroque, and whose presence can be guessed at behind Linnaeus's terminology and taxonomy. A key can refer to a methodology or a system, or be an aid to taking the right course. In his dissertation *Sapor medicamentorum* (1751), Linnaeus wrote that the doctor must never be content with empirical knowledge because "theory is to the art of medicine the key with which to access practical observations and these in turn furnish the proving room in which the medical theory is tested." Exclusivity is an essential component of the image: only the key holders are let in, and everyone else is excluded. The key opens up the house or, here, the human body. The hands are guarding it, keeping the enemies out although, with old age, they tremble. The symbolic key has been linked to Linnaeus himself: Carl Gustaf Tessin, who had hung a portrait of Linnaeus in his country home, Åkerö Manor, wrote on it in 1757 using block letters: "The Lord of Nature has given you the key to the Kingdom of Nature. Teach us to use all that which is transitory until we, in your company, are allowed to see that which endures always."[2]

The key to medicine is double—*duplex*—or, rather, consists of an inner and an outer part. Each key has five bits, set in opposite directions and so a total of twenty, that is two keys with two lots of five bits. In a letter to his student Johan Adolph Murray, Linnaeus drew what is surely just two differently sized keys instead of one "doubled" key.[3] The translation of *Clavis Medicinae Duplex* should use the plural; this is supported by what Linnaeus has to say in the opening remarks about the "two feet of medicine"—common sense and experience—and that the practitioner also needs "two adept hands." In his dissertation from 1766 on *De effectu et cura* (1766), Linnaeus developed a similar line of thought: one key is the empirical observation, and the other the dogmatic or rational hypothesis.

The number five plays a central role. In his 1771 lectures on *Notata subitanea* ("observed without preparation"), Linnaeus pointed out that there are five senses, five notes, and five vowels (we use a different number nowadays); surely it wasn't by chance that he dedicated *Clavis* to five contemporaries, all leading medical men. There are, he claims, five pathological states of the firm tissues and of the chemical composition of the body's liquid components, and five different forms of taste and smell as well as their opposites. The number five is integral to human beings with our five fingers and five toes; it could be that it somehow expresses a perfection of the organic form. As we will discuss shortly, Linnaeus also wanted to extend the four elements of the classical definition to include a fifth, a *quinta essentia*. It must have given him a measure of satisfaction that the ennobled form of his name, Linné, has just five letters.

However, two weigh in against five. There are two sects, or camps, in medicine: empiricists, who concentrate on case histories, and dogmatists—presumably the professor of theoretical medicine, Linnaeus, belonged among the latter. Reproduction and life depend on paired organisms, and polarity rules throughout: the male and female sex of plants (though the third, hermaphroditic form is also important); significant opposite poles include plant and animal, heaven and earth, hot and cold, sweet and sour, taste and smell, and so on. Analogous, or like with like, pairings are also important. In *Clavis*, he declared that "the cure for illnesses is illness." Linnaeus understood the world in terms of dualisms and analogies. God's plan for the world is sensible and hence straightforward, which is reflected again and again in nature. Linnaeus strove to establish a sustainable scheme only so that afterward he could devote himself to the details that show the manipulative skills of the Creator and the rich variability of nature *within* the fixed framework.

The fascination with pairs is there in the 1750s and '60s, when Linnaeus became ever more preoccupied with the bark-and-marrow concept, his own willful take on physiology, simple but versatile enough to be functional in many different contexts. "Marrow" signifies the soft, innermost, and feminine, and the "bark" is the hard outer layer, encapsulating, protective and nourishing, a masculine substance. The marrow provides continuity of life while the bark is the insulating agent. The marrow, if not controlled, will grow unstoppably, like life itself. The bark, while providing nutrients, can also kill life by turning it into stone or bone or bark. Linnaeus seems to base his thinking on an extrapolation from plants to animals in contrast with earlier analogies used to describe plants—for instance, flowers reproducing like the "pairing of two lovers." With time, his use of metaphors had grown more daring; at this stage, he is thinking about animals including man. In *Clavis*, the bark produces the taste-rich substances and the marrow is "unlocked" by scents.

The model for his thinking is, arguably, the Aristotelian duality between form and matter. Linnaeus introduced an important distinction: in his version, the heart comes from the father and the brain from the mother. His way of thinking about the division of functions is traditional, but he is still content to claim that rational thought is a feminine inheritance. For those who wonder how far he was prepared to take that idea, there is an answer of sorts in his 1767 letter to Daniel Tilas, head of the College of Arms, who had requested Linnaeus's family tree. Linnaeus cannot claim any great ancestors but said that both his wife Sara Lisa and he himself were firstborn children. He continued: "Often, I have wondered wherefore we human beings do not count family as coming from the maternal side because the Brain *sedes vitae* with its *viribus ingenii* come from the mother and not to speak of this would always be deemed *fallibile*." He obviously rather enjoyed teasing the officer of heraldry by confronting him with paradox that spelled ruin to his College, and the entire House of Nobility, and a total reversal of established sexual norms. At the same time, his remark was the logical outcome of the marrow-bark hypothesis he had been pondering—here, Linnaeus is a feminist.[4]

His introductory statements in *Clavis* set the dominant chords: the body is a pneumatic machine under the control of Life; movement wears the body parts while the right ways of living and dieting restore them. The mouth is the site of vitality, the nose a gateway to the animalistic. The spinal cord vibrates and thus radiates vital powers. Life is localized to the cord; he states, in line with an irrepressible old prejudice, that at tail-bone level, the spinal cord is closely connected with the genitals. "The seat of the Vestal principle, the fire inherited from the ancestral Mother, as no life is generated outside of the body." That sentence sums up aspects of his ideas on the continuity of life, the impossibility of the original act of reproduction, and a never-extinguished spark of life—an eternal life on Earth.[5]

But how does the spark of life get around? To Linnaeus and his generation, living bodies, indeed all the material world, was composed of the four elements earth, water, air, and fire—in that order—and in different combinations, the elements are universal constituents. Linnaeus sought a fifth element, one able to flow around as well as through all the elemental constructions and give them life and spirit. He called this fifth element *electricum*. *Clavis* is relatively silent on this idea, but the theme turned up in the lecture series *Notata subitanea*. Electricity had been a known phenomenon since antiquity and had become a general talking point in the eighteenth century, once people had heard about the invention of the electricity machine and Benjamin Franklin's discovery of atmospheric electricity. Together with many others, Linnaeus took an interest in electrotherapeutics, famously practiced in Holland by, among others, Pieter van Musschenbroek. In 1754 Linnaeus presided at the dissertation *Consectaria electromedicina* (1754) presented by the respondent Pehr Zetzell; Zetzell actually ended up running electrotherapeutic clinics in Stockholm and Uppsala—a quite sizable business even though the outcomes were debatable.

It is worth adding that Linnaeus was not alone in thinking of electricity as a new life force and designating it as the fifth essence or element.[6] For instance, electricity was the subject of a dissertation presented in 1761, after supervision by J. G. Wallerius: *Chemical remarks on the lightning strike at the Roy. Castle in Upsala*. Lightning had struck the castle the previous year, and the study carefully accounted for the different types of damage, proving that, in many cases, a bolt had penetrated objects without causing any visible cracks. The governor's harpsichord was one example: the keyboard lid was closed, but inside the instrument, fifteen strings had been cut. The rooms were filled with a sulphureous odor, and the observer equated the power of electricity and that of lightning, in line with Franklin but with certain reservations. The chemical implicated in both manifestations had been determined: it was sulphur.

Linnaeus referred to *electricum* as a substance but also as a property, *electricitas*. One principle for separating the two would be to see *electricum* as the material of the vital force. However, the effect was the opposite:

this invisible power kept all possibilities open. *Electricum,* he reckoned, had probably more to do with smell than with taste, so in 1755 he asked his friend François de Sauvages in Montpellier to test the effects on the nervous system of refreshing and objectionable plant smells: "Certainly, the matter is settled, were these smells to affect electricity." Sauvages provided the desired result: bad odors suffocate the electrical fluid. Now, Linnaeus felt justified in linking *electricum* to the "higher organ" or nervous system—brain, spinal cord, and peripheral nerves—a compound structure ventilated by breathing in *electricum.* Whoever wanders in a sweetly scented garden feels well, as Swedenborg had put it in *Oeconomia regni animalis.* Linnaeus, of course, agreed and, hence, pleasantly scented plants were likely to be more electrically active.[7]

He argued that the findings on electricity added up to a fatal strike at the old ideas about the vital spirits, and instead spoke of "something fine" in the air, a phenomenon of light that could later be taken to mean oxygen. He emphasized how critical *electricum* is to continued life: "If we come upon air which is not electric and breathe it, we die in that very instant. The more electr. air we get into ourselves, the merrier we are." This compares strikingly with Scheele's findings about the fire-sustaining gas or "fire air," eventually identified as oxygen, and published in the early 1770s. Linnaeus's lectures confirm that he knew of the young apothecary at Upsala Wapen. "Like a dry wick can be made to burn by applying oil, so it happens with us old people, whose dry brains are incited by stimuli such as to make their vital flame burn more strongly, which can clearly be seen when old men use *spirituosa.*" He suspected an affinity with alcohol and referred to his burlesque thesis from 1761 called *De inebriantia*: old men were becoming intoxicated by all the electricity they consumed in the pub.[8]

Another characteristic of this vital essence is the speed with which it travels—1,000 kilometers in a minute, he declared in 1771 to his amazed students. That year, his German student Paul Dietrich Giseke went back home across the sea and could observe that passengers with lung problems seemed to recover in the purer, more electric sea air; however, lack of patients made him reserve judgment. Linnaeus apparently at no point thought that *electricum* would join other substances to form chemical compounds.[9]

Thinking about electricity could seduce him into lyrical writing: "We are therefore like electrical lights with which God has illuminated this, his theater. We stand here as does the sea in quick illumination, glittering like snow crystals in the sun. We have been granted the honor to be the lights in the palace of God. . . . We also reflect the Creator's brilliantly shining

Majesty, *duplicat luce*. When we have burned out, He no longer has use for us. Then, he takes us away, lets others come in our stead. Thus, Nature judges us, *contra* Theology."[10]

Electricity acts but in hidden ways. Johann Beckmann, on a visit to Linnaeus in 1765–66, thought his host's ideas about electricity peculiar.[11] Despite the theory's close relation to vitalism, it is noteworthy that *Clavis* ends with a quotation from the Roman philosopher Lucretius, who unflinchingly insisted that all knowledge is acquired through our senses: "It will be found that knowledge about reality is based on our senses and that our senses cannot be refuted. If they are untrue then all our knowledge is also false."[12] Linnaeus, however, mistakenly ascribed it to Ovid, perhaps thinking him a more suitable source than the materialist Lucretius. In international editions of his work, it was reverence for the author that the mistake was not corrected.

Linnaeus was a sensual man. Even though the table-packed *Clavis* comes across as pedantic, it is actually an ode to sensuality. Another pair of guards stand watch at the gates to the temple of the body: our paired perceptions of taste and smell, which cannot be tricked. Already in *Diaeta Naturalis*, he singled out taste and smell as of critical importance to health.[13] "Nature chooses between tasting and smelling substances"; the olfactory sense is the highest, as the nose is more loftily placed in the face than the mouth. Of course, perceptions can be difficult to pin down, and their historical roles are correspondingly difficult to describe. Traditionally, the senses have different status, and the hierarchy was usually ordered so that sight comes first, followed by hearing, then the sense of smell, taste, and touch.[14] Commissioned by Collegium Medicum, Linnaeus was engaged to work on the new *Pharmacopoea svecica* and, in that context, as in the rest of his activities, he argued in favor of *medicamenta simplicia*—single medicines—rather than *composita*. A sworn enemy of mixtures, he believed that they ruined the clarity of the relationships between the body and the administered substance. He never supported works that promoted the notion of *Natura per Artem* (Nature through Art), which was used by those who favored complexity. Johan Haartman, for instance, used that Latin tag for his successful "traveling pharmacy" (1765), which recommended various mixtures.

Can illnesses be classified along principles similar to the classification of plants? The ordering of plants was allegedly based on an "artificial" system, and the search for a natural one was ongoing—are illnesses more amenable to ordering on some natural principles? Or are afflictions of the human body social constructs (as we would put it now)? Perhaps types of

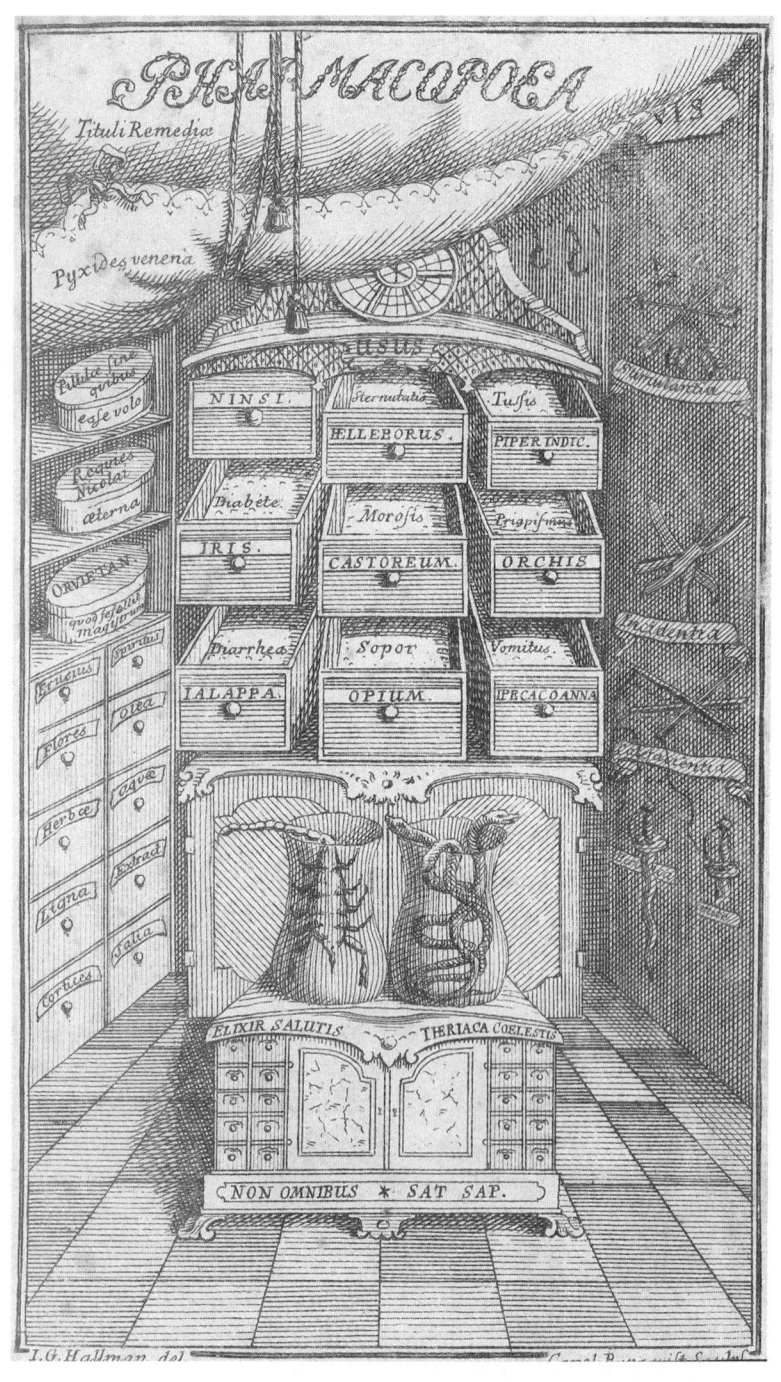

FIGURE 47. Frontispiece to *Materia Medica* by Johan Gustaf Halman. It shows a pharmacy interior with its shelves and jars including jars with snakes and a scorpion. Royal Library, Stockholm.

seeds are the cause of illness, or else, as Plato had already hypothesized in *Timaeus*, very small animals living parasitically on the body? What of the possibility that there is really one core illness which varies, is expressed in different ways, and is transformed in the process? The opportunities were opening up for a natural history of illness, which would also meet the demand for distinct unitary definitions needed by experimentalists for research and clinicians for treatments. Which route to choose to reach that goal: the inductive or the deductive, generalized arguments or rational systematization? This analysis of the physician's problems had been published about a hundred years earlier by the highly regarded doctor Thomas Sydenham ("the English Hippocrates"). Linnaeus, who normally went by the symptomatology, made *febres* into a class defined by any manifestation of raised body temperature. But he couldn't resist speculating on the role of the environment and the nature of the contagion. Could it be that the setting was noxious and the cause contagious? One problematic aspect of this line of thought is the lack of a defined relationship between the illness "itself" and the patient's symptoms or the observed syndrome.

One route was to classify plants according to their "medical powers." In *Fundamenta Botanica* (1736), Linnaeus wrote: "Plants which agree as to genus also agree in their *virtute* [property]. Plants which are close in their natural order also stand close to each other with regard to property." A little later, he went on to say: "Plants act either through the scent as it affects the nerves or the taste as it affects the bodily fluids."[15] When correctly understood, taste and scent offer the basis for a natural system in botany. A natural system of plants can be built on their tastes and smells, probably to be confirmed by chemical analyses.

When Linnaeus considered the senses as channels of information, he was sometimes inconsistent: the dietician and medic stressed smell and taste, but the natural historian looked first at visible characteristics. On several occasions, he praised himself for having set up a "sexual system" using only "the unaided eye"—no technical, expensive equipment required, which actually made his approach popular. It appealed to all kinds of people, from women, normally excluded from science, to solitary wanderers like Rousseau. In medicine too, Linnaeus wanted to avoid using instruments and rely on the human senses, which were surely all we needed—what the Creator has given us ought to be sufficient to understand His Creation and celebrate its originator. To measure anything in particular was unnecessary, but it was essential that all senses were alert, as if at the dawn of Creation.[16] As he grew older, he increasingly had to turn to senses other than sight. As he wrote in 1775 as part of an expert report on species descriptions: "I however like best descriptions of such plants which have some rare quality of structure, beauty, scent, taste, strength; others would be better left to have their own treatises."[17] Linnaeus was sniffing and tasting in the hope of getting the hang of nature's innermost patterns.[18] *Clavis* discusses mostly plants because people should only use vegetation-derived medicines; the foreword to *Clavis* speaks precisely about this, and the carpet of abundant plant life that the Creator has rolled out. The letters he wrote to Bäck about the new *Pharmacopoea svecica* of 1775 are informative on the range of medications based on animal products that Linnaeus wanted to discard in favor of herbal medicines. In one of his proposals, lists of appropriate plants occupy twenty-three folio pages; animals and minerals get one page each, including, for instance, leeches, millipedes, musk, and bezoar stones.[19] He urged that medicines should be derived from native sources, as this would be cheap, practical, and patriotic. Overall, his emphasis on the senses, his praise of plant-derived foods and "natural" medicines make it obvious that the aging Linnaeus believed in a vegetarian vision of paradise, in which true forms of nature and human beings still existed, and all was for the best.[20]

If there is a correspondence between that provision of nature and medicine, then the natural order of plants ought to match the order of illnesses, such as they had always been since the beginning of time. This was a criterion for "naturalness": because illness is a real, existing entity, it had surely been around since the Creation. If its stability could be proven and also such a close match with nature, it could be evidence of how the Creation was meant to function. Inherent in his thought is the possibility of a total synthesis of nature and medicine. The practical problems were how to delimit an illness and also the number of illnesses. His friend and medical authority Sauvages thought forms of illness reached a high number; in taxonomic matters, he quite clearly didn't belong to the "lumpers" but to the "splitters." Linnaeus never seemed to have taken this onboard and was forced to come to a Hannibal-like halt outside the gates, also in the incomplete attempt at an analysis of *Clavis*.

Linnaeus had dedicated his work to five of the leading physicians of his day: Siegfried Albinus, Albrecht von Haller, Nils Rosén von Rosenstein, Gerard van Swieten, and François de Sauvages, who was frail and unwell. All of them except Rosenstein seem to have ignored the honor or at least avoided any public statements, which is quite understandable. Linnaeus wrote to Bäck: "Would only My D. Brother go through this work and show my demonstrations, I hope that she were to be more comprehensible"; "more comprehensible" sounds rather like a response to objections. Perhaps these were dealt with because, a few months later, Linnaeus explained: "It is to me sufficient satisfaction that I have received from My D. Brother and Arch. Rosensten his approbation for my little *Clavis Medicinae*." When Bäck spoke of his dead friend's difficulties with *Clavis*, he formulated it well: "It is an *Ilias in nuce*, but a hard nut to bite into before the kernel is reached. He admitted to much labor on that book, and said that Medicine will need a man's span of life before the hidden becomes known. His audience found everything easy when he spoke to it himself." (*Ilias in nuce* means "*Iliad* in a nutshell"—a compressed version of an extensive work.)[21]

Although Linnaeus wanted the opinions of five stars among medical professionals, *Clavis* had been written for students; "to leave such a book to *Med. Studiosi* no one had thought of before." His faithful pupil Sven A. Hedin wrote: "The draft which the Master has composed under the name *Clavis medicinae* is to those who have not heard him lecturing perhaps likely to be considered a closed book, sealed until read at a future time when some may understand how to solve the problems it sets out; enough does however come through beneath its unbroken seal to conclude *Clavis* to be a work of an exceptional Genius." That is: Hedin understands nothing but doesn't hold back on his praise.[22]

Many testified to Linnaeus's ability to make his science come alive. He wrote this about *Clavis* in one of his autobiographical essays: "Happy the disciples who heard him demonstrate this." However, there are very few extant sets of notes from his lectures on *Clavis* in 1769 and 1772, which hardly supports what Hedin wrote.[23] In *Diaeta Naturalis*, Linnaeus stressed that he was addressing the "untaught" who were undamaged and open to words of ancient wisdom.

Clavis can be seen as a vision of a medical utopia as well as an attempt to cast new light on both medicine and science. He had opened up perspectives of future insights before—for instance, in the foreword to *Fauna Svecica* (1746, 1761): "Oh how happy are not those who, a few centuries hence, will experience this science when it has reached its perfection and so allows them lives more blissful than ever." In a letter to his friend Bäck in 1763, he set out five conditions that must be met: people should learn from history; base their diet on healthful things; realize the role of poor diets in pathological change; make more enlightened study of the semiotics of physical signs; analyze every source of *materia medica*, including its own nourishment. Besides, it was utterly wrong to use composite medicaments: "I do not hesitate to call it a madcap barbarism in the present light of science."

Words such as *barbarism* and *Enlightenment* recur when he wrote in this vein.[24] The key-metaphor adds an image of finally being able to open up what had long been locked away. It would have been easier to support Linnaeus's claim of having offered a basis for a new science if his ambitions for *Clavis* had been achieved and established a sound way of living, indicated the right medicine for the correctly specified medical conditions, created a natural system for classifying plants, founded a new kind of physiology, and identified electricity as the answer to the enigma of life. His belief was comparable to how he saw his binominal nomenclature in botany—"It was the same as placing the pendulum in the clock; through his doing this, botany sprang back into life . . . an entire new and natural procedure."[25] An exclusive profession of medical practitioners, bound by the Hippocratic oath and the high aims of the art of medicine, was part of his utopia, which cast out the charlatans in the trade.[26]

Linnaeus declared that he wanted to create a plan for medicine, which should be completed over time—more or less the same intention as for *Systema Naturae*, which was meant to function as a long-term, shared project in natural history. In *Clavis*, he sought to create the same new impetus in another field; the stakes were higher still. A substantial task remained: to make the second key fit more precisely and finally capture the Natural System. Entailed in this aim was to create a methodological approach that would make the systems for nature and medicine coincide. Both *Systema Naturae* and *Clavis* outline the core concepts for later generations of investigators, and both present charts and tables (from Lat. *tabulae*) and maps for future completion. The first edition of *Systema Naturae* was only a few dozen pages long, but the last one authorized by Linnaeus ran to more than a couple of thousand pages; *Clavis* could have gone the same way.

Both works tend to reductionism, rather like the dream of a universal language interlinked with a universal order. Both works were also intended to teach and address the open-minded beginner—the innovator.

CHAPTER THIRTY-FIVE

A Philosopher of Science or a Scientist?

THE YOUNG LINNAEUS WAS CURIOUS, observant, and too sharp-eyed to need a microscope; at the age of twenty-three he had formulated concepts of everything and won over the world with his charm, vitality, physical energy, simple messages, adventures in Lapland, and faith in the words of the Bible. The old Linnaeus was weighed down with honors, exhausted, prone to illness, fixated on the Fates, speculative, searching for the new key to unlock nature in his ideas about marrow/bark, smell/taste, and electricity, and sensing an unstable world in which *Systema Naturae* would never be complete. There was so much to ponder and nature was almost slipping out of his grasp.

Normally, people focus on the young Linnaeus and his youthful ways, and forget about the thinking that preoccupied him as he grew old. It is, however, important to recognize the old man's traits in the young one as well as the modern and innovative sides of the old Linnaeus, who had the wit to be interested in new findings such as the enigmatic polyps, the invisible world, and the mysteries of electricity—but understood these phenomena as an old magus would.[1]

His Danish student Fabricius noted down that Linnaeus had said that, during his lifetime, he did not dare to publish observations that risked violent reactions from the clergy.[2] His library at Hammarby contained several books on the occult. In his later years, he sometimes conceded that the secrets of nature might be accessible only to a select few and he thought about himself as one of the illuminati. By naming nature, Linnaeus had exerted a magic power of his own; "His encyclopedic mind, interest in utility and obvious glances at hermetic and alchemical concepts fall into the same category.

FIGURE 48. The wild-strawberry girl, Ulrika Charlotta Armfelt. Painting by Nils Schillmark (1745–1804). National Gallery, Helsinki.

His thinking was close to the Rosicrucianism of the seventeenth century."[3] In his day, occult undercurrents flowed in many directions through Sweden, partly driven by the writings of Emanuel Swedenborg. Linnaeus used few biblical references but, even so, they show a marked increase in the tenth and twelfth editions of *Systema Naturae*. All of them were taken from the Old Testament (the lack of New Testament references had nothing to do with religious divergence) as was often seen in learned texts. The primary difference between Linnaeus and those who believed in esoteric orders like the Rosicrucians was "his obvious lack of interest in Jesus."[4]

Linnaeus was searching for a grand, coherent explanation of the world and saw it in terms that bring to mind old fertility myths. In his lecture on "The Creator's Intentions for Nature's Work" in 1763, he described the very first steps, or generations, of nature's development: "Thus clay is the daughter of the sea," and "therefore, sand is the daughter of the air," so "therefore, soil is the daughter of the plants"; more followed in this variant on an alchemical genealogy.[5] The third part of *Systema Naturae*, published in 1768, was more or less contemporary with *Clavis* and opened with a ceremonial introduction in which Linnaeus invoked Moses, Thales of Miletus, and Seneca (and no one else) in support of his concept of all things beginning in water: "The moist, cold, passive, receiving water was made fruitful by the dry, warm, active, and generative air, became so pregnant with twins—masculine salt and feminine, appealing soil." Through this union, different salts and soil types were bred. Still looking far back in time, Linnaeus spoke next about the function of "the fathers of the stones" and then about "the mothers of the stones."[6] The chemist J. G. Wallerius commented in 1760: "The same Linnaeus as ever is now in labor to produce his mineralogical conceptus, as he wants to persuade the whole world that he alone is *mineralogus* as well as *botanicus*; he is driven by one part pride and one part shameless selfishness."[7] The man himself expressed it more soberly: "The generation of stones he understood better than anyone else, knew crystals and how they are prepared from salts, silicates from Crete confirmed the withdrawal of the water, also proved the four *terrarum ortum*; besides this he did in the veritable method in the realm of Stones *cum lapidifico calcareo*."[8] The dream of somehow shouldering responsibility for all nature was irresistible, and mythical interpretations tempted him away from limited scientific explanations. At this stage, he introduced *Systema Naturae* with references to God and claimed divine sanction before starting on the downward path to the system. In the wider history of science, people such as Bacon, Newton, and Swedenborg were also examples of similar, abrupt combinations of myth and science.

Was Linnaeus a Swedenborgian? These two men were the most internationally famous Swedes of their age and, on top of that, they were related and knew each other or, at least, were aware of each other. Reuterholm's home in Falun was a place where their paths may have crossed. Both were originally interested in mineralogy, and both were mechanistic thinkers who later turned to vitalism and religion; both were absolutists. It was Linnaeus who proposed Swedenborg's candidature in a membership election to the Academy of Sciences. He had read Swedenborg's book *Regnum animale* and referred to him in *Lachesis* and elsewhere.

There was a crucial difference between them: to Linnaeus, everything was located on this Earth and measured in the human life span. Swedenborg believed in an afterlife, admittedly in a world very much like our earthly one. Swedenborg turned to heaven, Linnaeus to Earth.[9]

Vitalism, materialism, mechanism, animism in different combinations; life as an element, a substance, an organization, a divine spirit, a growth force? When Linnaeus declared growth to be a fundamental characteristic of stones, he seemed to base it on the hermetic tenet that all nature is alive. Where exactly did he stand in relation to the rest of society, to "academic philosophy" (for want of a better definition) and to conventional thinking, established views, and alternative ideas? He had been taught the scholastic style of reasoning at senior school and, in educated circles, no one was unaware of the new Cartesian philosophy. The main school textbook, *Svicerus compendium*, is subtitled *Aristotelico-Cartesiana*. Actually, Linnaeus only rarely cited Descartes, though there is one exception in the earlier version of *Fundamenta Botanica*: "When I first examined nature, I found that she resisted against what the authorities had laid down, which is why I threw aside all prejudice and became a *scepticus*."[10] Also, in the same vein: "I set out to search around the entire kingdom and to myself seek out what I could do. I discovered a common sense free of prejudice, ready to take, with care, one step after another as findings and trials showed, will soonest find the door open to knowledge of nature."[11] Sometimes, his thinking resonated with Bacon's ideas but, on the other hand, he did not refer to Christian Wolff, the contemporary big name in academic philosophy. Wolff systematized most things and was—or so it can seem—someone who shared Linnaeus's opinions. True, Linnaeus didn't quote Locke either. Seemingly, he didn't pay much attention to any of the dominant philosophers of his time and stuck to his own thought system. He looked for older ideas about achieving wisdom: stoicism, hermeticism, and alchemy.

These are difficult questions, but Linnaeus was indeed a "philosopher"—someone who thought about the world in general terms. At the time, those who investigated nature were regarded as philosophers, and one of his most important publications was called precisely that: *Philosophia Botanica*. It was a book of rules specifying how to describe and understand a phenomenon as difficult to pin down as a plant, a flower—a form of life. His work on taxonomy grants him a place in the history of philosophy; in the ten-volume work *Routledge Encyclopedia of Philosophy* (1998), he is one of just five Swedish entries—the others are Axel Hägerström, Anders Nygren, Karl Olivecrona, and Emanuel Swedenborg—and Linnaeus is given more space than anyone else. Possible others, such as Erik Gustaf Geijer, Christopher Jacob Boström, Ingemar Hedenius, or Anders Wedberg are excluded. Notably, Linnaeus is not named in corresponding Swedish encyclopedias.[12]

We have noted that his philosophical home territory is hard to define: Was he a scholastic (essentialist), empiricist (a follower of Bacon), Cartesian (mechanist), Wolffian (systems fanatic), Platonist, Rosicrucian, stoic, or perhaps a self-made philosopher? It is difficult to place Linnaeus in a strictly defined Enlightenment tradition, but the best fit is probably in the middle period with its realistic appreciation of nature and cult of utilitarianism. Notably, his years in Holland left no trace of what might be taken as "an influence of the Enlightenment." We need a wider definition for his lifelong ambition to reach out beyond the confines of academe. He sometimes used light-metaphors (e.g., in his speech to the royal couple in 1759)—but so did the illuminati. Even though Linnaeus knew of the great names of the Enlightenment, it seems very unlikely that he read their books.

Linnaeus, a clergyman's son, had left the literal faith behind but sought correspondences between his own ideas and the Bible's words; with time, his interpretations grew evermore tentative and self-serving. His mindset changed a great deal between 1735 and 1770; the Enlightenment, which cast a faint light also into Svartbäcken, had had some effect, but what mattered more was his inner drive, the will to understand and to peep into nature's council chamber. We question his lines of thought but ought to be impressed by his determination to test and reconsider. Wisdom does not necessarily increase with age, but growing old can liberate thought from conventions. Linnaeus's mind continued an untiring search for the new, which is perhaps especially impressive—this is also in sharp contrast to the not unusual image of a provincial Linnaeus with a one-track mind.

Linnaeus arguably took on board elements of the entire development of natural history and was also on his way into the Romantic world of analogies, polarities, and forces. It doesn't follow that he should be labeled as, say, a Rosicrucian—a movement he had barely heard of—but rather that he had been influenced in his youth by writings of a similar kind, which combined occultism with humanitarian obligations and belief in one's own place among the elect.

The old Linnaeus kept the faith in his mission, his God-given task to teach people to see and know themselves in many defined ways. Much changed over the years in both the form and the content of his work; *Systema Naturae* rolled on, of course, but with adjustments and additions of sometimes weighty material (a new human species; a new kingdom of nature). Also, as he had begun with identifying an inner strength and virtue—optionally, taste and smell—rather than shape and structure, he relied on chemical criteria and also on magical properties. All is then woven into patterns laid down in number mysticism.

CHAPTER THIRTY-SIX

The Back of God and God's Footsteps

LINNAEUS HAD GROWN UP during the rule of orthodoxy in Sweden. He sat in church as a child, listening to father Nils giving a sermon from the pulpit, but close at hand was also the powerful emotional message of pietism, with its almost poetic approach to nature. At home, Johan Arndt's *Sanna Christendom* (True Christianity) was part of the family reading, as was Christian Scriver's *Själaskatt* (The Treasure of the Soul) and, on the basis of stylistic similarities, it has been claimed that such literature influenced him. He often returned in later years to the old metaphor about "the Book of Nature"; if read in the right spirit, it would bring to light the secrets to complete the Book of Revelations. Arndt in particular was likely to inspire such revelatory nature study, and it was an exegesis that Nils Linnaeus would approve of. In his Växjö school, young Linnaeus studied the Bible, hymnbook, and catechism. The Lutheran form of Protestantism busied itself with ordering your days as well as the purity of your mind, which had to be examined before you were allowed to travel abroad with all its temptations. While the philosophy of the Enlightenment probably came closest to whispering into his ear, the expansion of the natural sciences steered him, as we have seen, in a very personal direction toward the union of the Creation narrative with ideas drawn from elsewhere. Linnaeus was no purist and probably not "deeply religious" in the accepted sense but instead "naturally religious" not only because he was born and brought up within a religious worldview but because of his personality, perception of the grandeur of nature, and awareness of social norms. In any case, he was not orthodox Christian. Some people even told him to his face that they thought him an atheist, but later generations have described him as gentle and pious. Johan Otto Hagström, a doctor, wrote to Wargentin in 1773: "This *vulgi* notion has traveled so far that quick botanical youngsters have asked me, 'have not profs. And. Celsius and von Linné been infected with *Atheisterie*? Or at least *indifferentisterie*?'" At the time, atheism would have had you fired from teaching posts.

The charge of atheism might have been supported by Linnaeus's rare church attendance; he actually cared little for the core Christian dogmas. Looking beyond the popular physicotheological line of thought—its argument being that nature, so lovely and well ordered, must have a creator—his library held no theological works and no exemplary reading matter, not even a Bible. These might of course have been misplaced but hardly lost without a trace. With a few exceptions, there seem to have been no religious practices at home in Svartbäcken and Hammarby.

This did not prevent Linnaeus from making frequent references to the Bible; his writing is full of them although almost exclusively from the Old Testament books such as Genesis, commandments about what one is to eat or not, hymns in praise of the Creation in Psalms, and a handful of passages about the punishing God and stories suited for *Nemesis divina*. He showed little or no interest in the Gospels, ideas of salvation and eternal life, or in the Final Judgment and the Apocalypse. Jesus made few appearances in Linnaeus's form of religion; it is hard to work out whether he was a good Christian by the standards of his time.

Returning to the Book of Nature: it is an old belief that if you know your Bible and its teaching, you will want to study nature. In that sense, the development of the natural sciences took place within the church and with its support. There was no antagonism, but with time, the balance shifted toward making nature studies self-referential. The pious entomologist Jan Swammerdam's *Biblia Naturae* (Boerhaave supported the posthumous publication of this work in 1737) could represent an entire genre: physico- or natural theology. This became dominant during Linnaeus's formative years and was embraced by many eminent men: Christian Wolff, Willian Derham, Linnaeus himself (e.g., his talk on "Notable Features of Insects"), and later Pehr Högström, Johan Fredrik Krüger, Christian Lesser, and Anders Wahlström. Linnaeus's attributes to God tell a story: Creator, Founder, Builder, Ruler, Majesty—but not Savior. Consequently, the innermost structure of his thinking has been said to be physicotheological even though he never wrote anything about it. He had no time for the theologians' style of Bible studies: his God showed himself in nature, so that is where His intentions should be studied and interpreted—the Work praised its Master. The introductory words in the later editions of *Systema Naturae* are often quoted: "I saw the infinite, all-knowing, and all-powerful God from behind as he went away, and I grew dizzy. I followed his footsteps over nature's fields and saw everywhere an eternal wisdom and power, an inscrutable perfection."[1]

If an atheist asks a historian if there is just one example of how religion has contributed positively to science, the reply might be "Yes, look at Linnaeus." Physicotheological reasoning reveals the intricate interaction between the two; there are plenty of examples. Sometimes, contempt is expressed for these views, which are analogous to the functional arguments for intelligent design. As Linnaeus put it: "When the reindeer walks his foot clicks. I wondered thereat, and sought the reason. When I asked, all answered that he is created thus by Our Lord, and then I asked how Our Lord brought it about that it clicked." Linnaeus followed up: he investigated the joints, failed to find the answer but was, as usual, curious and persistent. By raising fundamental questions and the need to complete revealed truths, religion urged science on. Linnaeus's teaching was vigorous proof of his religious awareness, despite the potential conflicts. No one seemed surprised when two of his pupils, Browallius and Mennander, both professors of natural history, were appointed to bishoprics; innumerable clergymen have combined natural history with theology.

Whoever studies in the Book of Nature ought to be *curiosus*, or curious in both senses of the word. As a concept, it has lost some of its positive shine by the professionalization of science, and other meanings like "wonderous," "mysterious," and "awe-inspiring" have also faded. Linnaeus and his followers, however, emphasize *Curiositas naturalis*, also the title of the 1749 dissertation that the theological faculty tried to censor; such inquiries entailed interference with their sphere of interest and the word synonymous with hubristic. This upset Linnaeus, who wrote in one of his autobiographies: "Henceforth, Linnaeus never trusted any clergymen." He was his own priest, at least as far as he dared. He cultivated nature study as a theology, keen to employ biblical phrases such as "consider the lilies." There must be no doubt that minds should focus on the green pages of nature's book rather than the gray ones of learned works; by doing this, we fulfill our duty to God and come to understand our place in the Creation.

To Linnaeus, Genesis was the source of his thought. In his early medical handbook *Diaeta Naturalis*, he told of how, once after being ill, he had been overwhelmed by the beauty of nature. It was a life-affirming, and possibly life-changing, experience: "Wherever you look there is greenery which restores and quickens the eyes, wherefore the Creator has made the entire Earth green; here play flowers of divers form and *coleur*, which excite and delight mankind. The trees swing their leaves and make a pleasing noise, the birds chime in, the insects whirl through the air and settle here and there like small paintings or *pocader*, yes, everywhere are the traces of the indescribable Creator." In linguistic terms, it is, of course, possible to translate the Creation into Latin as Genesis or Natura. From Linnaeus's first sight of it to his last, he was filled with its luminosity; to him, it was especially important that the Creation was that essential beginning of all things. This might appear self-evident, but that is hardly the case; to Darwin, a hundred years later, one way of approaching such questions was to give only limited and hence incomplete explanations, avoiding the very first moments. Darwin's discussions are open-ended while Linnaeus's are closed off inside firm frameworks of place and time. Darwin's worldview was unstable, Linnaeus's stable—at first. Later, he changed his mind a little.

It does not mean that he took the Bible literally, as we have seen. Only in his distaste for everything papist was he truly in line with pure Lutheranism. In his writings, he went well beyond the Book, as for instance in his fundamental sentence "We count this day as many species as were created in the beginning." Here, he has locked the interpretation of the words in the Bible into what his science required. He stated that "all living things come from the egg," which meant that there was no original generation, and he used the model of Paradise for a variety of rather eccentric discourses. He successively moved away from a traditional biblical chronology of a 6,000-year timetable and on several occasions expressed his wonder at the sheer age of creation. An example from Helsingborg in Skåne, which he reached in 1749: "It makes me giddy as I stand on this height and see below me how long eras have come and gone like the waves in the strait and left behind only these much-worn traces of the past, now only whispering as all else has fallen silent."

He needed the Creation story as a starting point for his scientific work, as a takeoff board but also a stable platform; if only Paradisial ground is there to stand on, other things can remain uncertain. The Creation offered a framework for a sensibly arranged world; true, he later partially dismantled that structure. It took a few decades for him to distance himself from an apparently literal-minded faith (though it wasn't) to allow for something that approached a concept of evolution (though it wasn't, quite). He still envisaged a stable system but came to accept that new genera and species could appear within it.

Without a well-intentioned Creator, everything could "go *confust*." There are those who think of his passion for orderliness as a neurotic trait but, for one thing, it seems anachronistic. Bach surely can't be classified as a neurotic even though, like Linnaeus, he was a builder searching for sound foundations in order to create the mightiest of buildings. We should instead look to Linnaeus's background in Baroque aesthetics as well as to the rationalism of religious orthodoxy. Another crucial ingredient in his thinking was the nation-building that began during Sweden's period as a Great Power. One of the reforms entailed organizing a central administration for regional soldier recruitment. Linnaeus approved of its orderly subdivision of the country: he saw parallels between it and his hierarchy of classes, orders, genera, species and their varieties, and the five levels of geographical, military, and philosophical thinking.[2]

The Book of Nature was meant to match and complete the Book of Revelations. In Swedish science, as elsewhere, there had been plenty of work done on the zoology and botany of the Bible. Olof Rudbeck Jr. had penetrated deeply into ancient languages and more recent natural history, looking for clues to the real substance described as manna that rained down over the desert. Olof Celsius, Linnaeus's first mentor in Uppsala, wrote the *Hierobotanicon*. One of Linnaeus's achievements, his investigatory journeys through Sweden, had been so successful that the university, and especially the theological faculty, supported Fredrik Hasselquist's expedition to the Holy Land; that it would go very wrong had been, as ever, impossible to predict. Samuel Ödmann, in the generation after Linnaeus, would make wide-ranging studies of biblical philology with an express focus on natural history, as did Linnaeus's famous successor, Carl Peter Thunberg. A long series of dissertations, starting in the seventeenth century and continuing into the nineteenth, were on subjects in this academic field—a kind of hybrid of theology, philology, and biology. It is quite surprising that Linnaeus seemed not to care very much about the biblical flora and fauna. He preferred to spend his time on the utility of the Swedish flora and the new world of plants at the Cape of Good Hope.

One of Linnaeus's less well-known writings reflects his sense of being among the elect, as he imagined it in 1768: "To grasp *principium* or *fontem* is of God. But what makes it begin or end, of that we know nothing. Maybe there is no beginning or end? All things must be defined, or else we do not know what each is; but to define God is hard. We see some have painted him, mostly in churches, to look like an old man, which is a most unlike and miserable simile of such a great Majesty. . . . Moses, of whom was written in the Book that he was granted the grace to see God. [He] was commanded to hide so that he would see Him and when it happened, he had to exclaim with the wonder of it and say: Oh, how almighty eternal, etc. Hr. Archiater has been granted the same grace to see God albeit *a posteriori*."[3]

To sum up, with some additions: the Bible was seen as a source of knowledge about nature. The dependence on the Bible was obvious in many ways as it defined ideas about creation, hierarchy, and chronology. Nonetheless, Linnaeus vigorously stressed the importance of personal study in the Book of Nature. He interpreted the Fall in terms of sex but never mentioned the expulsion from Paradise. He seems to have been as uninterested in the Gospels and any ideas of individual salvation and life after death. The Flood turned into discussions of withdrawal of water, the constancy of species was questioned, the creational timetable expanded, and new human species were included in *Systema Naturae*. Elis Malmeström, a bishop-to-be, thought Linnaeus's religiosity was "of the Enlightenment, in its generalized, somewhat lackluster character."[4] Linnaeus needed an ultimate cause, a Creator, but was mentally attracted to successive processes. In a smoking break, Johann Beckmann asked Linnaeus if he thought animals had souls, and his teacher replied, rather reluctantly, that the soul is nothing but an electric spark. However, one had better take the theologians on trust in such matters—and beware of them, never try to dispute with them. His pupil still insisted that Linnaeus loved religion.[5]

In 1774 Linnaeus was asked to take on a theological task, namely, to join King Gustav III's Bible Commission as its expert on the natural sciences with the specific mandate to identify the plants and animals referred to in the Holy Book. The commission had twenty-two members, including Rosén von Rosenstein and Wargentin, who were to deal with matters of medicine and astronomy (chronology), respectively. In discussions about the Flood, Linnaeus suggested that "there was surely room enough for the animals onboard that ship of Noah's" but expressed reservations about the space for animal feed. His idea about how Adam and Eve covered themselves with banana rather than fig leaves was not appreciated. He retired relatively quickly, claiming health reasons. Again, the general conclusion was that "Linnaeus showed no greater understanding of the uniqueness of Bible texts."

But some passages tempted him to speculate; in this record kept by the librarian Erik Mikael Fant (Fant dates it to 1774), he spoke about the Creation of Eve as described in Genesis 2:22:

> Herr Archiater and Knight von Linné has his own thought concerning the bone from which woman was created. He believes that the medulla is the very element in all things and that it can propagate *ad infinitum*. A tapeworm for ex. is only combined from many worms of which each and every one, once he has become fully grown, projects a medulla

from which a new one grows, and so on. The vertebrae in a human body contain such a medulla and they end in four small bones which make up the coccyx or tailbone. Had the Creator not snipped it off or stoppered it then, or so von Linné would have it, another human would have grown out of the rump of the previous one. Therefore, he says, Our Lord removed the fifth bone and from that one, while it still contained medulla, created Eve. Thus, flesh filled the bone instead so that there was no further exit for the medulla and consequently reproduction in this manner was prevented. Linné has made inquiries to professor Aurivillius and Councilor Michaëlis if the Hebrew word [*tsala*] which is used here would not in the oriental Dialects mean something which upheld this interpretation, but both answered no. Otherwise is it untrue, as some still claim, that men have one rib less although that would seem necessary if Adam was to have lost one when Eve was created.[6]

A few sheets of Linnaeus's working notes are kept in the Linnean Society library. He matched a column of quotes from the Bible with his comments in each instance; the subjects are mainly plants and animals, but he had also tested meanings of the difficult sentence "And the spirit of God moved upon the waters." Linnaeus made a brief, concise entry: "*Aër vitalis electricus.*" At this point, it seems that the traditional, anthropomorphic representation of God has finally been abstracted into a life-giving and ordering concept—which is a remarkable shift in religious emphasis. If *electricum* is the very principle of life, then God must have something to do with electricity. His academic colleagues were spared having to confront this alarming identification, but if Fant is right, Linnaeus even so "had much stirred them with hypotheses and paradoxes that tasted little of seriousness." But to Linnaeus himself, all this was presumably seriously meant. From his seat on the Bible Commission, he wrote in 1772 to the great orientalist Johann David Michaelis in Göttingen, asking if the fifth coccygeal bone—the site where the spinal cord is "closed in"—did not agree better with the Hebrew word *tsala*, which would support the Creation of Eve from Adam's tailbone rather than one of his ribs. The reply from Michaelis was a mildly expressed negative.

Linnaeus preferred preaching his own gospel as well as avoiding conflict. He felt that the theologians ought not waste their time on arguing with "small heretics" and "the simple-minded should not be ensnared." He appeared to want calm more than rebellion. The analysis quoted above contains no surprises but has the advantage of being Linnaeus's own, complete with its contradictions. Here, he comes across as precisely what he warned against: "naturalist."

In 1779 Bäck spoke of his dead friend and necessarily stressed the purity of his faith: "He worshipped Religion and did not attempt to investigate its secrets," and went on to compare Linnaeus with Boyle and Newton, who bowed every time the name of God was used. But, even then, no mention was made of the individual having an eternal life after this one. Anyway, Linnaeus would not have expected it, at least not in the traditional religious version.

However, he apparently has gained eternal life of another kind.

CHAPTER THIRTY-SEVEN

Nemesis Divina

THE OLD LINNAEUS left a testament: *Nemesis divina* (Divine Retribution) was intended to enlighten his son but, at some point, disappeared from the home by an unknown route. It set out a religious view based on the natural sciences or, perhaps better, a scientist's religion ruled by cause and effect. Linnaeus called his attempt a *theologia experimentalis*, an experience-based, empirical, and Bacon-inspired theology suffused with ideas about morality's ecology and a final equilibrium, guaranteed by the order created by the Almighty.

This collection of notes ended up in the ownership of Olof af Acrel, who gave it to a doctor in Kalmar, O. C. Ekman, who sold it to Uppsala University library in 1845. The library had the sheets bound, a simple binding and in a different order from the 1968 edition of the book. A censored selection had been published early on, but a later, more inclusive edition by Elias Fries came out in the 1850s. The text aroused international interest, and it was published in English, German, French, Italian, and Dutch translations. There were ideas for a film based on its dramatic, often dark tales. A choral work by Daniel Börtz was given the title *Nemesis divina*, as was a classic (arguably) LP album by the Norwegian black metal band Satyricon. The collection of tales has either been published or been analyzed by, among others, Elias Fries, Oscar Levertin, Knut Hagberg, Elis Malmeström, Lars Gustafsson, Wolf Lepenies, Georges Bataille, and M. J. Petry.[1]

Closest in time is a work from 1758 by the Danish clergyman Frederick Christian Friess, to which Linnaeus himself referred. He had moved on to the common land of morality, where the human fates he recounted had dark prospects. *Nemesis divina* is in some ways like Plutarch's comparisons of life stories with encouraging or—mostly—warning examples from history. Linnaeus's heritage of ancient Greek and Roman thought has been traced in *Nemesis divina*. Research has also been done on his emphasis on character, and the work has been analyzed in terms of an Enlightenment-influenced successor to Leibniz's *Théodicée* "on God's goodness." As the historians of ideas fill in a rich background, ponderous elements are added to the image of Linnaeus. Many of the episodes can be placed within a long tradition of storytelling; he chose them from classical literature, the history of warfare, and, above all, from events involving his contemporary colleagues. Some of his sources were in 1730s Uppsala; the confrontation with Rosén might well have been the starting point for the entire collection although he found stories from as far away as the Stenbrohult of his childhood.

In the autumn of 1767, his lectures on diet contained this passage: "At this age [21–28] one should consider *nemesis divina*, the greatest I know of and have observed. I have collected in excess of 300 examples thereof in my time. You may be assured that God lets nothing pass unapprised.... Without morals humankind is the most repulsive tribe of animals ever created." Note that his cited number of examples was more than double that printed in "the complete edition" and also that, in this lecture, he addressed young men in their twenties and probably thought sex was a particular threat to their moral fiber. These lecture notes suggest the date when Linnaeus began to work on this theme, which was also when his son Carl reached the dangerous twenties. The book was dedicated to Carl: "At one time I doubted that God cared for me. / Many long years since and have taught me what I now leave for you.... Should you not believe the Holy Book, trust experience. / I have set out these few *casus* I remember. / See yourself reflected in them and pay attention.... I have named no names as I wanted to shield them." He added a warning against reckless publicity: "Listen well, tell nothing, blacken no one's name and honor."[2]

It seems natural to think of *Nemesis divina* as the summation of an embittered old man's experiences, but as we know, the ideas can be traced back to the 1730s and *Diaeta Naturalis*, as for example: "An honest cause will always win in the end, God is a just judge. Unhappiness will be your lot, you who have not learned that God judges first all there is in this world. I never believed that such a just God would abide sinners. If you do what is right you will never have to beg for mercy, never fear a court. To act rightly is this: do not to flatter yourself or believe your own case to be the best."[3]

The final words of *Diaeta Naturalis* testify to active Bible study: "Everything I attempted went badly whenever my intention was revenge, so I changed and left all in the hands of God. Then all went well." This could be the birth of his belief in nemesis: namely, that he abstained from revenging himself on Rosén. The book's philosophy is focused on a fair balance between crime and retribution: like for like, a tooth for a tooth. He had collected around a hundred stories, essentially about God's fair and just rule of the world (though one could conjecture an anti-nemesis, a list of cases in which due retribution didn't happen). It is a philosophy based on order and neatness, backed by a pedagogic line in warnings, even threats. Or is *Nemesis divina* a commentary on existence, compiled in the fear of death? No angels, no salvation, no eternal life.

As far as is known, Linnaeus never attempted to give his collection of stories a structure. It can be done, and the exercise has been carried out by publishers and critics, who have tried various ways—alphabetical order (Elis Malmeström), by definitions of people in *Systema Naturae* (M. J. Petry), by the Ten Commandments (Carl Gustaf Rollin). Reasonably, one could sort the stories by the date they were written, the type of crime, or gender, or social class, or group. If the latter was picked, about a quarter were linked to the university and Linnaeus's closest circle. A few examples, chosen at random:

The first comes from Linnaeus's farm, Säfja: Jaen Jaenson, "a feeble lad" married "a quick housemaid." They had no children, but Jaen realized that his wife was unfaithful with a man at a neighboring farm. In 1770 she was distilling aquavit when there was a fire. It caught in her clothes and caused burns on her thighs and genitals. For eight days, she "screamed to high heaven," and after six weeks, she was dead. "It was said she slept with 2 brothers. If that were true, then God's vengeance must be done." This is a late story, recorded with a stated year and the names of those involved; the point is that adultery is a serious crime, which a vengeful God will punish. (We are not told what Linnaeus, the master of the estate and a doctor, may or may not have done.)

The next story is called "Joseph, whom I trusted." Joseph was "exceeding *libidinosus*," and the story recounts his careless affairs and his punishment, which was gonorrhea; it plagued him until "a merciful God" pardoned him and he was well again, "though had intimations all his days." We can guess who "Joseph" was: probably Johan Browallius, who was mentioned in a letter sent by Linnaeus in 1739 as having been punished in this way. God was indeed merciful in this instance, possibly because "Joseph" had a bad conscience.[4]

A complex case, which aroused concern among many of the academics, was centered around the university's bursar Peter Julinschiöld. The story follows his rise and fall, and his punishment, which was bankruptcy.

"Malicious tongues had it he was unmarried and ate in their house, owned half-share in Mrs. Rosensten with her husband the other half." What had to happen, did: "Julinschiöld must after his *cessio* via Georgi beg from *consistorio*."[5] The theologian Petrus Ullén, who had been to a disputation party, turned up intoxicated at a meeting of the consistorium, where he attacked Johan Ihre, who didn't reply. At the next meeting, Ihre explained that he had kept silent because Ullén had been inebriated. All of which was duly entered into the minutes. Piqued by the accusation, "an altercation began" after which Ullén suffered ill-health, never recovered, and finally died.[6]

Above all, two bad behavior themes recur in *Nemesis divina*: infidelity and financial malpractice. There are many quotations from the Old Testament and some from the New Testament. Retribution strikes hard on adulterous or debauched women, who are shown no mercy: "Fate is the judgment of God against which no bets can be taken." But, just as some colleagues are false, some women are noble:

> The housemaid Maja Hierpe was the most virtuous, kind, lovely, willing, best. Her mother and siblings died in 1773 from the infectious spotted fever. She served at my house and asked me leave to go home for the burial of her mother. Every one of us, the entire household, begged her in tears not to do so, as she might be infected and die. But no heavenly or earthly power could prevent her; she said she would otherwise have done what was not good. She went, returned infected; we moved to the country, she was placed *in nosocomio* so as not to infect the house but with an accord that if she died she was not to be dissected. Professor Sidrén, *praefectus nosocomii*, promises me by everything holy not to anatomice her. She dies there; Sidrén dissects her head. He had been uninfected although there have been many *in nosocomio* with the illness, but now he is infected and very ill with it.[7]

Theft was not a category of crimes punished by Nemesis, except in the case of Broberg the gardener, who stole bulbs and willfully disobeyed Linnaeus himself. Frivolity, infidelity, and fashion-crazed folly were punishable, and the collective as well as individual could be sentenced. The tsunamis created by the Lisbon earthquake in 1755 raised questions about the existence of a good God: an edge of Europe cracked and the reverberations reached the Nordic countries. "Half the Earth quaked and He was revealed, able to feel, hear, and show mercy for the miserable though they may be heretics"; God was good as He didn't permit everyone to go under.

The editor of the 1878 volume, Thore M. Fries, declared that Linnaeus put less faith in auguries and portents than most people. However, some material of this kind is included. The vicar Carl Gustaf Rollin retold a story recorded by one of Linnaeus's students: "Bringing a few plants from East India which I had collected from Grill's office, I visited the archiater one day. I encountered the archiater as he left the lecture hall, stated my errand, and was kindly received. We proceeded to his study. When we had entered his museum, he looked at the table and the chair which stood by it, and on which he was in the habit of sitting when studying, and said in a loud voice, speaking clearly: 'Well, now! Sitting there, are you, Carl? No matter, stay in peace, I shall not disturb you.'—I dared to ask: 'Who is being addressed in this manner, Sir?' Linnaeus replied: 'From time to time, I have the sensation of sitting there (he pointed to the chair) deep into my work.'"[8] Was the old Linnaeus holed up in his attic room, talking to his younger self?

The tenet of equilibrium, a *Nemesis ecologica*, recurred on 6 November 1765 when Linnaeus, in his capacity as inspector at the Småland nation, mentored his fellow member Andreas Neander on the theme of virtue and happiness. The text tells a great deal about the way he lived and how he felt about it:

> You seeth, Dear Sirs and County fellows, that when a human being walks in the bright sunlight, she has a double shadow which follows her close on her heels however she turns and wherever she goes. To these two shadows I would liken two human aspects: Virtue and Happiness, which constantly follow, direct, and lead her. The darker shadow is Happiness, the lighter is Virtue. However cloudy the day, still these two shadows will be there whether they are seen by us or not. The two human servants, I said, follow us whether we walk and process forward, or we rush and drive forward across the theatrical stage that is our world. The two are quite different to mind and being. Happiness, angel of this world, is lively, flighty, excessive, dizzy, and unstable, and should we hand her the reins the drive will go daringly over sticks and stones.... Virtue, angel of God, is on the other hand God-fearing, restrained, careful, thoughtful, and steady, and if we hand her the reins, she will drive with caution and care, attentive to every stone.

Of course, Neander—who was hardworking, poor, and prone to ill-health—chose Virtue as his angel. To reach harmony, men and women had to choose, as he had done, the narrow road.

CHAPTER THIRTY-SEVEN

FIGURE 49. Linnaeus, aged 67. Painting by Per Krafft Jr., 1774. The art collection of Uppsala University.

Another story from the university nations: the curator at Småland nation and Linnaeus were talking together in front of a window; Linnaeus asked who the elegant gentleman in the street might be—That is magister ***—Linnaeus: Had he not been engaged to N. N.'s daughter?—Yes, that's right—Linnaeus: And seduced her?—Yes.—And he has now abandoned her? L. is now greatly saddened: You are surely to live for longer than I, and you will observe that nothing goes well for magister ***. How right L. was: twenty years later, the dead magister was hauled out of his own well.

Nemesis divina can be interpreted in many ways and seems fit for some purpose at every turn—it is the force of Old Testament justice and retributive sentences of an eye for an eye, and of the revenge themes of popular beliefs as well as of scientific and especially biological model systems based on cyclic processes and equilibrium reactions. It is manifest in the list of the seven mortal sins but also in the gossipy anecdotes about colleagues at the university.

Philosophically, it belongs among the many attempts to deal with the theodicy problem, assembled by the German philosopher G. W. von Leibniz under the overarching question of how all the misery that afflicts the world is permitted by an almighty and good God? Is he either not almighty or not good? Because Linnaeus himself didn't publish *Nemesis divina*, we don't know how he wanted it to be structured typographically or how he related it to other, similar texts.

The book contains a large selection of words of wisdom; one is "God sees and hears everything"—no one can escape. Linnaeus actually formulated his own purpose: "This must be the highest [task?] to be aware of that which is seen by God who knows all and forgets nothing."[9] Did he imply that he was beyond nemesis?

One last question: Where would Linnaeus have placed himself? Perhaps he would have categorized himself with the sin of pride. For someone who had learned to recognize every plant and insect, the punishment was forgetfulness. It is almost there to read in his last work: his nemesis was *Amnesia*, the loss of memory that comes with old age. But what had he done to deserve such pain?

CHAPTER THIRTY-EIGHT

Solomon on Growing Old

THE DISSERTATION ENTITLED *Senium Salomoneum* (Old age according to Solomon) was up for public discussion in 1759. It is a version of a Swedenborgian interpretation of Bible texts. Chapter 12 in Ecclesiastes is explicated in medical and social terms: "The sun or the light" is equated with rational thought, "the keepers of the house" with human arms and hands, and "the grinders" with our teeth. The opening line, "Remember now thy Creator in the days of thy youth," touched on a central Linnaean precept, as did the subdivision of life into different stages. Linnaeus went through the lines of the text, identifying the signs of old age: people lose their teeth, suffer constipation ("the doors shall be shut in the streets"), legs and knees grow shaky and one's gait uncertain ("they shall be afraid of that which is high"); it is when "desire shall fail" and "the pitcher [is] broken at the fountain" with the urinary bladder, which "[is] broken at the cistern." Finally: "Then shall the dust return to the earth as it was." For the young respondent Johan Pilgren, barely twenty and a medical student, this naturalism made it an alien thesis to defend, even as an academic exercise. It was also an oddity in that Linnaeus, for once, had dared to carry out an exegesis of a biblical text. He was fifty-two years old at the time but preoccupied by the conditionality of life, in the most fundamental and darkest sense.[1]

This passage in a 1756 letter to Wargentin suggests that the dissertation could be read as an autobiography: "Now it has happened to me as to the mole, who ere he dies is permitted to open his eyes and see the vanity of the world. Here, the heavy time in which you die by degrees is all that remains to me: the time when senses are dulled and limbs numbed; when sight, hearing, gait, teeth all die ahead of your final moment . . . and all lust ceases."[2] By the way, it would seem that "the mole" was one of his favorite self-analogies.

Illnesses are recurring themes in his correspondence with Bäck. In an example from 1754, he wrote about fever: "I have had a miserable Christmas as on the Eve, my wife fell ill with the current malign fever and has been quite badly affected. It costs to be married, with your wife ill and many small children. Let us see if I escape the illness, as I cannot take severe blows." Pity not Sara Lisa but Carl? In his 1757 dissertation on the so-called Uppsala fever, he told his audience: "The wives of Herr *Praeses*, of myself and numerous others, who here in Upsala have suffered from this disease, have had to pay dearly for their learning about it." The causes, he explained, were unsound air and miasma. One should avoid living in damp premises or near moldy cellars or privies. He warned about the university pond (Kamphavet) and the city ditch. The recommended treatment was to administer emetics (e.g., ipecacuanha), chinchona bark (quinine), and calming drugs. The same fever forced him to take to his bed between 2 May and 24 June in 1764.[3]

In February 1772, he wrote to Bäck on a serious subject: "My time is nigh and my fate must be a stroke as my head will wobble and I risk harm, my feet will stumble as if someone drunk, mostly on the right side." The theme of the transformation of the body is a specter that Linnaeus often came back to; apart from influence from the Baroque period, the idea can be traced back both to Ovid's *Metamorphoses* and the popular image of the Ladder of Ages. He piled up symmetries between the ages of man and the times of the day or the months of the year, leaving the impression that the human life span follows a relentless timetable. Walking along the road of life, we can at all times know how near we are to the end. In his analysis from 1764, *Diaeta per Scalam aetatis humanae* (Order of life and the different ages of humankind), he explained: "Old age entails an almost complete change of life.... All that has previously flourished grows dull and fades, wherefore the old are not rarely affected by illness.... The substance of the marrow stiffens and thus all functions of life will grow slack. Teeth are corroded to fragments, loosen, and fall out. The head goes parched and bald. Life's golden fountain dries out and all desire is lost." In *Metamorphosis humane* from 1767, he listed the four last phases of the twelve that signal human old age: first, decline (women stop menstruating); aging (gray hair begins to dominate; and names vanish from memory); old age (sexual desire is weakened); and old age frailty (you become like a child). Next, specific characteristics are added to each stage: "Old age. Teeth eventually come out, one after the other. While in the young, teeth are well-embedded in alveolar sockets, they now creep out and lose their hold.... It is a time of frailty. Few reach this stage, but if they do, they often are like children again. 'All sooner or later hurry toward the same goal / all steer a course to this place as here is our last home.'" The quotation is from Ovid's *Metamorphoses*. Always, spooked by his own damaged jaws, he mentioned teeth, a characteristic of the mammalian order.

Before his sixties, his teeth had mostly gone, his hair had turned gray, and his gait unsteady, as described by J. G. Acrel: "When in his 50th decade, he had begun to shuffle his feet forward rather than lifting them." He leaves you with the impression of one who hurries through the timetable of aging. Linnaeus often complained and not just because he tended to hypochondria—various things troubled him from early on. This, from 1751, is taken from his small florilegium of pain: "A rare illness, of the kind I would rather have examined in another body, has rendered lard of me ever since I was in Stockholm; she returns morning and evening, plaguing me for 5 hours at a time, as if a nail were driven into my left temple." Linnaeus envisaged the approach of death in 1763: "I am building a house for my widow and fatherless children." And, in 1764: "I have never been closer to death than this time; the Lord alone knows if I shall ever gain strength. I came through the harshest fight with death and I am quickening." By 1767 he was on his way out: "It is the most pleasing of gospels to learn that my D. Brother will once more see his servant alive, a state that will not last long. *Oleum omne consumtum est in lampade vitae.*" "For as long as oil is still in my lamp"—an expression we will see again.[4]

Aged before his time but much venerated, he also attracted a kindly lack of respect. Since at least 1765, Anders Falck had spoken of him as "the old boy."[5] In 1765 Johann Beckmann wrote a nuanced pen-portrait of the old—well, at least 58-year-old—Linnaeus: "As it was a holiday when I arrived in Upsala, the Herr Archiater resided in his country manor but was daily expected back. In the bookstore, I caught sight of an elderly, not very large man with dusty shoes and stockings, a long beard and an old green coat on which was pinned the order of the Polar Star. I was not a little surprised to be told that this was the famous Linnaeus. . . . Because he had come on foot from the country, he was sweaty and hurried homeward, taking me with him."[6]

Beckmann noted that "I heard something remarkable" every time he had the honor of conversing with Linnaeus. He had been told about Peter Forsskål the explorer, about culturing pearls and eating wild strawberries against gout, shown the horn of a Chinese rhinoceros and Linnaeus's collection of insects. He learned the ins and outs of the younger Haller's criticism of Linnaeus as well as Siegesbeck's and, among many other observations, was treated to his host's rank-order of his nineteen most important contributions to botany. On 31 October 1765, Beckmann was granted the special privilege of smoking a pipe of tobacco in Linnaeus's study while they talked extensively about electricity, the soul and its relevance to polyps, the marrow/bark ideas, various publications in Holland, the hydra, and all interspersed with anecdotes about Sauvages, Osbeck, card games, Jacquin, Münchhausen, Salvius, Rolander, Anton Martin, Laxman ... in summary, Beckmann had taken on the role of a Boswell or an Eckermann.

Linnaeus made no allowances for rank or age. Jacob Wallenius, born in 1760, a pupil at the Uppsala cathedral school who later went to work in Pomerania, recalled how Linnaeus took the time to show the boy his museum and teach him the ABCs of nature. Once, when Linnaeus and Jacob's father were discussing this and that over glasses of wine, two elegant Dutchmen arrived bearing gifts. They wished to meet the famous man but got a cool reception. Linnaeus showed them a few things and then left them. However, when two female relations of the anatomist Adolph Murray visited Uppsala in 1772, "it seemed the old boy sprung to life; he showed them many remarkable items and amused them so much they declared themselves exceedingly entertained. How the old boy addressed them he won them over and they him. Beauty will always tempt even the old to speak easily."[7]

By the late 1760s, Linnaeus's status as an international celebrity was such that faked interviews were published. The following tissue of damned lies was penned by a traveler called Joseph Marshall:

> Upsal is a considerable town, on a branch of the lake Meler; there is nothing in it so worthy of notice, as the famous Sir Charles Linnaeus, the head of the university here.... Accordingly, I wrote a card to him.... I had a most polite and obliging answer, requesting my company to spend the evening that night in his rooms at the college.... The conversation not being in English but in French I did not sufficiently recollect the train of argument so as to venture a repetition of them.
>
> Upon the whole, I never had a more agreeable or instructive evening; for besides our conversation Sir Charles Linnaeus shewed me part of his cabinet of natural history.[8]

The claim to have conversed with Linnaeus in French is enough to give the whole piece away as falsehood from beginning to end.

In 1772 an Icelander visited: "In the afternoon, traveled some six miles to Hammarby, the home of the archiater and knight Linné. There, we met his daughters ... and family, with the exception of the son who was in Uppsala. He then conducted us up a steep rock on a hillock near his house;

there was a small pavilion built of stone which he called his study.... He declares that all of Sweden had once been underwater but has abandoned the view that it is possible to determine how much the water level declines each century.... We inquired about the wallpapers [in the house, showing images of plants]: were they not damaged by damp? He agreed it was the case but that they must be there so that he can see them for as long as he lives. The fire still burns brightly in him and shines through his eyes; he is a short, stout man, not especially beautiful. His speech is cool, indifferent but still pleasant if a little stubborn—*enfin*: like a Dutchman." Fabricius, too, wrote about Linnaeus's eyes: "His eyes are the most beautiful I have ever seen; true, they are rather small but glowing; there is something penetrating about them which I have never found in anyone else. Seemingly, I cannot recall their color but they often met mine with glances such as if he wanted to see through my entire soul."[9]

In the late winter and early spring of 1772, Linnaeus was apparently still a vigorous man. His tone sounds brisk in a letter dated 27 March, unusually enough addressed to Sara Lisa: "My dear wife, I believe my last letter arrived properly; in it, I requested horses and a carriage so that I can travel from here by Maundy Thursday. The snow quality is such I doubt if traveling by sledge is possible. Be so good as to see to it that they are here in time. Should you wish to order me to buy something, let me know it. I believe that Faeton would be preferable to the carriage as the slopes are slippery. Be well My D. Wife. Carl Linné."[10] As Faeton was probably a horse, it proves that he was still able to ride.

An end to one chapter of his life was set out in his famous speech as the departing rector of the university. The ceremony took place in the cathedral on 14 December 1772: "Hence, Gentlemen, allow me to say a few words in honor of my dear Nature study, which most hold in contempt as a mere curiosity. I regret that I lack the cosmetic of eloquence and so must portray her naked, as she was created. She is the one who presents all the Masterpieces of the Almighty; who lets us see the back of our eternal Creator; who pleases and entertains us tirelessly; who gives us all that can be had; who teaches us to open our eyes, full of attention and willingness to learn." Then he began a tour of the many halls of Nature's temple. His speech ended with emphatic praise addressed to the students: "None of You, indeed a single one, have come to me for the slightest of misdeeds throughout my tenure as rector. No shouting, indeed not the smallest noise has been heard in the streets at night. No clamor or agitation anywhere. No one has been reported in disguise. No rows, no injustices between You." You wonder who Linnaeus is really praising, the students or their rector.

Then, the coda: "But at this time of the cold Clash of the Stars [the winter solstice], in this cold cathedral at my cold age, Your patience, Gentlemen will be cooling and demand that I finish."

There was a follow-up to the speech. The next morning, a delegation of the curators from all the student nations visited Linnaeus: they wanted him to have the text printed—"preferably, for the common people, converted to the native tongue." It was a request Linnaeus neither could nor wanted to reject, although he regarded the address as "composed in haste during some of my days of ill-health and unpolished in its formulation."[11]

On other occasions, he would use a darker tone. When, in the early 1770s, he introduced the dietetics course to the students, he said the following:

> You come here to learn wisdom; the first degree of all wisdom is to know thyself. Indeed, what splendid animals we are, for whom all else in the world has been created. We have been generated in a frothing passion in an obscene place. Hatched in a canal between feces and urine, we have tumbled ahead into the world through the most contemptible of triumphal arches. We are thrown naked and trembling onto the earth, more forlorn than any other animal. We are brought up amongst folly like apes and monkeys. Our daily task is to concoct from foods our abominable dung and stinking urine, we are doomed to end our days as cadavers which stench most vilely. Such are we all, from the beggar to the purple-clad. Thus: we are the brood of Sodom, beautiful on the outside but full of ashes.[12]

He wailed in innumerable letters; this is one of them, written in 1770, to his old friend Mennander:

> For a whole month I have been severely ill and never thought of anything other than valediction as the evening of my life has arrived and the night must come; nonetheless, it has pleased God to spare me for another few hours as I am now able to walk about if only I could gather the strengths which, at my age, hardly are easily gained. Now I have remarked from all the conditions that my time is well neigh and that my fate must [be] Stroke as my head will wobble and I risk harm, my feet will stumble as if someone drunk, mostly on the right side. This does not seem odd to me. I have now gained an age that not one 9th of 100 newborns reach.

The last act had begun, a period as dreary as his years in Holland had been fast-moving. He announced in 1773: "I am now bedridden with

[386] CHAPTER THIRTY-EIGHT

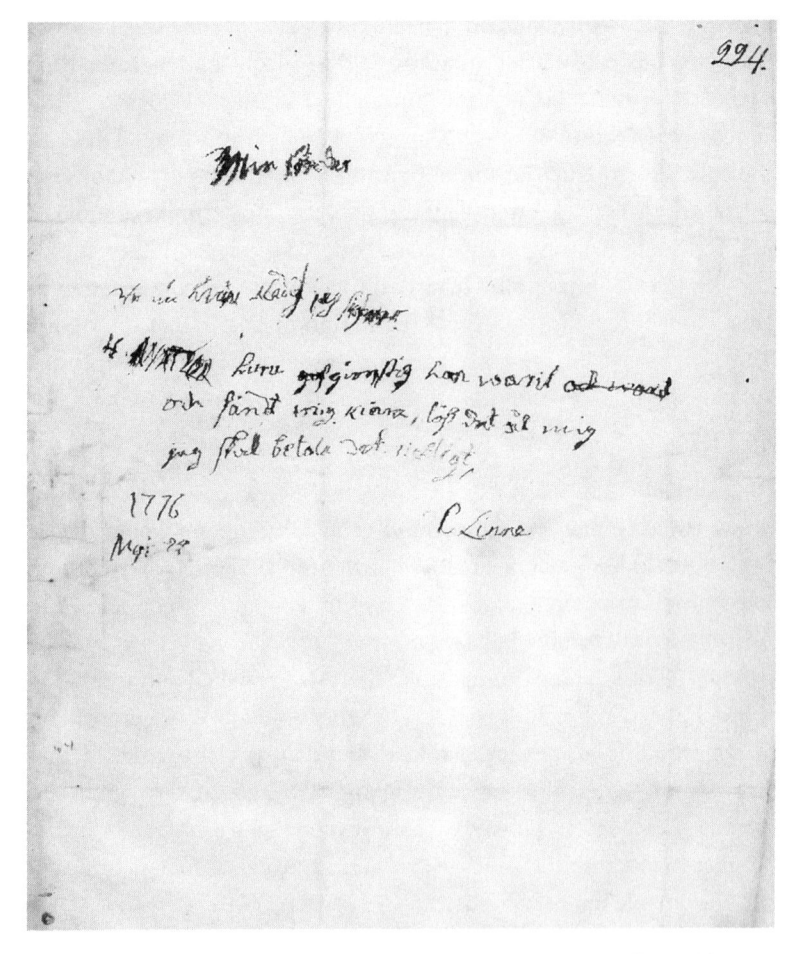

FIGURE 50. The old man's writing is shaky, in thought as well as style. Letter to Abraham Bäck. The text reads, "Dear Brother. H? how he has favored me and sent. dear. Lice. It for me. I shall pay properly. 1776 March 24. C. Linné." Uppsala University Library.

dolore ischiadico and have been here 2 days. No more strength to write." His capacity for work could no longer be relied on: "*Systema Naturae* in a new edition I may never have ready to be printed except for the first part on Animals. I have already had a fair copy made except for the second class of the Birds which is not yet complete." If this information were true, it would be an interesting document to unearth.

When Albrecht von Haller's enormous collection of letters began its run of publication, Wargentin was becoming alarmed on his own account. He asked that his earlier letters be burned, writing to Haller in Swedish—a language Haller understood; however, it was expensive for him to print

Swedish text. Wargentin also seemed to write on Linnaeus's behalf, with particular reference to "a tender concern, not so costly, for our honest and frank Linné, who would by You, Sir, be harshly exposed by publication of his letters. I know as well as anyone his weaknesses; which person exists who has not his own? However, his many merits ought to conceal them."[13]

In 1775 the chemist Torbern Bergman wrote to Wargentin:

> Linné drinks and smokes tobacco like a healthy man but cannot speak, at least not say the *polysyllaba* rightly; also, cannot walk or write legibly; memory and right thinking are lost. He has however retained his passion for money and wishes at last to be inspector ararii as by that is gained 1,100 th.c [thalers, in copper coinage]. It is a delicate *casus*: on the one hand, it is impossible for him to do battle with such a difficult commission, and on the other, the risk is too great. Were he a needy man, it would not be to wonder at, but he is, as is well known, the wealthiest Professor here and is furthermore enjoying a double income (an instance not seen since the first day of the Academy) without being capable of lecturing, examining, and completing any other part of his work, so in my opinion he should stay content.

"Passion for money"—speaking in memory of Linnaeus, Bäck defended his friend: "The noblest of metals gladdened his sight and why should he not collect it, as he collected all else in nature?"

On 29 November 1776, Wargentin wrote to a French correspondent about Linnaeus's miserable condition: "Despite this, he remains wondrously merry and in a good mood, and is at work all day long in the company of young doctor Sparrman, who has returned from the Cape of Good Hope, bringing a large treasure of new plants." The son of Linnaeus's old friend Mennander visited in March 1776: "The old chap seemed infused with new life, accompanied me himself through all his orangeries, and also showed me, among other birds, a Cassowary, recently presented to him by the Royals."[14]

Sparrman wrote an opaque note to Georg Forster on 27 March 1777: "I wrote to you in my last [letter] that tho' Linnaeus keeps so much alive, he lives for himself only, moves very little about his rooms, and is dead to mankind, the soul being, I believe, sometimes out of the body. Three weeks ago, he knew me and spoke to me (one must also guess what he means, as he sometimes knows not of his own children). Pray, why do Mr. *Banks and Solander long to hear of his death?*" Indeed, what about Banks and Solander—and why ask? John Rotheram, later appointed professor in Edinburgh, wrote this in Swedish to Linné Jr.: "Banks is a strange man, Solander likewise. They received me quite coldly as they do everyone from Upsala once they have learned that the caller is among the Professor's friends or acquaintances." There were several plausible reasons for this hostility, not least rivalry, as Banks and Solander were working on a new edition of *Species Plantarum*; also, Linnaeus had sent Banks a letter "which so offended him that in a rage he tore it to pieces." Ever since this event, he had held a grudge against Linnaeus, but neither Rotheram nor later investigators have been able to find out what was said in the letter. We will return to what has been called "the Solander issue."

C. C. Gjörwell reported on 10 June 1777 on the state of health in the world of learning: "Arch. Linné is now like a child and thus the son has become a man. The old one sits there, smiling a little, asks for the names of plants. However, young Murray is a rising star, shining with a strong light in Upsala. He will become to anatomy a new Rosenstein." Seemingly, a new generation was already knocking on the door. This is Ihre, writing about Linnaeus on 8 September 1777: "The ornament of our academy ... now utterly returned to childhood: he vegetates more than lives, cannot complete an entire sentence, only a few monosyllables which also don't work."[15] In 1823 Adam Afzelius recalled the last days: "As soon as the warmer season had begun, Linné was taken out into the countryside ... where he remained the whole summer and was daily carried outside as soon as the weather permitted ... up to his museum wherein he was aroused by happy memories and enjoyed several hours at a stretch looking through his collections of dearest treasures and was always carried away from there in a much jollier mood." The same reporter noted the old man's state of health in spring 1777, observing with his cool, unflinching doctor's gaze: "His limbs and organs, notably the tongue, lower extremities, and urinary bladder, were all paralyzed. His speech was incoherent and most often incomprehensible. ... Of his organic life, only his breath, digestion, and blood circulation were still in a reasonably good state. ... He had even forgotten his own name and seemed for most of the time to be without awareness of either the past or the present."[16]

When the snow cover made sledging possible in December 1777, Linnaeus ordered the stable lad to take him to Säfja, where he stayed for far too long. He was carried into the house and, sitting by the fireplace, he puffed contentedly on his pipe. It grew dark outside and started to rain, which may have brought on the end more swiftly. His condition anyway took a turn for the worse; all he could take by mouth was beer, and he drank it with such intensity that "it was no use to take the vessel away from his mouth for as long as a drop was left." Death arrived at eight o'clock in the morning on 10 January 1778, soon after the death of his adversary Haller and in the same mournful year during which Rousseau and Voltaire passed away.[17]

On 30 December 1777, his son wrote about his father's last days: "My father has this time had a most dangerous attack of convulsions, starting in his face and then to the entire body such that I feared each respiration would be his last; I had been left alone with him as no *medici* were in town; once more, this state has now been overcome but God alone knows for how long." Then, just ten days later, the message of Linnaeus's death, "the most grievous event," was sent to Bäck on 11 January; it led to a discussion of the cause of death, which they decided was not a stroke, but inflammation of the bladder. Several letters were exchanged with Bäck about where to find an autobiography, which Bäck needed for the "parentation"—the funeral celebration of a dead parent—and about the funeral ceremony in general. Bäck, a prominent and successful doctor, wrote a full account of Linnaeus's medical history. In his younger days, he had suffered from tonsillitis and toothaches; in middle age, from severe pain involving half his head and also painful gout "which he believed himself able to medicate with wild strawberries." During the long hours of sedentary work on *Species Plantarum*, he had pangs of pain from passing kidney stones. In 1774 a sudden, dangerous episode, like a flash of lightning: he lost consciousness and had a stroke during a lecture in the academic gardens. He believed it to be a warning of imminent death but recovered when a batch of plants from Surinam was brought to him. The sight of the specimens well preserved in aquavit "gave him what was like a new life." He examined all of them and later wrote up the results in the 1775 dissertation *Plantae Surinamenses*. However, his fatigue returned and grew so serious he could no longer walk or speak. He was helped to get up the hill to his collections and "sat there alone for many hours, as was his previous habit." At the time, he made an entry in his diary, noting that "Linnaeus limps, can hardly walk, speech is disorderly, and he can hardly write." He wrote to Bäck several times and set down with trembling hand his protestations of friendship "as if it were for the last time." Next, another stroke was followed by a right-sided paralysis. His senses much weakened, a three-day fever ensued. It was accompanied by "painful dysuria and the water passed mixed with pus until a quiet death put an end to all his misery."

There are several sources of information about the pathological picture, but only a few modern attempts at diagnoses. We have noted that Linnaeus spent a month or more in bed with a fever before midsummer

1764, and a hundred years ago, Karl Hedbom wondered about the cause of his attacks of shivering: was it typhus or the "spotted fever" variant or some other febrile condition?[18] He had presumably been infected by the "Uppsala fever" that had been so prevalent the previous year. In his own house, five people had fallen ill with it at the same time, but they began to recover as soon as they had been moved out into the yard. Linnaeus explained to his students that "no ditch, pond, or swampy ground should be close to houses. Turks have houses with windows set into the roofs."[19] Birger Strandell and Patrick Sourander are the only ones who have compiled an overview of Linnaeus's often unclear complaints that he is getting old and nearing the end of his life—these started in 1751. Their diagnoses are: paroxysmal hemicrania with migraine attacks, several strokes leading to right-sided hemiparesis (weakness or paralysis of musculature on one side of the body), and dysphasia (language disorder). His history also included autoscopy (perceiving your body or face from a distance): Linnaeus was once described as seeing himself seated at his desk. Idiopathic autoscopy and similar hallucinations may be a characteristic of certain egocentric or narcissistic personalities. The serial strokes were probably due to arterial hypertension, but this isn't sufficient information to explain all his recorded signs and symptoms.

Rotheram, a family friend who was present at Linnaeus's deathbed, noted that there was an elm tree, which "the old man himself had planted for employment to his coffin. It fell over in the summer of 1775 in the garden at Hammarby where it had grown." On Linnaeus's death, a sealed envelope with his funeral instructions was opened, but all it said was that he was to have a sheet for his shroud, a white nightcap on his head, and then be put into the coffin. Rotheram confirms that this was done: "thus he lies in his grave." Linnaeus's autobiographies, which were published much later, were used in the parentation presentations to keep his memory alive. In the version completed in 1750, he had already written in the introduction: "After my death, this document is to be submitted first to Prof. Beronius, who has promised to present at my parentation, and the presentation I wish that my wife will have printed if she is well at the time; then it will, as soon as Pr. Beronius has it copied out for the Academy of Sciences in Stockholm, of which my wife will request that they do for me, as they have promised, the most honorable parentation and ask them to print the speech."[20] All this was later crossed out. Besides, Archbishop Beronius had himself passed on before he could fulfill his promise.

Linnaeus had left the following exact instructions: "1. Place me in the coffin unshaven, unwashed, undressed. Wrap the body in a sheet, then quickly close the coffin up so that my frailty is not seen; 2. Have the big bell ring out but not the others nor in the chapels, at the farm, or the hospital, but well in Danmark parish church; 3. Have a service of gratitude said both in the Storkyrka and in the Danmark churches to thank God who hath given me so many years and all blessings; 4. Have men from my county nation carry me to the grave and give each a small medal inscribed with my image; 5. Do not serve anyone food and drink at my funeral and do not receive condolences." His instructions were followed, except for item 5: the students from Småland nation were treated to free food and drink.[21] He made a few further points: "Do not have me taken to Drottningholm and put my medallion on the grave and let it be etched into him: *Princeps botanicorum, natus 1707 majus 13 mortus* . . ." That he didn't want to be taken to royal Drottningholm might have been due to his view that its miasmatic air was a health risk, or a last radical gesture to distance himself from the court. Linnaeus's wish to have his funeral bells rung from the local church proves that he regarded Danmark as his home parish.

What do these wishes say about the individual's attitudes to life and death, and what of the general views held by his contemporaries? For the matter-of-fact wishes of a medical man, they can be compared with the 1792 testament of the doctor Johan Otto Hagström: "In this town, no arrangements with horses and carriage should be made, no bells should be rung as suchlike are papist inventions. No washing must be carried of my dead body, as it will be the home of worms. Shrouding can be the simplest, an old sheet, and the coffin made with pine boards, a linen or cotton cap to be put on my head. In my house no room must be clothed in black, just as the burial must be free of garnishes. I wish to be buried in a cemetery, not in the church, so that my corpse should not with its stench trouble the living."[22] There are other examples of what might be "medical men's misanthropy"—take Christofer Carlander's instructions for managing his death: "When I by the grace of God have been taken home to my father's house, let the cadaver lie in a bed for a day and a night so that it is well stiffened. No bells, but the minister will get his burial fee as is proper. Finally, dispatch the cadaver to its last rest, and that should be that."[23]

Jean Eric Rehn has said that he drew a shrine for his friend Linné, but the drawing has not been found. Another architect may have been charged (probably in receipt of an anonymous contribution from Gustav III) with creating the 1798 Linnaeus memorial in Uppsala cathedral; perhaps the proposals by Rehn were used as a model.

An anonymous poetic admirer, thought to be the secretary of the university, a Gustaf Flygare, who was with Linné the younger in London at the time when Linnaeus was dying, had a dream—an apotheosis of sorts—and wrote it down. He had seen, as if floating in the sky, Linnaeus on a wonderful, lovely meadow. Nature herself, the Indescribable, was enthroned there: "Although this Goddess is always good, it nonetheless seemed Linné's arrival enhanced the gentleness of her countenance and suffused it with a pleasure which in a mortal would be called joy." It then seemed that Linné was recognized as belonging among the first rank of the students of nature: Aristotle, Pliny, Dioscorides, Boerhaave, and von Haller.

Linnaeus's will, dated August 1770, favored his daughters by stating that his precious herbarium was to be sold and the income from the sale shared between them. The widow was left in charge of the properties, including the military obligations, while the son was given the collection of minerals and the library. Were he to use his position at the university and the library as intensively as his father "he would have as much wherewithal as all his sisters together." An additional clause stated that Captain Bergencrantz, Elisabeth Christina's husband and a known wifebeater, should not benefit from the inheritance.[24]

Thore M. Fries grows a little hesitant when discussing the issues raised by the various and sensitive ultimate concerns. Linnaeus left drafts of two different messages about his death, one according to the formulaic "It has pleased the Almighty God to end the lifespan of . . ." while the other, according to Fries, was "obviously written during his last years of weakness, and consists of a longer piece of writing including a brief, dry biography and a few, rather too boastful comments." We are not told what these were or what his source was. However, Linnaeus's son apparently wanted to change the conclusion that his father had died from a stroke, but the widow forbade any autopsy. There is an unattributable note saying that she feared that professor Adolph Murray had planned a dissection to take place within a week of the funeral, which was why she kept the keys to the church safe in her own purse.

The funeral, which took place at six o'clock on 22 January in Uppsala cathedral, a contrast with the spring birthday of the man whose life they were celebrating. The Book of the Dead states: "Of the Bells, ring the Big Bell alone, black clothing, and black coverings."[25] The route to the cathedral was lit by the lanterns accompanying the twenty-one carriages in the funeral procession.

Jonas Hallenberg, who was among those present, remarked: "Such a crowd of people as on this occasion no one can ever remember having seen in Upsala church." Another witness, Adam Afzelius, wrote: "It was a dark and silent evening, its darkness only broken along the roads through the city where the slowly marching procession lit it with torches, lights, and lanterns." Another comment in the same vein: "Archiater Linné's funeral at 6 o'clock in the evening was made well *éclair* along the road to the church by the lanterns that accompanied twenty-one carriages." The officiant was the dean of the cathedral and County Småland fellow Eric Hydrén. "Given the absence of relatives at the ceremony walked, next to the son, the fellows from the nation." This could be understood as Sara Lisa not attending, perhaps because she and her daughters were preparing the meal to which the men from Småland nation would be invited; alternatively, there was a custom at the time which didn't allow female relatives to be present at the funeral service.[26] The account for the funeral has an entry of 90 thalers in copper, not including the cost of the meal.

Jonas Sidrén, Linnaeus's colleague in the Faculty of Medicine, joined all the occasional poets: "He has been the jewel of Sala / He was well liked by all men of nobility / He gained everything and more / Will we ever see his like? / No, He was a masterpiece / We will never see his like."[27] Arguably, not great poetry but a true summing-up. Despite Linnaeus's instructions, we have been given the impression of the funeral as an appropriately staged social event. Still, the anti-Linnaeus camp follower Adolph Murray commented that the funeral meats were "pitiable."[28]

Few professors are likely to get a mention in a monarch's speech from the throne, but when Gustav III addressed the parliamentary estates in October 1778, he spoke with almost personal feeling: "I have lost a man who did his Country as much honor as a worthy citizen, and he truly was rightly famous in the whole world. Uppsala will long remind itself of the fame which the name of Linné brought to its university."[29]

Science, too, will long remind itself of his contributions.

EPILOGUE I

Family Life 3

MOTHER AND CHILD

LINNAEUS'S ESTATE CONSISTED of his country properties and, above all, the wealth of scientific material in his collections, his wide-ranging correspondence, and costly library. The inventory was reassuring, but the matter of his will remained unclear. The document "A cry from inside my grave to my wife, so dear to me in my time" established that his son "should not take charge of the herbarium but it may pass to a future son-in-law who is a *Botanicus*." It undeniably sounds like a "Failed." However, the son was given the other collections in his care to spare his mother the responsibility.

The relationship between mother and son had been poor for a long time. Johann Beckmann commented: "The son fears his parents more than he loves them, and his mother has always detested him intensely. On one occasion, when he had already been appointed professor, she slapped his face. When his father asked her to consider that he was a professor, she replied: 'I would slap him even if he were an archiater.' This was the reason why the father had rented a room for him in the house where I stayed, and also why the son did not like to be in the company of his parents." Linnaeus's student Fabricius wrote: "It was most remarkable that she hated her only son. He could not have had a greater enemy in the world than his own mother." Fabricius himself got on well with Linné Jr.: "It must be said of him that he had not inherited his father's lively manner but the great knowledge he had acquired through constant practice in botany and learning from his father's many and excellent observations, which he must have found in the older man's manuscripts, will make him a fine teacher at the Academy." Adolph Murray—Linné Jr.'s rival—remarked in

FIGURE 51. Carl von Linné Jr. Painting by Jonas Forsslund. Photo by Mikael Wallerstedt. The art collection of Uppsala University.

April 1778: "If the lad is thought ornery, it is also a fact that he has an old witch to deal with, who persecutes and hates him excessively."[1]

Linné Jr. inherited his professorial chair from his father; the authorization had been drafted in 1769, but his induction took place first in 1777, on 27 October. To install your son as your academic successor was not unique; for instance, Rudbeck Jr., a perfectly competent man, succeeded Rudbeck Sr. Anders Berch cited his failing eyesight as a reason for a temporary leave, and his son stepped in as a locum. Linné Jr. was awarded an honorary doctorate in the 1765 round of nominations, at the instigation of the crown prince but with his academic "promoter" Samuel Aurivillius stating explicitly that it was based on the father's achievements—"meritis summi viri"; it is not hard to work out how the son felt. Envious bystanders didn't miss this opportunity: at the 1766 "Disting," the early spring market in Uppsala, you could buy a pamphlet allegedly written by "von Linné the younger" and entitled "Thoughts of a child: On the possibility of becoming a doctor without knowledge and a professor without a head."[2] The following year, Linné Jr. wrote to Bäck: "I am convinced that both Bergman and Bergius said at my lecture concerning my findings: 'It is surely unfounded, poor chap. He knows no better; he has heard his father say so, as a charlatan would have, and with his superficial knowledge and fluttering spirit, tried next to trick people roundly,' and more of the same kind; I know them all well enough and let them not worry me anymore; so, they remain *ingrata*—let it be; they cannot with such utterances diminish my Father's reputation; nor will I have them eat me."[3] Early in 1780, his tone is somewhat less cocky: "It is not written in my Fate that I should win at home, so I shall try to be happier among foreigners. I have not yet grown too weary, my passion for Natural History cannot easily cool."[4]

Sure enough—it did not take him long to set out for foreign places. His journeys between 1781 and 1783 deserve a chapter of their own: he went to England, where he met Joseph Banks and was present at the death of Daniel Solander; to France to meet Buffon; to Holland and then, very successfully, to Denmark. The only letter left from this period was sent from London in 1781 to his mother and sisters, and in it, he asked for money. He followed this up in a challenging, perhaps characteristic style: "However, to tell of the country where I now live, I have sought to think of something to amuse my D. Mother and sisters. I have to speak of the women here; they are, most of them, quite appealing; they are quite white, well-shaped, move with ease and freedom, and quite merry; they never bind themselves, don't use corsets, often have their hair free and unpowdered until they are 16 years of age." The ladies at home are not likely to have found this information purely entertaining, as it for one thing appeared to repeat his father's criticisms of women's artificial fashions.

Stoever, too, reported that Linné Jr. had the misfortune to be the target of his mother's hatred. Hedin took a mildly contrary line: "The relation between the son and his father's household was never broken; surely one could sense the maternal love and wish to support a firstborn son." Certainly, there is more to be said about Linné Jr. and his career, friends (for one thing, he inherited not only his father's academic appointment but also the close contact with Bäck), as well as his personality and work as a scientist.

FAMILY LIFE 3: MOTHER AND CHILD [397]

FIGURE 52. Tahitian Omai, Sir Joseph Banks, Daniel Solander. Oil painting by W.J.S. Brown after William Perry. Royal Academy of Sciences.

In the summer of 1783, Linné Jr. was struck by fever and gallstones, which weakened his constitution a great deal. Soon after his return to Sweden, he had a fatal stroke; it took four hours for his life, which had lasted just a little more than forty years, to be extinguished, only a few months after the death of his friend Solander in London. The Linnaean male line died with him, and on the announcement of his death, the family coat of arms was broken at the House of Nobility.

Daniel Solander has been mentioned several times: his story is interwoven with the Linnaeus family and, in later years, with Linné Jr. in particular. The question of whether he was also a biological family member has been raised and should be given an airing here even though it seems impossible to answer it definitively. The critical event would have taken place during Linnaeus's travels in Lapland. On arriving in Piteå on 15 June 1732, he stayed for a week, spending two nights in the rectory with the then curate (later vicar) Carl Daniel Solander and his nineteen-year-old wife, Magdalena Bostadia. A little more than eight months later, on 19 February 1733, Magdalena gave birth to Daniel. In 1750 the youth enrolled as a student at Uppsala University and soon became one of Linnaeus's favorite pupils as well as an occasional lodger in his house. "Not that I constantly lodge in Herr Archiater Linnaei house, but visit every day so that I almost spend more time there than at home in my proper quarters." Linnaeus mentioned Solander on several occasions and made it quite clear that he thought of the young man as a successor and also—as was Solander's own expectation—as a son-in-law. To marry the right person was one route to an appointment, and his choice was Linnaeus's daughter Lisa Stina. As it turned out, his love was not reciprocated, the rejected Solander took himself off to England, and cut all links with Lisa Stina's father—or, at least, no longer replied to his letters. He didn't keep in touch with his mother either, a silence that is quite hard to explain. Several unopened letters from his mother were found among his belongings in his London home after his death in 1782.

This is the parentage story in brief: it can be extended but also questioned. The established facts in the case are few but suggestive; some, however, do not support a father-son relationship. Nothing relevant is said in the Lapland travel diary, Linnaeus and Magdalena didn't know each other—she was quite newly married—and they had had little time together to procreate. The reason why Solander and Lisa Stina did not marry was that she didn't care for him.

Linnaeus had really tried to recapture Solander. At first, the young man responded warmly, but from 1763 on, there were no replies at all. Linnaeus's letter dated 15 August 1761 begins with a wish to come to terms: "My honest advice was perhaps ill received." He had thought it would have been good for Solander to be at home, as market conditions (for university jobs) were favorable. He went on to list some thirty-five news items and rounded off with: "Everyone in the house sends the very warmest greetings." He wrote again on 11 December the same year: "My dearest Sir, the affection and enjoyment I felt for my D. Friend long before this has been conducive to make me wish Your interests were to be conjugated with the achievement of our shared interests. . . . Your Mother, who dearly wishes for Your best, as indeed do I, has now reached the years when Her *Catamenia* [menstruation] should cease, but she continues to suffer from *heamorrhagia uteri* [uterine hemorrhage], which at her years has a most

FIGURE 53. Sara Lisa as an adult. Drawing in black crayon, probably by Sven Höök (d. 1787). Museum of Natural History, Göteborg.

dangerous *prognosie*."[5] Why didn't Solander write in reply? Was it because he had nothing to say, or because the matter was so sensitive, he wanted to bury it in silence? In 1763 some event must have taken place; his mother wrote to him several times, Linnaeus and Wargentin tried to mediate, but she only ever received three letters.

Then, five years later, a letter arrived from Brazil, dated 1 December 1768. His lack of any sign of life is admitted—"Not now to repeat the causes of why [I have not written] for such a long time"—but not explained, although there is an unmistakable allusion to his love for Linnaeus's daughter Lisa Stina.[6] As we have noted, just about ten years later, Sparrman wondered why Banks and Solander apparently longed to learn that Linnaeus had died.

After the death of her son, Mrs. Solander tried to get in touch with his friend Joseph Banks; she evidently wanted to know how her son's last years had been. Banks replied that he "had left out such Anecdotes which are as unfamiliar to Yourself as some are to me"—not a very helpful formulation—but we can guess that Banks was referring to Solander's years in England. Of course, family conflicts rarely have only one cause; perhaps the schism between Linnaeus and Solander was not primarily over Lisa

Stina but might have had more to do with Linnaeus's way of presuming to steer Solander back to Uppsala, or to a professorial post in Petersburg that at one point was open to applicants.

Once the men of the family were gone, Linnaeus's widow Sara Lisa took control at Hammarby. She had shared her life with her husband for forty years, and her irritable state of mind was surely understandable: he traveled most summers, kept nocturnal hours of work, and attracted hordes of students in dirty boots. However, it wasn't her husband but her children that engaged her most: from giving birth with its feared risks to both mother and child, to bringing up her daughters, with all the attendant hopes and disappointments. By the time Fabricius and Beckmann described her, she had reached her fifties, and they were around half her age. This is Fabricius: "She was a large woman, strong, despotic, selfish, and utterly lacking in culture. She often drove the joy away from our get-togethers. As she was quite unable to share in our talk, she also did not care for our company. In such circumstances, the children's education could only be wanting. They are kindly but rough children of nature, without the good manners that a careful upbringing could have given them." By then, they were either dead, or had left home.

It has been said that Linnaeus had never been happy in his homelife, a belief based on a mistaken statement by Fabricius, who wrote: "In his family he [Linnaeus] was not completely happy." Within the family, however, there is a tradition of defending Sara Lisa. Tullberg, a descendent, explained that Fabricius, who was a foreigner, knew no better and, besides, Sara Lisa was fed up with foreigners.[7] It should be stressed that Linnaeus himself never complained about his marriage or expressed any wish for another relationship, in his autobiographies or elsewhere. There is a comparison to be had in his son's lectures in a series entitled *Mores*: "If one were to find a wife who well understands how to manage a household and is nimble in her behavior, it would be a happy outcome but, should either be missing, the outcome is likely to be troubling. To have a learned wife is not the best, as an educated woman will always mock her husband whenever she is able to go further into a matter than a man could and also, because of their studying, the housekeeping will be disorderly."[8] Sara Lisa was an efficient and orderly housewife who supported her staff. Linnaeus comes across rather like an absentminded professor–type, but he was not a docile spouse. If he had been bothered by his wife's uncultured ways, he would presumably have seen to it that his daughters were better educated.

Besides, these social judgments were unfair. Linnaeus was the celebrity, and Sara Lisa was considered of no account because she is uninterested in botany—though she actually joined other ladies on a botanical excursion in 1747. However, she did know how to run a household in the style expected of her.[9] During her many years at Hammarby, she also developed the skills of an estate manager. The charcoal drawing of her kept at Hammarby is a portrait of a woman with authority and, perhaps, a sense of humor.[10] Despite much scrutiny of the record of the Linné family burials, it has proved impossible to determine the location of Sara Lisa's grave.[11] She survived her husband by twenty-eight years and died in her nineties in 1806. Her youngest daughter, Sophia, took over at Hammarby. In 1879 the state bought the house.

In this context, attention should be given to another relative: the eldest daughter, Elisabeth Christina or Lisa Stina, born in 1743, who was taught as well as loved by Daniel Solander. They were regarded as a couple, but Lisa Stina apparently didn't want to marry him.[12] News that she had married in 1764 presumably reached Solander and could partly—but only partly—explain his silence with regard to Linnaeus. Of course, she might have heard of the speculations concerning Daniel's parentage. Her marriage to Lieutenant, later Major, Carl Fredrik Bergencrantz was unhappy. Bergencrantz abused his wife and had "a closer understanding with his housemaid Maja, a widely notorious whore"—in Linnaeus's words.[13] Linnaeus forbade Bergencrantz's presence in the house; Lisa Stina and her daughter Sara Lisa, born in 1766, moved back to the home of her parents. She died early, in 1782—the same year as Daniel Solander.[14]

What happened to Linnaeus's collections, the extensions of his personality? In brief: their fate was obscure. The testaments were unclear for a start, and the muddle grew worse after the death of Linné Jr. Gjörwell wrote to Olof Knös on 6 December 1784 that he had learned from a piece in the newspaper *Dagligt Allehanda* of the sale of Linnaeus's collections and asked: "Is this true? Shall such dishonor and loss now afflict our literature at this enlightened time?" The report was true—the collections had been bought by a young Englishman, James Edward Smith. His purchase consisted of an herbarium containing some 14,000 plants, 45 glass display cases with birds, 158 dried fish specimens, 3,198 insects, 1,564 *Conchyler* [shells], a large number of corals, 2,424 mineral specimens; added to the *naturalia* were around 1,600 books, 3,000 letters, and an unsorted pile of manuscripts and other papers. Since then, the birds have disappeared, the minerals were sold in 1796, and a number of medical books were donated to Uppsala University library—but all the rest has, since 1970, been stored in a strong room constructed under the home of the Linnaean Society in Burlington House, London. There, the collections are kept safe.

EPILOGUE II

Linneanism

THE CULT THAT had formed around Linnaeus had already taken root before his death but burst into full flower quite soon afterward. His influence can be seen in innumerable names and book titles, statues and jubilees; he is everywhere, engraved on medals, remembered in the names of streets and girls. He had profound effects on science and academic thought in general; all this adds up to what we here call *Linnaeanism*. Gustav III, who knew well how to play the strings of national history, donated his royal castle garden to Uppsala University so that they could construct a new academic garden; he also funded the building of a botanical temple in beautiful neoclassical style. It was opened in 1807 and given the name Linneanum. At the same time, there was a collection of money to pay for Niklas Byström's marble statue of Linné, draped in a Roman toga. The national treasure Tegnér wrote a poem for the fiftieth anniversary of the Swedish Academy: "Swedish honor conquered new roads / in the unseen lands of thought: Linné / stood there, joyous with victory among flower banners / like them, as innocent, as lovable and as artless."

There were, of course, those who hit out against the cult. The astronomer Daniel Melanderhjelm's long essay from 1783 about Uppsala University lingers thoughtfully on Linnaeus's rare achievement: he had succeeded in combining fame and a respected name with keeping an eye on finances and had ended up respectably well-off. This was possible because of two conditions, which existed then but would never be repeated: botany was so undeveloped that everything was new, and nature study "was seen as entertaining and suitable to every understanding." Linnaeus had an easy time of it, in other words, because people enjoyed novelty, as ever. True, Newton's law of gravitation was also new but too difficult to go down well in nontechnical circles. Instead, everyone jumped at the possibility of seeing "in Sweden, a rose from the Cape, an amaryllis from Asia, monkeys and snakes from Africa, and parrots from America, all of which were rated as such marvels that no one except Linné could display."[1] This was a very tendentious hypothesis: botany, like astronomy, had a long history (consider the two Rudbeck professors), in Uppsala and elsewhere, but Melanderhjelm was determined to protect his science and fend off exposure and aggression from without.

The idea had a long life. Thomas Thorild, ever bad-tempered, wrote from Uppsala in 1787: "One arrives down from beautiful hills and fields to this contemptible city, looking as if dumped in a hole in the ground. One believes at first that its constructor wished to conceal the truth; then, preferred to suffocate it slowly. He has gained his purpose! / All alive in this place can be brought into one single *Classis*: Louts and Fools. Seen in this *Systema Naturae*, as surely as everything else that is human: the able are *monstra*, or *praeter naturam*, no spark of Glory! No glimpse of Sociability: a desert! / I sigh for Stockholm. To live here is as costly as there. I rediscover Swedishness: Cheat and prey on others; ineptitude and credence. / Here people *bow* more haughtily, and also more slavishly than in any other place in Sweden. This is the effect of the *Majesty of Pedantry*—Lund rises far above this." The "Majesty of Pedantry" is a reference to Linnaeus or, possibly, to Thunberg, who succeeded him in 1784. All of it is, of course, very unfair, but it remains a rightly memorable lamentation.[2]

Internationally, Linnaeanism was a success, with the 1788 founding of the London Linnean Society, which held the great collections in its care. The society's coat of arms was modeled on Linnaeus's own: an egg in the middle and three fields, colored black, green, and red—one for each of the three realms of nature. In 1790 a bust of Linné was formally installed in Jardin des Plantes in Paris; many other cities followed suit.

We will confine this narrative to a few statements and names from the international scene. When Björnståhl visited Rousseau in 1770, his host exclaimed: "You know then my Master and teacher! . . . In this book [*Philosophia Botanica*] there is more wisdom than in the greatest folios."[3] Rousseau was an influential promotor of Linnaeus in France, although still ready with some criticisms in *Les Confessions*: he said that "his Master" was "excessively studied in herbaria and gardens, and not sufficiently in nature." Goethe admired Linnaeus: "Apart from Shakespeare and Spinoza, I know of no one who has made such a strong impression on me."[4] He stressed that they were both morphologists—however, studying Linnaeus seems not to have taught him botany. Such long-distance influence could be complex, as Walter Wetzel pointed out in his essay on Linnaeus and Goethe.[5] Jeremy Bentham, Immanuel Kant, who discusses the Linnaean system in the context of his own logic, and Michel Foucault were others who responded in their own ways—Foucault took his friends along to show them Hammarby.[6]

Then, of course, there is Charles Darwin. "It is very possible that Darwin would not have been able to envision his ancestral tree of life without the Linnean hierarchy," suggested the creator of sociobiology, Edward O. Wilson. The two great naturalists can be compared without any attempt at rank-ordering: they were separated by a hundred years, by different national cultures, and by family wealth (the modest minister's son grew up in a meager countryside; the doctor's son enjoyed his family's wealth); Linnaeus published early and Darwin late; one became a professional and a careerist, the other stayed an amateur (in the best sense) and somewhat withdrawn; Linnaeus wrote compendia, Darwin narratives—but both were buried in the nation's grandest churches, both traveled early in life, both were natural historians rather than biologists, both suffered from various forms of malaise—and so on. And both were epoch-making.

In Holland, where Linnaeus's methodological works were published, there was an early interest in his lines of thought. Adriaan van Royen and Jacob Gronovius were working in Leyden, Johannes Burman was professor of botany in Amsterdam, Evert Jacob Wachendorff was in Utrecht, as were Maarten Hottuyn and the eminent David de Gorter. The Dutch-born medic Gerard van Swieten and Nicholas Joseph Jacquin were in Austria and Johann Anton Scopoli in Tyrol. In England, Linnaeus's work had "a solid victory," and was studied eagerly by John Ellis, Peter Collinson, Philip Miller, John Hill, Joseph Banks, Erasmus Darwin, Daniel Solander, Jonas Dryander, and James Edward Smith. Erasmus Darwin took the Swedish scholar's writing on board with enthusiasm: "Botanic Muse! Who in this latter age / Led by your airy hand the Swedish sage, / Bade his keen eye your secret haunts explore / On dewy dell, high wood, and windling shore."[7] In France, Linnaeus was one initiator of what has been called the birth of systematization, and followed by François Sauvages, Antoine Gouan, Antoine Laurent, Bernard Jussieiu, Buffon, and Michel Adanson. In Germany and Switzerland, there were both supporters and attackers; among the former were Christian Gottlieb Ludwig, Johann Gesner, and Albrecht von Haller. Haller had an objective mind and could be generous in his judgments of Linnaeus's work—as in this quote from *Bibliotheca Botanica* [translation from Latin]: "Linné has begun one of the greatest revolutions and has almost in all entirety reached his goals. Highly gifted, with an incisive mind and equipped with an extraordinary imagination as well as a systematic mentality, Linné elaborated a new system for plants by setting his great intellectual ability to work."[8]

The international network is still intact: in *Letters to Linnaeus*, published by the Linnean Society (2009), sixty-odd, mostly still living people write to Linnaeus and express their deep respect for "Dear Carl," "Dear Baron," or "Your Excellency." Edward O. Wilson stressed the threefold importance of Linnaeus's work: the systematic principles, the naming, and the inspiration to examine biodiversity because he argued for an inventory of the natural assets on the still relatively unknown Earth. All the writers, including Wilson, pointed out that Linnaeus counted around 18,000 species and that, by now, we know of about 1.8 million and counting; most of them are still not properly described. The task is enormous, but our planet needs an *Encyclopedia of Life*.

There were critics abroad as well as at home. The anatomist Pieter Camper attracted attention when he turned down an honorary membership in the London Linnaean Society because, he said: "I look upon Linnaeus as a mere Cataloger, and the most superficial naturalist I ever knew. He offered in this century little honor to that science."[9]

A lively snapshot of the big outside world, taken inside one of its temples of science: On 14 February 1770, Johan Hinrik Lidén attended a meeting in Paris of the French Academy of Sciences, and among the speeches, one was directed against Linnaeus. It was given by the botanist Michel Adanson:

> Hardly had the auctor begun to read before a general murmur arose. M. Guettard at first interrupted M. Adanson and took to defending Hr Linné; he accused Adanson of having quoted the other's words incorrectly and without sufficient accuracy explicated the experiments he indicated as in his own favor, and then asked several sharp questions; the respondent replied angrily and continued his reading anew; soon after, he was interrupted by M. Duhamel's interjection which lasted for quite a while; he frequently referred "le Diable," a reference which led me to inquire of Hr De la Lande if Satan himself had been voted in as a Member? . . . The irritation increased and finally the quarrel grew truly violent and general; I perceived no one who stood by Hr Adanson, except for M. Buffon, who in any case was quite moderate. . . . M. de la Condamine fell asleep during this quarrel even though there was much said so loudly that he, even though deaf, could still hear it. M. D'Alembert was mute and listened to the dispute leaning his learned head in his hand. M. De la Lande was laughing with me. Finally, the

clock rang out for 5, the members rose, and Hr Adanson must leave his Anti Linnean lecture before he had had time to complete it. De la Lande asked me afterward what was gained by such a "row" and received the reply "nothing."[10]

Meanwhile, nature study had been divided into green fieldwork and white laboratory work. In 1802 Gottfried Treviranus and Jean-Baptiste Lamarck created the concept "biology" in order to include later experimental methods, as distinct from the observational investigations. If such distinction were applied, Linnaeus would be a natural historian, not a biologist. His followers and successors, including Darwin, also would not count as biologists.

Individual aversions, like Thorild's, are of little account compared with the accolades of scientists such as Thunberg, Swartz, Acharius, Wahlenberg, Agardh, and Fries, all committed in word and action to Linnaeus. There are many instances of their work making special contributions to the completion of *Systema Naturae*; Thunberg, less concerned with the Swedish flora, added the world of plants from Japan and South Africa while Agardh, Wahlenberg, and Fries spoke and wrote about Linnaeus's greatness. In 1800 the Uppsala *Societas pro historia naturali* was founded; in 1803 it was renamed the Zoophytolithic Society and, from 1807, the Linné Institute; in 1834 the Linnéan Foundation was established in Stockholm. The man had become a national icon and had to be treated accordingly. Adam Afzelius said this, on the publication of Linnaeus's autobiographies and other writings in a suitably dignified format: "To my mind, all this should be Swedish and the most beautiful conceivably achievable in our country. This, as Linné was not only a Swede but a most unusual Swedish Man, who acquired for his Native Country a previously unknown learned knowledge, which aroused the admiration of his fellows and successors, and left us the great memory of him, not only in Sweden but in all of Europe, yea, on all the continents of the world, and for time eternal."

Be that as it may, as a nation, we also love C.J.L. Almqvist's vision of the modest little wild rose, the "flower of Swedish poverty." However, Almqvist, too, was influenced by the Linnaean heritage. Swedish culture is often supposed to have begun with Strindberg. It is true that he coined the term "buttonology"—scholastic sorting of buttons with holes, and without, with round or oval loops—in order to shame historians who didn't accept his vision of history but instead spent their time arranging stone axes in long rows, in the style of Linnaeus.[11] Strindberg was really after

FIGURE 54. Linnéträdgården (Linné garden), 1864. Photo by Emma Schenson (1827–1913). Uppsala University Library.

Oscar Montelius; he regarded Linnaeus and Swedenborg as the original teachers: "Those who dare call him the dry systematizer have not found the master where he should be looked for." Strindberg recommended the travelogues, the letters, some of the lectures, and Linnaeus's overarching vision—"Linné is closer to the Pythagoreans and Swedenborg than one might have assumed and hence we await his return."[12] Characteristically, all of this was published in the celebration year 1907 and sure to irritate maximally.

The Linnaean tradition has evolved into different forms, and its variants can be combined and recombined. Just as the word *nature* can be interpreted in innumerable ways, so too the name Linnaeus can be linked to many things, from ecological interactive systems to imperialistic human power-grabs. If the Swedish work ethic is a manifestation of "Luther inside us," our relationship with nature reflects "Linnaeus inside us."

It is indisputable that Linnaeus has transfused green blood into Swedish veins. At the beginning of the previous century, the statistician Gustav Sundbärg averred that Swedish people are especially gifted in science, share an innate, special feeling for nature, but lack self-awareness.[13]

Appendix XVI of the *Report on Emigration* is a series of aphorisms and begins by establishing two facts about the Swedish mentality: (1) We take a profound interest in nature but do not include human beings; (2) We lack a sense of nationhood (presumably the first item is linked to the Linné celebrations in 1907, and the second to emigration issues). "Love of nature is deeply rooted in our people and was given a huge awakening by Linné. Linné determined the direction of our people's interest for centuries."[14] The historian of ideas Gunnar Eriksson observed: "It is surely significant that [the popular tale of] Nils Holgersson speaks almost exclusively about the nature and the fauna of Sweden. Wherever is the Sweden made by human beings and where are its people?"[15]

Nature sensibility is a diffuse concept, to say the least. The list of those who have this quality, and discovered it in Sweden and Swedish identity, can be made quite long: Sten Selander, Knut Hagberg, Gunnar Brusewitz, Rolf Edberg, Carl Fries, Harry Martinson, Björn von Rosen, and Gustav Sundbärg—this list contains only nonacademics. The Linnaean inheritance can be traced in people who have explored Sweden, in the main a mixed crew of painters and writers: Ehrensvärd, Elias Martin, Marcus Larson, Kilian Zoll, Almqvist, Strindberg, Liljefors, Bergh, Prins Eugen, Johan Tirén, Helmer Osslund, Peterson-Berger, Lubbe Nordström, and Lars-Erik Larsson.

Linnaeus's impressionistic notes on what he saw, vivid writing, realism, eye for detail, and exactitude as well as his Old Testament citations, parallelisms, and list-making have found ways into writing as diverse as by Ödmann, Fischerström, Linnerhielm, C. U. Ekström, Erik Axel Karlfeldt, Erik Rosenberg, and Dag Hammarskjöld—and the Swedish Tourist Association's annuals.

Well spotted: Linnaeanism seems purely male so far. There is every reason to include Vivi Täckholm, Harriet Hjorth, Elin Wägner, Karin Berglund, Kerstin Ekman—not least works by Ekman such as "The Masters of the Forest" (*Herrarna i skogen*), "See the Flower" (*Se blomman*, with Gunnar Eriksson), and "Gubba's Meadow" (*Gubbas hage*). But we are not just talking about being a keen gardener or potted plant enthusiast: here, we trace his influence in great literature, in essays, nature writing, and, of course, poetry.

Compulsory plant collection was introduced into the senior school by the new regulations of 1856, in the old humanities curriculum as well as the new science-based one; extra points were awarded for rare plants. The emphasis on teaching botany did not go down well with everyone. Frans G. Bengtsson railed: "Damn the lot of them, most of all Linné! was

FIGURE 55. Statue of Linnaeus by Fritjof Kjellberg in Humlegården, Stockholm. Chicago has a similar statue. The four female figures on the base symbolize respectively botany, zoology, mineralogy, and medicine. Uppland Museum.

what I thought. There is no telling how much damage that man has caused with his extermination of plants. In Sweden, not a single herb is left in peace because of the idiotic botanizing tradition he started."

Linnaeus's approach to botany was embraced by the Swedish school system, not only as an element in teaching nature studies or biology but also as a model for instilling orderly, disciplined thinking in all forms of learning, like models such as Latin conjugations and vocabulary lists. It also taught opportunism: how to get away with using collections made by older siblings.

The celebrations over the years offer insights into the changes in his image, and Henrik Björck has studied precisely that: "In 1807 Linné was above all a Smålander and, to a degree, a Swede. By 1907 he had become Swedish and only secondarily from Småland. . . . By 2007 he was all things to all men: he belonged to Småland and to Sweden, was a scientist but also much more—an international celebrity, the first ecologist, a visionary, a universal genius able to work across scientific disciplines, an experimentalist with a strongly results-oriented ethos. He seemed like a shape-changing cartoon figure—or an ink blot, interpreted almost at will at different times."[16]

It is rare that an individual can be held solely responsible for a tradition or a school of thought. We must try to establish a sensible relationship with the broad historical narrative as well as event by event. The best way is to meet the human being, flawed as well as creative. Linnaeus is charming, but we have to accept his other sides as part of the bargain; he also offers us so much of the story of his life. He showed that individuals, to varying degrees, can affect history, including the present. No one can escape the past, not even if you are a Linnaeus. The best we can do is get to know him, look at his words, and inquire into his views on the state of nature and other things.

ABBREVIATIONS

Bref Letters and other writings (*Bref och skrifvelser*)
CFM C. Fabricius, manuscripts
DN *Diaeta Naturalis* 1733
FaS *Fauna Svecica*
FlS *Flora Svecica*
FB *Fundamenta Botanica*
Fries Thore M. Fries, *Linné* 1–2
ID *Iter Dalecarlicum*
IL *Iter Lapponicum*
KB The Royal Library (Kungliga Biblioteket)
KI Karolinska Institute (Karolinska Institutet)
KIB Karolinska Institute University Library (Karolinska Institutet Universitets Bibliotek)
KVAH Acta, Royal Academy of Sciences (*Kungliga Vetenskapsakademiens Handlingar*)
LC Linnean Correspondence
LK *A Linnaean Kaleidoscope* 1–2
L Linneaus
LN *Lachesis Naturalis*
LS Linnean Society of London
LT Linnaeus in "Learned Messages" (*Lärda tidningar*)
LUB Lund University Library (Lund Universitets Bibliotek)
ND *Nemesis Divina*
SKVA Writings by Carl von Linné published by the Royal Academy of Sciences (*Skrifter af Carl von Linné utgifna af Kungl. Vetenskapsakademien*) I–V, 1905–13
SLÅ Annual Report by the Swedish Linnaean Society (*Svenska Linnésällskapets Årsskrift*)
SLS Swedish Linnaeus Society (Svenska Linnésällskapet)
SN *Systema Naturae*
UUÅ Umeå University (Universitetet Umeå)
UUB Uppsala University Library (Uppsala Universitet Bibliotek)
UU ILH Uppsala University Linnaean Institute Collections (Linnéska institutets handlingar)
VA Selected dissertations
VAS Selected dissertations and other publications (*Valda avhandlingar och skrifter*)
Vita *Caroli Linnaei* I–V

NOTES

A reference apparatus such as the present one is a distillation of many years of textual study; it follows that it could become almost endlessly long. I have imposed necessary limitations, in contented awareness that there are excellent Linnaeus bibliographies available for readers with special interests.

Here, I had originally planned to include references to Linnaeus's own work and other primary sources, but the list has been expanded to cover other relevant literature, while not always giving page numbers—the simple, if sad reason being that to investigate all these books before listing the endnotes would take almost as long as writing this biography in the first place.

Preface

1. United Nations, Intergovernmental Science-Policy Platform on Biodiversity and Ecosystem Services (IPBES)

Who Was He?

1. Otto Sylwan in *Svenska litteraturens historia* (History of Swedish literature), 1919, 352.
2. *Vita*, 86.
3. *Vita*, 144ff.
4. Beckmann, *Tagebuch*.
5. *Vita*, 24.
6. Linné to C. F. Mennander 26.1.1762.
7. *Bref* (Letters), I:5, 191.
8. Von Hofsten, SLÅ, 1948, 98. Cf. Vila 2018, regarding Enlightenment France and the illnesses of intellectuals.
9. In Johann Beckmann, *Tagebuch*, 1911. See the medical journal *Läkartidningen*, 2013, 110.
10. *Anteckningar* (Notes), 1823, 90f.; cf. *Vita*, 158ff.
11. Malmeström, 1974, 358.
12. *Bref* (Letters), I:6, 206.
13. LT, 166.
14. *Vita*, 150ff.
15. Lindroth, SLÅ, 1979/81, 156 (orig. DN 12.5 1957).
16. *Classes plantarum, praefatio*.

Part I. A Great Man Can Come from a Small House, 1707–1741

1. *Critica Botanica* (transl.)—Much attention has been paid to Linnaeus's childhood by his biographers; Fries devotes more space to it than to other stages in his subject's life. The home neighborhood is described in many studies by Assar M. Lindberg; Ingrid Wallerström has made the boy Carl, his mother, and the local landscape and villages the focus of three books.

Chapter One. The Guardian Tree

1. *Vita*, 145.
2. *Vita*, 90.
3. See Per Martin Jörgensen in SLÅ, 2004–5; about Linnaeus's family—the ancestral Ambjörn—most recently in *Släkt och hävd* (Family and Tradition), 2017:3.
4. *Vita*, 155.
5. Samuel Linnaeus to Jonas Hallenberg, 1778. See SLÅ, 1931.
6. LN, 72.
7. Quotation from Ovid's *Epistulae ex Ponto* III.
8. Rogberg, 1772 (introduction).
9. Foreword comments on Stenbrohult, from Gunnar Eriksson's history of Swedish botany (on Wahlenberg).
10. Linné to Roberg, 1729.
11. See Broberg, SLÅ, 1972–74.
12. *Vita*, 91.
13. *Vita*, 59.
14. SLÅ, 1931, 1.
15. *Metamorphosis humana*, 1767.
16. LN, 48.
17. DN, 168.
18. *Iter ad exteros*, in *Iter Dalecarlicum*.
19. *Vita*.
20. *Vita*, 96.
21. Broberg, SLÅ, 1972–74; K. Rob Wikman, *Lachesis and Nemesis*.
22. LT, 62.
23. LN, 40.
24. LN, 46f.
25. LN, 150.
26. In *Carl von Linné* by Oscar Levertin, 1906, 28; cf. Hagberg, 1951.

Chapter Two. Studies in Växjö and Lund

1. *Vita*, 41.
2. *Vita*, 59f.
3. *Svenska arbeten* (Swedish work), 1880, 1:2–6.
4. *Vandrande scholares* (Wandering *Scholares*) by Gustaf Sivgård.

5. M. B. Swederus, *Pedagogisk Tidskrift*, 1880.
6. SLÅ, 1960, 96. The herbal has been published twice: in 1957 by Telemak Fredbärj; in 2009 by Torbjörn Lindell (facsimile and a translation into English).
7. DN, 175.
8. *Vita*, 92.
9. Samuel Linnaeus, SLÅ, 1931, 119.
10. Broberg, SLÅ, 1972–74.
11. LS, Linné Jr., Manuscript of *Diaeta*, 164f.
12. *Vita*, 55.
13. *Vita*, 156.
14. *Vita*, 34; cf. also Westling, 2007.
15. *Manuscripta medica* 1:173.
16. SN, 10th ed.; Lindroth, 1967, Part I:713ff. See also Linnaeus to Wargentin, 1760, in *Bref* (Letters), I:2, 229f., 236.

Chapter Three. The Academy in Uppsala

1. Mats Westerberg, *Lefvernes beskrifning*, 1973, 19.
2. *Bref* (Letters), I:5, 48f.
3. *Vita*, 156.
4. *Bref* (Letters), I:1, 110f.
5. SLÅ, 1953.
6. Linnaeus to Stobaeus, 25.3.1730, in *Svenska arbeten* (Swedish work), 1880, 2:22.
7. Linnaeus to Stobaeus, 8.11.1728, in *Svenska arbeten* (Swedish work), 1880, 2:12.
8. DN, 202.
9. Linnaeus names Rudbeckia. After Afzelius, 138ff.
10. Copied in *Bidrag till en Lefnadsteckning öfver Carl von Linné* (Contribution to the life of Carl von Linné), II:107ff.
11. DN, 179; see scholion 114.
12. ND, 176. For Greta Benzelia story, see Schück (1917); also *Gyllene äpplen* (Golden apples), 2:1013ff.
13. *Praeludia sponsaliorum plantarum*, in SKVA, IV.
14. SLÅ, 1931, 157.
15. *Acta Litteraria Upsaliensis*, 1736.
16. Hesselius, in Nils Jacobsson, SLÅ, 1938.
17. Wallin, in Fries notebook X:69. Cf. *Georg Wallin* by Tor Andrae. Stockholm 1936.
18. After B. Hildebrand, 1939, 205.
19. *Bröloppsdiktning* (Wedding poems) by Stina Hansson, 2014.
20. UUÅ, 1907.
21. Cf. below for source regarding Reaumur's duck-rabbit.
22. Linnaeus to Stobaeus, in *Svenska arbeten* (Swedish work), 1:2.
23. Ibid.
24. *Carl von Linnés ungdomsskrifter* (Youthful writings by Carl von Linné), 1.
25. SLÅ, 1964:13.
26. On Linnaeus's library, see Grape, *Caroli Linnaei Bibliotheca medica*. Publ. by T. Fredbärj, VA, 223. On Linnaeus as a Rosicrucian, see A. J. Cain, *Linnean*, 1992.

27. Uggla V:27, in LS.
28. *Speech*, 1779, 15.
29. *Vita*, 98.
30. UUB, W35: 17.

Chapter Four. In a Mythical Landscape: Lapland

An edition of the travelogue about the journey in Lapland was first published as a translation into English; the title was *Lachesis Lapponica* (J. E. Smith, 1811). Fries prepared the most important edition (1913), with its many appendices. The Skyttean Foundation's edition in three volumes was followed up by a volume of essays entitled "So why does Linné travel?" (*Så varför reser Linné?*). This section is mainly based on the Fries edition. Linnaeus's "Account" (*Relation*)—with his expenses claim—submitted to the Society of Sciences is printed in "Letters" (*Bref*) I, 1:334ff.

1. DN, 215.
2. Linnaeus to Lars Roberg, dated Svappavari g.s. (in Lapland) 14.8.1732.
3. IL, 1913, 5.
4. Ibid.
5. In print by K. B Wiklund, SLÅ, 1925: 92.
6. *Flora Lapponica*, translated by Th. M. Fries, 353. Cf. *Kollegieanteckningar* (Student notes), 45.
7. *Bref* (Letters), I:5, 250.
8. Comment, 362.
9. Fries, 1903, 1:108.
10. *Vita*, 101 ("fatigue"). See Knut Hagberg, *Carl Linnaeus* (1958), 60.
11. *Så varför reser Linné?* (So why does Linné travel?).
12. DN, 192f.
13. IL, 52f.
14. Cf. IL, 52f. See also an account of the *furia* in *Relation* (Account), 349. When Linnaeus approached the ferry station at Älvkarleby to cross the Dal River, he is reminded of Rudbeck's version of the Charon myth (*Karon*, 10): Linnaeus was traveling in a mythical landscape.
15. IL, 64f. "Fåglarnas budskap" (The message of the birds) by Gunnar Eriksson, in *Historiens vingslag* (The wing-beat of history), Stockholm, 1988. Also in *Flora Lapponica*, § 163 (on Andromeda); on *Sceptrum Carolinum*, § 143.
16. *Kung Karls spira* (The moor-king—Pedicularis sceptrum-carolinum), printed in SKVA, IV:243–59, 1908.
17. *Flora Lapponica* (Andromeda 10.6 resp. 4.7). On color blindness, see Nelson, 2010.
18. IL, 97. Cf. *Relation* (Account), 350f., in DN, 48: "When I reached the high hillsides it was as if given new life by the clean winds, when I went down it seemed to me that I had lost something, but then, back up there, I was enlivened."
19. *Vita*, 66 (clouds), resp. 103 (in Falun).
20. IL, 93.
21. DN, 165.
22. DN, 167.
23. Carl-Otto von Sydow, in SLÅ, 1972 and 1974, 58 and 48.

24. See Sven Widmalm, 1990, chap. 4.

25. *Flora Lapponica*, 278.

26. IL, 59 (take a piss); 44 (farmed countryside); 49 (flour and milk).

27. IL, 119f.

28. Fries, 1:137ff. and 160f.

29. SLÅ, 1962, 19 ("all three realms"); *Bref* (Letters), I:6, 240ff. (Gyllengrip).

30. The Academy of Sciences' adherence to an academic, Latin-based ideal approach is contrasted with the new ideas. Cf. Tessin's diary (*Dagbok* 6.7.1757), publ. 1824. Also, Linnaeus to Gyllengrip, 1.10.1733.

31. The planting theme is explained in *Linné och växtodlingen* (Linné and plant cultivation) by M. B. Swederus. The Gothic understanding of climates is illustrated in Linnaeus's dedication to Renmarck (ed. P. Ekerman of *De praestantia orbissviogothici*, 1747) and in the dissertation *Morbis ex hyeme* (Diseases in the winter), 1752. Cf. *Linné och vintern* (Linné and the winter) by T. Fredbärj, in SLÅ.

32. *Berättelse om En till Lappland och Norrsiöen giord resa, under sommar månaderne 1736* (The story about a journey made to Lapland and the Northern Sea during the months of summer) by Jonas Meldercreutz, publ. by Jouko Vahtola, Rovaniemi, 1984, 28ff. He listed his writings in a letter to Gyllengrip: "mss af mig propria Minerva elaborerade" (mss. by me propria, Minerva elaborated), among others, *Lachesis Lapponica* "in Swedish."

33. Clearly, issues of power balance come into this; the contrast between older-style heroics and this more skeptical approach to history is worthy of analysis in terms of ideological changes. On the subject of large-scale attempts to acclimatize, see *Nature, the Exotic, and the Science of French Colonialism* by Michael A. Osborne, Bloomington, IN, 1994.

34. Koerner, 1999, 72; Gibson, 2015, 80.

35. K. B. Wiklund: "Linné at no point states into which of these four groups [of human variants, according to *Systema Naturae*] the Lapps should be counted and, at the time of his travels in Lapland, he had evidently not arrived at any opinion about the extent to which their characteristics were determined by race but rather assumed that they were related to their lifestyle." In SLÅ, 1925, 77.

Chapter Five. Diaeta Naturalis

1. Linné to Cronhielm, ca. 1733, *Bref* (Letters), I:5, 317.

2. DN, 17 (from To the Reader).

3. Willa apinian.

4. "Medical primitivism" is meant to imply a stress on a healthy lifestyle with plenty of greens and exercise, and a skeptical attitude toward "modernity"—and so on.

5. Several of these themes are carefully exemplified by von Sydow. On Swedish eighteenth-century (notably its late half) ideas about the climate, see *Klimat och karaktär* (Climate and character) by Carl Frängsmyr.

6. *Flora Lapponica*, 229.

7. DN, 128. Cf. similar statements, e.g., "about a temporary bed in the wilderness," in *De lecto in desertis extemporaneo*, one of the essays submitted to the Society of Sciences. Printed in *Bref* (Letters), I:1.

8. DN, 56.

9. DN, 59.
10. DN, 131, scholion 81.
11. DN, 134.
12. DN, scholion 83.
13. See, e.g., DN 46, 48, 56f., 61, 74, 78, 80, 82, 85, 89, 91.
14. *Morbi mentales*, 117.
15. DN, Food for the first human beings. Similar comments in *Kollegieanteckningar* (Student's notes), 1907, 168 and 198ff.
16. DN, 35.
17. DN, 35, 123, 130, 132, 133.
18. DN, 190, 199.
19. DN, 59f.
20. DN, 122. On Linnaeus's "Sami persona," see Koerner, 64ff.
21. DN, 206.
22. SLÅ, 1931, 157.

Chapter Six. In the Mountains and under the Ground: County Dalarna

The Journey in Dalarna (*Dalaresan*) was first published by Ewald Ährlin. The second version (1953), was edited by A. Hj. Uggla, who added *Iter ad foedinas* and exhaustive comments; in 2007, Roger Jacobsson edited and added new comments in a third edition.

1. Hedin, *Minne af Linné fader och son* (Memories of Linné, father and son), 43.
2. Mennander, letter 94.
3. *Bref* (Letters), I:5, 313f.
4. *Vita*, 67f. (the conflict with Rosén).
5. According to C. Forsstrand. SLÅ, 1919, 50.
6. Copy of ID.
7. ID, 103.
8. ID, 147.
9. ID, 89f.
10. ID, 122.
11. ID, 146.
12. ID, 148.
13. ID, 25f.
14. ID, 67.
15. UUB Mss. N 971, 39a.
16. Ibid. 971, 39.
17. UUB N 551, 145a.
18. *Den Swenske patrioten*, 1735:8.
19. *Den Swenske patrioten*, 1735:16.
20. *Vita*, 194f.
21. *Iter ad exteros*, 313.
22. Conflict about Sara Lisa.

23. *Vita*, 194f.
24. Cf. Fries, 1:272.
25. Linnaeus to Mennander, 1738.
26. *Bref* (Letters), I:5, 313ff.
27. ID, XIIIf. Foreword to *Flora Lapponica* (1737).
28. C. O. von Sydow in SLÅ, 1962, 18. This work is not noted in the list of qualifications in Linnaeus's letter to Gabriel Gyllengrip 1.10 (g.s.), 1733. K. B. Wiklund listed additional texts linked to the journey. SLÅ, 1925, 62ff. Uggla refers to Acrel in *Försvunna Linné-handskrifter* (Mss. by Linné that have disappeared). SLÅ, 1952, 83. The entire title, according to S. A. Hedin (1808, 96): *Iter ad superos et inferos in campiis elysiis et aula subterranea ubi Regio Hyperboreorum et Regnum Plutonis cum descriptione beatorum et infelicium susceptum Anno 1732 & 1734*. Hedin comments: "In which He speaks frankly of his Lapland travels and his stay at Fahlu [where] the Mineworkers' wretched state was described in a Roman manner and a style that was all his own." It would seem that Acrel and Hedin had read the text.
29. In *Svenska arbeten* (Swedish work), 1:2.
30. These suggestions are easily seen in *Gyllene äpplen* (Golden apples), I:560ff.
31. *Vita*, 84.

Chapter Seven. In the Land of Tulips

1. *Iter ad exteros*, 318.
2. ID, 321.
3. *Iter ad exteros*, 328.
4. Ibid.
5. About the hydra, see Uggla, 1946; Lars Forsberg, 2004.
6. *Vita*, 104.
7. Printed in 1767; see G. Broberg and U. Marken, 1990.
8. Linnaeus yoiked in Dalarna. Cf. diary notes, publ. by Uggla, SLÅ, 1935.
9. See *Kort beskrivning av Lappland* (A brief description of Lapland) by Nicolaus Örn, publ. by Leif Lindin, Sture Packalén, and Ingvar Svanberg, Umeå, 1982; also Svanberg's additional notes on Örn's biography, in *Rig*, 1984:9ff. Linnaeus refers dismissively to *Reise Beschreibung von Lappland und Botnien* (Travelogue from Lapland and Bothnia) by Joh. G. Schellern, Jena, 1727; cf. Wiklund, 91, for judgment on Schefferus's work.
10. *Iter ad exteros*, 333; examples used by Broberg in *The Swedes*, ed. Kristian Gerner, 2014.
11. DN, 47.
12. DN, 193.
13. DN, 199.
14. DN, 192.
15. *Abraham Bäck's Diary*, 1742.
16. DN, 172, 178.
17. LN, 129.
18. LS III.6., Students' notes, 231.

19. LS, 232.
20. DN, 50 (smell).
21. DN, 85.
22. DN, 100f. (tobacco).
23. DN, 101.
24. DN, 102.
25. DN, 127 (smoking).
26. Tersmeden, 2; Lindeboom, 1968; Broberg 1975, 257ff.
27. Linnaeus's version. *Vita*, 70, 107-11.
28. UUB Mss. N 971.
29. Summary of text after translation by Heller, *Taxon* 17 (Dec. 1968).
30. N 971, 308b.
31. *Vita*, 84.
32. Tersmeden, 1914, 2: 39f., 137f.; the date seems strange.
33. LN, 137.
34. See Johan Nordström, SLÅ, 1954-55.
35. Travel diary, 1745. KIB Ms. 40:2.
36. *Vita*, 106.
37. *Bref* (Letters), Nov. and Dec. 1736, I:5, 255f., LUB.
38. *Vita*, 110ff.
39. KB C XI.1.9, 117f.
40. Linné Jr. remembers Paris.

Chapter Eight. Nature's Order 1

1. UUB Ms. N 971.176v, 12.12.1735.
2. KB Mennander, Hagströmer library.
3. Nordenskiöld, *Lychnos*, 1940, 5f.
4. A. J. Cain in *Archives of Natural History* 19 (1992).
5. Edward Heron-Allen, *Barnacles in Nature and Myth* (London, 1928).
6. *Guds werk och hwila* (God's work and rest), Stockholm, 1705, 191ff. and index. Cf. Spegel on paradoxes, also in *Spegels Guds werk och hwila* by Bernt Olsson, Lund, 1963, 364ff.
7. See *Järven—filfrassen-frossaren* (The wolverine—the glutton or quickhatch) by G. Broberg, *Lychnos*, 1971-72, 181-216. Quotes from *Föreläsningar öfver djurriket* (Lectures on the animal kingdom), Stockholm, 1913, 27 and 35.
8. *Pan Europaeus* and *Manuscripta medica*, Linnean Society, London. DN 1733: 201. UUB Westin, 35-39: corresponding Mss. KB (manus C. F. Mennander).
9. IL, 1913, 99f.
10. KVAH, 1740.
11. Re the printing of SN, see "Linné and Gronovius" by Johan Nordström, SLÅ, 154-55. *Anonymous Linné och den sjuhövdade hydran* (The anonymous Linné and the seven-headed hydra), publ. by A. Hj. Uggla, Uppsala, 2004.
12. G. Broberg, "The Dragon Slayer," *Tijdschrift voor Skandinavisiek*, 2008, 29:1-2.
13. Lecture on the animal kingdom, 1748.
14. *Vita*, 84.
15. Uggla, 1956, 221.

Chapter Nine. A Stockholm Interlude

1. List published by Uggla, SLÅ, 1953, 127.
2. "an elderly, not very large man with dusty shoes and stockings, a long beard and an old green coat on which was pinned the order of the Polar Star" (*einen etwas etas bejahrten, nicht grossen Mann, mit bestauten Schuen und Strümpfen, langem Barte und einem alten grünen Rocke, worauf das Ordenszeichen hieng*). Beckmann, *Tagebuch*, 48.
3. Beckmann, *Tagebuch*, 130.
4. *Vita* 113f., cf. 73.
5. See Carl Forstrand, 1915.
6. Linnaeus to Sauvages, *Lettres inédites*, Alais, 1860.
7. *Stockholms Post-Tidningar*, 1739, 40 (21.5).
8. Introduction to appendix IX, Fries.
9. Linnaeus to KVA, Summer 1761 (I:2, 243ff.). Cf. Bengt Hildebrand, 1939, 226; also Sten Lindroth,1967.
10. KVA protocols for the years 1739, 1740, 1741, publ. by E. W. Dahlgren, Uppsala, 1918.
11. Ibid.
12. Quote from Nordström, *Lychnos*, 1940, 429.
13. 26.11.1740. KVA protocols for 1739, 1740, and 1741, publ. by E. W. Dahlgren, Uppsala, 1918.
14. KVA protocols for 1739, 1740, and 1741, publ. by E. W. Dahlgren, Uppsala,1918.
15. Almanac, after Fries, 2:95.
16. *Wetenskapsakademiens Handlingar* (Acta of the Academy of Sciences) 1741, § 5, § 12, § 13, § 16, § 20, § 28.
17. "Tal om märkwärdigheter uti insecter" (Talk: On notable features among the insects). Held in October 1739, in *Auditorio illustri*, Riddarhuset (*House of the Nobility*, Stockholm). Published (probably) by Royal Academy of Sciences (Kungliga vetenskapsakademin).
18. *Vita*, 116.
19. The appointment issue is extensively discussed by Fries and Annerstedt. *Orbis*, publ. in transl., VA, 1952, 12.

Part II. At the Height of the Ages of Man, 1741–1758

1. DN, 93.
2. *Vita*, 129f.

Chapter Ten. Uppsala and Enlightenment

1. *Bref* (Letters), I:4, 151. Seminar essay by Margareta Eriksson, on UUB S 139a, at UU ILH (around 1988). For interesting gossip about Uppsala academics, see Olle Bergquist, 1965–66.
2. History of UU by Lindroth.
3. Celsius, see Lindroth in *Lychnos*, 1956–57.
4. Linné Jr. in *Blomsteruret* (The flower clock). For the theme "Linnaeus in Uppsala," see the books by Bertil Gullander (1978) and Helena Harnesk (2006).
5. Gjörwell.

Chapter Eleven. Three Programmatic Speeches

1. About necessity, *Skrifter* (Writings), 2.
2. Uggla, SLÅ, 1941.
3. Journey of exploration, *Skrifter* (Writings), 2.
4. Speech on habitability, etc., *Skrifter* (Writings), 2.
5. *Botaniska utflykter* (Botanical excursions) by Elias Fries, 1843, 1:185.
6. Thore M. Fries, comment, *Skrifter* (Writings), 2.
7. Tore Frängsmyr, 1969, chap. 4.
8. *Fundamenta Botanica*, § 132.
9. *Bref* (Letters), I:4, 109.

Chapter Twelve. Provincial Travels on Behalf of Parliament

1. Cf. Uggla, SLÅ, 1932.
2. Heckscher, SLÅ, 1942.
3. Protocol, 24 April.
4. At Lidköping.
5. 28.6.1746.
6. Fries 1:appendix XIV.
7. On Linnaeus as an ethnobiologist, see Svanberg, 2005. On "botany for females," see Broberg, 1990–91.
8. The protocols are in the Linnean Society, London. Texts used here are published by Bertil Gullander and Sigurd Fries.
9. *Bref* (Letters), I:2, 98; Nina Sjöberg, *Linné och riksdagen*.
10. Utg. 1884, nr 806, April 1744.
11. Kalm to S. C. Bielke, 12.11.1745, in *Brev*, 1960, 105.
12. *Bref* (Letters), 1.4.1747.
13. *Bref* (Letters), 160, 4.11.1747.
14. Drover's rattle (Sw. *Märrskramla*).
15. LT, 31f.
16. Hallenberg, in *Historisk tidskrift för Skåneland* (The history of Skåne Bulletin), 1913. See Jan Dahlin and Annika Tergius, *Gröna arkiv*, 2007.
17. Foreword, *Wästgöta resa*.
18. On the number of local species, see *I Linnés hjulspår runt Skåne* (Following Linné's wheel tracks around Skåne) by Sven Snogerup and Matz Jörgensen, 1997.
19. "To slaughter the people." Erik Rydbeck's speech on the usefulness of cultivation, 1762. Transl. in Swederus, 1907.

Chapter Thirteen. A Language in Which Everything Matters

1. *Vita*, 113. Linné's "wallet" from the Lapland journey has been kept, with his personally drafted Finnish phrasebook inside; it included "Where is the coaching inn?"
2. SLÅ, 1947, 72.
3. Cf. Heller in *Taxon*, Dec. 1968, 679ff.

4. *Bref* (Letters), I:1, 212.
5. Malmeström, 1925, 92.
6. Margit Abenius, SLÅ, 1935.
7. *Wästgöta resa*, 1747, 275.
8. Bäck's speech in memory of Linnaeus.
9. Linné to Laxman, 12.3.1764, *Bref* (Letters), I:8, 180.
10. Ährling, 2, 95.
11. *Skrifter* (Writings), 4, 43f.; see *Sponsalia plantarum* (not previously printed).
12. Ibid.
13. For more examples, see Broberg, 2008.

Chapter Fourteen. Flora et Fauna Svecica

1. Kalm to Bielke, 4.6.1745, *Bref* (Letters), 1960, 89.
2. Ibid.
3. L 3134, 10.10.1762, see Eriksson (1986) on the concept "flora"; Erdtman (1945) for an overview of the flora in some Swedish regions.
4. *Vita*, 138, 53.
5. LT, 300.
6. Transl. A-M. Jönsson.

Chapter Fifteen. Family Life 1: Scenes from a Marriage

1. SLÅ, 1932.
2. *Vita*, 117f.
3. On restoration, see Viktor Edman, SLÅ, 2008.
4. Beckmann on Linné Jr., *Tagebuch*, 122f.
5. Tullberg, 1878, 62.
6. SLÅ, 1931, 122.
7. Fries, 2:224. Fries, notebook IV: 133ff. See UUB N 947. Voltemat's anecdotes, 151ff.
8. Death of Johannes: D. Solander to E. G. Lidbeck 25.3.1757; Linné to Bäck 11.11.1757. See *Bref* (Letters), I:5, 34.
9. General comments on "apparent death," see Broberg, 2008.
10. On the local "Disting" market, to Bäck (Feb. 1754).

Chapter Sixteen. In the Garden, at Herbations, among the Collections

1. On thermometers: see Nordenmarck's summary (1936 and 1937); see also previous discussion by Fries in NT 1897, 658ff.
2. Eric Sefaström in *Swenska samlingar* (Swedish collections); J. A. Lindblom in *Anteckningar* (Notes), 1925, 37.
3. Annerstedt, 3:1, 147. Linné's proposals, etc., see *Bref* (Letters), I:1, e.g., 149ff., 160ff., 196ff., and 213ff.
4. 1.5.1750, July 1753.
5. Transl. Fries, 1:96ff.

6. Linnaeus to KVA 28.9.1756, *Bref* (Letters), I:2, 206.
7. ND, 192.
8. *Bref* (Letters), I:1, 196ff.
9. *Bref* (Letters), I:3, 107.
10. *Skånska resa*, 234f.
11. L 3579, 9.4.1765.
12. Linnaeus to Bäck, 13.7.1753.
13. 1913, 310.
14. On the elephant and sperm whale, see Widmalm, 2018. For authoritative work on Swedish *naturalia* collections, see the thesis by Yngve Löwegren; other sources include the works by Lars Wallin and Torleif Ingelöv.
15. *Bref* (Letters), I:4, 249.
16. UUÅ, 1916, 11.
17. *Vita*, 120.
18. Ehrström, SLÅ, 1946, 69f.
19. Acrel, 1796.
20. Lindblom, 1925, 37.
21. Ödmann, 1925, 1:286f.
22. Melanderhielm, 1783; see Annerstedt on the history of Uppsala University, app. V, 1913, 137f.
23. *Bref* (Letters), I:7, 138f.

Chapter Seventeen. Ex Cathedra

1. Linné to the university *consistorium*, October 1768, *Bref* (Letters), I:1, 224. A "declaration concerning the state department of education" gives many details of academic teaching at the time. See LT, 173.
2. Useful to the state, see the chapter "Entrepreneur and Economist."
3. Useful to the student, see Uggla, 1940.
4. *Bref* (Letters), I:4, 210.
5. Acrel, 1796, according to Hildebrand, 1939, 298f.
6. *Beskrifning öfwer stenriket* (A description of the mineral realm), publ. by Carl Benedicks, UUÅ, 1907.
7. Foreword to *Fundamenta Botanica*, publ. by Lars Bergqvist, after Osbeck, 1748, 55.
8. Discussing the animal kingdom, Linnaeus says, "That a human being is an animal cannot be denied by anyone, as she has a heart." The feminine pronoun may reflect the premodern feminine gender of the word for "human being," which is neutral in modern Swedish, or his old-fashioned Småland dialect.—Trans.
9. Lectures, 1913, 9f.

Chapter Eighteen. What Is More Precious Than Life, More Pleasing Than Health?

1. Erik Eurén, *Dagbok 1732–1762* (Diary), 1927, 24f.
2. S. A. Hedin, 1808.

3. *Lachesis Naturalis*; also in students' notes from about 1748, publ. by A. O. Lindfors, UUÅ, 1907. A later series was published in 2009 by Broberg under the title *Notata subitanea*, 1772.
4. *Kollegieant* (Students' notes).
5. DN, 76 T.
6. *Bref* (Letters), I:3.
7. DN, 99.
8. Malmeström, 1926, 115; LN 88.
9. *Vita*, 144f.
10. See Nils Uddenberg, *Linné och mentalsjukdomarna* (Linnaeus and mental illnesses); Fredrik Berg, *Linnés Systema morborum*. On vegetarianism, see Colin Spencer, *The Heretic's Feast*.

Chapter Nineteen. Academic Amusements

1. Bo Lindberg, introduction to Bergquist and Nynäs, *Linnaean Kaleidoscope*.
2. Fredbärj, SLÅ, 1966.
3. SLÅ, 1940, 14.
4. Annerstedt, 3:2, 240f.
5. Even Krok (1925) thinks the respondents (at a disputation) carry most of the responsibility.
6. Acrel, Speech, 1796.
7. In the series "Selected dissertations," 84 have been published.
8. LT, 82.

Chapter Twenty. Appetite for Work, Weariness, Communication

1. T. Tullberg, Family tree, from Fries, 2:247.
2. Linné to Bäck, 3.4.1761.
3. Linné to Jacquin.
4. SLÅ, 1956–57, 162.
5. Banks, on David Solander.
6. *Nattens historia* (The history of the night) by G. Broberg, 125ff.
7. Cf. Staffan Müller-Wille and Isabelle Charmantier.
8. See Spencer Savage, 1940.
9. *Vita*, 79.
10. *Bref* (Letters), I:7, 138f.
11. *Bref* (Letters), I:7, 145.
12. Cf. Malmeström's thesis, 90ff., and biography.
13. *Bref* (Letters), I:4, 115.
14. *Bref* (Letters), I:4, 133.
15. *Bref* (Letters), I:5, 41.
16. Eva Nyström, 2007.
17. *Bref* (Letters), I:5, 270.
18. *Hallers Netz* (Haller's network), ed. Stuber, Hachler, and Lienhard, 2005.

19. Eva Nyström, 2007; Anfält, 2005.
20. *Bref* (Letters), I:6, 221f.
21. 134; *Bref* (Letters), I:3, 293, 332, 113.
22. Gronovius to Ph. Miller.
23. Linnaeus to Bäck, 3.4.1745, *Bref* (Letters), I:4, 55.
24. *Bref* (Letters), I:6, 271f.
25. J. O. Hagström to Linné, 1.1.1767, *Bref* (Letters), I:6, 271f.
26. Quoted from Ährling, 1878, 103f.

Chapter Twenty-One. When Linnaeus Wrote, Salvius Printed, and Tessin Bought the Books

1. Schück, *The History of Publishers' Bookshops in Sweden*, 1923; cf. Bo Bennich Björkman in *Bokens vägar* (The ways of the book), 1998.
2. Linnaeus to KVA, *Bref* (Letters), I:2.
3. Schück, on Lars Salvius.
4. L 3580, 9.4.1765.
5. L 5889, 3.1.1769.
6. *Bref* (Letters), I:5, 204.
7. Unidentified letter.
8. *Incrementa botanices* (1753), LT, 182.
9. *Bref* (Letters), I:4, 217.
10. Blunt (Matisse), 56.
11. See Broberg (1990) and Isabelle Charmantier. Among other aspects of natural history, some of its delight derives from the illustrations. Again and again, Linnaeus complains that he could never compete with the splendid but expensive illustrated tomes. His choice was to produce "epistemic images" defined as "translating abstract epistemological priorities into concrete pictures. . . . Epistemic images make collective empiricism possible." Linnaeus uses this kind of pedagogic images, based on numbers, shape, position, and proportion in several contexts. Cf. Lorraine Daston.
12. L 3275, 22.7.1763; Sylwan, 1901.
13. L 3617, 15.7.1765.
14. LT, 183, 355, 359.
15. Adam Afzelius to C. P. Thunberg, London, 26.11.1798; UUB G 300a.
16. SLÅ, 1918, 158.
17. *Bref* (Letters), I:8, 158. Lagerberg's letters aren't worth publishing!
18. *Bref* (Letters), I:4, 194, to Ulla Tessin.
19. Linnaeus to Charlotte de la Gardie, 1.9.1762, *Bref* (Letters), I:5, 355f.
20. LT, 290; *Auctores botanici* (1759).

Chapter Twenty-Two. Linnaeus, "the Sexualist"

1. Desmond King-Hele, *Doctor of Revolution*, 1977, 112; Gibson, 82ff.
2. E. da Costa.
3. *Naturahistoriens studium vid Åbo universitet* (The study of natural history at Åbo University) by O. Hjelt, 1896, 60. See Stearn in *Notes and Records of the Royal*

Society 40, 1982; and *The Flamingo's Smile: Reflections in Natural History* by S. Jay Gould, 1985.
  4. DN, 111, 162, 169.
  5. DN, 111, 162.
  6. Mats Ola Ottosson, SLÅ, 2010.
  7. *Lychnos*, 1953, 246.
  8. J. Sahlgren, 1924.
  9. SLÅ, 1950–51, 167ff.; *Bokvännen*, 1952:5.
  10. GP, 3.6.1950.

Chapter Twenty-Three. Curiosity-Driven Research

1. To be curious—*curiosus*, possibly in the sense of being a little odd—may sound like the opposite of being a serious scholar, but, by now, curiosity-driven research is seen as praiseworthy. The word *curious* has a long, contrarian history. Augustine called it *concupscentia oculorum*, or "the eyes' pleasure," which concentrates our interest of the world around us and, by distracting us from the true contemplation of God, leads to both delight and pride. In Linnaeus's time, the word was eagerly debated. See "Curiosity killed the cat?" (a loose transl. of *Nyfiken i en strut*) by G. Broberg, including its references.
  2. *Bref* (Letters), I:7, 148.
  3. Early foreword to *Fundamenta Botanica*.
  4. Printed in *Skrifter* (Writings), 2.
  5. *Nödföda* (Food in a famine), 1756: Linnaeus wrote to the king on this matter; see *Bref* (Letters), I:1, 4ff. Svanberg uses "food in a famine" in several articles.

Chapter Twenty-Four. Nature and Culture

The subject is elusive, as the concepts have several different definitions. One theme that will not be discussed here—although it is in the chapter on the Lapland journey—is the impulse to see landscapes in classical terms and identify aspects of it with classical imagery. Such euhemerism is seen in the names for apes: pan, sphinx, satyrus, diana, faunus, hamadradyas, oedipus, and midas. Molluscs are named doris, thetys, nereis, and the genus *Papilio* (butterflies) is named consistently after, for example, heroes in the Trojan war, the nine Muses, and so on.

  1. On equilibrium, 1978.
  2. LT, 68ff.
  3. 293ff.
  4. Stauffer, 241.
  5. From "Tal om den beboeliga," 1743 (Speech on the habitable world).
  6. DN, 52.
  7. Cf. Carina Lidström, "Resenärer om fågelskjutningens moral" (Travelers on the morality of shooting birds). ND, ID (Orsa).
  8. Heller on euhemerism.

Chapter Twenty-Five. Entrepreneur and Economist

1. Lisbet Koerner's *Nature and Nation* focuses on Linnaeus as an economist and emphasizes Linnaeus's lack of understanding—or indeed knowledge—of Adam Smith's ideas of laissez faire markets. Linnaeus didn't regard trade as potentially leading to mutual gain and was, if anything, suspicious of traders. His views were based on the utility of using national resources, and he attempted to extend them by experiments in acclimatization. Koerner categorizes this as "Linnean cameralism."
2. Notes, in *The Journey in Dalarna*, 4 August.
3. *Bref* (Letters), I:1, 3ff.; I:1, 295ff.; Koerner, 131, ref. to Bäck in memory of L, 78.
4. Koerner's thorough analysis.
5. Karin Johannisson, *Det mätbara samhället* (To measure society), 1988; also Leif Runefelt, *Historisk Tidskrift* 2004:2.
6. C. G. Tessin, 1995, 992f.
7. *Bref* (Letters), I:2, 27.1742; cf. no date [1755]. *Bref* (Letters), I:2, 191f. on the deaths of reindeer and the consequences for the Sami. Earlier, to Gyllengrip 1.10 1733, *Svenska arbeten* (Swedish work), 1880, 2:85, item 14: "Vain would be any attempt to persuade in this matter either the Lapps, the newly arrived, or the farmers, as they will never abandon the habits of their ancestors."
8. Sjöberg, 2007, 48ff.
9. LT, 221f.
10. LT, 238.
11. SLÅ, 1956–57, 162ff.; LT, 71.
12. SLÅ, 1946, 70.

Chapter Twenty-Six. To Describe the World

1. Herman Richter, 1959; Sven Widmalm, 1988, 1990.
2. DN, 116, 60.
3. *Bref* (Letters), I:2, 73.
4. This, according to a protocol, 1760 (Koerner, 120; no identified source).
5. *Bref* (Letters), I:1, 232.
6. *Philosophia Botanica*, § 17.
7. Linné to Wahlbom, 8.12.1752, in *Anteckningar* (Notes), 1823.
8. *Kollegieanteckningar* (Students' notes), around 1750, 78f.
9. On Kalm's idea about acclimatization, etc., see Martti Kerkkonen, 1989.
10. *Bref* (Letters), I:11.
11. *Bref* (Letters), I:2, 105f.
12. Cf. Sörlin, Nyberg-Hodacs, and Leos Müller; on the East India Company, see Kjellberg, 1964, 196.
13. Hodacs et al., 2018, 12.
14. *Bref* (Letters), I:2, 29.4.1749.
15. Bäck, 1779, 45.
16. *Bref* (Letters), I:2, 180.

Chapter Twenty-Seven. Nature's Order 2

1. *Bref* (Letters), II:11, 1757.
2. *Philosophia Botanica*, § 132.
3. *Philosophia Botanica*, § 77.
4. See also Broberg, 1975, 74ff.
5. *Vita*, 171. Cf. Weinstock.
6. Hoquet, 2007.
7. Darwin, *Origin of Species*, ed. R. C. Stauffer, 354.
8. *Nature*, 9.9.1999.
9. *Bref* (Letters), I:4, 140.
10. Cf. Broberg, 1975, 66ff.
11. The "worm letter" to Linnaeus's mother-in-law, *Svenska arbeten* (Swedish work), 1880, 2.
12. *Philosophia Botanica*, § 79.
13. Speech on habitable land.
14. *Philosophia Botanica*, § 251.
15. *Vita*, 167.
16. *Bref* (Letters), I:4, 154.
17. On the reform of naming, see William T. Stearn, 1959; Koerner, 1999, chap. 2; and John Heller, *Huntia*, 1964:1. Bengt Jonsell writes that 1753 was probably one of Linnaeus's happiest years. Linnaeus (*Vita*, 167) penned the famous words: "*Trivialia nomina* were unheard of before; but he appended such to all plants. It was the same as placing the pendulum in the clock; through his doing this, botany sprang back to life as now the names could easily be remembered and spoken of and written as previously was done with definitions." On the concept "fauna," see Mats Rydén, 1992.
18. Schiebinger, L. "Why Mammals are called Mammals etc." *American Historical Review* 98:2 (1993).

Chapter Twenty-Eight. Homo sapiens

1. KB, X 508:2. Notably, the many pages the eminent philosopher G. Agamben devotes to Linnaeus's delight in monkeys are full of unacknowledged quotes from my thesis.
2. *Philosophia Botanica*, § 206; cf. Nickelssen, 2006, 3:23.
3. After Broberg, 1975; also Broberg and Christensen-Nugues, SLÅ, 2008. Cf. Hoquet on *Homo nocturnus*, 2007.
4. *Flora Lapponica*, transl. Fries (1907), 302.

Chapter Twenty-Nine. Honors

1. *Bref* (Letters), I:7, 179.
2. Stoever, 1792, 117–18.
3. *Bref* (Letters), I:4, 35.
4. *Vita*, 153f.

5. Tessin, "En gammal mans bref till en ung prins" (An old man's letter to a young prince), in *En gammal mans utvalda bref til en ung prins* [An old man's selected letters to a young prince], by Carl Gustaf Tessin, 1785.
6. Wallenberg, *Min son på galejan* (My son onboard the galley), 1781.
7. Unidentified source.
8. *Bref* (Letters), I:2, 265.
9. Gjörwell, letter, 2.12.1809; Brinkman archive 2, 369.

Chapter Thirty. Among Students and among Senior Academics

1. Linné to Bäck, 9.5.1766.
2. *Bref* (Letters), I:5, 10f.
3. *Bref* (Letters), II:362f., 13.
4. *Bref* (Letters), I:5, 157.
5. Bo Lindberg, *Den akademiska läxan* (The academic lesson), 2018.
6. *Bref* (Letters), I:4, 97ff.
7. *Bref* (Letters), I:4, 97ff.
8. *Nemesis divina*, 46f.; C. G. Spangenberg, 2007.
9. Fries 2:226f.

Chapter Thirty-One. Family Life 2: Hammarby

1. *Svenska arbeten* (Swedish work), 1:2. The entire letter forms a small autobiography. As for what he paid for Hammarby, different costs are quoted: *Vita*, 128, states 80,000 thlrs; here the sum is halved.
2. Karin Martinsson and Helena Backman, 2013.
3. Linné to Bäck 9.5.1766; Beckmann, *Tagebuch*, 96ff.
4. Sigurd Wallin, SLÅ, 1963, 52; G. Björnhag, 2009. Kelley Rourke has made music of the cows' names. SLÅ, 2013.
5. *Bref* (Letters), I:5, 185.
6. Linné to Bäck, 25.7.1766.
7. Fabricius's "nyckelharpa"; see Stoever, 1792.
8. Hildebrand, 1939, 301.
9. G. A. Reuterholm; cf. *Skånska resa*, 7.6, at Tungbyholm.
10. Fabricius, as in Stoever, 1792.
11. SLÅ, 1946, 71f.
12. LN, 47.
13. *Bref* (Letters), I:8, 118f.
14. On Adam Afzelius, see Olsson, 1949, 69.
15. *Bref* (Letters), I:2, 210.
16. Tullberg, 1919.
17. SLÅ, 1949, 69f.
18. Kerstin Ekman writes in *Gubbas hage* (Gubba's meadow), 35: "It was published in her father's name"—but that isn't right; nor is Thore Fries quite right when he comments in *Skrifter* (Writings), 2, 321: "One might however reasonably assume

that it is highly probable or, rather, certain, that the here cited essay in fact flowed from Linné's own pen."

Chapter Thirty-Two. Friends and Enemies

1. SLÅ, 1961, 42.
2. 38. L3757.
3. L 4228.
4. Erik Müller, SLÅ, 1921.
5. *Bref* (Letters), I:4, 324. "My little wife" is too free; the Latin is *uxor mea*.
6. *Bref* (Letters), I:5, 57.
7. *Bref* (Letters), I:5, 148ff.
8. After Fries, notebook IV, 68.
9. *Bref* (Letters), I:5, 165–70.
10. *Bref* (Letters), I:5, 344.
11. Ibid.
12. On "nightcap," see *Bref* (Letters), I:5, 28.
13. Ihre, 187, UUB.
14. *Vita*, notes in his own hand, 61.
15. Fries, 2:145.
16. Fries, 2:142.
17. SLÅ, 1958, 139.
18. Linné to KVA, Nov. 1764, *Bref* (Letters), I:2, 268; letter 19.11.1764 in *Bref* (Letters), I:3, 142ff.
19. *Bref* (Letters), I:2, 229.
20. The examples show Wargentin's role as a diplomatic intermediary between Haller and Linnaeus.
21. Sigrid Lejonhufvud, Lovisa Ulrika, and Carl Gustaf Tessin, 1920, 62. Cf. *Bref* (Letters), I:4, 281f.
22. Notata sub.1771, 230.
23. *Bref* (Letters), I:4, 221f.
24. *Bref* (Letters), I:4, 24.
25. *Bref* (Letters), I:5, 94.

Chapter Thirty-Three. Problems

1. In the eighteenth century, a popular image of being as a great chain was sustained by the core idea of nature as "plenitude," a rich source. God was perfect and thus couldn't have left anything incomplete—hence, nature as an integral whole presupposes an infinite number of creative beings. See also Lovejoy's *Great Chain of Being*.
2. Broberg, "The Broken Circle," in *Quantifying Spirit* (1990).
3. On an addition to nature's kingdoms, see *Bref* (Letters), I:2, 289.
4. Usually, this is associated with the African springhare (*Yerbua capensis*) because it runs on its hindlegs, and not to the American opossum (*Opossum didelphis*)—Kalm had brought one of these home to Åbo; it ate his slippers. Cf. Shillito, 1978.
5. *Philosophia Botanica*, § 77.

6. See Uddenberg, 2007; more recently M. Lidén and J. Kårehed in "Linné och det naturliga systemet" (Linné and the natural system), *Svensk botanisk tidskrift*, 2018, 111:2. Even more recent: Kees van Putten's overview "Three Eighteenth-Century Attempts to Map the Natural Order: Johan Hermann-Georg Christoph Würtz—Paul Dietrich Giseke," *Early Science and Medicine*, 2019, 24:33–89.

7. *Bref* (Letters), I:5, 126f.

8. See *Svenska arbeten* (Swedish work), 1:20, 25f., 33, 45; *Bref* (Letters), II:2, 3–8; S. Müller-Willie, *Natural History and Science Overload: The Case of Linnaeus*, 2012; Daniela Blei, "How the Index Card Cataloged the World," *Atlantic* 2017, 1:12.

9. Linn Soc., Linnaeus pat. Introductions, no date. Transl. from Latin.

10. SLÅ, 1956–57, 160f.

11. Svanfeldt, 399f. Karin Johannisson, "Science in Retreat," *Lychnos*, 1979–80.

12. *Bref* (Letters), I:1, 217.

13. Linnaeus's valuable London contact John Ellis wouldn't accept that fungi were animals. See, e.g., *A Selection*, 1821, 2:216f.

Chapter Thirty-Four. A New Synthesis?

Hofsten's *Linnaeus's Conception of Nature* (1957) is a notable attempt to chart Linnaeus's theoretical thinking. See also Lindroth, Stevens, and Müller-Wille. However, none of them have focused on *Clavis*.

1. Jewel in medicine: *Vita*, 167f.

2. Tessin, *Stenstilar* (Block lettering), 1771.

3. See *De effectu et cura*, § 14. Linnaeus to Murray 24.1.1765 [1766], at UUB G 152.

4. KB, *Autografsamlingen*, Linnaeus to D. Tilas, 24.11.1767.

5. The far from reliable Sven Hedin, a pupil of the elderly Linnaeus, wrote in his *Handbok för pracktiska Läkarevetenskapen* (Manual for the practical science of medicine), 1796, 29: "Linné, and with him also many Learned men have assumed that the body is ruled by two *principia vitalia*. The one he called *Soul* and the other *Nature*. The occupations of the Soul are all of its free will. Nature acts through an unceasing continuation of the functions of the body according to mechanical laws over which the Soul has not within its influence any immediate concern."

6. See T. Fredbärj's introduction to *Elektriskt-mediciniska satser*; also Zetzell's account in UUB: *Nordin* 111, 77–215. For an earlier take on the idea of a fifth element, see Otto Sibum, *Das fünfte Element: Wirkung und Deutungen der Electricität*, 1987, on the work by Johan Heinrich Winkler, 1745.

7. Linnaeus had actually read this in Swedenborg's writings; cf. Broberg, 2008, 133ff. On Sauvages, see Fredrik Berg, "Linné och Sauvages," *Lychnos*, 1956. On the effects on the human body of *electricum*, its enlivening effect on the old, see the abstract of the dissertation *Odores Medicamentorum*, 1752; Drake, *Linnés disputationer*, 146ff.

8. Quoted from KI, *Föreläsningar över Clavis*, 5, 6.

9. Giseke on electricity at sea.

10. Electric light: Broberg, 1975, 144.

11. Beckmann, *Tagebuch*, 120f.

12. *De rerum natura*, 4:478f, 485.

13. DN, 172ff.

14. Cf. Robert Jütte, *A History of the Senses: From Antiquity to Cyberspace*, 2005, 61ff.
15. *Fundamenta Botanica*, § 337, 361.
16. Cf. Gerlinde Hövel: "*Qualitates vegetabilium,*" "*vires medicamentorum*" und "*oecononomicus usus plantarum*" *bei Carl von Linné (1707–1778)*, 1999, 175–88.
17. Linné to J. C. Wilcke, June 1775, *Bref* (Letters), I:2, 358.
18. Hövel, 62ff. The questions raised above were aired in Chr. D. Wilcke's *Diss. de usu systematis sexualis in medicina*, 1764. The conclusion expresses the hope that the sexual system can be found to agree with the criteria based on taste, smell, and color, and hence be seen to form a natural order (312ff.).
19. Some passages in *Clavis* favor "animal products"—e.g., bone marrow, butter, cream, spermaceti, egg yolk (V:2), and animalistic ones: egg white, deer-horn jelly (XI:1), some animalistic substances like umber, musk, zibet (XIV:5, last), and game meat: venison—but these are only 4 out of 66 groups. Earlier, in 1749, Linnaeus proposed in a letter to Bäck that about a quarter of animalistic *materia medica* should be eliminated and followed up by asking, "May not others be excluded also?" (*Bref* [Letters], I:4, 110). Cf. C.-J. Clemedson, "Some selected items from Linné's *materia medica*." From *Utur stubbotan rot* (Out of a stubborn root), 1978. The work on *materia medica* can be seen to start in the early essay *Caroli Linnaei Pharmacopaea Holmiensis*. See "Chosen Dissertations" (*Valda avh.* nr 18) and continue to "Draft to M" (*Utkast till Materia Medica*), 1761. VA 19, publ. by T. Fredbärj, 1954, and go on from there. Quote from *Autografsamlingen*, KB.
20. Cf. G. Broberg, "Mat och levnadskonst: Arkiater Linné ger goda råd" (Food and the art of living well: Good advice from the Arkiater Linné), *Lychnos*, 2005.
21. UUB G 152 (Linné to Murray); *Bref* (Letters), I:5, 139; Hjelt, 1907, 139; 3.4.1766, *Bref* I:5, 142; *Minne* (In memory), 87f., 123. *Ilias in nuce* (the *Iliad* in a nutshell) is a phrase used by, among others, Roman emperors in important notifications to the Senate.
22. Sven A. Hedin, *Minne af Linné fader och son* (Memories of Linné, father and son), 1, 123.
23. Manuscript, KI Föreläsningar (Lectures); see also UUB D 809, 1769. The latter compilation is well done and clearly set out but is in Latin and so very close to the printed version and hence less helpful.
24. Linnaeus to Bäck, 1.3.1763, *Bref* (Letters), I.5, 108. The letter was actually about the reform of veterinary medicine. Peter Hernquist was recommended in the letter as the right man for the job, a good choice as it turned out.
25. *Vita*, 167.
26. Linnaeus's faith in his own superior ability can become comical: "Profess. Sauvage's eyes grew so weakened that he could neither read nor write; has corresponded worldwide on the matter of his eyes and tried everything. He writes that what I prescribed for him has now fully restituted his sight. I have now myself forgotten what it was and inquired from him how he had proceeded." Linnaeus to Bäck, 23.6.1765, *Bref* (Letters), I:5, 136.

Chapter Thirty-Five. A Philosopher of Science or a Scientist?

The ethnologist K. Rob Wikman's *Lachesis and Nemesis: Four Chapters on the Human Condition in the Writings of Carl Linnaeus* (1970) is an in-depth study of the work of the young and the old man, and includes manuscript material. The book is rich in ideas, but the chronology lacks clarity. The older Linnaeus thought along lines that could be characterized with a term used by Theodor Adorno and Edward Said, as his *Spätstil* (late style).

1. Broberg, 1975, 59ff.; P. F. Stevens and S. P. Cullen in "Linnaeus, the Cortex-Medulla Theory, and the Key to His Understanding of Plant Form and Natural Relationship," *Journal of the Arnold Arboretum*, April 1990, 179–220.
2. Stoever, 1780, 280f. Cf. Linnaeus on the Creator's intentions.
3. Cain, 1992.
4. Sahlgren, 1922.
5. Published and translated by A. Hj. Uggla, SLÅ, 1947.
6. For the mothers of soil, see SN, 12th ed.; 3, 1768.
7. See Annerstedt, 3:2, 448n.
8. *Vita*, 152.
9. L. Bergquist and G. Hillerdal, *Smålandsakademien*, 2007, 104ff. and 126ff.
10. *Ungdomsskrifter* (Youthful writings), 93ff.
11. Quote from Johan Browallius's already mentioned magazine *Den Swenska patrioten* (The Swedish patriot), 1735:8, text probably by Linnaeus.
12. Stevens, 1998. True, the article is entitled "Carl von Linnaeus," but it is otherwise trustworthy.

Chapter Thirty-Six. The Back of God and God's Footsteps

Elis Malmeström's thesis and biography are important sources on the topic of Linnaeus's religiosity, which has also been considered in several essays—for example, Erland Ehnmark's *Linnaeus and the Problem of Immortality*, in Kungliga Humanistiska Vetenskapssamfundet, Lund, 1951–52. The author Gunnar E. Sandgren, who also grew up in the now rebuilt rectory in Stenbrohult, reacted vigorously when asked in an interview about the portrayal of Linnaeus as a "good Christian" (Lars Ola Borglid, *Linnés landskap*, 1999). Sandgren has written the play *Linnés tebuske* (Linnaeus's Tea bush).

1. Quote from Malmeström, 215f.
2. *Philosophia Botanica*, § 155.
3. About Linnaeus's sense of being "an elect," see, e.g., the autobiographies Linnés ("Gud Himself had led him with his almighty hand"), in *Vita*, 145f.
4. Malmeström, KHÅ, 1925, 23.
5. On whether animals have souls, see Beckmann, *Tagebuch*, 1911, 104.
6. E. M. Fant; Linné appointed to the Bible Commission; for plants mentioned in the Bible, see *Bref* (Letters), I:1, 275ff.

Chapter Thirty-Seven. Nemesis Divina

1. Petry.
2. From the Dedication of ND, re his son.
3. DN, 195.

4. ND, 147f.
5. ND, 148f.
6. ND, 193. An entertaining parody of ND: Bengt Hallgren's *Guds finger i Uppsala*, 1981.
7. ND, 142; cf. 99 (see ghost hunting).
8. ND, 664; cf. Rollin 70; Oliver Sacks, 2015, 262ff. on autoscopy and Linnaeus.
9. *Vita*, 173.

Chapter Thirty-Eight. Solomon on Growing Old

Fries writes extensively on Linnaeus's last years, as does Afzelius in Notes (*Anteckningar*), 1823, 226ff., but there are no other studies on the subject.

1. Transl. in VA.
2. *Bref* (Letters), I:2. Linnaeus to Wargentin, 1756.
3. Linné in 1757, to Bäck about the Uppsala fever.
4. Oil in the lamp, to Bäck 19.6.1767, *Bref* (Letters), I:5, 151.
5. *Lychnos*, 1965–66. Falck, like most students, seems to have called any professor "an old boy"; Linnaeus was "the old boy in Svartbäcken."
6. Beckmann, *Tagebuch*, 48.
7. Jacob Wallenius, 2016; letter to J. A. Murray, in Fries, 2:83.
8. Joseph Marshall, *Travels . . . in the years 1768, 1769, and 1770*, 1773, 2:312–29.
9. Hannes Finnson, 82f. Two other visits in CFM 22.3.1776, 556f; CFM 1, 303.
10. Linnaeus to Sara Lisa, SLS, UUB.
11. *Bref* (Letters), I:5, 202.
12. Introduction to diet ideas. VA 35 (1960).
13. Wargentin to Haller, 7.3.1775; Nordenmark, 1939, 66; the quote "Linné drinks" in Nordenmark, 1946, 47.
14. SLÅ, 1941, 116.
15. SLÅ, 2000–2001, 125.
16. Afzelius, *Anteckningar om sig sjelf* (C.L., notes about himself), 1823, 223.
17. Fries 2:402f.
18. Karl Hedbom, SLÅ, 1918.
19. Notata sub, 256.
20. *Vita*, 58.
21. Fries, 2:407, after an original in the National Museum; no one, however, has found it since.
22. SLÅ, 2004–5, 147f.
23. In *Nyaste samling af svenska anekdoter* (The most recent collection of Swedish anecdotes), 2, 1875, 40f.
24. *Bref* (Letters), I:1, 267ff.
25. SLÅ, 2014, 181.
26. Matts Floderus, SLÅ, 1919, 113.
27. Dagligt Allehanda, 25.91778.
28. Fries's notebook X:9.
29. Fries, 2:411, 1903.

Epilogue I. Family Life 3: Mother and Child

Inventories after the deaths of Carl and Sara Lisa Linnaeus, in SLÅ, 1950-51, 1953, 1954-55, and 1962. Tomas Tullberg on Sara Lisa, her grave; see C. Backman, in SLÅ 2002-3, also SLÅ 2006; Linné Jr.'s letters to Abraham Bäck, in SLÅ 1956-57 and 1958.

1. Cited after Tullberg in SLÅ, 2006, 34f.
2. Annerstedt, Appendix IV, 213.
3. SLÅ, 1956-57, 141.
4. SLÅ, 1958, 75.
5. Uggla, in SLÅ, 1954-55.
6. Letters Duyker-Tingbrandt, 283f.; *A selection*, 1821, II:2.
7. Tullberg (1878), 58ff.
8. Linné Jr., *Dietetics*, 1761, 55v.
9. Cf. Christina Backman, "Sara Lisa writes," SLÅ, 2006.
10. Manktelow, 2007, 41.
11. Christina Backman, SLÅ, 2002-3, 154ff.
12. Strandell, "Linné's descendants," SLÅ, 1979-81.
13. N. E. Landell on *Läkaren Linné* (Linné the doctor), 214f.
14. This talented woman's unhappy fate is the core narrative in Christina Wahlén's novel *Den som jag trodde skulle göra mig lycklig* (I thought he would make me happy), 2013.

Epilogue II. Linneanism

1. Annerstedt, Appendix V, 137.
2. *Harmens diktare* (The poet of hurt), 2, Stellan Arvidson, Stockholm, 1993, 373-77.
3. *Bref* (Letters), I:3, 240.
4. SLÅ, 1963.
5. *Goethe and Linné, Contemporary Perspectives on Linnaeus*, 1985, 135ff.
6. Didier Eribon, *Michel Foucault*, 1991, 192.
7. *The Loves of the Plants* by Erasmus Darwin, 1791.
8. In Stafleu, 1971, 248.
9. In Eva Nyström, *Biblis*, 48, 2009-10.
10. UUB X 399.
11. "De lycksaliges ö" (The island of happiness), short story in *Svenska öden och äventyr* (Swedish lives and adventures) by August Strindberg, 1907, 382f.
12. *En blå bok* (A blue book) by August Strindberg, 1907, 1, 332f.
13. *Det svenska folklynnet* (The mind of the Swedish people) by Gustav Sundbärg, 1911.
14. Supplement XVI, *The Report on Emigration*, 24.
15. Introduction, *Linnean apostles I*, 2010.
16. *Konsten att minnas Linné* (The art of remembering Linné), by Henrik Björck, SLÅ, 2018.

SOURCES AND LITERATURE

"it should be perfectly possible to write an exhaustive and satisfying study of Carl Linnaeus if the author were a very learned humanist, a practicing doctor as well as thoroughly versed in the history of medicine, and an everyday Latin speaker, who carried out research in ethnography as a hobby . . . and, last but not least, a reasonably good poet." This multitalented biographer will presumably never be found, especially as he or she would do well to be professionally qualified in the full range of biological subjects, have a working knowledge of epistemology, and so on (Hagberg 1939; foreword). The list should also include being a decent graphologist, an intellectual multitasker, and an archive scholar with a fox's patience and tracking ability. In practice, anyone writing about Linnaeus has to be selective and decide to emphasize some themes rather than others: on aspects of medicine, say, or on geology and mining technology, or plant characterization and the philosophy of botany. The present work is, above all, a life story and might of course have included more biographical material. There is a wealth of sources, as must be clear by now.

Linnaeus steered, to a remarkable extent, how posterity would perceive him by composing autobiographies, planning a collection of his letters (not completed), and, whenever the context allowed, pointing out his own contributions. As for biographers, he considered four or five candidates: Johan Browallius, Carl Fredrik Mennander, Magnus Beronius, Abraham Bäck, and, possibly, Richard Pulteney. Linnaeus's autobiographies have been drawn together into a single volume entitled *Vita Caroli Linnaei* (1957; an earlier version had appeared in 1823). It seems safe to assume that Linnaeus's own voice is heard in Johan Browallius's long introduction to the "book of patronage" (dated Falun, 15 February 1735). The early twentieth-century initiative to publish Linnaeus's large correspondence was energized by the end of the century when its digitization began, a still ongoing project. It was problematic from the start; Linnaeus expressed his qualms in a letter to Gjörwell in 1769, and Bäck said in his speech "In Memoriam" (1779, 69): "It would appear probable that his learned correspondence would never be seen in print, such is the dearth of publishers in this nation"—a reference to the death of Lars Salvius. In the present work, the extensive text collections have been selectively used, as has the material held by the Linnean Society of London. Its archive contains most of the material, but the university library in Uppsala (UUB) also has a wealth of documentation; an overview is in the library's publication for the 2007 centenary *Låt inte råttor och mal* (Let moth and vermin not destroy). The sheer scale of the task makes a fair representation impossible, a caveat that also applies to the plentiful variety of notes taken at Linnaeus's lectures, which not infrequently include biographical observations but are scattered among many libraries.

A few sets of lecture notes have been published: on dietetics (LN; Lindfors, 1907), the animal kingdom (Lönnberg, 1913), *Fundamenta Botanica* (Bergquist, 2007), and the *Notata Subitanea* on dietetics (Broberg, 2000). The *Diaeta Naturalis* collection compiled by Uggla (1957) contains a wealth of information. When Lindfors pulled together the LN series, he had no access to the manuscript notes, and the gaps this caused were not detected until the London Society sent the scripts (A. Hj. Uggla, KVA Annual for 1957, 396–97). The rough student notes used to complete the publication were made between 1748 and 1753, and are, of course, unrepresentative of what preoccupied Linnaeus two decades later. So much written evidence has been accumulated about Linnaeus that he is, at least in many ways, the best documented Swedish individual in the eighteenth century; the list includes his letters and large library, lecture texts and students' notes, essays, articles and dissertations, visitors' descriptions, house inventories, family recollections, and innumerable portraits. Uggla's articles in SLÅ deserve special mention for their discoveries of new material. Linnaean family traditions were recorded by several relatives, most notably his great granddaughter's son Tycho Tullberg, professor of zoology (SLÅ 1918, 1919).

Let us return to the correspondence: a wide-ranging selection was compiled by James Edward Smith (1821; new edition in 2012), and letters exchanged with, among others, Sauvages, Haller, Scopoli, and Gunnerus. Ewald Ährling pulled together a seemingly endless collection of notes from the correspondence with Tessin, the Academy of Sciences, and Linnaeus's siblings. Having already completed work on a basic organization of Linnaeus's letters in 1885, Thore M. Fries was ready to begin in 1907 his ambitious *Bref och skrifvelser* (Letters and other writings), which reached ten published volumes. Uppsala University is now publishing it on www.alvin-portal .org together with a wealth of other Linnaeus-linked documentation—documents, books, and images.

An enormous amount of literature has been written about Linnaeus: *Bibliographica* has Hulth's and Basil Soulsby's later compendium from the British Library catalogue (1935). This work was continued with the so-called Linnaeus Link—www .linnaeuslink.org—a site that brings together links to important internet sites—for instance, Uppsala University library's Waller and Hammarby collections. The latter includes the books in the young student's library, donated by the Linnean Society in London. Tycho Tullberg compiled a portrait catalog in time for the 1907 centenary, and more supplements have been added since. A few other outstanding aids to Linnaeus studies: *Linnaeus i Lärda tidningar* (ed. Ove Hagelin, 2007), a set of articles by Linnaeus written for *Lärda Tidningar*, and *A Linnaean Kaleidoscope 1-2* (ed. Carina Nynäs and Lars Bergquist, 2016), a survey of Linnaeus's 186 dissertations.

Ten Biographies of Linnaeus:
A Quick and Hence Unfair Overview

The frontispiece of *Lebens des Ritters Carl von Linné* by Dietrich H. Stoever (1792) carries the inscription "Deus creavit, Linnaeus disposuit" (God created, Linnaeus ordered). As is often the case, this first major biography proved influential on the later ones and launched many of the handed-down anecdotes, most of which had been collected by Giseke. Of course, Stoever's *Life* was not the first in absolute terms—the introduction lists 34 previous Linnaeus texts, starting just two years after his death— but Stoever's version, with its more than 700 content-packed pages, came to be seen as the definitive work.

It is useful to read the early biographies, composed when the material still felt recent. Published in 1808, Sven A. Hedin's *Minne af Linné fader och son* (Memories of Linné, father and son) is factually unreliable but responsive to what his contemporaries wanted to know. He cites the aphorism on nature's continuity—*Natura non facit saltus*—and discusses Linnaeus's vision of development in nature being like a chain that began with a link between man and monkey (see 99ff.). Hedin also wants to persuade us of Linnaeus's interest in the brain and its relationship to Gall's studies of functional locations in the brain, suggesting that Linnaeus shelved his observations to avoid a row. However, no record of any brain research has been found, unless one is prepared to include Linnaeus's speculations about the "medulla." Hedin belongs to the school of biographers who feel that everything their subjects did was by definition excellent; "We ought to mention that striving for glory was the strongest of *Linnaei* passions" (32).

Ewald Ährling (1837–1888) must be mentioned. His daywork was to teach at the gymnasium in Arboga, but with admirable energy inspired by his interest in nature, he trailed Linnaeus around Sweden and all the way to England and the Linnean Society. He shared the contemporary fascination for amassing autographs and manuscripts, which was another driver of his enthusiasm; he found traces of his quarry everywhere, and copied everything. Ährling's large collections (held at KVA), were a feast waiting for future scholars. Fries found them (cf. below), examined, and had published the journeys to Lapland and County Dalarna, and began the intricate sorting of the letters. Ährling corresponded with Gustaf Lindström at the Royal Academy of Sciences (KVA), but their exchange ceased—far too soon—when Ährling suffered a stroke that ended his plans for using his collections as a basis for what might have been a wide-ranging biography (cf. SLÅ, 1931:27; 1932:3; 1937:1).

Research on Linnaeus must begin with a study of the 900-page biography by Thore M. Fries (1903)—two volumes, which testify to the author's extensive knowledge of his sources, botanical knowledge, and deep admiration for his subject. Fries (1832–1913) had a remarkable father in professor Elias Fries, a mycologist and founder of systematic knowledge about fungi. Thore Fries, too, was an Uppsala professor of botany and an outstanding lichenologist, who, like his father, served as the rector (president) of Uppsala University. Even so, his fame rests on the Linnaeus biography, which was anticipated in Fries's *Bidrag till en lefnadsteckning 1–8* (UUÅ, 1893–98; Contributions to a Life, 1–8), a volume of about 500 pages with 10 appendices (cf. Rutger Sernander in SLÅ, 1932; also, a bibliography). It is, on the whole, equivalent to part 1 of the biography. In the 1903 introduction, Fries writes: "As my apology, and in my defense, I may be allowed to mention that I, enlivened by the heritage of my father's profound appreciation of the King of Flowers, have during more than three decades, when at leisure from other more immediately demanding tasks, collected material for just such a work as this. I certainly realize that, through further researches, such material as I have presented could well be augmented, but increasing old age forces me to conclude, without further delay, the work so long prepared as a small offering and a token of my adoration of and gratitude to the memory of the Master." This summary of his motives begins with stating positive values, leaves some room for potential criticism, but ends by declaring that any objections will be seen as negligible.

Fries's biography has virtues also seen in the great works by some of his contemporaries, all of whom were capable of covering a great many pages: Söderbaum on Berzelius, and Annerstedt on the history of Uppsala University—others include

Schück, Lamm, Warburg, and Hjärne. Later, Fries edited and had published Johann Beckmann's *Tagebuch*, much used here, as well as the first volumes of Linnaeus's correspondence and classical writings, partly translated and deciphered in conjunction with KVA—five volumes in total. Fries's wonderful edition of *Iter Lapponicum* was published in 1913, the year he died. The sum total of his work is hugely impressive; students on the lookout for quotable lines by Linnaeus will often find them already extracted by Fries.

Many episodes had to be disentangled; the alleged conflict between Linnaeus and Rosén is just one example. Fries devotes twenty-odd pages to it (170ff.), but a reprimand is due for the way he censored to avoid embarrassment; he dismisses some of the myths but sidesteps other aspects. Citing cases from *Nemesis Divina* (in part 2), he excludes the Kyronius story because "the word used [ass] should not be reproduced in print." As for a reference to a Parisian female, he fails to note or comment on the phrase "lies herself with him" ("him" being a *maître d'hotel*). Fries adds in a footnote: "As for the punishments awaiting frivolous and unchaste behavior, a number of examples are offered, the lewdness of which however being such as to make them impossible to quote in print." Linnaeus was fascinated by *sexualia*, but Fries makes him out to be impeccably proper and a guardian of respectable morals. Here is another case in point, from 1731, in Fries's version: "Quite without warning, an event occurred that made him [Linnaeus] decide to resign from his employment as tutor in Rudbeck's home." The event in question was that "G. B, a hussy, had taken up residence in the Rudbeck house and she pushed her presence to such an extent that Mrs. Rudbeck acquired a small hatred of *Linnaeum*, as the minx did not keep her children neat enough" (Fries, vol. 1:76). Linnaeus "decided to resign"? It seems more likely that he was sacked for his affair with G. B. Elsewhere, Fries refers proudly to the praise heaped upon Linnaeus in *Lärda tidningar*, apparently unaware of who had penned these favorable judgments (Fries, vol 2:102–3).

Oscar Levertin's incomplete biography (1906) was later extended by the publication of a chapter entitled "With Linné in Holland" (SLÅ, 1929); it focused on Linnaeus's thesis and disputation in Harderwijk. Levertin waves Sweden's banner whenever possible: "That day, 24 June 1735, was also a day of victory for the Swedish coat of arms, as the young man from Småland took his place at the cathedra ... to defend his doctoral dissertation, his blazing vitality and genius made him far from the least of Swedish conquerors." Levertin's descriptions of Leyden and Hartecamp are as typical of the grand style of his time, but his book is beautifully written and views the subject from a psychological perspective.

The theologian Elis Malmeström, bishop in Växjö, had his own ideas about *Nemesis divina*, which he made the central theme of his thesis (1926) as well as his Linnaeus biography (1964), to which he gave the subtitle *Geniets kamp för vetenskaplig klarhet* (The struggle of genius for scientific clarity). Malmeström, too, was eager to defend Linnaeus's morality; here he is on the subject of *Nuptiae plantarum*: "That his system is *sexualis* has however no relationship with the modern cult of sexuality nor with a systematic approach to sexuality. In Linnaeus's mind, the matter in hand is natural and linked to the title of his work in a way similar to publications like *Flora suecica* and *Fauna suecica, Pluto svecicus, Lachesis naturalis*, and *Nemesis Divina*" (in *Linnaei väg till vetenskaplig klarhet*, Malmö, 1960:33). In 1735 Linnaeus, however, consistently compared the marriage rites of plants and people—using words like "house, bridal chamber, wedding, nuptial rites, bed, reproduction, hermaphroditism, males and females, brothers, mothers, close relatives"—and although he later pruned his range of expressions, the original idea stuck in his mind, "probably to no one's joy or edification." Had he wished to clean up his system and eliminate the entire lot of human imagery, it would have been a simple exercise. Malmeström can't get over this (1960:55)!

Knut Hagberg has had a major influence on Linnaeus's image, internationally as well as at home. There was academic resistance to his essays, and notably to his biography of Linnaeus, published in 1939 and in a slimmed-down edition in 1958. His chief opponent was Sten Lindroth, at the time assistant professor of the history of ideas and epistemology, soon to become a full professor and elected to the Swedish Academy. The heat of debate was fueled by their clashing political ideas: Hagberg wrote columns for Sweden's leading conservative paper, Lindroth for the liberal one. The tone was learned, though, with headlines such as "Vivat Linnaeus, pereat Lindroth." Hagberg, an arts journalist, didn't base his work on primary sources, preferred Linnaeus's travelogues, and could come across as an arch know-it-all; he often introduced his arguments with "It is a matter of fact that." Throughout, his biography focuses on his own special interest: birds.

Sten Lindroth used another, modern tone in his long, critical article "Linné—legend och verklighet" (Linné—legend and reality; in *Lychnos*, 1965–66). It was based on firsthand readings of the core texts but also influenced by current research. The British zoologist Arthur Cain had launched the notion of "the scholastic Linnaeus," quoting Arthur O. Lovejoy's observation that Linnaeus hadn't made a single discovery. Julius Sachs made the case that Linnaeus had stuck to the dry list making of taxonomy and ignored the growing interest in experimental physiology; the argument was taken on board by Bengt Lidforss, to whom all this ordering was mere hackwork. Lindroth, sounding rather Wiggish, pointed out how dated some of Linnaeus's thinking was, but his bracingly critical view of his subject would change. Later in his life, Lindroth contributed to the Linnaean anniversary symposium in 1978, and his *Svensk lärdomshistoria* (History of Swedish Knowledge; four volumes published in 1975–81) contains a section that unapologetically presents a very positive image of Linnaeus.

In this context, it is relevant to quote from the article by the Nestor of the history of ideas, Arthur O. Lovejoy, on "The Place of Linnaeus in the History of Science" (*Popular Science Monthly*, 1907, 498, 500): "No naturalist of this century, and few naturalists of any period, have so universal a popular reputation, or are, by so nearly common consent, given a place among the immortals not far removed from Copernicus, Galileo.... Yet, when seriously scrutinized, Linnaeus's position in the history of science is a peculiar one. With his name, there is commonly associated no epoch-making hypothesis, not a single important discovery, not one fundamental law or generalization, in any branch of science"; admittedly, "he was an unsurpassed organizer, both of scientific material and of scientific research; he introduced form and order, clearness and precision, simple definitions and plain delimitations of boundaries into sciences previously more or less chaotic or confused, or impeded with cumbrous and inappropriate categories and terminology" (500).

Michel Foucault, an *enfant terrible* among historians of ideas, may well have been influenced by Lindroth in his *Les mots et les choses* (1966). Lindroth, who appreciated

Foucault's brilliance, still wouldn't allow the French philosopher to defend his thesis in Uppsala. In Lindroth's eyes, Linnaeus was an old-fashioned collector given to scholastic classifications, traits, which made him alien to his contemporaries, while Foucault defined the entire mid-eighteenth century as spiritually and factually Linnaean.

In Sweden, many relied on a work that became, in the nicest sense of the phrase, a coffee-table book: Wilfrid Blunt's *The Complete Naturalist* (1971; two editions published in Sweden). In a very British way, it is an agreeable read; it is also very well illustrated—perhaps too well, as a large proportion of the images was copied from Thornton's *Temple of Flora* (1795), a splendid work that is incongruous with Linnaeus's simple, Småland-style world of plant drawing. Indeed, Blunt is also the author of *The Art of Botanical Illustration*. In a critical mood, one might add that Blunt's approach has something of the tourist about it: he devotes about a third of the book to Linnaeus's travels and his journey through Lapland in particular (Blunt tells the young man off for cheating with his travel expenses). Still, this biography has provided good service for years, and later editions incorporate William T. Stearn's pedagogic introduction to Linnaean botany.

The last entry in this lineup is Lisbet Rausing-Koerner's *Linnaeus: Nature and Nation* (Harvard University Press, 1999). A history of science theme has been embedded in the biography: a presentation analyzing the Linnaean taxonomy and binominal naming system. Another strand is a discussion of Linnaeus's views on national economics, a subject that Koerner sees as the mainspring of his life's work. Her approach is effective and offers a suitably current, if somewhat materialistic, perspective on the man. The main objection would be that the reforms Linnaeus backed, although in vain, must be seen as related to the widespread poverty at the time; no one can foresee the future.

Koerner is very well read in Linnaean studies, and also born and raised in Sweden (her family has a very successful industrial background). Her insights into the period and the country are those of a knowledgeable native, and the book currently has the status of being the most up-to-date and so most influential on ongoing research.

The production of academic papers exploded prior to the 2007 celebrations of Linnaeus's birthday. The biographically themed publications include books by, for instance, Marita Jonsson and Kerstin Berglund, and by Nils Erik Landell and Torbjörn Lindell; Nils Uddenberg's *System och passion* (in collaboration with Helen Schmitz and Pia Östenson) is both beautiful and readable, and fits into this group although, apart from a modicum of biography, Uddenberg concentrated on Linnaeus the botanist. Several major works by German, French, Italian, and Finnish authors were also published for the 2007 centenary—they are listed in fifty pages of SLÅ (2008). At the same time, a growing interest in the explorations by Linnaean students and their role in globalization was reflected in publications by, among others, Sverker Sörlin, Kenneth Nyberg, and Hanna Hodacs. Staffan Müller Wille's study contained many new observations on Linnaeus's way of working—his "paperwork." Gunnar Eriksson's and Bengt Jonsell's many diverse and rich compilations must also be mentioned.

There are so many other contributions that deserve a mention. For instance, there are at least fifty-odd examples of music written in honor of Linnaeus, many of them occasional pieces but also longer works. The first opera was James Plumtree's *The Lakers—A Comic Opera in Three Acts* (1798), thought to be rather risqué for playing around with ideas from Erasmus Darwin's *The Loves of the Plants*. Other operas include Mogens Christensen's *Systema Naturae* to a libretto by Sanne Björk (2001)—a "botanical requiem"—and Jonas Forsell's *Trädgårdsmästaren* (The gardener) to a libretto by Magnus Florin (for other works, see SLÅ, 2013:166f.). The trend of building fictional narratives around historical people is illustrated in the novels about Linnaeus by Rune Pär Olofsson, Magnus Florin, and Gunnar E. Sandgren, and also in books about the apostles, which include Kerstin Ekman's nonfiction story about Bjerkander, Pär Wästberg's novel about Sparrman, and Monica Braw's about Thunberg. Ann-Mari Jönsson has written about Linnaeus and Sara Lisa in *Silverbrevet: En roman från Linnés Uppsala* (The silver letter: A novel set in Linné's Uppsala; 2017). Maria Gripe's popular story for young adults *Tordyveln flyger i skymningen* (The dung beetle flies at dusk; 2018) should definitely be included here.

Last but not least: the invaluable resource that is *Svenska Linnésällskapets Årsskrift* (Annual Report by the Swedish Linnaean Society) or SLÅ for short. Its hundredth edition was published in 2018; given the number of double volumes of between 100 and 250 pages, the Annual Reports have a total number of pages on the order of 15,000. The content has widened over the years, from its original focus on Linnaeus to natural history in general. It goes without saying that it has covered an amazing range of topics and rare material, notably because several manuscripts were first published in an Annual Report. Arvid Hj. Uggla made an outstanding contribution during his time as the editor, the outcome of many visits to the Linnean Society of London, whose magazine *The Linnean* must be included here.

Lists of unprinted material and more recent printed literature are set out below. The abbreviated selection of Linnaeus's own writings can be followed up in the complete bibliographies by, respectively, Hulth, Soulsby, and Sandemann-Olsen.

Manuscript Material

UPPSALA UNIVERSITY LIBRARY

Browallius, S36; S139a
D 809 Lectures on *Clavis* 1769
D 1103k 1767
D 1454 Anders Tidström
Fries's notebooks
G 300a Afzelius to Thunberg
Ihre 187
Letters G152a
N 947 Voltemat's anecdotes
N 971 Andreas Browallius's diary

THE ROYAL LIBRARY / *KUNGLIGA BIBLIOTEKET*

Autograph collection Tilas X 508
Manuscripta Mennandria

HAGSTRÖMER LIBRARY

Abraham Bäck's diary
Lectures on Clavis medicinae

LONDON, LINNEAN SOCIETY

Linné Jr. Lectures on dietetics
Linné Jr. Travel diary
Manuscripta medica 1, 3

Cited Writings by Linneaus

Linné, Carl von. *Diaeta Naturalis.* 1733. Publ. by Arvid Hj. Uggla. Stockholm, 1957.
Linné, Carl von. *Systema Naturae.* Ed. 1, Leiden, 1735; ed. 2, Stockholm, 1740- ; ed. 4, Paris, 1744; ed. 6, Stockholm, 1748; ed. 10, 1 and 2, Stockholm, 1758–59; ed. 12, 1 to 3, Stockholm, 1766 and 1768.
Linné, Carl von. *Hortus Cliffortianus.* Amsteladami, 1737.
Linné, Carl von. *Flora Svecica.* 1745, 2nd ed. 1755. Translation, Stockholm, 1986.
Linné, Carl von. *Öländska och gotländska resa* [Journey to Öland and Gotland]. Stockholm, 1745.
Linné, Carl von. *Fauna Svecica.* 1746, 2nd ed. Stockholm, 1761.
Linné, Carl von. *Wästgöta resa* [Journey to Wästgöta County]. Stockholm, 1746.
Linné, Carl von. *Amoenitates Academicae seu Dissertationes Variae* [Academic pleasures of his various dissertations]. Stockholm, Leipzig, Erlangen, 1749; 1790.
Linné, Carl von. *Skånska resa förrättad.* 1749 [Journey in Skåne concluded 1749]. Stockholm, 1751.
Linné, Carl von. *Museum Tessinianum.* Stockholm, 1753.
Linné, Carl von, Jr. "Brevväxling" [Exchanges of letters]. To Bäck, 1 and 2, 1756–57, 1758. Publ. by Arvid Hj. Uggla in SLÅ.
Linné, Carl von. "Skaparens afsikt med naturens verk" [The Creator's intention with the work of nature]. 1763, public. Swedish translation by Arvid Hj. Uggla. SLÅ, 1947.
Linné, Carl von. *Anteckningar om sig sjelf* [Notes about himself]. Publ. by Adam Afzelius. Stockholm, 1823.
Linné, Carl von. *Svenska arbeten* [Swedish work]. Publ. by Ewald Ährling I–II:1. Stockholm, 1879–80.
Linné, Carl von. *Skrifter* [Writings]. Translated by Th. M. Fries et al., publ. by the Royal Swedish Academy of Sciences (Kungl. Svenska Vetenskapsakademien), 1–5. Uppsala 1905–13.
Linné, Carl von. *Bref och skrifvelser* [Letters and writings]. 1907–.
Linné, Carl von. Dietetics. *Lachesis Naturalis.* Notes in UUÅ 1907.
Linné, Carl von. *Föreläsningar öfver djurriket* [Lectures on the animal kingdom]. Publ. by Einar Lönnberg. Uppsala, 1913.
Linné, Carl von. *Iter Lapponicum.* Publ. by Th. M. Fries, 1913; also publ. by the Skyttean Foundation, Umeå, 2005.
Linné Carl von. *Valda avhandlingar* [Selected dissertations]. In translation from the Latin. SLS 1–84; Uppsala 1921–.

Linné, Carl von. *Iter Dalecalicum*. Publ., incl. *Iter ad exteros*, by Arvid Hj. Uggla. Stockholm, 1951.
Linné, Carl von. *Örtabok* [Herbal]. Edited by Telemak Fredbärj. Stockholm, 1957. English translation by Torbjörn Lindell.
Linné, Carl von. *Vita Caroli Linnaei* [Life of Carl Linnaeus]. Edited by Elis Malmeström and Arvid Hj. Uggla. Stockholm, 1957.
Linné, Carl von. *Nemesis Divina*. Publ. by Elis Malmeström and Telemak Fredbärj. Stockholm, 1967. Translated into English and edited by M. J. Petry. Dordrecht, 2002.
Linné, Carl von. *Om sättet att tillhopa gå* [About the way to go together]. Göteborg, 1969.
Linné, Carl von. *Om jämvikten i naturen* [On equilibria in nature]. Translated by Anders Piltz. Edited by Gunnar Broberg. Uppsala, 1978.
Linné, Carl von. *Philosophia Botanica*. Translated by Stephen Freer. Oxford, 2005.
Linné, Carl von. *Föreläsningar över Fundamenta Botanica* [Lectures on *Fundamenta Botanica*]. Notes by Pehr Osbeck. Publ. by Lars Bergqvist. Stockholm, 2007.
Linnaeus in *Lärda tidningar*. Publ. by Ove Hagelin. Stockholm, 2007.
Linné, Carl von. *Musa Cliffortiana*. 1736. Translated by Stephen Freer. Introduction by Staffan Müller Wille. Lichtenstein, 2007.
Linné, Carl von. *Clavis Medicinae Duplex*. Translated by Birger Bergh. Commentary by Gunnar Broberg, Bengt Jonsell, and Bengt I. Lindskog. Stockholm, 2008.

Secondary Sources

Abenius, Margit. "Allt är viktigt" [Everything matters]. SLÅ, 1935.
Åbonde, Anders Johansson. *Drömmen om svenskt silke* [The dream of Swedish silk]. Stockholm, 2016.
Acrel, Johan Gustaf. *Tal, om läkevetenskapens grundläggning och framväxt vid rikets äldsta lärosäte i Upsala* [Lecture on the foundations of the medical science and its growth at the oldest seat of learning in this nation: Uppsala]. Stockholm, 1796.
Afzelius, Adam. *Tillägg i Linné Egenhändiga anteckningar* [Additions to notes made by Linné]. Stockholm, 1823.
Agamben, Giorgio. *The Open: Man and Animal*. Berkeley, 2004.
Ahlström, Otto. "Linnés mikroskop." SLÅ, 1948.
Ährling, Ewald. *Carl von Linnés brefväxling*. Stockholm, 1885.
Aldén, Gustaf. "Linné som student och inspector." In *Festen till Carl von Linnés minne* [The feast in memory of Carl von Linné]. Uppsala, 1878.
Anderman, Tomas. *Herr Archiatern på Hammarby*. Uppsala, 1999.
Anderman, Tomas, and Hans Norman. *Linnés Sävja*. Uppsala, 2007.
Anfält, Tomas. "Linnés nätverk" [Linné's network]. In *Så varför reser Linné* [So why does Linné travel?]. Umeå, 2005.
Annerstedt, Clas. *Uppsala universitets historia*. Appendies 3 and 5. Uppsala, 1912–13.
"Annual Report, Swedish Linnaean Society." Svenska Linnésällskapets Årsskrift, 1918–.
Banks, Joseph. *The Banks Letters*. Edited by Warren Dawson. London, 1958.
Beckmann, Johann. *Tagebuch: Schwedische reise in den Jahren 1765-66*. UUÅ, 1911.
Berg, Åke. Publ. of "Herbationes Upsalienes." SLÅ, 1950–51.
Berg, Fredrik. *Linné's Systema morborum*. UUÅ, 1957.

Berg, Fredrik. "Hygienens omfattning i äldre tider" [The extent of hygiene in the past]. *Lychnos*, 1962.

Berglund, Karin. *Linné: Han som såg allt* [Linné: The man who saw everything]. Stockholm, 2006.

Bergqvist, Lars, and Carina Nynäs. *A Linnaean Kaleidoscope 1-2*. Stockholm, 2016.

Bergquist, Olle. "Anders Falck's Letters to His Brother in Petersburg." *Lychnos*, 1965-66.

Bendyshe, Thomas. "On the Anthropology of Linnaeus." In *Mem Anthrop. Soc. of London*, 1863-64.

Benz, Ernst. *Theologie der Electricität*. Mainz, 1970.

Benzelstierna, Gustaf. *Censorsjournal*. Uppsala, 1884.

Björck, Henrik. "Konsten att minnas Linné: Firandets former vid jubileerna 1807, 1907 och 2007" [The art of remembering Linné: Celebration formats at the anniversaries in 1807, 1907 and 2007]. SLÅ 2018.

Björnhag, Göran. *Linnés kor* [Linné's cows]. Uppsala, 2009.

Blunt, Wilfred. *The Complete Naturalist*. London, 1971.

Boerman, Albert. "Linnaeus Becomes Candidatus Medicinae at Harderwij." SLÅ, 1956-57.

Bondestam, Maja. "När svensken gick in i puberteten" [When the Swedes entered adolescence]. In *In på bara huden: Festskrift till Karin Johannisson* [Close to the skin]. Nora, 2010.

Broberg, Gunnar. "Den unge Linné speglad i några hittills obeaktade document" [The young Linné, as reflected in some previously overlooked documents]. SLÅ, 1972-74.

Broberg, Gunnar. *Homo sapiens L*. Uppsala, 1975.

Broberg, Gunnar. "Linnaeus and Genesis." SLÅ, 1978.

Broberg, Gunnar. *Petrus Artedi in His Swedish Context*. Proc. Congr. Eur. Ichthyology V. Stockholm, 1985.

Broberg, Gunnar. "The Broken Circle." In *The Quantifying Spirit*. Berkeley, 1990.

Broberg, Gunnar. *Den linneanska bildvärlden* [The Linnaean world of images]. Conferences, Roy. Acad. Learning. (Kungl. Vitterhetsakademien). 1990:23.

Broberg, Gunnar. "Fruntimmersbotaniken" [Botany for females]. SLÅ, 1990-91.

Broberg, Gunnar. *Gyllene äpplen 1-2* [Golden apples]. Stockholm, 1992.

Broberg, Gunnar. "Mat och levnadskonst" [Food and the art of living well]. *Lychnos*, 2005.

Broberg, Gunnar. "Naturen gör inga hopp" [Nature does not jump]. In *Filosofiska citat: Till Svante Nordin*. Stockholm, 2006.

Broberg, Gunnar. *The Dragon-Slayer*. Tijdsrifft voor Skandinavistiek, 2008.

Broberg, Gunnar. "Liv, död och romantik skendöden" [Life, death, and the Romantics: The appearance of death]. In *Til at stwdera läkedom* [A study of healing]. Lund, 2008.

Broberg, Gunnar, and Charlotte Christensen-Nugues. "Homo sapiens: 250 years as an Animal and a Moral Being." SLÅ, 2008.

Broberg, Gunnar, and Ulf Marken. *I skogar, på berg och i dalar* [In forests, on mountains, and in valleys]. In *Linné i dikten* [Linné in poetry]. Uppsala. 1990.

Browallius, Johan. *Den Swenska patrioten*. Stockholm, 1735.

Browne, Janet. *The Secular Ark: Studies in the History of Biogeography*. London, 1983.

Bryk, Felix, ed. *Linnaeus im Auslande*. Stockholm, 1919.
Bryk, Felix. *Linné als Sexualist*. Stockholm, 1951.
Burman, Lars. *Vältaliga studenter* [Eloquent students]. Uppsala, 2013.
Busser, Johan. *Utkast till beskrifning om Upsala* [Draft of a description of Upsala], 2. Uppsala, 1769.
Buttimer, Anne, and Tom Mels. *By Northern Lights: On the Making of Geography in Sweden*. London, 2006.
Byrne, Angela. *Geographies of the Romantic North*. New York, 2013.
Cain, A. J. "Was Linnaeus a Rosicrucian?" *Linnean*, 1992.
Charmantier, Isabelle. "Carl Linnaeus and the Visual Representation of Nature." In *Historical Studies in the Natural Sciences*, 2011, 41:4.
Charmantier, Isabelle, and Staffan Mülle-Wille. "Carl Linnaeus's Botanical Paper Slips." *Intellectual History Review*, 2014, 24:2.
Christensson, Jakob. *Petrus Artedi* (in preparation).
Dal, Björn. "Grågökar och Papilioner: Om Carl Clercks fjärilsverk Icones insectorum rariorum" [The ash-colored cuckoos and the butterflies: On Carl Clerck's study of butterflies]. SLÅ, 1984–85.
Darwin, Charles. *Charles Darwin's Natural Selection: Being the Second Part of His Big Species Book Written from 1856 to 1858*. Edited by Richard C. Stauffer. Cambridge, 1975.
Daston, Lorraine. "On Epistemic Images." In *Vision and Its Instruments*, edited by Alina Payne. University Park, PA, 2015.
Davidsson, Åke. *Till vänskapens lov* [In praise of friendship]. Göteborg, 1971.
Dintler, Åke. *Lars Roberg*. Uppsala, 1958.
Djurberg, Vilhelm. *När det var anatomisal på Södermalms stadshus i Stockholm 1685–1748* [When the council chambers in Södermalm housed a dissection room]. Stockholm, 1927.
Drake, Gustaf. "Linnés försök till en inhemsk teodling" [Linné's attempts to create a native tea plantation]. SLÅ, 1927.
Drake, Gustaf. "Linné och pärlodlingen" [Linné and the cultivation of pearls]. SLÅ, 1930.
Dunér David. "Naturens alfabet: Polhem och Linné om växternas systematic" [The alphabet of nature: Polhem and Linné on a systematic approach to plants]. SLÅ, 2008.
Dunér, David. *Människan maskinen* [Man and machine]. In *Tankemaskinen* [The thinking machine]. Nora, 2012.
Du Rietz, Rolf. "Om tryckningen av Species plantarum 1753" [On the printing of Species plantarum]. SLÅ,1965.
Duris, Pascal. *Linné et la France (1780–1850)*. Geneva, 1993.
Ehnmark, Erland. "Linnétraditionen och naturskildringen" [The Linnean tradition and describing nature]. SLÅ, 1931.
Ehnmark, Erland. "Dygden och lyckan" [Virtue and happiness]. SLÅ, 1944.
Ehnmark, Erland. *Linnaeus and the Problem of Immortality*. "Annual report for the foundation for studies in the arts." Humanistiska vetenskapsamfundets årsberättelse. Lund, 1951–52.
Ehrensvärd, Gustaf Johan. *Dagboksanteckningar* [Diary notes]. Stockholm, 1878.

Ehrström, Robert. "Minnesanteckningar om Linné" [In memory of Linné]. SLÅ, 1946.
Ekstedt, Olle. *Linné i Växjö.* Växjö, 2014.
Engel, Hans. "Linnaeus' Voyage from Hamburg to Amsterdam." SLÅ, 1936.
Erdtman, G. "Linnés Flora svecica." SLÅ, 1945.
Eribon, Didier. *Michel Foucault.* Translated by Betsy Wing. Cambridge, MA, 1991.
Eriksson, Gunnar. *Botanikens historia i Sverige intill år 1800.* Uppsala, 1969.
Eriksson, Gunnar. "The Botanical Success of Linnaeus." SLÅ, 1978.
Eriksson, Gunnar. "Carl von Linné." SBL, 1980:23.
Eriksson, Gunnar. "Fåglarnas budskap" [The message of the birds]. In *Historiens vingslag* [The wing-beats of history]*: Festskrift till Allan Ellenius.* Stockholm, 1988.
Eriksson, Gunnar. "Olof Rudbeck d.y.'s Campanula serpillifolia alias Linnaea." In *Lärdomens bilder* [Images of learning]*: Festskrift till Gunnar Broberg.* Stockholm, 2002.
Eriksson, Gunnar. "Barndomens botanik" [The botany of childhood]. SLÅ, 2015.
Fabricius, Johan Christian. From Stoever, 1792.
Fåhraeus, Olof Immanuel. "Om perlfiskeriet och Linné" [On pearl fishing and Linné]. Translated for KVA meeting, 1859.
Finnsson, Hannes. *Stockholmsrella.* 1772 [?]. Translated by A. Hj. Uggla, 1935.
Floderus, M., and C. Forsstrand. "Linnés ättlingar" [The descendants of Linné]. SLÅ, 1919.
Ford, Brian J. "The Microscope of Linnaeus and His Blind Spot." *Microscope,* 2009, 57:2.
Forsstrand, Carl. *Linné i Stockholm.* Stockholm, 1915.
Forsstrand, Carl. "Linné och familjen Reuterholm" [Linné and the Reuterholm family]. SLÅ, 1919.
Forsstrand, Carl. "Uppsala på Linnés tid" [Uppsala at Linné's time]. SLÅ, 1924.
Foust, Clifford. *Rhubarb: The Wondrous Drug.* Princeton, 1992.
Frängsmyr, Carl. *Klimat och karaktär* [Climate and character]. Stockholm, 2000.
Frängsmyr, Tore. *Geologi och skapelsetro* [Geology and creationism]. Uppsala, 1969.
Frängsmyr, Tore. *Wolffianismen i Sverige* [Wolffianism in Sweden]. Uppsala, 1972.
Frängsmyr, Tore, ed. *Linnaeus: The Man and His Work.* Canton, MA, 1983; 2nd ed., 1994.
Franzén, Olle. "Hur Linnébilden formades" [How the image of Linné was formed]. SLÅ, 1963.
Fredbärj, Telamak. "Carl H. Wänman, en misskänd Linnélärjunge" [Carl H. Wänman, a misunderstood pupil of Linné]. SLÅ, 1966.
Fredbärj, Telamak, and Jan Knöppel. "Kofsan på Linnaei tid och idag" [Kofsan (rocky landscape in southern Sweden) in Linné's day and nowadays]. SLÅ, 1953.
Fries, Elias. *Botaniska utflykter 1* [Botanical excursions]. Stockholm, 1843.
Fries, Nils. "Linnés resedagböcker" [Linné's travel diaries]. SLÅ, 1966.
Fries, Nils, ed. *Linnés språk och stil* [Linné's language and style]. Stockholm, 1971.
Fries, Robert. "Linné i Holland." SLÅ, 1919.
Fries, Thore [Theodore] M. *Bidrag till en lefnadsbeskrifning 2* [Contribution to the description of a life]. Installment 2 of 8, published between 1888 and 1893 for professorial inaugurations. UUÅ.
Fries, Thore [Theodore] M. Linné: Lefnadsteckning [Linné: A Biography]. Vols. 1 and 2. Stockholm, 1903.

SOURCES AND LITERATURE [449]

Gerner, K., ed. *The Swedes and the Dutch Were Meant for Each Other*. Lund, 2014.

Gertz, Otto. "Sperling, Stobaeus, Linné och Leche, och de första undersökningarna av Skånes flora" [Sperling, Stobaeus, Linné, and Leche, and the first studies of the flora in Skåne]. SLÅ, 1926.

Gibson, Susanne. *Animal, Vegetable, Mineral: How Eighteenth-Century Science Disrupted the Natural Order*. Oxford, 2015.

Giertz, Martin. *Svenska prästgårdar* [Swedish rectories]. Stockholm, 2009.

Gjörwell, C. C. "Samlingar" [On collections]. In *Samlingar*. Skåne, 1875, 3:8.

Goerke, Heinz. *Carl von Linné Arzt, Naturforscher Systematiker*. [Carl von Linné—doctor, scientist taxonomist]. 2nd ed. Stuttgart, 1989.

Göransson, Sven. *De svenska studieresorna och den religiösa kontrollen* [The Swedish travels for studies abroad and religious control]. UUÅ, 1951.

Granit, Ragnar, ed. *Utur stubbotan rot* [Out of a stubborn root]. Stockholm, 1958.

Grape, Anders. "Abraham Bäcks utländska studieresa" [Abraham Bäcks's studies abroad]. SLÅ, 1937.

Grape, Anders. "Linné, Abraham Bäck och Pharmacopaea svecica av år 1775" [Linné, Abraham Bäck and Pharmacopaea svecica of the year 1775]. SLÅ, 1946.

Grape, Anders. *Ihreska handskriftsamlingen 1-2* [Ihre's manuscripts]. Uppsala, 1949.

Gullander, Bertil. *Linné och Uppsala* [Linné and Uppsala]. Stockholm, 1978.

Gustafsson, Åke. "Linnés Peloria." In *Utur stubbotan rot* [Out of a stubborn root], ed. Ragnar. Granit. Stockholm, 1978.

Gyllensten, Lars. "Carl von Linné." In *Författarnas litteraturhistoria*. Stockholm, 1984.

Hagberg, Knut. *Carl Linnaeus*. Stockholm, 1939 and 1958.

Hagberg, Knut. *Carl Linnaeus: Den linneanska traditionen* [Carl Linnaeus: The Linnean Tradition]. Stockholm, 1951.

Hagelin, Ove. *Georg Dionysius Ehret and His Plate of the Sexual System of Plants in Linnaeus' Own Copy of "Systema Naturae."* Hagströmerbiblioteket at Karolinska Institute (publication), Stockholm, 2000. See also *Biblis*, 2000:10.

Hagelin, Ove, ed. *Linnaeus i Lärda tidningar* [Linnaeus in scholarly journals]. Hagströmerbiblioteket at Karolinska Institute (publication). Stockholm, 2007.

Hagström, Johan Otto. *Brev från åren 1755-1785* [Letters]. Linköping, 1993.

Hallenborg, Carl. "Carl Hallenborgs Anmärkningar till Carl von Linnés Skånska resa" [Carl Hallenborg's Remarks on the subject of Carl von Linnés travels in Skåne county]. In *Historisk tidskrift för Skåneland* [Journal of the history of Skåneland], 1911–1913.

Hamberg, Erik. "Carl von Linnés stjärngossar och hans undervisning hösten 1747" [Carl von Linné's star pupils and his teaching during the autumn of 1747]. SLÅ, 2019.

Hansen, Lars, ed. *The Linnean Apostles*. Vols. 1–10. Whitby, 2015.

Hansson, Stina. *Svensk bröllopsdiktning under 1600- och 1700-talen* [Swedish wedding poetry in the seventeenth and eighteenth centuries]. Göteborg, 2011.

Harnesk, Helena. *Linné och Uppsala* [Linne and Uppsala]. Stockholm, 2006.

Heckscher, Eli. "Linnés resor: Den ekonomiska bakgrunden" [Linné's journeys: The economic background]. SLÅ, 1942.

Hedin, Sven A. *Minne af von Linné fader och son* [In memory of Linné, the father and the son]. Stockholm, 1808.
Hedlund, Emil. "Johan Rothman." SLÅ, 1936.
Heller, John. "The Early History of Binomial Nomenclature." *Huntia*, 1964:1.
Heller, John. "*Hortus Cliffortianus* by Linnaeus." *Taxon*, 1968:17, 6.
Heller, John. *Classical Poetry in the Systema naturae of Linnaeus*. Trans. Proc. Am. Ass. 102, 1971.
Hildebrand, Bengt. *Kungl. Svenska Vetenskapsakademiens förhistoria* [The early history of the Royal Swedish Academy of Sciences]. Stockholm, 1939.
Hirell, Anders. *Den svenska matsvampens historia* [The history of edible mushrooms in Sweden]. Stockholm, 2013.
Hjelt, Otto. *Linné som läkare och naturforskare* [Linné, the doctor and the scientist]. 1907.
Hodacs, Hanna. "Linneans Outdoors." *Br. J. History of Science*, 2011:44.
Hodacs, Hanna, and Kenneth Nyberg. *Naturalhistoria på resande fot* [Natural history while traveling]. Stockholm, 2007.
Hodacs, Hanna, Kenneth Nyberg, and Stéphane van Damme. *Linnaeus, Natural History and the Circulation of Knowledge*. Oxford, 2018.
Hofsten, Nils von. "Systema Naturae: Ett 200-årsminne" [At the 200th anniversary]. SLÅ 1935.
Hofsten, Nils von. "Linnaeus anno 1748." SLÅ, 1948.
Hofsten, Nils von. *Linnaeus's Conception of Nature*. Kungl.Vetenskapssocietetens årsbok, 1957.
Holm, Åke. *Specimina linneana* [Linnaean specimens]. UUÅ, 1957.
Hoquet, Thierry. *Buffon / Linné: Eternels rivaux de la biologie?* Paris, 2007.
Houdt, Toon van, and Kris Delacroix. See *Neulatinisches Jahrbuch*, 1999.
Hövel, Gerlinde. *Qualitates vegetabilium, vires medicamentorum und oeconomicus usus plantarum bei Carl von Linné (1707-1778)* [Carl von Linné on the qualities of vegetables, medicinal potencies and economic uses of plants]. Stuttgart, 1998.
Hult, Olof T. "Om Linné och 'den osynliga världen'" [About Linné and the "invisible world"]. SLÅ, 1934, 1935.
Hulth, J. M. "Studenten Carolus Linnaeus' bibliotek." In *Studier tillägnade Isak Collijn* [Studies in honor of Isak Collijn]. Uppsala, 1925.
Hylander, Nils. "Linnaea / . . . / och andra växtsläkten uppkallade efter svenskar" [Linnaea . . . and other plant species named after Swedes]. SLÅ, 1967.
Ihre, Thomas. *Abraham Bäck*. Stockholm, 2012.
Ingelög, Torleif. *Skatter i vått och torrt: Biologiska samlingar i Sverige* [Treasures from the wet and the dry worlds: Biological collections in Sweden]. Stockholm, 2013.
Jacobsson, Nils. "Magister Andraae Hessellii anmärkningar" [Mag. Andraae Hesselii remarks]. SLÅ, 1938.
Jacobsson, Roger. *Så varför reser Linné?* [So why does Linné travel?]. Umeå, 2005.
Jarvis, Charlie. *Order out of Chaos*. London, 2007.
Johannisson, Karin. "Vetenskap på reträtt" [Science in retreat]. *Lychnos*, 1979-80.
Johannisson, Karin. *Det mätbara samhället* [To measure society]. Stockholm, 1988.
Jonsell, Bengt. "Linnaeus at His Zenith." *Symp. Bot. Ups.*, 2005, 33:3.
Jonsell, Bengt. "Apostlarnas resor och gärningar" [Travels and finding by the disciples]. In *Ljus över landet* [Light above the land], edited by Paul Hallberg. Göteborg, 2005.

Jönsson, Ann-Marie. "Linnaeus's 'Svartbäckslatin' as an International Language of Science." SLÅ, 2000-2001.
Jönsson, Ann-Marie. "Retorik—inte Erotik. Linné och Lady Monson" [Rhetoric and not eroticism]. SLÅ, 2006.
Jönsson, Ann-Marie. "Brev till Jungius" [Letters to Jungius]. *Humanitas*, 2017.
Jörgensen, Matz. *Lund på Linnés tid* [Lund at the time of Linné]. Lund, 1999.
Jörgensen, Per M. "Om Linnés okända norska ursprung" [About Linné's unknown Norwegian origins]. SLÅ, 2004-5.
Juel, Oscar. *Hortus linnaeanus*. Uppsala, 1919.
Jütte, Robert. *A History of the Senses*. Malden, 2005.
Kalm, Pehr. *Brev till Friherre Sten Carl Bielke* [Letters to Count S. C. Bielke]. Publ. by Carl Skottsberg. Åbo, 1960.
Kant, Imanuel. *Anthropologie in pragmatischer Hinsicht* [A pragmatic review of anthropology]. Translated by V. L. Dowdell, 1996.
Kerkkonen, Matti. *Pet and His American Journey*. Helsinki, 1959.
Kiöping, Nils Matsson. *Kort beskrifning uppå trenne resor* [Short descriptions while on three journeys]. Wisingsborgh, 1667.
Kjellberg, Sven T. *Svenska ostindiska kompanierna 1731-1813* [The Swedish East Indian companies]. Malmö, 1964.
Knapp, Sandra, and Wheeler Quentin, eds. *Letters to Linnaeus*. London, 2009.
Koerner, Lisbet. *Nature and Nation*. Cambridge, MA, 1999.
Krok, Thorgny. *Bibliotheca botanica svecana*. Uppsala and Stockholm, 1925.
Landell, Nils Erik. *Trädgårdsmästaren Linné* [Linné, the gardener]. Stockholm, 1997.
Larson, James Lee. *Reason and Experience*. Berkeley, 1971.
Larsson, Tord, ed. *Den gustavianska bibelkommissionen* [The Gustavian Commission on the Bible]. Lund, 2009.
Låt inte råttor eller mal fördärva . . . [Let moth and vermin not destroy]. *Linnésamlingar i Uppsala* [Linné's collections]. Uppsala, 2007.
Leide, Arvid. *Akademiskt 1700-tal* [The academic eighteenth-century]. Lund, 1971.
Lejonhhufvud, Sigrid. *Lovisa Ulrica och Carl Gustaf Tessin*. Stockholm, 1920.
Lenk, Torsten. "Linnés nordstjärna" [Linné's order of the Polar Star]. *Livrustkammaren* 1941:2.
Levertin, Oscar. *Carl von Linné*. Stockholm, 1906.
Levertin, Oscar. "Med Linné i Holland" [With Linné in Holland]. SLÅ, 1929.
Lidström, Carina. *Skrivare på resa* [Writer on his travels]. Stockholm, 2015.
Lindberg, Assar M. "Linden i Jonsboda" [The lime tree in Jonsboda]. SLÅ, 1955-56.
Lindberg, Bo. Introduction to *A Linnaean Kaleidoscope* (cf. Bergqvist and Nynäs, 2016).
Lindberg, Bo. *Den akademiska läxan* [The academic lesson]. Stockholm, 2018.
Lindberg, Bo. *Peregrinatio medica, Svenska medicinares studieresor i Europa 1600-1800* [Swedish medics and their educational travels in Europe]. Uppsala, 2019.
Lindblom, J. A. *Äldre svenska biografier 8* [Older Swedish biographies]. Uppsala, 1925.
Lindeboom, G. A. *Herman Boerhaave: The Man and His Work*. London, 1968.
Lindell, Torbjörn. *Carl von Linné: Den fulländade forskaren* [Carl von Linné: The complete scientist]. Lund, 2007.
Lindgärde, Valborg, and Elisabeth Mansén. *Ljuva möten och ömma samtal* [Sweet meetings and tender talk]. Stockholm, 1999.

Lindroth, Sten. "Naturvetenskaperna och kulturkampen" [The natural sciences and the culture wars]. *Lychnos*, 1957–58.
Lindroth, Sten. "Linné: Legend och verklighet" [Linné: The legend and the real man]. *Lychnos*, 1965–66.
Lindroth, Sten. *Kungl. Vetenskapsakademiens historia 1-2* [The history of the Royal Academy of Sciences]. Stockholm, 1967.
Lindroth, Sten. *Uppsala universitets historia*. Uppsala, 1977.
Lindroth, Sten. "Samtal om Linné" [Conversations about Linné]. SLÅ, 1979–80.
Linné, Carl von, Jr. *Blomsteruret* [The flower clock]. VA, 1971:60.
Linnel, T. "Några ord om Linnés Peloria" [A few words about Linné's Peloria]. SLÅ, 1932.
Löfling, Pehr. "Oration." Publ. S. Rydén and A. Hj. Uggla. *Norrlandica*, 1961:5.
Lövcrona, Inger. ". . . hans knif skär wackert up hennes stek" [. . . his knife neatly carves her steak]. *Kulturen*. Lund, 2007.
Lovejoy, A. O. "The Place of Linnaeus in the History of Science." *Popular Science Monthly*, 1907:71.
Lovejoy, A. O. *The Great Chain of Being*. Cambridge, MA, 1936.
Lovejoy, A. O. and George Boas. *Primitivism and Related Ideas in Antiquity*. New York, 1935.
Löwegren, Yngve. *Naturaliekabinett i Sverige under 1700-talet* [Nature specimens in Swedish collections during the eighteenth-century]. Lund, 1952.
Löwendahl, Björn (London-based dealer in rare books). Catalogue, 1973:4.
Lundmark, Bo. "Linnés samiska trumma" [Linné's Sami drum]. SLÅ, 1982–83.
Malmeström, Elis. *Carl von Linnés religiösa åskådning* [Carl von Linné's religious views]. Uppsala, 1926.
Malmeström, Elis. "Gudsnamn" [God's name]. In *Annual Report, History of the Church*. Kyrkohistorisk årsskrift, 1926.
Malmeström, Elis. "Linnés avhandling om ålderdomen, Senium Salomoneum" [Linné's dissertation on old age]. SLÅ, 1929.
Malmeström, Elis. "Carl von Linnés kyrkohistoriska ställning" [Carl von Linné's views on the history of the church]. SLÅ, 1942.
Malmeström, Elis. *Linnaei väg till vetenskaplig klarhet* [Linnaei way to scientific clarity]. Malmö, 1960.
Malmeström, Elis. "Linnés bruk av bibelord i Nemesis divina: En statistisk undersökning" [Linné's use of the words of the Bible in *Nemesis divina*: A statistical study]. SLÅ, 1963.
Malmeström, Elis. *Carl von Linné: Geniets kamp för klarhet* [Carl von Linné: The struggle of genius for clarity]. Stockholm, 1964.
Mantkelow, Mariette. "Linnés Hammarby: Ett blommande kulturarv" [Linné's Hammarby: A flowering cultural heritage]. *Svensk Botanisk Tidskrift*, 2001:5.
Mantkelow, Mariette, and Petronella Kettunen. *Kvinnorna kring Linné* [The women around Linné]. Uppsala, 2007.
Martinsson, Karin, and Svengunnar Ryman. *Hortus rudbeckianus*. Uppsala, 2005.
Mayr, Ernst. *The Growth of Biological Thought*. Cambridge, MA, 1986.
Meldercreutz, Jonas. *Berättelse om en till Lappland och Norrsiöen gjord resa under sommarmånaderna 1736* [1737] [A story about a journey to Lapland and the Bothnian sea in the months of summer]. Rovaniemi, 1985.

Mennander, Carl Fredrik. *Brevväxling* [Letters]. Publ. by Kaarlo Östterblaadh 1–3. Helsinki, 1939–42.
Müller, Erik. "Linné och Abraham Bäck." SLÅ, 1921.
Müller-Wille, Staffan. "Names and Numbers: Data in Classical Natural History." *Osiris*, 2017:32.
Müller-Wille, Staffan, and Isabelle Charmantier. "Natural History and Information Overload: The Case of Linnaeus Studies." *History and Philosophy*, 2012:43.
Mustelin, Olof. "Pehr Kalm om våra inhemska växters nytta" [Pehr Kalm on the utility of our native plants]. SLÅ, 1952–53.
Nelson, E. C. "The Enigma of the Skye-Blue Andromeda, or: Was Linnaeus Colour-Blind?" *Linnean*, 2010:26.
Nickkelsen, Kärin. *Draughtsmen, Botanists and Nature: The Construction of 18th Century Botanical Illustration*. Amsterdam, 2006.
Nilsson, Gunnar. "Linnaeus som medicus vid Amiralitet i Stockholm" [Linnaeus as Admirality *medicus*]. SLÅ, 1936.
Nisser, Marie. "Tessin får ett celebert besök" [Tessin has a celebrity visitor]. *Sörmlandsbygden*, 1966.
Nordenmark, N.V.E. "Anders Celsius, Linné och den hundragradiga termometern" [Anders Celsius, Linné and the hundred-graded thermometer]. SLÅ, 1937.
Nordenmark, N.V.E. *Pehr Wilhelm Wargentin*. Uppsala, 1939.
Nordenmark, N.V.E. *Frederic Mallet och Daniel Melanderhjelm*. Uppsala, 1946.
Nordenskiöld, Erik. "En blick på Linnés allmänna naturuppfattning" [A look at Linné's general ideas about nature]. SLÅ, 1923.
Nordström, Johan. "Linné och Gronovius." SLÅ, 1954–55.
Nyberg, Kenneth. See Hanna Hodacs.
Nyström, Eva. "Carl von Linnés korrespondens: Ett vetenskapligt nätverk från 1700-talet" [Carl von Linné's correspondence: An eighteenth-century network of scientists]. *Meddelanden från Postmuseum*, 2007:56.
Nyström, Eva. "Peter Camper och the Linnean Society." *Biblis*, 2009–10.
Ödmann, Samuel. "Om disciplinen och lefnadssättet wid Wexiö schole" [On discipline and the manner of living at the school in Wexiö]. In *Skrifter och bref* [Writings and letters]. Uppsala, 1925.
Ödmann, Samuel. *Skrifter och brev* [Writings and letters]. Vol. 1. Uppsala, 1925.
Olsson, Torsten. "En linnean om Linné" [A linnaean devotee on Linné]. SLÅ, 1949.
Örn, Nils. *Kort beskrivning av Lappland*. [A brief description of Lapland]. [1710] Uppsala, 1984.
Osborne, Michael A. *Nature, the Exotic, and the Science of French Colonialism*. Bloomington, IN, 1994.
Östlund, Michael, and Lars Forsberg. *Linné och den sjuhövdade hydran* [Linné and the seven-headed hydra]. Stockholm, 2004.
Ottosson, Mats Ola. "Hur Venus kom tillbaka till Linnéträdgården" [How Venus returned to Linné's garden]. SLÅ, 2010.
Pagliaro, H. E., ed. *Racism in the Eighteenth Century*. Cleveland, OH, 1973.
Petersson, Gunnar. *Prästgårdens trädgård* [The rectory garden]. Stockholm, 2010.
Platen, Magnus von. "Den svenska resan" [The Swedish journey]. In *Naturligtvis, festskrift till Gunnar Eriksson*. Uppsala, 1981.

Platen, Magnus von. *Svenska skägg: Våra manshakor genom tiderna* [Swedish beards: Our male chins through the ages]. Stockholm, 1995.

Porter, Roy, and George S. Rousseau. *Gout: The Patrician Malady*. New Haven, 1998.

Rausing, Lisbet. "Underwriting the Oeconomy: Linnaeus on Nature and Mind." Supplement to *History of Political Economy*, vol. 35 (2006). See also Koerner.

Reuterholm, Axel. *Dagbok 1738–1739* [Diary]. Stockholm, 2006.

Richter, Herman. *Geografiens historia i Sverige till 1800* [The history of geography in Sweden until 1800]. Uppsala, 1959.

Ritterbush, Philip. *Overtures to Biology: The Speculations of Eighteenth-Century Naturalists*. New Haven, 1964.

Rogberg, Samuel. *Historisk beskrifning öfver Småland* [Historical description of County Småland]. Karlskrona, 1770.

Rollin, Carl Gustaf. *Nemesis divina*. Stockholm, 1857.

Rydén, Mats. "Om termen fauna" [About the word *fauna*]. In *Festskrift till Bengt Odenstedt*. Umeå, 1992.

Rydén, Mats. *Botaniska strövtåg: Svenska och engelska* [Botanical rambling: Swedish and English]. Uppsala, 2003.

Sacks, Oliver. *Hallucinations*. London, 2012.

Sahlgren, Jöran. "Linnés talspråk" [Linné's spoken language]. SLÅ,1920.

Sahlgren, Jöran. "Linné som predikant" [Linné as a preacher]. SLÅ,1922.

Sahlgren, Jöran. "Linnés bildspråk" [Linné's use of imagery]. *Annual Report by the Society of Sciences / Vetenskapssocietetens årsbok*, 1924.

Sandemann Olsen, Sven-Eric. *Bibliographia discipuli Linnaei* [The bibliographies of Linnaeus's disciples]. Copenhagen, 1997.

Santesson, Lillemor. *Tryckt hos Salvius* (Printed at Salvius's). Lund, 1986.

Savage, Spencer. *Synopsis of the Notations by Linnaeus and His Contemporaries in His Library*. London, 1940.

Schiebinger, Londa. "Why Mammals Are Called Mammals: Gender Politics in Eighteenth-Century Natural History." *American Historical Review*, 1993, 98:2.

Schivelbusch, Wolfgang. *Paradiset, smaken och förnuftet: Njutningsmedlens historia* [Paradise, taste, and common sense: The history of pleasurable drugs]. Stockholm, 1982.

Schoeps, Hans Jachim. *Philosemitismus im Barock* [Philosemetism in the Baroque]. Tübingen, 1951.

Schück, Henrik. "En skilsmässoprocess från 1700-talet" [Divorce in the eighteenth century]. In *Från det forna Uppsala* [From the Uppsala of the past]. Stockholm, 1917.

Schück, Henrik. *Den svenska förlagsbokhandelns historia* [The history of publishers' bookstores in Sweden]. Vols.1 and 2. Stockholm, 1923.

Schück, Henrik. *Lars Salvius: Minnesteckning* [Lars Salvius: In memory]. Stockholm, 1929.

Sefström, Eric, ed. *Swenska samlingar 1763–65* [Swedish collections]. Stockholm, 1763–65.

Selling, Olof. "Från Linnés ungdomstid och hans småländska hembygd" [From Linné's youth and his Småland neighborhood]. SLÅ, 1960.

Sernander, Rutger. "Hårleman och Linnaei *herbationes upsalienses*." SLÅ, 1926.

Sernander, Rutger. "Linné som uppsalastudent" [Linné as a student in Uppsala]. SLÅ, 1929.
Shiebinger, Londa. "Why Mammals Are Called Mammals." *American Historical Review* 1993:98.
Shillito, J. F. "Linnaeus: Zoology in the Last Years." SLÅ, 1978.
Sivgård, Gustaf. *Vandrande sholares*. Lund, 1965.
Sjöberg, Nina. *Linné och riksdagen* [Linné and parliament]. Stockholm, 2007.
Skottsberg, Carl. *Pehr Kalm: Levnadsteckning* [Pehr Kalm: A life]. Stockholm, 1951.
Skuncke, Marie-Christine. *Carl Peter Thunberg: Botanist and Physician*. Uppsala, 2014.
Smith, J. E., ed. *A Selection of the Correspondence of Linnaeus and Other Natural Historians*, 1–2. London, 1821.
Snogerup, Sven, and Matz Jörgensen. *I Linnés hjulspår runt Skåne* [Following Linné's wheel tracks around Skåne]. Stockholm, 1997.
Sörlin, Sverker. *Rädslan för svaghet* [Dreading weakness]. Stockholm, 2014.
Sörlin, Sverker, and Otto Fagerstedt. *Linné och hans apostlar* [Linné and his disciples]. Stockholm, 2004.
Soulsby, Basil. *A Catalogue of the Works of Linnaeus*. 2nd ed. London 1933, 1936.
Sourander, Patrik. "Linnaeus and Neurology." SLÅ, 1978.
Spangenberg, Carl Gustav. "Linné som domare" [Linné as a judge]. In *De lege*, edited by Peter Asp. Uppsala, 2007.
Stafleu, Frans. *Linnaeus and the Linneans: The Spreading of Their Ideas in Systematic Botany 1735–1789*. Utrecht, 1971.
Stearn, William T. *Introduction and Appendix to Species Plantarum 1–2*. London, 1956–58.
Stearn, William T. "The Background of Linnaeus's Contribution to the Nomenclature." *Syst. Zool.* 1959, 8:4-22.
Stearn, William T. "The Origin of the Male and Female Symbols." *Taxon* 1962:11, 4.
Stevens P. F. "Linnaeus." In *Routledge Encyclopedia of Philosophy*. Vol. 5. London, 1998.
Stockholms weckoblad, 1739–40.
Stevens, P. F., and S. P. Cullen. "Linnaeus: The Cortex-Medulla Theory." *Journal of the Arnold Arboretum* 1990:71.
Stoever, D. H. *Lebens des Ritters Carl von Linné* [The life of the Knight Carl von Linné]. 1792. Hansebooks, 2017.
Strandell, Birger. "Linnés ättlingar" [Linné's descendents]. SLÅ, 1979–81.
Strandell, Birger. "Linnés lärjungar" [Linné's pupils]. SLÅ, 1979–81.
Stuber, Martin, Stefan Hächler, and Luc Lienhard, eds. *Hallers Netz: Ein europischer Gelehrtenbriefwechsel zur Zeit der Aufklrung* [Haller's networks: Exchanges of letters between scholars during the Englightenment]. Basel, 2005.
Svanberg, Ingvar. "Deras mistande rör mig så hierteligen: Linné och hans sällskapsdjur" [Loss of these truly touches my heart: Linné and his household pet animals]. SLÅ, 2007.
Svensk etnobiologi 1–3. Stockholm, 2001–7.
Swederus, Magnus. *Botaniska trädgården i Upsala 1655–1807* [The botanical garden in Upsala 1655–1807]. Uppsala, 1877.

Swederus, Magnus. "Linné och växtodling" [Linné and the cultivation of plants]. UUÅ, 1907.

Swenska vetenskapsakademiens protokoll för åren 1739, 1740 och 1741 [Swedish Academy of Sciences, minutes of meetings]. Ed. E. W. Dahlgren. Stockholm, 1918.

Sydow, Carl Otto von. "Den unge Linnés författarskap" [The writings of young Linné]. SLÅ, 1962.

Sydow, Carl Otto von. "Rudbeck d.y. och hans dagbok från Lapplandsresan 1695" [Rudbeck Jr.'s diary from his travels in Lapland]. SLÅ, 1968–69, 1970–71.

Sydow, Carl Otto von. "Linné och Lappland." SLÅ, 1972–74.

Sylwan, Otto. *Från stångpiskans dagar* [From the days of the ponytail]. Stockholm, 1901.

Tersmeden, Carl. *Memoarer 2* [Memoirs 2]. Ed. Nils Sjöberg. Stockholm, 1914.

Tessin, Carl Gustaf. *En gammal mans utvalda bref till en ung prins* [An old man's selected letters to a young prince]. Publ. by A. J. Nordström. Stockholm, 1785.

Tillgren, Josua. *Linné och brännvinet* [Linné on aquavit]. Lund, 1960.

Tinland, Franck. *L'homme sauvage* [The wild man]. Paris, 1967.

Tobin, Beth Fowkes. *The Duchess's Shells*. New Haven, 2014.

Torgny, Ove. *Med Linné genom Sverige* [With Linné through Sweden]. Stockholm, 2006.

Tullberg, Tomas. "Sara Lisa von Linné: En biografisk studie." SLÅ, 2006.

Tullberg, Tycho. *Linné-porträtt*. Uppsala, 1907.

Tullberg, Tycho. "Linnés Hammarby." SLÅ, 1918.

Tullberg, Tycho. "Familjetraditioner om Linné" [Linnaean family traditions]. SLÅ, 1919.

Uddenberg, Nils. *Linné och mentalsjukdomarna* [Linné and mental illness]. Stockholm, 2012.

Uddenberg, Nils, Pia Östensson, and Helene Schmitz. *System och Passion: Linné och drömmen om det naturliga systemet* [System and passion: Linné and the dream of nature's system]. Stockholm, 2007.

Uggla, Arvid Hj. "Kilian Stobaei instruktion för en forskningsfärd i Skåne 1729" [Kilian Stobaei instruction for an investigative journey in Skåne 1729]. SLÅ, 1932.

Uggla, Arvid Hj. "Linnnés almanacksanteckningar för 1735" [Linné's almanac notes]. SLÅ, 1935.

Uggla, Arvid Hj. "Linnés tankar om den akademiska ungdomens fostran" [Linné's thoughts on the education of academic youths]. SLÅ, 1940.

Uggla, Arvid Hj. *Linnés adelskap* [Linné's knighthood]. Artre et marte, 1941.

Uggla, Arvid Hj. *Den sjuhövdade hydran i Hamburg* [The seven-headed hydra in Hamburg]. Festschrift for Harald Nordenson. 1946.

Uggla, Arvid Hj. "Skaparens afsikt med naturens verk: En promotionföreläsning av Linné 1765" [The Creator's intention for the work of nature: A lecture at a disputation by Linné]. SLÅ, 1947.

Uggla, Arvid Hj. "Linné och linneanerna" [Linné and his followers]. In *Ny illustrerad litteraturhistoria* 2. Stockholm, 1956.

Uggla, Arvid Hj. "Linné d.y. och hans brev till [Linné Jr.'s letters to] Abraham Bäck." SLÅ, 1956–57, 1958.

Vartanian, Aram. "Trembley's polyp." *J. History of Ideas*, 1950:11.

Vila, Anne C. *Suffering Scholars: Pathologies of the Intellectual in Enlightenment France*. Philadelphia, 2018.

Virdestam, Gotthard. "Linné och Stenbrohult." SLÅ, 1928.
Virdestam, Gotthard. "Omkring några brev av Samuel Linnaeus" [On some letters by Samuel Linnaeus]. SLÅ, 1931.
Wagner, Florian. *Die Entdeckung Lapplands* [The discovery of Lapland]. Kiel, 2004.
Wallenberg, Jacob. *Min son på galejan* [My son onboard the galley]. Göteberg, 2014.
Wallenius, Jakob. *Einige Begebenheiten meines Lebens* [Some stories from my life]. Greifswald, 2016.
Wallin, Olle. "Bibliotheca botanica." *Biblis*, 2014:65.
Wallin, Stig. "Urkunder rörande familjen Linné" [Search of documentation concerning the Linné family]. SLÅ, 1950–51, 1953, 1954–55, 1962.
Weinstock, John, ed. *Contemporary Perspectives on Linnaeus*. Austin, TX, 1985.
Westerberg, Mats. *Lefvernes beskrifning* [Description of a life]. Publ. by Magdalena Hellquist. Västerås, 1973.
Westling, Håkan, ed. *Linné och de lärde i Lund* [Linné and the scholars in Lund]. Lund, 2007.
Widmalm, Sven. "Gravören och docenterna" [The engraver and the academics]. In *Kunskapens trädgårdar, festskrift till Tore Frängsmyr* [The gardens of knowledge, in celebration of Tore Frängsmyr]. Stockholm, 1988.
Wiklund, K. B. "Linné och lapparna" [Linné and the Lapps]. SLÅ,1925.
Wikman, K. Rob. *Lachesis and Nemesis*. Åbo, 1970.
Wiman, Lars Gösta. *Abraham Bäck*. Uppsala, 2012.
Windahl, Annika, and Roger Tallroth. "Beständit menuett och polska: Om musik och dans i familjen Linné" [Always minuet and polska: On music and dancing in the Linné family]. SLÅ, 2017.
Wiséhn, Eva, *Images of Linnaeus: Medals, Coins, Banknotes*. Stockholm, 2011.
Wittrock, Veit. *Linnaea borealis*. Stockholm, 1907.
Wraxal, Nathanel. *En tur till Stockholm 1774* [A trip to Stockholm]. Translated by T. Nyman. Stockholm, 1963.
Wright, John. *The Naming of the Shrew: A Curious History of Latin Names*. London, 2014.
Zachrisson, Sune. *Rabarber: Från laxermedel till delikatess* [Rhubarb: From laxative to delicacy]. Sörmlandsbygden, 1995.
Zennström, P. O. *Linné Sveriges upptäckare, naturens namngivare* [Linné—the man who discovered Sweden and named nature]. Stockholm, 1957.

INDEX

"About the Necessity of Investigative Journeys in Our Native Country," 154
About the way to become together (Om sättet att tillhopa gå), 250–51
Academia, 223–24
Academia Naturae Curiosorum, 306
Académie des sciences (Paris), 118, 135
Academy of Sciences: application to, 131; censor for, 332; competition to find the best ways of fighting infestations in fruit trees, 307; contacts with Ekeblad de la Gardie, 165; founding of, 77–78; Gregorian calendar, adoption of, 149; illustrators working under the aegis of, 241; Kalm's journey to America supported by, 276; Latin-based approach idealized by, 417n30; Linnaeus's proposal of membership for Swedenborg, 361; membership in, 232, 327; motto of, 76, 243; origin of, 134–37, 147; questions about the country, publication of memorandum with, 158; revision of *Fauna Svecica*, proposal for, 183; role in Swedish education, 314; Skåne, eagerness to see, 168; Swedish language and, 177, 239; the Tärnström case and, 277; thermometer purchased from, 197; translation of works in Latin, 182–83; yearly scientific mission abroad, 272–73
Acharius, Erik, 242, 406
Acrel, Olof af, 373
Acrel (or Acrell), Johan Gustaf: descriptions of Linnaeus, 4–5, 382; on Linnaeus as a lecturer, 208–9, 229; on Linnaeus leading his students in outdoor excursions, 204; on Linnaeus's authorship of disputations, 225; mixing of rabbit and hen, observation of, 343; text on journey to Lapland, reading of, 419n28
Adanson, Michel, 341, 404–6
Adolf Fredrik (king of Sweden), 177, 332
Adonis stenbrohultensis, 23
Adonis Uplandicus, 53

Adorno, Theodor, 433
Afzelius, Adam, 275, 325, 388, 392–93, 406
Agamben, Giorgio, 429n1
Agardh, Carl Adolph, 406
aging: description of, 380–82; descriptions of Linnaeus, 387–88; dulling of senses, 345, 355; Linnaeus's mind and, 363, 433; Linnaeus's self-perception of, 385–86; sources on Linnaeus, 435; the transformation of the body and, 381–82; the young versus the old Linnaeus, 359
Ährling, Ewald, 192, 234, 277, 438–39
Åkerman, Anders/Andreas, 149, 271–72, 295
Akrell, Fredrik, 319
Albinus, Bernhard Siegfried, 103, 109, 356
Aldrovandi, Ulisse, 41–42, 125
Alexander the Great, 84
Almqvist, C.J.L., 406, 408
Alströmer, Claes, 290
Alströmer, Jonas, 135
Amoenitates academicae (Academic Amusements), 224–25
amphibians, 337. *See also* frogs
anatomy, 47, 212, 222
Andersen, Hans Christian, 6
Anderson, Johann, 107
Andromeda caerulea, 70
Animalia composita, 286
Animalia per Sveciam observata, 54
animals: apes, 294–95; birds (*see* birds); compassion for, 260–61; earthworms, 40; frogs (*see* frogs); human beings as, 83; hybridization in the natural system for, 343; learning from, 84–85; moral sense of, 201; natural system for ordering, 342; opossum, slipper-eating, 431n4; pets (*see* pets); plants and, distinction between, 286; souls of, 370; at the Uppsala garden, 199–200. See also *Fauna Svecica*
Annerstedt, Claes, 197, 439
Anthropomorpha, 294–95
ants: "Notes on the Practices of Ants," 137, 140

[459]

apes, 294–95
"apparent death," 36–37
appearance of Linnaeus. *See* descriptions
aristocracy, contempt for, 86
Aristotle, 33–34, 56, 121, 282, 392
Armfelt, Ulrika Charlotta, 360
Arndt, Johann, 56, 365
Artedi (Arctedius), Petrus: advice about places to stop for the night on the coastal road, 64; death of, 114–15, 231; influence on Linnaeus, speculation about, 124; low profile of, 58–59; method in ichthyology worked out by, 121; the siren included in work by, 128; the unicorn/narwhal identified as *Monodon* by, 124
atheism, accusation of, 343–44, 365–66
Auctores botanici, 246
Augustine, Saint, 427n1
Aurivillius, Samuel, 332, 371, 396
autobiographies of Linnaeus, 2–4, 437
awards. *See* honors

Bach, Johann Sebastian, 11, 368–69
Bäck, Abraham: background and life of, 328; biographer, consideration as, 437; on Clifford and the conflict between Linnaeus and Cramer, 116–17; collaboration with Linnaeus, 111, 222, 234, 281, 285–86, 328; corals, question of, 285; correspondence with Linnaeus, 157, 177, 198, 202–3, 230–33, 240–41, 285, 312–15, 317–18, 330, 333, 341, 348, 356–57, 381, 386, 389, 433n19, 433n26; correspondence with Linnaeus Jr., 396; correspondence with Rosén, 329; diary entries about Holland, 109; donation for foreign explorations, 276; on the friendship between Artedi and Linnaeus, 58; friendship with, 2–3, 109, 328–30, 348, 387; Holland, diary entries about, 109; on the improbability of seeing Linnaeus's correspondence in print, 437; on Linnaeus after his death, 247, 387; Linnaeus as part of the new generation in science, 131; Linnaeus's autobiography, delivering the text of, 2–3; on Linnaeus's collaboration with God, 305–6; Linnaeus's death and, 389; Linnaeus's moaning about becoming a courtier, 335; on Linnaeus's offering of alms to starving persons, 265; on Linnaeus's sorrow over the loss of disciples abroad, 277; on Linnaeus's writing style, 177–78; on the purity of Linnaeus's religious faith, 372; residence in Stockholm, 330; translation of Latin works into Swedish, proposal for, 182; tribute to Linnaeus, 3; warning to not mention trip to Stockholm to his wife, 193
Bäck, Anna Charlotta, 329
Bäck, Carl Abraham, 329–31
Backman, Johan, 46, 197, 216
Bacon, Francis, 56, 60, 83, 254, 361
bacteria, 340
Bagge, Peter, 267
bananas, 268–69
Banks, Joseph: correspondence with Mrs. Solander, 399; hostility to Linnaeus, 387–88, 399; Linnaeus Jr. and, 396; painting of, 397; reports from Cook's explorations, source of, 340; on Solander's work habits, 229; study of Linnaeus's works by, 404
Bartsch, Johann, 111–12, 116, 277
Bassaport, Madeleine, 118
Bataille, Georges, 373
Bauhin, Caspar, 41, 50, 201
beard(s), 87, 104, 131, 300, 382
Beckmann, Johann: conversations with Linnaeus, 382–83; description of Linnaeus, 2, 131, 382; description of Sara Lisa, 399; description of the Uppsala fire, 319; at Hammarby, 323–24; Linnaeus's ideas about electricity, peculiarity of, 353; on the plants named for Browallius, 235; on the relations between Carl Jr. and his mother, 394; religion, Linnaeus and, 370; *Tagebuch*, 440
Bedoire, Frans, 199
Bellman, Carl Michael, 263
Bengtsson, Frans G., 408, 410
Bentham, Jeremy, 404
Benzelia, Greta, 51, 185
Benzelius, Erik, 51
Benzelius, Mathias, 159
Benzelstierna, Gustaf, 172
Berch, Anders, 147, 249, 316, 395
Berch, Anders, Jr., 249–50, 395
Berch, Christer, 152
Bergencrantz, Carl Fredrik, 234, 392, 401
Bergh, Sven Richard, 408
Bergius, Bengt, 220, 332

Bergius, Peter Jonas, 332, 396
Berglund, Karin, 408
Berglund, Kerstin, 442
Bergman, Torbern: as an anti-Linnaean, 332; faculty at Uppsala, member of, 148; on the judge's bench at Uppsala, 316; letter to Wargentin on the elderly Linnaeus's condition, 387; Linnaeus Jr. and, 396; lover of natural history, described as, 236; *Physisk beskrifning öfwer jordklotet* (Physical description of the earth), 271–72; prize from the Academy of Sciences, winner of, 307
Bergquist, Carl, 241
Berlin, Andreas, 277
Bernigeroth, Martin, 261
Beronius, Magnus Olai (archbishop of Uppsala), 3, 390, 437
Bible, the: absence of in Linnaeus's library, 366; authority of in Stenbrohult parish of Linnaeus's youth, 20–21; bananas, as a presumed source of information about, 268; botany of, 369; "consider the lilies" in, 34; final words of *Diaeta Naturalis* and, 374; geography, as a source on, 271; Linnaeus quoting/frequent references to, 215, 366; Linnaeus's faith in the words of, 359, 363; Linnaeus's science and, 368–70; Linnaeus's use of in "On the Growth of Habitable Parts of the Earth," 154–56; Linnaeus's writing style influenced by, 177; misuse of words from, warning about, 254; the study of nature and, 366–70. *See also* Creator/Creation/role of God; religion
Bible Commission, 370–71
biblical philology, 369
Bielke, Sten Carl, 135, 181, 268, 327
binomial nomenclature, 8, 129, 226, 255, 291–92
biographies of Linnaeus, 438–42
biology: creation of the concept of, 406; double name convention introduced into, 291; ethnobiology, notes on from Lapland journey, 65; recognition of "hybrid" as classic moment in the history of, 284; sexual system of, 247, 250; significance of *Systema Naturae* for, 119; taxonomic base of, 346 (*see also* taxonomy project)

birds: classification of, 59–60; dissection of, 214; drawings of, 42, 49; migratory, 27; pelican, old tale about, 124; as pets, 199–200; short-eared owl, drawing of, 67; song of, 22, 25, 104, 260–62; stuffed in Linnaeus's collection, 50; swallows spend winter underwater, belief that, 128; in Sweden, 154; work on classifying, 59–60
birth/birthplace, 17–18, 20, 27–28
Biurman, Georg, 160
Björck, Henrik, 410
Björk, Eric, 137
Björk, Sanne, 443
Björk, Tobias, 117
Björnståhl, Jacob Jonas, 244, 275, 403
Blackburne, Anna, 246
black henbane, curious effects of, 34
Blackwell, Elisabeth, 246
Blake, William, 6
Blom, Carl Magnus, 286
Blunt, Wilfrid, 79, 242, 442
Boerhaave, Herman: dislike of Linnaeus, 112; in the first rank of students of nature, 392; as an iatrochemist, 213; illness of and farewell to, 113; *Institutiones medica*, 57; as learned celebrity from Holland, 103; meeting, 111–12; notice of Linnaeus by, 47; as a patron/supporter, 111–12, 141; portrait of in Linnaeus's grand salon, 235; publication of Swammerdam's work supported by, 366
Boëthius, Jacob, 87
bog rosemary (*Andromeda polifolia*), description of, 69–71
Böhme, Jakob, 35
Bontius, Jacobus, 126
books: hiding the books at Uppsala from the Russians, 45; illustrations and drawings in, 241–44; library borrowing while at Lund, 41–42; Linnaeus's library while a student at Uppsala, 57–58; literary works provided by Nils Linnaeus, 33–34; medical texts, Linnaeus's library of, 56; number of written by Linnaeus, 238; publication of (*see* publication/publishing); reviews, 245; Stobæus's library, 39–41; target readership of Linnaeus's, 245; texts owned while a grammar student, 33–34
Börtz, Daniel, 373

Boström, Christopher Jacob, 363
botany: Book of Herbs kept while at school, 1725–1727, 34–35; definition of, 255–56; "herbations" as excursions focused on, 203–6; introduction to, 97; irritation at restructuring of plant nomenclature by Linnaeus, 129; lectures on, 134; Linnaeus's approach to teaching, 410; Linnaeus's reputation as a student at Uppsala in, 49; medicine and, 220; Nils Linnaeus's interest in, 23; as the people's science, 245; taxonomy (*see* taxonomy project); teaching, concerns about, 408, 410; women attracted to, 245–46. *See also* reproduction
Boyle, Robert, 372
Brandes, Jan, 278
Braw, Monica, 443
Bredberg, Sven, 108
Bref och Skrifwelser (Letters and Writings), 246
Broberg, Lars, 199, 376
Brodersonia, Christina, 18
Bromelius, Olaus (or Olof), 34, 181, 292
Bromell, Magnus von, 242
Browallius, Carl Fredrik Andreas, 54–55, 95–96, 100
Browallius, Johan: academy membership, Linnaeus's support for, 136; as biographer, 437; bishopric, appointment to, 367; correspondence with, 104; Dalarna, journey to, 93, 101; description of Linnaeus's room in Rudbeck's estate, 59; destruction by fire of letters to, 233; ghost writing for Linnaeus, 96–97; as go-between for Linnaeus and Sara Lisa, 98–99; on the intellectual status of Linnaeus, 119; as "Joseph" in *Nemesis divina*, 375; "Journey in Dalarna," plans to publish, 101; Lapland, journey to, 80–81; naming of plants for, 235; as part of the new generation in science, 131; political beliefs held by, 132; Sara Lisa Moraea, flirtation with, 99–100, 117, 132, 375; Siegesbeck, conflict with, 100
Brown, W.J.S., 397
Brusewitz, Gunnar, 408
Bruyset, Jean Marie, 342
Bryk, Felix, 242, 251–52
Bryson, Bill, 248

Buffon, Georges-Louis Leclerc, Comte de, 283–85, 396, 404–5
Burman, Johannes, 112, 404
Burser, Joachim, 201
Busser, Johan Benedict, 147, 150
"buttonology," 406
Byström, Niklas, 402

Cain, Arthur, 441
Cajanus, Daniel, 20
Cajanus, Gustaf, 208
cameralism, 269
Camerarius, Rudolf Jakob, 54
Camper, Peter, 405
Caps, the, 31, 132, 141, 192, 265, 312–13
Carlander, Christofer, 391
Carl Linnaeus Notebook, 34
Carlsdotter, Stina, 315–16
cataloging, method of, 340
celebrity, faked interviews and, 383
Celsius, Anders: as colleague, 111; cover illustration for *Arithmetica, or the Art of Counting*, 151; French expedition to the north, participant in, 61, 64; on his deathbed quip about the possibility of eternal life, 148–49; as inspector at Uppsala, 310; nephew of Olof, 47; as part of the new generation in science, 131; ranking of candidates for professorial chairs at Uppsala, 141; reduction in water levels following the Flood, observation of, 156; support for Linnaeus's expedition to Lapland, 64; on teaching of medicine at Uppsala, 49; thermometer, manufacture of, 197; as the university's most brightly shining light, 152
Celsius, Olof: "Captain," playful designation as, 306; correspondence with, 116–17, 129; on *Flora Lapponica*, 65; *Flora uplandica*, 181; friendship with, 47; *Hierobotanicon*, 369; lodgings for Linnaeus, assistance in finding, 59; *Praeludia* dedicated to, 53; ranking of candidates for professorial chairs at Uppsala, 141; on the sexual relationship of plants, 54; species of local flora, figure for, 204; theology examination required for study abroad, examiner for, 103
Celsius family, 44
censorship, 130, 171–73
Cesalpino, Andrea, 23

Cesnecopherus, Johannes, 213
Charles XII (king of Sweden), 18, 30, 131, 305
childhood: biographers on, 414n1; Book of Herbs kept while at school, 1725–1727, 34–35; education (*see* education); grammar school, 30–33; Linnaeus's thoughts about, 28–29; medicine, playing at, 23–24; plants, interest in, 23, 25; religion, 365; Stenbrohult, growing up in, 22–24, 29
Christensen, Mogens, 443
Christina (queen of Sweden), 160, 219, 272
Cicero, 219, 331
Classes Plantarum, 111, 129, 341
classification: of birds, 59–60; of illnesses, 221–22, 353, 355–56; of mammals (*Mammalia*), 210, 280; of man (*Homo sapiens*), 254; of minerals, 96–97; of nature, systematic (see *Systema Naturae*; taxonomy project); of plants according to their "medical powers," 355; sexual system of, 59, 123, 128, 130, 137 (*see also* sex)
Clavis Medicinae Duplex, 11, 346–53, 356–58
Clerck, Carl, 232, 242–43
Clewberg, Carl, 92
Clifford, George: apes in the *vivaria* of, 294; dedication of *Hortus Cliffortianus* to, 113–14; favorite of, competition for, 116–17; irritation at restructuring of plant nomenclature by Linnaeus, 129; Linnaeus as garden manager for, 112–13, 198; London trip for Linnaeus, paying for, 117; staying with, 99, 114, 116, 118, 324
climate: dietetics, included in, 213; foreign plants and, 267–68; Gothic understanding of, 417n31; interconnectivity and interdependencies in nature and, 258; lives of men shaped by, 84, 86–87, 132, 297; of the mountains, 75, 78; zones of, 87, 155
Colden, Jane, 246
collections/collecting: compulsive/obsessional, 5, 202; compulsory in the senior school, 408; devoted list-maker and orderly hoarder, 178; fate of following death of Linnaeus, 401; insects, 5, 35, 50; maintaining the university and royal, 202–3; minerals, 92; plants, 35; popularity of, 201–2

Collegium Medicum, 353
Collin, Johannes, 27
Collinson, Peter, 404
continuity: in Darwin, 283; derivation of concept of, 281; fascination with, 282; of life, letter about, 288; of life, marrow providing, 350; of life, sentence summing up aspects of ideas on, 351; in a map model of the orders, 341; in nature, aphorism on, 439; in nature, the egg as symbol of, 123
Cook, James, 340
Corallia baltica, 285
corals: animal-plant boundaries and, 282–83; as animals, development of understanding of, 285; classification of, 290; collection of, 328, 401
Cornelius Agrippa, 56–57
Cornelius Nepos, 33
correspondence/letters: conditions of professional, 234; from the family, 233; of introduction, 234; number of correspondents, 233; professional networks via, 232–33; publishing, 233–34, 437–38. *See also* Bäck, Abraham, correspondence with Linnaeus; Mennander, Carl Fredrik, correspondence with; Roberg, Lars, correspondence with; Salvius, Lars, correspondence with; Sauvages de Lacroix, François Boissier de, correspondence with; Stobæus, Kilian, correspondence with; Tessin, Carl Gustaf, correspondence with; Wargentin, Per Wilhelm, correspondence with
Cosmographical Society, 271
Creator/Creation/role of God: electricity, God and, 371; Eve, Linnaeus on the creation of, 370–71; Genesis as the starting point for Linnaeus's thought/work, 121, 154–56, 281, 367–68; on God's judgment, 374–75 (see also *Nemesis divina* (Divine retribution); God's plan for the world is sensible and straightforward, 349–50; hybridization and the laws of nature, 342–44; "know thyself" means realizing that you are the aim of Creation, 299; nature and, union of, 365; perfection of and science, difficulty of reconciling, 338. *See also* Bible, the; religion
Crichton, Alexander, 5

crickets, 164, 324–25
Critica Botanica, 129, 296–97
Cronhielm, Gustaf, 83, 91, 101
Cui bono?, 255, 274, 279
culture: as a factor shaping the lives of men, 84–86; natural form of medicine and, 82–83; nature and, combinations of, 260; nature and, opposition of, 87, 257; Sami, ruining, 78; Sara Lisa as lacking in, 324, 400; Swedish, 15, 406; variants of species created by, 121
Curiositas naturalis, 367
curiosity, 253–55, 427n1
"Curious Features of Insects," 153–54, 177–78, 259–60, 299

Dahle, Lars, 94
Dalarna, journey to: account of, 93–97; availability of the account of, 173; commission for, 92; entertainment breaks during, 97–99; personnel for the expedition, 92–93; praise of, 95–96
Dalarna resa, 173–74
Dali, Salvador, 5
Dalin (later Dahlin), Olof von, 107, 132–33, 177, 203, 327
Darwin, Charles: creation of coral reefs, first to explain, 225; Down, comparison of Hammarby to, 320; earthworms, fascination with, 40; journey on *The Beagle*, 81; Linnaeus, comparison with, 404; Linnaeus as inspiration for, 259, 283–84; marriage, thoughts on, 189; methods of, concept of "biology" and, 406; "nature doesn't jump," reference to, 283; questions of Creation, approach to, 368
Darwin, Erasmus, 248, 404, 443
data management, 229–30
death: burial spot, 1; causes of, 389–90; dream on the night of Linnaeus's, 392; fate of collections following, 401; the funeral, 392–93; funeral instructions, 390–91; instructions for managing, examples of, 391; interest of Banks and Solander in Linnaeus's, 387–88, 399; of Linnaeus, 388–89; self-perception as close to, 382; the will, 392, 394
De curiositate naturali, 230, 253–54
De effectu et cura, 349

defining by exclusion, practice of, 210
De Geer, Charles, 242, 281, 327
De inebriantia, 352
De la Gardie, Catharina Charlotte, 246
De la Gardie, Eva Ekeblad, 165, 246
Deliciae Naturae, 216, 246
De Peloria, 284
depression, 230–31
Derham, William, 366
Descartes, René, 57, 119, 254, 257, 362
descriptions of Linnaeus: by Acrel, 4–5, 382; appreciation from a professor of literature, 1; by Bäck, 3; by Beckmann, 2, 131, 382; eyes described by Fabricius, 384; by Frondin, 4; by Queen Lovisa Ulrika, 333; as an older man, 382, 387–88; by Rothman, 4; self-portraits and self-praise, 1–2, 4, 6–7; size/height, 19–20; as summation of multiple images, 9–11; temperament, 28. *See also* portraits of Linnaeus; professional career/accomplishments
Diaeta Naturalis, 82–90; on American Indians, 72–73; animals portrayed as moral exemplars in, 300, 325; city and countryside, differences between, 260; coffee and tea are always to be drunk hot, 145; compiled by Uggla, 437; on God's judgment, 374; health, critical importance of taste and smell to, 353; intended audience for, 357; introduction to, melancholy and bitterness in, 89–90; on Lapps never having seen a snake, 77; life-affirming/changing experience of the beauty of nature, 367–68; observations of folk customs in foreign countries in, 272; precursor to *Paradoxa* list included in, 126; sex drive, observations linked to, 249; vegetarianism discussed in, 115; views from Lapland journey in, 67–68; work on beginning in 1733, 91
Diaeta per Scalam aetatis humanae (Order of life and the different ages of mankind), 381
dictionary of natural history, 342
dietetics: drinking and alcoholism, 216–17, 219, 324; eating practices, 213–16, 324; harsh judgment of contemporary Swedes' practices, 219; iatrochemists

in the seventeenth century and, 213; introduction to the course on, 385; journey to Lapland and, 72; lecturing on, 89, 207–8, 211–12, 219–21; luxury, ridding the nation of, 266; path to wisdom and, 299; six main elements of, 83; subject matter of and focus on, 213, 221; texts on, 58; warm drinks, dangers of, 217

Dillenius, Johann Jacob, 117, 235

Diogenes, 83–84

Dioscorides, 392

Döbelius, Johan Jacob, 42

Dodoens, Rembert, 41

double name convention (binomial nomenclature), 8, 129, 226, 255, 291–92

drawing, Linnaeus's abilities at, 242

dreams: contempt for, 64–65; about the murder of Caesar, 261; on the night of Linnaeus's death, 392; at the outset of the expedition to Lapland, 65; of primitive humans in a state of innocence, 73; Swedenborg's erotic fantasies, 109

Drelincourt, Charles, 287

Dryander, Jonas, 404

earthworms, 40

East India Company, 272–73, 277

economics: Linnaeus's lack of knowledge about, 427–28n1; Linnaeus's optimism about, 265–66; Linnaeus's pronouncement on the importance of, 136; proposals (mostly unrealistic) for addressing, 266–70; travels through Swedish landscapes and, 77–78, 102, 265

Edberg, Rolf, 408

Edison, Thomas Alva, 5

education: disputations and dissertations, 223–27, 425n5; focus shifting away from the sciences in, 345; geography, early teaching of, 271; at grammar school in Växjö, 30–33, 365; introduction to the course on dietetics, 385; Linnaeus's approach to botany, 408, 410; Linnaeus's engagement in and opinion of, 314–15; at Lund University, 37–42; moving from Lund to Uppsala, 42–44; at Uppsala University, 45–60 (*see also* Uppsala University). *See also* teacher/teaching

Education Commission, 148

egg(s): every being has its origin in an, 120–21, 281, 368; in Linnaeus's coat of arms and seal, 123, 403; *Mammalia* as class that does not grow in, 210; number laid linked to size of animal, 259; as a representation of the world, 123; seed and, similarities of, 52, 121, 123, 226; seed and bud of plants, difference in, 288; seed as the egg of the plant, 55; sorting birds by number laid, 59; sperm and, combination of, 55–56

Ehrenholm, Lisa, 99

Ehrensvärd, Augustin, 134

Ehrensvärd, Carl August, 335

Ehrensvärd, Gustav Johan, 408

Ehret, Georg Dionysius, 122, 129, 241, 318

Ehrhart, Friedrich, 204

Ehrström, Anders, 269, 324

Ekeberg, Carl Gustaf, 145

Ekeblad, Clas, 170

Ekerman, Petrus, 141, 176, 224–25, 310

Ekman, Kerstin, 408, 430n18, 443

Ekman, O. C., 373

Ekström, C. U., 197, 408

electricum/electricity, 351–53, 371

electrotherapeutics, 351

Ellis, John, 282, 340, 345, 404, 432n13

Elvius, Johan, 55

Emporelius, Erik, 92

enemies/rivals/criticisms, 230, 331–32; Adanson, 405–6; Bengtsson, 408, 410; "buttonology," 406; Camper, 405; dismissal of Linnaeus's work as mere collecting and organizing, 441–42; noisy opponents, dealing with, 237; Siegesbeck, 100; the "Solander issue," 387–88; upstart, criticism as an, 96. *See also* Bergman, Torbern; Haller, Albrecht von; Ihre, Johan; Rosén von Rosenstein, Nils; Solander, Daniel; Wallerius, Johan Gottschalk

entrepreneur, Linnaeus as, 276

equilibrium, 373, 377, 379

Eriksson, Gunnar, 408, 442

Erik XIV (king of Sweden), 333

"eternity machine," possibility of building a, 136

Eugen (prince of Sweden), 408

Euhemerus/euhemerism, 77

Eurén, Erik, 211

INDEX

Exanthemata viva (Live Eruptions), 340
extinction, species at risk of, ix

Fabricius, Johan Christian, 320, 323–24, 359, 384, 394, 399–400
Fagerstedt, Otto, 277
Faggot, Jacob, 136, 158
Fahlstedt, Ingel, 92
Fahrenheit, Daniel, 195
Falck, Anders, 382, 435n5
Falun, Sweden: birth of first child in, 187; description of mines in, 95; laughing sickness in, case of, 267; partying (without Linnaeus) in, 99; post as medic in, 44, 90; return to, 118; 1734 as the year of, 92–93
family life: at Hammarby (country manor house) (*see* Hammarby); members of the household in Uppsala, 193; at the prefect's house in Uppsala, 186–87. *See also* romance
famine: of 1733, 76; of 1756, 255; of 1773, 309
Fant, Erik Mikael, 370–71
"fauna," introduction of the term, 292
Fauna Svecica, 181–83, 185; checking proofs of second edition, 228; frontispiece of, 184; gifts to sponsors of, 234; kinds of people who live in Sweden, account of, 297; Linnaeus called "The Second Adam" in reference to, 305; publication of by Salvius, 240; quote from foreword to, 357; the wolverine included in, 126
feminist, Linnaeus as, 350
Fischerström, Iwan F., 408
Fjellström, Johan, 159
flattery: appreciation of, 7; ingratiating, 50–51; not heard in the north, 65; of royal recognition, 333
"flora," introduction of the term, 292
Flora Dalecarlia, 93
Flora Lapponica: Boerhaave's refusal to write an introduction for, 112; descriptions from Lapland journey incorporated into, 65; frontispiece for, 62, 242; inventory of plants in Lapland as an international contribution, 129; metaphors in the foreword to, 178; mushrooms, inedibility of, 216; passages on the Fury and Andromeda in, 69; printing plans for, 101
Flora oeconomica, 265

Flora Svecica, 15, 181–85, 207, 240–42, 244–45
Florin, Magnus, 443
Flygare, Gustaf, 391–92
food: cooking and cookbooks, 219–20; evolution of the human diet, 85; at Hammarby, 324; kitchen interior, drawing of, 218; recommended eating practices, 213–16; suggestions for, hunger and, 265; vegetarianism, 216; warm/hot drinks, dangers of ingesting, 217. *See also* dietetics
foreign, suspicious of the, 264–66
Forsell, Jonas, 443
Forsskål, Peter, 149, 382
Forsslund, Jonas, 395
Forster, George, 340, 387
Forster, J. R., 340
Foucault, Michel, 79, 404, 441–42
Fowles, John, 194
Fraga vesca, 303
Franklin, Benjamin, 351
Fredrik I (king of Sweden), 30
Fredrik the Great (king of Prussia), 333
French, dislike of, 82, 87, 170, 264, 296
Freud, Sigmund, 248
Friedrich, Adolph, 162
friends, 327–31. *See also* Bäck, Abraham; Härleman, Carl; Höpken, Anders Johan von; Lovisa Ulrika (queen of Sweden); Mennander, Carl Fredrik; Tessin, Carl Gustaf; Wargentin, Pehr Wilhelm
Fries, Carl, 408
Fries, Elias, 155–56, 373, 439
Fries, Thore M.: anecdote on the naming of plants for Browallius, dismissal of, 235; attribution of essay to Linnaeus, 430n18; auguries and portents not taken seriously by Linnaeus, 377; battle with Aurivillius, 332; Beckmann's *Tagebuch*, editing and publishing of, 440; biography of Linnaeus, 439–40; commitment to Linnaeus, 406; complications of Linnaeus's private life not addressed by, 192; on the death messages from Linnaeus, 392; desperation, 1733 as a time of, 91; drawing from Lapland journey, 75; dutiful attender at meetings, Linnaeus as, 312; *Iter Lapponicum*, publication of, 440; letter from Celsius referring to a "new method," 54; letters, compilation

of, 438–39; Linnaeus's flexible attitude toward facts unremarked on by, 80; on Linnaeus's theory of all species being present at the Creation, 156; on Linnaeus's writing style, 177; loyalty to the king as a "good" quality, 313; on the marriage celebration for Linnaeus and Sara Lisa, 137; politics, minimizing references to, 312; publication of Lapland travelogue, 416; publication of Linnaeus biography, ix; publication of Linnaeus correspondence, 234; the rivalry/conflict with Rosén, 50
Friess, Frederick Christian, 373
frogs: children catching, 28; as a member of the "ugly family," 203; mythical figures and, 68, 70; potential illness associated with, 216; vocalizations by, 104, 171
Frölich, Charlotte, 165
Frondin, Birger "Berge," 4, 189–92, 197
Frondin, Elias, 189
Frondin, Erik, 189, 191
Fuchs, Rutger, 162
Fundamenta Botanica: belief that all species today were present at the Creation as core concept of, 155; botany, definition of, 256; Cartesianism in, 362; classifying plants by their "medical powers," 355; echoes of Descartes in foreword to, 57; lecture notes, mentioned in, 210; print run planned for, 100; publication plans for, 101; quotes from the introduction and postscript to, 53–54
Fundamenta Medicinae, 58
Fundamentum fructificationis, 342

Galen, 34
Galileo Galilei, 296
Gall, Franz Joseph, 439
gardens: application for a post at the botanical garden in Uppsala, 47–48; botanical in Skåne, 170; Clifford's, 113–14; dangerous, keeping Samuel out of, 35; experimental, 268; *Hortus Upsaliensis*, 225–26; Linné garden, photo of, 407; as living libraries of plants, 194; locations of significant, 194; in London, 117; Lund as a city of, 38; of Nils Linnaeus, 18, 23, 26, 28; Paradise, Linnaeus's garden known as, 306; at Råshult, 22–23; rectory, 22–23; Royal Academic Garden in Upsala, statue of Venus in, 249–50; at Uppsala (*see* Uppsala University Botanical Garden)
Geijer, Erik Gustaf, 363
Geisler, Johan Tobias, 55
Gemmae arborum, 287–88
Genera morborum, 221
Genera Plantarum, 15, 116–17, 141, 242–43, 342–43
genus: emphasis on, 342. *See also* taxonomy project
geography: climatic zones, 272; early teaching of, 271; Linnaeus's learning about, 271–72; research effort in (*see* travel)
Gesner, Johann, 404
Gesner (or Gesnerus), Conrad, 125, 306
Gibson, Susannah, 79
Gieses, Albert, 271
Giseke, Paul Dietrich, 341, 352, 438
Gjörwell, Carl Christoffer: on the aging Linnaeus, 388; correspondence with, 437; on Döbelius, 42; on flattery, 7; Linnaeus's foreign correspondence, as potential publisher of, 233; sale of Linnaeus's collections, inquiry regarding, 401; traveling clothes of, 152
glowworms, 168
God. *See* Creator/Creation/role of God
Goethe, Johann Wolfgang von, 326, 403–4
Goffman, Erving, 307
Gorter, David de, 404
Gorter, Johannes de, 110
Gothicism, 72
Gotland, 225
Gotländska resa, 173
Gouan, Antoine, 404
Gourlie, Norah, 66
Gregorian calendar, adoption of, 149
Gripe, Maria, 443
Gronovius, Jacob, 404
Gronovius, Jan Frederik: America, travel to, 273; assistance/support from, 111, 112, 116; association of Linnaeus with *Linnaea borealis*, 15, 235; "Colonel," playful designation as, 306; irritation at restructuring of plant nomenclature by Linnaeus, 129; letter of introduction for Linnaeus, 234; visiting, 118; warning against challenging Seba over the Hamburg hydra, 127

Grundtvig, N.F.S., 217
guinea pig, illustration of, 227
Gustafsson, Lars, 373
Gustafsson, Mikael, 321–22
Gustav III (king of Sweden): absolute monarchy reinstated by, 130–31; anonymous contribution for the Linnaeus memorial in Uppsala cathedral, 391; Bible Commission of, 370; birth of, 165; contributions to Uppsala University in commemoration of Linnaeus, 402; coup by, 309, 313; donation of royal garden to Uppsala University, 402; friendship with, 335; opulent image of the country under the rule of, 264; references by Linnaeus to, 135; remarks on the death of Linnaeus, 393
Gustav Vasa (king of Sweden), 134
Gyllenborg, Carl, 141, 186, 200, 234, 323
Gyllengrip, Gabriel, 76, 78–79, 92, 111, 159

Haak, Dirk, 123
Haartman, Johan, 247, 353
habits: bad, the upper class as the source of, 110, 217; to be noted by explorers, 273; of gifted men, 228–29; lives of men shaped by, 28, 84–85, 297; nature and, 85; of the Sami, commitment to, 428n7; as second nature, 219; sticking to simple, 83
Haeckel, Ernst, 258
Hagberg, Knut, 66, 80, 373, 408, 441
Hägerström, Axel, 362
Hagström, Johan Otto, 236, 366, 391
Halenius, Engelbert, 230, 311, 313
Hallenberg, Jonas, 392
Hallenborg, Carl, 173–74
Haller, Albrecht von: accusation of atheism conveyed by spokesman for, 343; "Colonel," playful designation as, 306; correspondence with, 98–99, 438; death of, 389; dedication of *Clavis Medicinae Duplex* to, 356; *Die Alpen*, 102; irritation at restructuring of plant nomenclature by Linnaeus, 129; number of letters in the archives of, 232; portrait of hung upside down in Linnaeus's grand salon, 235; praise for Linnaeus's work, 404–5; publication of collection of letters, 386; publication of letters from Linnaeus without permission, 233; "The Second Adam," Linnaeus referred to as, 305; Wargentin as mediator between Linnaeus and, 332
Hallman, Johan Gustav, 195–96, 354
Hammarby: crickets at, 325–26; escaping the Uppsala fire at, 319–20; etching of, 319; as a farm, 320; life at, 320; obligations as member of landed gentry, 318; portraits of three daughters hung on the walls of, 322; purchased by the state in 1879, 400; purchase of, 317–18, 430n1; rebuilt and decorated with portraits, 318–19; visitors at, 383–84; wallpaper in Linnaeus's bedroom and study, 321
Hammarskjöld, Dag, 408
Härleman, Carl: disturbing letter from, 230; drawings for the remodel of the prefect's house, 187; expansion of the botanical garden in Skåne by, 170; friendship with, 312, 327; journey to Skåne, support for, 168; objection to clearing land by slash-and-burn, 172; as patron, 111; portrayed as heroic figure in New Year's greeting to, 7; post-fire rebuilding, 147, 309; restructuring of the garden at Uppsala, 195; warning about the new style of clothing worn by the young, 206
Harvey, William, 121, 199, 251
Hasselquist, Fredrik, 28, 114, 174, 276–78, 291, 369
Hats, the: characterized in a flippant pamphlet, 313; formation of, 30–31, 132; Linnaeus and, 111, 282, 312–13; mercantilism of, 240; silkworm cultivation, medal struck for, 149; Uppsala University and, 148; war against Russia waged by, 309
health: critical importance of taste and smell to, 353; height of residence and, 317; illnesses (*see* illnesses); importance of and practice related to, 212–17; "medical primitivism" as an approach to, 417n4; mental (*see* mental health); riding as beneficial for, 160; visiting the country house and, 317; wild strawberries for (*see* wild strawberries). *See also* dietetics; medicine; near-death experiences
Heckscher, Eli, 159
Hedbom, Karl, 390

Hedenblad, Petrus, 92
Hedenius, Ingemar, 363
Hedin, Sven A.: biography of Linnaeus, 438–39; on *Clavis*, 357; on *Iter*, 419n28; on Linnaeus as lecturer, 211; on the relationship of Carl Jr. and his mother, 396; 1733 as a time of desperation for Linnaeus, 91; on Soul and Nature in Linnaeus's approach to medicine, 432n5
Heister, Lorenz (Laurentius), 306
herbations, 203–6
Hermann, Johan, 338
Hernquist, Peter, 433n24
Herodotus, 271
Hesselius, Andreas, 54
Hiärne, Urban, 242
hierarchy: the Bible and, 370; derivation of the concept from three basic principles of nature, 281; imposed on nature, 9; in Linnaeus's taxonomy, 369; Linnean, Darwin and, 404; in the map model of nature, 341; in nature, difficulty of maintaining notion of, 282; ordering humanity with a model of, 300; of the senses, 353
Hildegard of Bingen, 246
Hill, John, 404
Hippocrates, 56, 212
history: ideology and approaches to, 417n33; natural (*see* natural history); pansexual view of, 249
Hjorth, Harriet, 408
Hodacs, Hanna, 442
Hoffman, Martin, 88–89, 106–7, 119
Hoffmann, Friedrich, 58, 213
Hofsten, Nils von, 5
Högström, Pehr, 366
Höijer, Anna Ovena, 216
Holberg, Ludvig, 133, 177, 179
Holgersson, Nils, 408
Holland: almanac kept during the journey to, 104, 109–10; Artedi, death of, 114–15; Boerhaave, interaction with, 111–13; celebrities in science and medicine from, 103; Clifford's garden, manager of, 113–14; club for scientists in Leyden, 116–17; freedom to worship in, 108; hydra affair, 107, 127; illness contracted while in, 27–28; improvement in health upon leaving, 117–18; interest in Linnaeus's line of thought in, 404; no "influence of the Enlightenment" from his years in, 363; sinful goings-on in, 109; travelogues from, 108–9; widening of horizons while in, 272

Homo sapiens, 294–300; introduction of, the two-name classification system and, 291; selection of as name for human beings, 254, 293, 298

Homo troglodytes, 125
honors: *Academia Naturae Curiosorum*, election to, 306; knighthood and change of name, 230, 250, 306–7; official titles, 307–8; Order of the Polar Star, 230, 382; professional pinnacle reached by Linnaeus, 305–6
Höök, Gabriel, 26, 31, 104
Höök, Sven, 399
Hoorn, Johan, 57, 251
Höpken, Anders Johan von: friendship with, 312, 327; on Linnaeus's scientific theology, 254; political face-making by, 111; portrait of at Hammarby, 318; recognition of Linnaeus's professional status by, 305; Royal Academy of Sciences, member of, 135; on the "utility of plenty," 266
Hoppius, Christian Emanuel, 295
Horace, 263
Horn, Arvid, 30–31, 77, 132–33, 312
Hortus Agerumensis, 130
Hortus Cliffortianus: Bartsia in, 111; erotica in, 248; four continents of the world represented in the frontispiece of, 271; frontispiece, designing of, 242; frontispiece, drawing of, 105; gold standard in publication (illustrations, binding) for, 241, 243–44; Hamburg hydra in the frontispiece of, 107; imported plants listed in, 271; introduction to, 113; Linnaeus in the frontispiece of, 306; thermometer in the frontispiece of, 197; writing style of, 176
Hortus Uplandicus I–IV, 53, 56
Hortus Upsaliensis, 195–97, 225–26, 235
Hottentots, 87–88, 251, 297–98, 300
Hottuyn, Maarten, 404
Hübner, Johann, 271
Hulth, J. M., 438
human beings: as animals, 83; tapeworms and, 287; two factors shaping the lives of, 84. See also *Homo sapiens*

human sexuality, 250–51
Humble, Johannes, 47
hunting, 260
hybrids/hybridization, 284–85, 290, 342–43
hydra in Hamburg, 107, 127
Hydrén, Eric, 393
Hydrén, Lars, 152
Hyltén-Cavallius, G. O., 171

iatromechanical school of medicine, 111
Ihre, Johan: attacked at a meeting of the consistorium, 376; as brilliant linguist, 152; description of the aged Linnaeus, 388; election of inspector, votes received in, 310; farm sold to Linnaeus, 318; flippant characterization of, 313–14; as part of the new generation in science, 131; ranking of candidates for professorial chairs at Uppsala, 141; rivalry with, 331; wealth of, 148
illnesses: aging and, 382 (*see also* aging); ague contracted in Holland, 27–28; causes of, 381, 390; classification of, 221–22, 353, 355–56; gout, 303, 382, 389; laughing sickness caused by saffron, 267; Linnaeus's medical history, 389; in 1718, 25; Uppsala fever, 148, 192, 381, 390
Inebrantia, 263
Ingeborg in Mjärhult, 167
insects: bees, 86, 137, 140, 255; collections of/collecting, 5, 35, 50; crickets, 164, 324–25; "Curious Features of Insects," 153–54, 177–78, 259–60, 299; eating, 215; *Furia infernalis*, the "monster" insect, 40, 68, 339, 416n14; *Miracula insectorum*, 8, 40; mosquitoes, 81; "Notes on the Practices of Ants," 137, 140; overflowing of records and the emergence of entomology, 338; transformations of, 290
Instructio perigrinatoris, 274
Insulin, Stephan, 272
interconnectivity, 258, 285
Isidor of Seville, 290
Iter ad Superos & Infernos ("Travels in heaven and hell"), 101–2
Iter Lapponicum: Andromeda episode described in, 107; classical motifs in, 179; drawing of Andromeda threatened by a dragon in, 71; drawing of a Sami woman, 75; drawing of a short-eared owl in, 67; evaluations of, 79–80; language in/writing style of, 65; on a list of planned publications, 130; primitivism in, 296; publication of, 172, 440; read by Meldercreutz, possibility of, 79

Jacquin, Nicholas Joseph, 228, 404
Jaenson, Jaen, 375
Jews: in Hamburg, 104; worship by, 108
Johan af Wingård, 161
Jonsell, Bengt, 429n17, 442
Jönsson, Ann-Mari, 443
Jonsson, Marita, 442
Jonstone, John, 125
"Journey in Dalarna," 80, 101, 418
"Journey in Lapland," 93
Julinschiöld, Peter, 375–76
Jussieu, Antoine de, 118
Jussieu, Bernard de, 118, 285, 306, 404

Kalm, Pehr: appointment to a chair without sitting for examinations, 223; discouraged by Linnaeus from publishing account of 1742 journey, 172; herbaria of, number of specimens in, 291; on Linnaeus's Biblical account of Creation, 155; "Major," playful designation as, 306; plan to search at "Swedish latitudes" for useful plant and animal specimens, 267–68; printing of *Flora Svecica*, support for, 181–82; realization of the extent of "biodiversity" upon landing in America, 337; slippers eaten by opossum, 431n4; travels by, 275–76
Kalmeter, Johan Olof, 324
Kant, Immanuel, 155, 404
Karlfeldt, Erik Axel, 408
Karl XI (king of Sweden), 61, 202
Karl XII (king of Sweden), 134
"key" metaphor, 348–49, 358
Kiellman, Tiburtius, 115
Kiesewetter, Gottfried, 172, 238–39
Kiöping, Nils Mattson, 156, 293
Kjellberg, Fritjof, 409
Kjellgren, Johan Henric, 7
Klein, Johann Theodor, 299
Klinckowström, Johan Mauritz, 202
Klingenberg, Carl, 141

INDEX [471]

Klingenstierna, Samuel, 119, 131, 148, 205, 327, 331
Klintberg, Bengt af, 171
knighthood, 230, 250, 306–7
Knös, C. J., 331
Knös, Olof, 401
Koerner, Lisbet, 159, 428–29n1, 442
Koskull, Anders, 104
Koulas, Samuel, 39–40, 42
Krafft, Per, Jr., 9, 378
Kraken sea monster, 128
Kramer, Johann, 116–17
Krok, Thorgny, 425n5
Krüger, Johan Fredrik, 366
Kuhn, Adam, 308, 320
Kvick, Åke, 21

Lachesis Lapponica, 63, 416
Lachesis Naturalis: dietetics, thoughts on, 212; on habits, 219; on music and dance, 110; near-drowning incident described in, 36; new material for teetotalers' propaganda provided in, 217; notes on sexual and erotic matters in, 251; Swedenborg, reference to, 361
Lagerberg, Anna Christina, 246
Lagerbring, Sven, 27, 39
Lagerlöf, Petrus, 219
Lagerström, Magnus, 200, 202
Lamarck, Jean-Baptiste, 406
La Mettrie, J. O. de, 230
Lamm, Martin, 179–80
land, steady expansion of, 155–56. *See also* water
Landell, Nils Erik, 442
Landell, Torbjörn, 442
Lange, J. G., 241
language: choice of for a speech to the royal family, 177; double name convention as contribution to, 291–92; Latin as the best for science, 176; Linnaeus's use of, 176–80 (*see also* writing style); naming items in Swedish in the royal collections, problem of, 202–3; reform of naming, significance of, 429n17; teaching Carl Jr. Latin, 188; translation of works in Latin, 182–83
Lapland: commission to journey to, 61, 63–64; contempt for the curate/schoolteacher in, 126–27; diary/travelogue describing the journey to, 63–72, 174 (see also *Iter Lapponicum*); discussion of journey to with Reuterholm, 92; evaluation of Linnaeus's travels in, 79–81; ideas adopted from journey to, 72–73; luggage, contents of, 64; melding of economics and natural history in, 77–79; pragmatic goals of the expedition, 73–75; proposals for changes to, 75–77; travels among the Lapps, 73–75
Lapps. *See* Sami, the
Larson, Marcus, 408
Larsson, Lars-Erik, 408
Laurent, Antoine, 404
Lawson, Isaac, 116–17
Laxman, Erik, 178, 383
Leche, Johan, 39, 183, 306
lecturer/lecturing: on botany in Stockholm, 134; on classification of illnesses, 221; climatic zones in botanical lectures, 272; content, examples of, 209–10; description of Linnaeus as, 208–9, 211; on dietetics (*see* dietetics); number of lectures per term, 207; popularity of Linnaeus as, 208, 219
Leeuwenhoek, Antonie van, 103
Leibnitz, Gottfried Wilhelm, 119, 374, 379
Lenngren, Anna Maria, 345
Lepenies, Wolf, 373
Lesser, Christian, 366
Levertin, Oscar, 29, 177, 373, 440
Levnadsordning vid en surbrunn, 303
Lidén, Johan Hinric, 108, 177, 405
Lidforss, Bengt, 441
Lieberkühn, Johann Nathanaël, 116–17, 129
Liedbeck, Eric Gustaf, 165–66
Liljefors, Bruno, 408
Lindblom, J. A., 204–5
Lindeboom, G. A., 112
Linder, Johan, 181, 292, 300
Lindfors, A. O., 217, 437–38
Lindroth, Sten, 80, 148, 441–42
Lindström, Gustaf, 439
Linnaea borealis: association of Linnaeus with, 11, 15; cultivation of, 332; Dalarna, observed in, 93; drawing of, 14, 236; as example of the two-name convention, 291
Linnaean/Linnean Society: London (*see* London Linnean [or Linnaean] Society); Swedish, 251, 443

Linnaean methodology/"Linneanism," 275
Linnaeus, Carl, Jr.: accusation of father's atheism, response to, 343–44; on Artedi, 58; birth of, 137, 187; career achievements, his father's position and, 395–96; cataloging information for *Systema Naturae*, role in, 340; death of, 397; death of his father, aftermath of, 392; dedication of *Nemesis divina* to, 374; on his father's visit to a madhouse in France, 118; his mother's hatred of, 394–96; installed as his father's successor at Uppsala, 316; last days of his father's life described by, 389; on the mouth-to-mouth resuscitation of his sister, 36; portrait of, 395; recruitment to be a demonstrator in the Uppsala garden, 199; on time, 149; upbringing/education of, 187–88
Linnaeus, Nils: background and career of, 18, 20–21; botany and gardening, knowledge of and interest in, 23; death of, 230; his son as a legacy of, 28; literary works provided by, 33–34; revelatory nature study approved by, 365
Linnaeus, Samuel: daughters of regretted not gaining from their uncle's knowledge, 189; education and children of, 27; on his brother's childhood play at medicine, 23–24; on his father's keeness for gardening, 23; intelligence of, 41; promise he would be kept out of dangerous garden, 35
Linnaeus, Sara Lisa (née Moraea): Bäck family, relations with, 329–31; birthdate, 99; courtship of, 97; death of, 400; drawing of, 399; family defense of, 400; as firstborn child, 350; flirtations by, 99; at Hammarby, 320, 323–26, 399–400; hatred for her son, 394–96; inheritance of, 186; letters to/from, 104, 137, 384; Linnaeus's funeral, not attending, 393; marriage of (*see* marriage); period of separation from Linnaeus, 97–99; shopping by, 193; skirmish between Browallius and Linnaeus over, 99–100; unflattering descriptions of, 324, 400; Uppsala fever contracted by, 381; wedding portrait, 139

Linnaeus family: absence of letters from, 233; Anna Maria (sister), 26; Bergcrantz, Carl Fredrik (brother-in-law), 234; Carl the Younger (son) (*see* Linnaeus, Carl, Jr.); Christina Brodersonia (mother), 18–19, 35–36; Elisabeth Christina (daughter, a.k.a. Lisa Stina), 188, 326, 398–401; Emerentia (sister), 27, 37; hatred of mother for her son, 394–96; Johannes (son), 188, 192; Linnaeus's will and, 392; Lovisa (daughter), 188; marriage of Linnaeus and Sara Lisa, 136–39; Nils (father) (*see* Linnaeus, Nils); rare mention of, 331; Samuel (younger brother), 18–19 (*see* Linnaeus, Samuel); Sara Christina (daughter, a.k.a. Sara Stina), 188–89; Sara Lisa (wife) (*see* Linnaeus, Sara Lisa [née Moraea]); Sara Magdalena (daughter), 188; siblings and extended family, overview of, 26–27; Solander and, 397–99, 401; Sophia (daughter), 36–37, 188–89, 192–93, 400; Sophia Juliana (sister), 27; widow Moraea (mother-in-law), 318. *See also* parents
Linnean cameralism, 428n1
Linnéan Foundation, 406
Linneanism, 402–10
Linné Institute (previously Zoophytolithic Society; *Societas pro historia naturali*), 406
Linnerhielm, Jonas Carl, 408
Locke, John, 148, 362
Löfling, Pehr, 182, 226, 277, 287–88, 291–92
London, England, 117
London Linnean (or Linnaean) Society: Åhrling and, 439; books donated by, 438; Camper's refusal of honorary membership in, 405; drawing from *The Linnean Herbarium* of, 236; founding of, 403; *Letters to Linnaeus* published by, 405; Linnaeus's working notes kept in, a few sheets of, 371; source material held by, 64, 401, 437; Uggla's visits to, 443
Louis XV (king of France), 177
Lovejoy, Arthur O., 441
Lovisa Ulrika (queen of Sweden): bananas presented to, 269; collections amassed by, 201–2; dedication of *Västgöta resa* to,

INDEX [473]

165; demonstration of Münchhausen's "discovery" sent to, 339–40; failed coup by, 333; friendship with, 327, 332–33; language skills of, 177; as patron, 111; references by Linnaeus to, 135; views on Linnaeus's sexual system not known, 248
Löwenhielm, Carl Gustaf, 314
Lucretius, 353
Ludwig, Christian Gottlieb, 404
Lund, Sweden, 37
Lundborg, Herman, 80
Lund University, 37–43
Luther, Martin, 20, 217, 407
luxury: debate about, 266; the primitive versus, 72; of the royal court, mixed emotions about, 333; slavery to pleasures and, 113
Lyell, Charles, 259

Magnus, Olaus, 61, 64, 80, 126
Malmeström, Elis, 370, 373, 375, 440–41
mammalian class: defining by exclusion in recognizing the need for, 210; fish species placed in, 128; launching of, 280; name of, reason for selecting, 292–93; teeth as a defining characteristic of, 382. See also *Homo sapiens*
Månsson, Arvid, 34
Manuscripta medica, 128, 159
Manuscripta Mennandria, 119
marriage: events of and leading to, 136–39, 189; rumors about, 189–92; as a union of convenience, 189
marrow-bark thesis, 287–90, 342, 346, 350, 359, 381, 383
Marshall, Joseph, 383
Marsigli, Luigi Ferdinando, 285
Martin, Anton, 383
Martin, Elias, 408
Martin, Roland, 286, 332
Martinson, Harry, 408
Martyn, Thomas, 340
Masson, Francis, 235
Materia Medica, 221
Maupertuis, Pierre de, 61
medicine: alcoholism, 217; animal derived, 433n19; bedding a child inside a slaughtered sheep, practice of, 37; benefits of traveling for doctors, 154;

botany and, 220 (*see also* botany); *Clavis Medicinae Duplex*, 346–53, 356–58; clinical work, status of, 221–22; country healers, visits with, 166–67; doctorate in, travel to Holland for, 103–4, 110–11 (*see also* Holland); iatromechanical school of, 111; illnesses, classification of, 221–22, 353, 355–56; "keys" to, 348–49; lecturing in, 211–12; medical texts, Linnaeus's library of, 56; medications, formulation and classification of, 222; mistakes in contemporary, 220; natural form of/medical primitivism, 82–83; number "five," central role of, 349; pharmacy interior, frontispiece showing, 354; plan for, creation of, 358; populist approach to education in, 220–21; schools of thought on, 212–13, 349; serious study of, human dissection and, 134; single (as opposed to composite) medicines, champion of, 353, 357; Stahlian, 38–39; study of at Lund, 38, 42; study of at Uppsala, 48–49; synthesis of nature and, 356; use of human senses in, 355–56; vegetation-derived and from native sources, preference for, 356; wild strawberries as (*see* wild strawberries); youthful determination to become a doctor, 34–35. *See also* dietetics; health
Melander, Eric, 313
Melanderhjelm, Daniel, 205, 316, 402–3
Meldercreutz, Jonas, 78–79, 152
Mendez da Costa, Emanuel, 248
Meniskans cousiner (The cousins of human beings), 298
Mennander, Carl Fredrik: biographer, consideration as, 437; bishopric, appointment to, 367; Browallius's "falsehood" revealed by, 100; correspondence with, 133, 385; defense of animals, vegetarianism and, 216; friendship with, 328; *Homo* in lecture notes of, 293–94; humans as animals, 83; lectures commissioned by, 53; Linnaeus's autobiography and, 2–3; as part of the new generation, 131; peacocks passed on to, 201; as pupil of Linnaeus, 91; Rosén married into family of, 92

mental health, 5–6
Merian, Maria Sibylla, 124, 246
Metamorphoses plantarum, 289–90
Metamorphosis humana, 186, 192, 305, 381
metaphors, use of, 350
methods: cataloging, 340; data management, 229–30; Linnaean methodology/"Linneanism," 275; natural, 281
Methodus, 124
Methodus avium sveticarum, 60
Michaelis, Johann David, 371
Miller, Philip, 116–17, 234, 404
mineralogy: importance to explorers, 273; Linnaeus's interest in, 56, 59, 93, 101, 273; sculpture icon, 409; subject of study, 38, 42–44
minerals: classifying, 96–97; collection of willed to son, 392; dowsing rods used to find, 170; lecture on, 170; origin of, 258; shells of crustaceans/mollusks and corals classed as, 215, 285; Stobæus's collection of, 38. *See also* stones
Miracula insectorum, 8, 40. *See also* insects
Momma, Peter, 238–39
Monson, Lady Anne, 246, 308
Monstrosi/monster, 297–98
Montelius, Oscar, 407
Montesquieu, Charles Louis de Secondat, Baron de, 87, 148
Montin, Lars, 286
Moraeus, Johannes, 97, 99
mouth-to-mouth resuscitation: by Linnaeus, 36–37; of Linnaeus, 36
Mozart, Wolfgang Amadeus, 5
Müller, Otto Friedrich, 286
Müller-Wille, Staffan, 442
Münchhausen, Otto von, 339–40, 345, 383
Mundus invisibilis (The Invisible World), 339
Munthe, Isac, 37
Murray, Johan Adolph, 347, 349, 383, 388, 392–95
Musa Cliffortiana, 106, 124
Museum Adophi Friderici: drawing of white-throated capuchin monkey, 336; drawings of fish, 334
Museum regis Adolphi Friderici, 241
Museum Tessinianum, 241
music written in honor of Linnaeus, 442–43
Musschenbroek, Pieter von, 103, 351
Mylius, Martin, 56

myths/mythology: Andromeda in Lapland notebook, 69–71; Charon myth, 416n14; classical imagery in species names, 427; description of nature and, 263; mobilized in support of present-day motives, 77; old wives' tales, 164; persistence of myths, 164; Samson, 50, 92; "the hydra in Hamburg," 107

Näcken (capricious water spirit), 36
naming of plants, 235, 246
Näsman, Reinhold, 92
natural history: alchemical genealogy of the first steps of development of, 361; battle against the old, 107; consequences of lacking knowledge about, 8; contributions to, 128–29; "crypto-zoology" addressed in *Paradoxa*, 124–27; curiosity and sensual pleasure in the study of, 253–55; debate between Buffon and Linnaeus, 283–85; declining interest in, 344–45; development from Genesis onward through continuous change, theory of, 155–56; dictionary of, 342; first Swedish book of, 125; global explorations of, 273–78; illustrations/"epistemic images" and, 426n11; "natural philosophy" of Linnaeus, 56–57; utility in the study of, 255
Naturaliesamlingar (Collections of *Naturalia*), 287
nature: appreciation of numerical order in, 11; cameralism and the attempt to rule over, 269; "chain of" and "ladder of," distinction between, 282–83; change in, status of continuity and hybridization in, 281–82; completeness, growing toward, 282; creating order in, fixed terminology and the double-name convention, 290–92, 429n17; growth as a fundamental characteristic of stones, 361–62; interconnectivity in, 258, 285; man and, division between, 257; meanings and descriptions of, 257–63; order of/systematic classification of (see *Systema Naturae*; taxonomy project); as "plenitude" in the eighteenth century, 431n1; religion and (*see* Creator/Creation/role of God; religion); sexual reproduction relied upon by, 128 (*see also* sex); study of in the service of

God, scientific theology and, 254–55; three principles showing the foundation of, 281; tripartite subdivision of, 340; universal laws of, 127–28
Nauman, Johan Justus, 227
Neander, Andreas, 311, 377
near-death experiences: drowning, 36; insect bite, 40–41; on three occasions, 41
Nemesis divina (Divine Retribution), 373–79; as a collection of moral fables, 179; compassion for the poor in, 26; execution of child-murderers, discussion of, 316; foreign policy fiasco, note on, 132; Fries's self-censorship in discussing, 440; Malmeström's defense of Linnaeus's morality in, 440–41; moving in the deep shadows where ghosts are lurking, 6, 9; the punishing God and stories from the Old Testament, passages on, 366; roll call of righteously punished sinners in, 132; story involving his brother and a fortune teller, 41; the supernatural addressed in, 171
Neptunist, 156–57
networks. *See* professional networks
Newton, Isaac, 205, 305, 361, 372, 402
Nietzel, Dietrich, 195, 197–98
Noctiluea marina, 263
Norberg, Matthias, 275
Nordenflycht, Hedvig Charlotta, 174
Nordström, Lubbe, 408
nostalgia/homesickness, 27–29
Notata subitanea, 349, 351
"Notes on the Practices of Ants," 137, 140
Notke, Bernt, 107
novels about Linnaeus, 443
Nutrix noverca, 134, 293
Nyberg, Kenneth, 442
Nygren, Anders, 362

Observationes in Regna Naturae, 156
Obstacula Medicinae, 220
occult, the: ghost story in Lund, 41; Linnaeus's interest in, 8, 359–61, 363–64; Näcken (capricious water spirit), 36; number mysticism, 364; spirits roaming at Stenbrohult, 27; supernatural phenomena, belief in, 171; witchcraft, 18, 164
Odhelius, Lorens, 286

Ödmann, Samuel: biblical philology, studies of, 369; on excessive drinking, 217; full revision of *Fauna Svecica* proposed by, 183; on herbations, 205; on the lake of Linnaeus's near-drowning experience, 36; Linnaean impact on the writing of, 408; Linnaeus's school years illuminated by, 31–32
Oeconomia Lapponica, 64–65, 75–76
Oeconomia naturae, 226, 258–59, 337
Olai, Benedictus, 213
Öland, 225
Öländska och gotländska resa, 178, 239
Olivecrona, Karl, 362
Olofsson, Rune Pär, 443
Omai, Tahitian, 397
"On Lapland and the Lapps" (*Om Lappland och lapparne*), 73
"On the Growth of Habitable Parts of the Earth," 155–56
Oratio de incremento, 258
Oration de increment telluric habitable, 267
orderliness, passion for, 368–69
Örn, Nils, 107
Osbeck, Pehr, 161, 205, 221, 268, 383
Osslund, Helmer, 408
Östenson, Pia, 442
Ovid: the Golden Age, vision of, 72; identification with, 70; Linnaeus's mental geography, as an element in, 65; Linnaeus's preference for, 177; *Metamorphoses*, 381–82; mistaken attribution of quote to, 353; motto above the bedroom door at Hammarby from, 318; mythology in descriptions of nature, 263; as "the poet," 68; shepherds, stories about, 73
Oxenstierna, Axel, 71
Oxenstierna, Erik, 170–71

Palmberg, Johannes, 34
Palmstruch, Johan, 14
Pan Europaeus, 126
Pan Svecicus, 167, 291
parents: career choice, response to, 35; description of, 19; father (*see* Linnaeus, Nils); mother, 18–19, 35–36
Paris, France, 118
Patriotic Society, 3, 102
pearl fishing, 267
Pedersdatter, Johanne, 18

Peterson-Berger, Wilhelm, 408
Petri, Laurentius, 325
Petri, Olavus, 177
petrified objects, 120
Petry, M. J., 373, 375
pets: cat, 200; childhood dog, 35, 37, 199; dogs defended against vivisection experiments, 199; list of in Uppsala, 200; parrot and Pompe the dog at Hammarby, 325; peacocks, 201; raccoon, 200; talking parrot at Uppsala, 200–201
Peyssonal, Jean André, 285
Pfeiffer, Johan August, 144
Pharmacopeia Svecica, 222, 328, 353, 356
philosopher, Linnaeus as, 362–63
Philosophia Botanica: color is of no value for definitions, 296; immutability emphasized in, 121; investigation of nature regarded as philosophy, 362; the marrow-bark thesis in, 288; *Methodus* reprinted in, 124; names, centrality of, 290; "natural system," suggestion of, 341; "nature abhors a vacuum" phrase crossed out in private copy of, 341; "nature does not jump" dictum found in, 281; "*nulla species nova*" phrase crossed out in private copy of, 342; piece of paper with Sophia's name glued into, 189; principles for any flora established in, 183; travels of great botanists, listing of, 273
physicotheological thought, 366–67
Pico della Mirandola, Giovanni, 298
Pilgren, Johan, 380
Pitton de Tourneforts, Joseph, 49, 53–54, 100, 235, 273
Plantae esculentae patriae, 265
Plantae Surinamenses, 389
plants: acclimatization of foreign, 267–68; animals and, distinction between, 286; black henbane, curious effects of, 34; bog rosemary (*Andromeda polifolia*), description of, 69–71; horticultural research, mostly crazy, 268–70; hunger and the need to raise level of knowledge about, 265; *Linnaea borealis* (see *Linnaea borealis*); "medical powers," classification according to, 355; naming of, 235; natural order of, order of illnesses and, 356; natural system of based on tastes and smells, 355; at night, 262–63; "power" or "virtue" of, 346; sexual reproduction in, reactions to Linnaeus's system of, 247–52; sexual system, drawing of, 122; tea bush, 144–45; transformations of, 290; word-and-number system for analyses of, 243. See also *Flora Svecica*; gardens

Plato, 283, 287, 355
Pliny the Elder, 271, 392
Plumier, Charles, 273, 318
Plumier, Ehret och, 321
Plumtree, James, 442–43
Plutarch, 373
Poe, Edgar Allan, 5
poet, Linnaeus as, 179
Polhem, Christopher, 73–74, 131–32, 216
Politia naturae, 258–59
politics: education policy and, 314; limited appeal of, 282; loyalty to the king, 313; party association, 282, 312–13
polyps, 272, 280, 285–86, 339, 359, 383
Pontins, Samuel, 108
Porphyry, 115
portraits of Linnaeus: age 67 by Krafft, Jr., 378; by Bernigeroth, 261; bust with Greco-Roman gods and goddesses, 10; in his normal clothes by Rehn, 262; by Roberg some time prior to 1735, 98; in Sami costume, 88; statue by Kjellberg, 409; wedding portrait by Scheffel, 138; the young Linnaeus, 24, 32
poverty: of grammar students at Växjö, 32–33; Linnaeus's food consumption shaped by upbringing in, 324; Linnaeus's self-description of his, 26; in Sweden, 264–65; at Uppsala, 46; "walking the parish" in Stenbrohult, 33
Praelectiones Botanicae Publicae, 53
Praeludia sponsaliorum, 53
Preussische Akademie der Wissenschaften, 135
primitivism, 72, 83, 296, 417n4
Printing Trade Society, 238
professional career/accomplishments: career options, 90; choice of career, 44; disputations and dissertations, 44; inventions, 8–9; number of books written by Linnaeus, 238 (*see also* publication/publishing); parental reaction to early career preferences,

35; summary of, 8; at Uppsala (*see* Uppsala University). *See also* honors
professional networks: correspondence/letters, 232–34; disciples/traveling students, 235–37; methods of strengthening, 234–35; need for, 231–32; variety of, 232
professional rivals. *See* enemies/rivals/criticisms
publication/publishing: of correspondence/letters, 233–34; "epistemic images," 426n11; fees and payments, 244–45; fine writing, ambitions to produce, 102; illustrations and drawings in, 241–44; of Linnaeus's works, 238–42; plans for, 100–102, 130; reviews, 245; in Sweden, 238–40; of travelogues, censorship and, 172–73. *See also* books; writing
Pulteney, Richard, 437

rabbit-hen chimera, 284, 343–44, 415n4
Rabenius, Olof, 315
racism, 299
Ramazzini, Bernard, 229
Råshult, Sweden, 17–18, 20–23
Ray, John, 49, 58, 125–26, 347
Réaumur, Ferchault de, 284–85, 343–44
Rehn, Jean Eric, 184, 241, 262, 391
religion: atheism, accusation of, 343–44, 365–66; Bible, references to, 360–61, 366; the Bible, zoology and botany of, 369; "Book of Nature," the Bible as, 365–67, 369–70; clergy, distrust of, 254, 367; examination in theology required before foreign travel, 103; on the freedom to worship in Holland, 108; influence on Linnaeus, 363, 365; Linnaeus as one of the elect reading the Book of Nature, 369; Linnaeus's view of, 7, 370; minor role in writings, 68; physicotheological thought, 366–67; science and, 367; sources on Linnaeus's, 434. *See also* Creator/Creation/role of God; *Nemesis divina*
reproduction: eighteenth-century theories of, 287–89; Linnaeus on (*see* marrow-bark thesis; sex)
Retzius, Anders Jahan, 345
Retzius, Nils, 39, 41–42, 159
Reuterholm, Axel Gottlieb, 92, 133, 267

Reuterholm, Esbjörn, 170
Reuterholm, Gustaf Adolf, 323
Reuterholm, Gustaf Gottlieb, 92–93, 99
Reuterholm, Nils: Academy of Sciences, role in founding, 78; *Classes Plantarum* dedicated to, 111; Dalarna, commission to journey in, 92; praise of Linnaeus for his travels in Dalarna, 95–96; proposal for Lapland by Linnaeus, 76; provincial travel project, idea for, 159; shifting dependence to, 89; spending Christmas with, 92; Wolff, impressed by, 119
rhubarb, 268
Rhyzelius, Andreas, 33
rivalries. *See* enemies/rivals/criticisms
Robeck, Johan, 231
Roberg, Lars: Artedi as student of, 58; correspondence with, 64, 66, 71–72, 102; death of, 186; egg and seed, thesis discussing the similarities of, 121; honoring the memory of, 154; illustrator, skill as a, 242; leave from professorial chair at Uppsala, 140–41; on the medical faculty at Uppsala, 48; as part of the street scene in Uppsala, 152; portrait of Linnaeus painted by, 98; *Spolia Botanica* dedicated to, 53; on youth in the kitchen, 219
Rolander, Daniel, 188, 192
Rollin, Carl Gustaf, 375, 377
romance: courtship of Sara Lisa Moraea, 97; period of separation from Sara Lisa Moraea, 97–99; skirmish between Browallius and Linnaeus over Sara Lisa, 99–100
romantic nature philosophy, 345
Rosen, Björn von, 408
Rosenberg, Erik, 408
Rosén von Rosenstein, Nils: apparent death, childhood experience of, 36; Bäck, correspondence with, 329; Benzelia scandal, role in, 51; Bible Commission, member of, 370; competition for professorial chairs at Uppsala, 140–41; conflict/rivalry with Linnaeus, 49–50, 92, 332, 374–75, 440; deal with Linnaeus regarding subject matter to be addressed, 186, 212, 221, 312; dedication of *Clavis Medicinae Duplex* to, 356–57; honor rescued by Fries,

Rosén von Rosenstein, Nils (*continued*) 192; management of Uppsala garden assumed by, 195; as part of the new generation in science, 131; reconciliation with Linnaeus, 331; Roberg, as a pupil of, 48; the Sami, question about, 63; Stobæus, as a pupil of, 39; student rowdiness, writing off, 311; tapeworm and polyps, paper on, 286; Uppsala, move to, 43; at Uppsala in the 1740s, 148

Rosicrucianism, 360, 363

Rosicrucian Order, 56

Roslin, Alexander, 9

Rostius, Christopher, 38

Rotheram, John, 387–88, 390

Rothman, Johan: co-author of *Hortus Agerumensis*, 130; dancing spirits at night, observation of, 27; decision to go to Uppsala as a student, contribution to, 43–44; gifts of books to Linnaeus, 57; *Hortus Agerumensis*, co-author of, 130; international standing of, 31; as mentor/contribution to development of Linnaeus's mind, 31, 37, 39, 111; notice of Linnaeus by, 47; visit from and time spent with Linnaeus, 104

Rothman, Johan Gabriel, Jr., 4, 189–90, 192

Rothof, Lorens, 264

Rourke, Kelley, 430n4

Rousseau, Jean-Jacques, 29; death of, 389; distrust of formal schooling, 188; era of, 317; positive comments on Linnaeus's writing style, 177; praise for Linnaeus, 403; primitivism influenced by, 72; use of senses, appeal of, 355

Roux, Frederic François Joseph, 32

Royal Academy of Sciences. *See* Academy of Sciences

royal court: collections, working on, 202; distancing himself from in funeral instructions, 391; interaction with, 332–33, 335–36; as patron of the Royal Academy of Sciences, 135. *See also* Gustav III (king of Sweden); Lovisa Ulrika (queen of Sweden)

Royal Printers, 238

Royal Society (London), 135

Royal Society of Sciences (Uppsala), 61, 63, 66, 71, 73–74, 80

Royen, Adriaan van, 306, 404

Rudbeck, Johan Olof, 70

Rudbeckianism, 89

Rudbeck the Elder, Olof: anatomical dissection theater built by, 151; *Atlantica*, 64, 123–24; botany, as professor of, 403; herbations arranged by, 203; son as successor to, 395; species inventory by, 48

Rudbeck the Younger, Olof: as academic successor to his father, 395; accusation against the Sami rejected by, 300; Benzalia scandal in the home of, 51; birds, illustrations of, 42, 49; botany, as professor of, 403; *Campus Elysii*, 41, 273; celebratory poem addressed to, 50–51; Charon myth, version of, 416n14; creation of the garden at Uppsala, 194–95; Gothic ideas on geography, 271; home-based science education for the family of, 188; *Hortus II* dedicated to, 53; illustration of *Linnaea borealis*, 15; illustrator, skill as a, 242; initial meeting with, 47; interest in the climate of Lapland, 75; journey to the north with Karl XI, 61; leave from professorial chair at Uppsala, 140; lecture on sexual biology attributed to, 250; Linnaeus as successor to, 142; manna, search for the real substance of, 369; as mentor/patron, 48–49, 58, 64, 69, 111; mythical references in writings of, 69, 73, 416n14; prefect's house built for, 186; professor of medicine at Uppsala, 47–51; pygmies of antiquity compared to the Sami, 77; residence of Linnaeus with, 59, 440; resources of Lapland praised by, 73; shifting dependence from, 89; third wife of, Linnaeus's condemnation of, 51, 53; writings on Lapland, 64

Ruskin, John, 217

Russian spy, suspected of being, 163

Ruysch, Fredrik, 103

Sachs, Julius, 441

Sahlgren, Jöran, 251

Said, Edward, 433

Salvius, Lars: anecdotes about in conversations with Beckmann, 383; correspondence with, 241, 245; donation for travel by Linnaeus's students, 276; on

INDEX [479]

fees and payments in the publishing industry, 244; *Flora Svecica*, printing of, 181–82; Linnaeus in *Learned News* published by, 254; as printer for Linnaeus, switching to, 172; publication of works by Linnaeus, 239–40; as Sweden's leading publisher, 238–41; *Systema Naturae*, printing of, 279; *Västgöta resa*, printing of, 161

Sami, the: celebration of, 82, 86–89; dressing like/telling stories about during journey to Holland, 107; endangering the culture of, 78; as enlightened but wild, 300; habits of, commitment to, 428n7; identification with, 88–89; ideological content of the Linnaean discourse on, 79; limited interest of the Society of Sciences in, 63; Linnaeus in dress of the, 88; Native Americans and, 296; as our teacher, 83–84; recalling while aboard a ship, 104; religion of, 68; running speed of, question regarding, 84; sorcerers, defense against rumors that they were, 299–300; travels among the Lapps, 73–75; woman on the mountainside, drawing of, 75

Samson, 50, 92
Sandel, Benjamin, 92
Sandgren, Gunnar E., 443
Sapor medicamentorum, 348
Sätherberg, Herman, 39
Satyricon, 373
Sauvages de Lacroix, François Boissier de: anecdotes about shared in conversations with Beckmann, 383; correspondence with, 133, 221, 438; dedication of *Clavis Medicinae Duplex* to, 356; high number of illnesses, belief in, 356; ideas for schooling his baby son, 187–88; plant smells, testing the effect on the nervous system of, 352; positive comments on Linnaeus's writing style, 177; support for Linnaeus from, 141; systematization, following Linnaeus in supporting, 404

scandal(s): appointment to the Lars Roberg chair as, 141; Browallius and, 99; Greta Benzelia, 51; Linnaeus's easygoing attitude on sex as, 100, 247–48; rumors about the Linnaeus household, 189–92

Scheele, Carl Wilhelm, 148–49, 352
Scheffel, Johan Henrik, 138–39
Schefferus, Johannes, 61, 64, 319
Schenson, Emma, 407
Schiebinger, Londa, 292–93
Schillmark, Nils, 360
Schmitz, Helen, 442
Schönberg, Anders, 318
Schück, Henrik, 51, 177, 179–80
Schultz von Schultzenheim, David, 332
science: fantasy and, drawing the boundary between, 125; Linnaeus and, 6–8; myth combined with, 361; new generation in Sweden of, 131–36; publicizing, information overload and, 338–39; religion and, 367 (*see also* religion); search for completeness/truth, difficulty of reconciling perfection of God's Creation and, 338; wit and literary learning as more valued than, 345
scientific rivals. *See* enemies/rivals/criticisms
scientific theology, 254–55, 259, 273, 275
Scopoli, Johann Anton, 404, 438
Scriver, Christian, 365
sea, fear of, 104, 163
Seba, Albertus, 103, 107, 114, 116, 127, 201
Sefström, Eric, 200
Selander, Sten, 408
Semina muscorum detecta, 123
Senium Salomoneum (Old age according to Solomon), 380
senses: acquisition of knowledge through, Lucretius on, 353; as channels of information, 355; hierarchy of, 353; Linnaeus as a sensual man, 353; plant smells, effect on the nervous system of, 352; tastes and smells as the basis for a natural system of plants, 355; use of in medicine, 355–56
Serenius, Jacob, 158
sex: advice about plants and, 35, 221; as a basis for classification, 59, 123, 128, 130, 137; human sexuality, text on, 250–51; human sexuality, transformations with age of, 192; man and woman, differences between, 300; plant sexuality, drawing of, 122; reactions to Linnaeus's interest in, 247–52; reproduction in plants, illustration from *Sponsalia*

sex (*continued*)
  plantarum, 52; reproduction of plants and, 54–56; reproduction of plants based on, 54–56; reversal of established sexual norms, 350
Shaw, George Bernard, 5
Sherard, William, 117
shoes: without heels, 84; wooden clogs, 174
short-eared owl, drawing of, 67
Sidrén, Jonas, 376, 393
Siegesbeck, Johann Georg, 100, 306
silicosis ("stone dust lung"), 95
silk from worms eating mulberry tree leaves, 268
Skåne: journey to, 168, 170–75; map of, 169
Skånska resa, 170–74, 230
Skjöldebrandt, A. F., 15
Skuncke, Marie-Christine, 8
Skyttean Foundation/Society, 66, 416
Sloane, Hans, 112, 116–17, 135
Smith, Adam, 266, 427n1
Smith, James Edward, 233, 337, 401, 404, 438
social conscience, 265–66
Society of Sciences. *See* Royal Society of Sciences (Uppsala)
Söderbaum, Henrik Gustaf, 439
Söderberg, Olof, 170–71
Sohlberg, Claes, 92, 97, 103–4, 112, 115
Solander, Carl Daniel, 398
Solander, Daniel: available light and work habits of, 229; on the death of Johannes (younger son), 192; hostility to Linnaeus, 332, 387–88, 399; Linnaeus Jr. present at the death of, 396; love for as a scientist, 236; painting of, 397; parentage of and relations with the Linnaeus family, 397–99, 401; as part of the street scene in Uppsala, 152; reports from Cook's explorations, source of, 340; study of Linnaeus's works by, 404
Solander, Magdalena (née Bostadia), 398–99
Somnus plantarum, 262
Sörlin, Sverker, 277, 442
Soulsby, Basil, 438
Sourander, Patrick, 390
sources on Linnaeus: autobiographies, 2–4, 437; biographies, 438–42; correspondence (*see* correspondence/letters); lecture notes, 437–38

Sparre, Ulla, 246–47
Sparrman, Anders, 213, 273, 387, 399
species: Buffon's critique of Linnaeus's definition of, 283; challenges of defining and changing role of, 342; classical imagery in names of, 427; concept of in *Systema Naturae*, 121; counted by Linnaeus, 405; estimated number of, 337–38, 405; extinction, at risk of, ix; generation of new through hybridization, 284–85, 290, 342–43; of local flora, number of, 204; "splitters or lumpers" on the issue of how to define a, 338, 356; of Swedish fauna, number of, 185; theory of rooted in Genesis, 155–56, 281. *See also* taxonomy project
Species Plantarum: absence of images in, 242; book production, considerations addressed in, 239–40; countries and gardens studied by Linnaeus, list of, 194; double name convention applied to all known species, 291–92; foundation of Linnaeus's fame, contribution to, 279; list of planned publications, on a, 130; new edition of, rivalry between Banks and Solander and, 388; on the number of plants, 337; pain from kidney stones while working on, 389; plants given names based on Swedish notables in, 235; three principles showing the foundation of nature, 281; working himself to death on, question of, 292
Spegel, Haquin, 125
Spolia Botanica, 21, 53
Sponsalia plantarum, 52, 226
Spöring, Herman, 141, 286
Spreckelsen, Johann Heinrich von, 107
Stauffer, Richard C., 259
Stearn, William T., 248, 291, 442
Steever, D. H., 37
Steinmeyer, Johann, 203
Stenbock, Magnus, 18
Stenbrohult parish, Sweden: birth and childhood residence in, 17–18; fostering of Linnaeus's qualities in, 21–23; Linnaeus's description of, 21; population and local authority in, 20; return visits to, 25–26, 91; spirits roaming at, 27; summers at, 35–36; Växjo, distance from, 30–31; "walking the parish" in, 33
Stobæus, Florentina, 43

Stobæus, Kilian: correspondence with, 49–50, 56, 126, 234; description of, 38–39; fact-finding in the field, inspiration for, 158–59; as mentor and patron, 38–44, 111; notice of Linnaeus by, 47; wishing to go back to, 46
Stobæus the Younger, Kilian, 43
Stockholm, Sweden: coffeehouses in, 132–33; Linnaeus in, 132–37
Stoever, Dietrich, 305, 396, 438
Stoics, the, 257
"stone dust lung" (silicosis), 95
stones: creation of rocks, 120; generation of, water and, 361; growth of, 55, 120–21, 282, 361–62; lifted by the sky, 127. *See also* minerals
Strabo, 271
Strandell, Birger, 390
strawberries, wild. *See* wild strawberries
Strindberg, August, 179, 406–8
suicide, 231
Sundbärg, Gustav, 407–8
Sundgren, Jöran, 315
supernatural phenomena. *See* occult, the
Svanberg, Seger, 78
Sven in Bragnum, 166–67
Svensson, Sigfrid, 171
swallows spend winter underwater, belief that, 128
Swammerdam, Jan, 7, 103, 366
Swedberg, Jesper, 87
Sweden: abolition of absolute monarchy and the Hats vs. the Caps, 30–31; the Age of Liberty, 130–31, 148, 175; divisions within the country, 265; love of nature in, Linnaeus and, 407–8; new generation in, science and, 131–36; orderly subdivision of, 369; poverty in (*see* poverty); public road network in, 160; the publishing industry in, 238–39; Skåne as showplace for, 170 (*see also* Skåne); small population of, 266; tea cultivation, attempt at, 145; wars against Russia and Denmark, 18, 309
Swedenborg, Emanuel: "About the Motion of the Earth and the Planetary Movements and Relationships," reference to, 82; academy membership, Linnaeus's support for, 136; the egg used as a representation of the world by, 123; erotic fantasies recorded while visiting Holland, 109; Linnaeus and, 361–62; myth and science combined by, 361; new generation of Swedish scientists, influence on, 131; observations on Amsterdam, 108; occult undercurrents driven by, 360; as the original teacher, 407; outdoor privy, seated in, 137; positive effects of wandering in a garden, 352; *Routledge Encyclopedia of Philosophy*, entry in, 362
Swedish Linné Society/Swedish Linnaean Society, 183, 251, 443
Swedish Medical Association, 123
Swedish Tourist Association, 408
Swieten, Gerard van, 356, 404
Sydenham, Thomas, 57, 355
Sydow, Carl Otto von, 80
Systema morborum, 252
Systema Naturae: aging and a diminished capacity to work on, 386; all things beginning in water, concept of, 361; arthritis hindering work on, 230; banana, name for, 268–69; begins with instruction to "know thyself," 299; biblical references/references to God in, 360–61; cataloging method for the new edition of, 340; daughters in, 193; double name convention applied in the tenth edition, 291; expansion through successive editions of, 279–80, 358; first appearance of in 1735, 101; frontispiece of 1760 edition, 306; the "Furia" mentioned in, 40; *Geographia naturae* as early title of, 119; geographic boundaries not indicated in, 273; grand design of illustrated in selected passages, 119–21; *Histoire naturelle* (Buffon) as the only rival to, 283; human population divided by continents in first edition of, 296; "key" metaphor used in presentation of sexual classification system, 348; Kraken sea monster included in, 128; as a long-term project, 338, 358–59, 364; Mantissa as supplement to, 339; manuscript taken on trip to Holland, 119; marginal annotations as method of storing data for, 229–30; *Methodus* as a printing on one sheet, 124; *Mundus invisibilis* (The Invisible World) as a commentary on the animal section

Systema Naturae (continued)
of, 339; naturalistic view of humanity in, 254; new human species included in, 370; numerical order in *The Well-Tempered Clavicle* and, 11; *Paradoxa* group of, 344; *Paradoxa* (zoological curiosities) taken up in, 124–27; printing of the first edition, 123–24; printing of the fourth edition supervised by Bäck, 328; publication, note regarding delays in work for, 241; published at Gronovius's expense, 116; religion in introductory words of later editions, 367; special contributions to the completion of, 406; species, concept of, 121; target readership for, 245; theory of spontaneous generation excluded from, 127–28; three principles showing the foundation of nature, 281; title page, 280; twelfth edition as crowning achievement, 347–48; Wallerius's criticisms of, 141; working on at Hammarby, 323

Täckholm, Vivi, 408
Taenia, 286, 289
tapeworms, 226, 286–89, 370
Tärnström, Christopher, 145, 277
taxonomy project: evolution of in successive editions of *Systema Naturae*, 279–81; feasibility of, 337–39; fixed terminology/nomenclature and the double-name convention in, 290–92, 429n17; map model for, 341; the micro level world and feasibility of, 339–40; names of classes, selection of, 292–93; "natural system" as a basis for, dream of, 341–42; orderliness reflected in, 369; ranks of the taxonomic system, 123; sexual system of classification used in, 123 (*see also* sex); species (*see* species); summary of, 128–29; travel and, 86; tripartite subdivision of nature and, 340. *See also* classification; *Systema Naturae*
tea bush, 144–45
teacher/teaching: disciples of Linnaeus, 235–37; faculty roles at Uppsala, 310–16; lectures, 207–10 (*see also* lecturer/lecturing); populist educator, role as, 245; populist medicine, 220–21; students, number of, 208; supervising dissertations and presiding at disputations, 224–27; tutoring at Uppsala, 50. *See also* education
Tegnér, Esaias, 402
Telander, Johan, 31
Tersmeden, Carl, 58, 115
Tessin, Carl Gustaf: bound copy of *Fauna Svecica* given to, 234; collections maintained by Linnaeus, 202; correspondence with, 279, 337, 438; dictionary of natural history project offered to Linnaeus, 342; friendship with, 312, 327–28; gold medal awarded to Linnaeus, 281; horticultural experiments by, 266; Linnaeus as the stoat or ermine in fairytale by, 306; mind of Linnaeus, description of, 78; as patron/supporter, 111, 133; portrait of at Hammarby, 318; portrait of Linnaeus in the home of, 348–49; support from Linnaeus for, 240; visiting in later life, 271; wife's interest in natural history, 246
theodicy problem, 379
thermometers, 195, 197
Thorild, Thomas, 403, 406
Thornton, Robert John, 442
Thunberg, Carl Peter: biblical philology, studies of, 369; contemptuous reference to, 403; contribution to *Systema Naturae*, 406; herbariums built by, 338; naming of a plant for, 235; number of dissertations supervised by, 224; on private enterprise in Holland, 108; travels by, 275, 277
Tilas, Daniel, 99, 350
Tillandz, Elias, 34
time, 149, 281
Tirén, Johan, 408
Tissot, Samuel-Auguste, 5
tobacco, 4, 32, 110, 191, 217–19, 383
travel: "About the Necessity of Investigative Journeys in Our Native Country," 154; characteristics of the ideal traveler, 274; to Dalarna (*see* Dalarna, journey to); dangers of and Linnaeus's sorrows over the deaths, 277–78; by disciples of Linnaeus, 274–78; financing, 276; to Holland (*see* Holland); "inland passport" required for domestic, 160; to Lapand (*see* Lapland); Linnaeus as

traveler, 171, 174; Linnaeus on, 274; London, England, 117; marketing of the travelogues, 173; Paris, France, 118; provincial, 158–75; Skåne, journey to, 168–75
Travels on Öland, 22
Trembley, Abraham, 272, 285
Treviranus, Gottfried, 406
Triewald, Mårten, 42, 134–36
Trolle, Stina, 35
Tullberg, Otto, 188
Tullberg, Tomas, 400
Tullberg, Tycho, 98, 438
twinflower. See *Linnaea borealis*

Uddenberg, Nils, 442
Uggla, Arvid, 129, 437–38, 443
Ullén, Petrus, 376
Ulrika Eleonora the Elder (queen of Sweden), 30, 250
Ulrika Eleonora the Younger (queen of Sweden), 28
Uppsala, Sweden: appearance from the south side, drawing of, 150; changes in circa 1764, 309; contemptuous description of, 403; fires in, 45, 147–48, 309–10, 319–20; geographic layout of, 149; upon Linnaeus's arrival, 45–46; move to in 1741, 142; post-fire restoration of, 147; street scene, elderly academics as part of, 152; as an unhealthy place, 147–48; Uppsala fever, 148, 192, 381, 390
Uppsala University: central university building, depiction of, 151; competition for two chairs at, 140–42; Faculty of Theology at, 254; history and facilities of, 45; judicial action at, 315–16; Linnaeus as faculty at, 154, 310–16; Linnaeus as rector, 316; Linnaeus as student at, 46–60; Linnaeus documentation held at, 437–38; medical faculty and curriculum at, 48–49, 311–12; move to, 42–44; pamphlet providing flippant overview of faculty at, 313–14; peak period during the Age of Liberty, 148–49; the prefect's house, residence in, 186–87; Society of Sciences, 310; speech as the departing rector, 384–85; students at, 45–46, 148, 152, 310–11
Uppsala University Botanical Garden: allowed to "run to waste," 42; animals as part of, 199–200; creation and care before Linnaeus, 194–95; dung from the university stables, fight over acquiring, 199; map of, 196; prospectus for, engraving of, 195; restructuring of and management by Linnaeus, 195, 197–99; thermometer, use in, 197
utility: botany as a study shorn of, 256; of centers of industry, 257; the Creation and, 255, 275; curiosity and, 255; from the "Earth," 209; of economics, 136; of etymology, 243; of keeping healthy, 160; of learning about herbs, 134; of measuring time, 149; of music, 262; "of plenty," 266; political prioritization of, 130; practical, 178, 255; of using natural resources, 428n1

Vaillant, Sébastien, 31, 54, 235
van Royen, Adriaan, 112, 118
van Swieten, Gerard, 116–17
Västgöta resa: dedication of, 165, 170; final passage of, 180; foreward focused on practical utility, 178; publication of, 172–73; title page, 161
Växjö, Sweden: population of, 30; Stenbrohult, distance from, 30–31
Virgil, 68–69, 263
Vita Caroli Linnaei, 2–3, 58, 114, 429n17, 430n1, 437
Voltaire (François-Marie Arouet), 330–31, 389

Wachendorff, Evert Jacob, 404
Wägner, Elin, 408
Wahlenberg, Göran, 21–22, 406
Wahlström, Anders, 366
Wallenberg, Jacob, 108, 306
Wallenius, Jacob, 383
Wallerius, Johan Gottschalk: electricity as the subject of a dissertation supervised by, 351; lecturing post in medicine won by, 43; mineralogical conceptus by Linnaeus, assessment of, 361; as part of the new generation in science, 131; professorial chair at Uppsala, candidacy for, 140–41; on rumors about Linnaeus's sexual teachings, 250; unfriendly relations with, 331–32; on the university faculty in the 1740s, 148
Wallerius, Nils, 148, 244

Wallin, Georg, 54, 103
Wandelaar, Jan, 105
Wänman, Carl, 223
Warg, Cajsa, 149, 218, 220, 324
Wargentin, Pehr Wilhelm: Bible Commission, member of, 370; complaint about the cost of travel to Stockholm, 330; on the condition of the elderly Linnaeus, 387; correspondence missing from, 233; correspondence with, 332, 380; Linnaeus's supposed atheism, letter raising, 366; mediation between Solander and his mother, attempt at, 399; as mediator for Linnaeus, 332, 431n20; as part of the new generation in science, 131; request that letters to Haller be burned, 386–87
Warmholtz, Carl Gustaf, 50
Wästberg, Pär, 443
water: all things beginning in, 361; decreasing levels of, 156, 166; Linnaeus's fear of the sea, 163, 171; steady expansion of land coinciding with reduction of, 155–56
Weber, Max, 292
Wedberg, Anders, 363
weddings, spring and, 55
Weibull, Martin, 43
Westbeck, Sacharias, 136
Westerberg, Mats, 46
Wetzel, Walter, 403
Widegren, David, 27
Wiklund, Karl Bernhard, 80, 417n35
Wikman, K. Rob, 171, 433
Wilcke, Johan Carl, 340
wild men raised by animals, 298
wild strawberries: drawing of, 302; medicinal qualities of, 303, 382, 389; regular consumption of, 323
wild-strawberry girl, 360
Willughby, Francis, 58
Wilson, Edward O., 404–5
Winge, Erik, 47
wisdom, 84, 298–300, 362–63
witches, witchcraft, and witch-hunting hysteria, 18, 164. *See also* occult, the
Wolff, Christian, 119, 362, 366
Wolffianism, 57

wolverine, 126
women: as botanists, 165, 245–46; Eve, creation of, 370–71; example of nobility in, 376; fashion for, views on the topic of, 264; interest in during provincial travels, 165; in Linnaeanism, 408; reaction to Linnaeus's sexual system, 247–48; on women's dress in Sweden and Holland, 108
wooden clogs, 174
work habits: data management, 229–30; daylight and, 229; depression and, 230–31; hours devoted to work, 228–29; publication, 241 (*see also* publication/publishing); volume and variety of work piling up, 231. *See also* professional networks
Wrangel, Carl Henrik, 201
Wrede, Elsa Beata, 246
Wrede foundation, 91
writing style, 1, 53; comparison of the Lapland and Dalarna travelogues, 94–95; electricity and lyrical, 352–53; in Lapland travelogue, 65; legacy of, 408; Linnaeus's standards for, 177; metaphors, use of, 350; mixed comments on Linnaeus's, 176–78; mythological references in Lapland travelogue, 68–72; as an old man, 386; poet, Linnaeus as, 179; in provincial travelogues, 162–63, 167–68; superlatives in letters, 178. *See also* language

Zeidenzopff, Ernst, 31
Zetzell, Pehr, 351
Ziervogel, Frederic, 201
Zimmerman, J. G., 343
Ziöberg, Magnus, 284
Zoëga, Johan, 320, 323
Zoll, Kilian, 408
zoology: beginning of modern/scientific, 125, 128 (*see also* taxonomy project); creation of orderly system for, 120–21; overtaking of botany in Linnaeus's work by, 279
zoophytes, 280, 282, 286
Zschotzscher, J. C., 31